Ablative Thermal Protection Systems Modeling

Ablative Thermal Protection Systems Modeling

Georges Duffa

AIAA EDUCATION SERIES

Joseph A. Schetz, Editor-in-Chief
Virginia Polytechnic Institute and State University
Blacksburg, Virginia

Published by the
American Institute of Aeronautics and Astronautics, Inc.
1801 Alexander Bell Drive, Reston, Virginia 20191-4344

American Institute of Aeronautics and Astronautics, Inc., Reston, Virginia

1 2 3 4 5

Front cover line drawing based on figure by Daniel Potter.

Library of Congress Cataloging-in-Publication Data

Duffa, Georges.
Ablative thermal protection systems modeling/Georges Duffa.
Pages cm – (AIAA education series)
Includes bibliographical references and index.
ISBN 978-1-62410-171-7 (alk. paper)
1. Ablation (Aerothermodynamics)–Mathematical models. I. American Institute of Aeronautics and Astronautics. II. Title.

TL574.A45D84 2013
629.4'152–dc23
2012043782

Dedicated to my wife Nicole, my children Sébastien and Céline,
and my grandchildren Abel, Jade, Eléonore, and Mila

CONTENTS

PREFACE

History

The hypersonic domain was born in the 1950s with the development of the first U.S. ballistic systems. This area was developed subsequently in the former USSR, then in Europe: the United Kingdom and France. The secrecy surrounding these activities make the publications of that time rare, and almost exclusively American. Until the late 1960s, the development of thermal protection systems for reentry was based mainly on an experimental approach for both the design of materials and their testing. This activity involved the development of major test facilities: plasma jets, free-flight wind tunnels, and flight tests. The volume of this activity was reinforced by the development of human space flight in the late 1950s. During this period of trial and error, researchers discovered the concept of ablative material. The physical and chemical transformations and gas injection resulted in an important protection from the re-radiation. This concept made extinct the use of a metal shield as a heat sink [1]. These concepts involved the development of knowledge of new phenomena related to heterogeneous physical and chemical reactions, the effect of gas injection in the boundary layer, and the phenomenon of pyrolysis. The radiative transfer problems were fairly well controlled by techniques developed in other areas of physics. Conversely, until the 1970s and beyond, the problems of laminar–turbulent transition, the setup and effects of parietal roughness, and the impact of particles remained largely unresolved problems, other than the development of correlations from the tests [2].

This first period is characterized by activity mainly based on engineering technology development, consuming important human and financial resources. The modeling effort was intense, but limited resources for the numerical calculation, in terms of both numerical analysis and computer resources, were such that it was necessary to wait until the late 1970s for calculation to become the third pillar of the field with modeling and testing.

With the advent of the shuttle projects in the 1970s—the Space Shuttle in the United States, Buran ("Snowstorm") in the USSR, and later Hermes in Europe—new problems appeared related to nonthermal equilibrium flow and heterogeneous chemical reactions, limited to the catalysis phenomenon. Indeed, the thermal protection developed for these reentry objects were intended to be reusable instead of ablative. It also should be mentioned that universities were solicited more than during the previous period, and the research activity spread to countries not previously present in the area, including Japan, Germany, and Italy.

This period saw a great development of numerical methods that would make the simulation code a central part of the design process. Such means would be used for mission simulation or testing, and then for preparing the tests themselves. In the 2000s this technique was applied to materials in the form of numerical calculation at the microscopic scale as well as taking a macroscopic approach. This was a necessity for progress in technological knowledge, which was constrained by the economic interests of the user to reuse existing materials. This approach has solved a number of issues, among those the ones listed previously. However, others, like erosion, are still waiting to obtain a physical solution. The domain of reentry and hypersonic systems consists of boom and bust cycles. At the time of this writing, the drastic reduction of resources allocated to this area makes it necessary to plan on a basic comprehensive program for the future.

Physical Basis

For one area discussed in this book, namely the modeling of phenomena, the sum of physical knowledge is broad. It includes (from microscopic to macroscopic) atomic physics, thermodynamics, gas kinetics, radiative transfer, physical and chemical reactions (both homogeneous and heterogeneous), fluid mechanics, and turbulence applied to previously studied areas, such as rough walls. There are, of course, many books on each of these areas, but very little literature on the synthesis of ablation problems [3–5]. The particularity of the area is undoubtedly the strong coupling between various physical domains. An overview of it is given at the end of the first chapter.

The objective of this book is to develop physical skills in the areas cited, applied to the modeling of thermal protection. The main basic equations are simply stated, recalling their physical origin, if necessary, and the main results are presented with significant concern that the text be light. For this, long demonstrations are given in the appendices.

References

[1] Sutton, K., and Gnoffo, P. A., "Multi-Component Diffusion with Application to Computational Aerothermodynamics," AIAA Paper 1998-2575, Seventh AIAA/ASME Joint Thermophysics and Heat Transfer Conference, Albuquerque, NM, June 1998.

[2] Walberg, G. D., and Sullivan, E. M., "Ablative Heat Shields for Planetary Entry. A Technology Review," ASTM/IES/AIAA Space Simulation Conference, Sept. 1970.
[3] "Tutorial on Ablative TPS," NASA Ames Conference Center, Moffett Field, CA, Aug. 2004.
[4] Hurwicz, H., Kratsch, K. M., and Rogan, J. E., "Ablation," *AGARDograph*, No. 161, edited by R. E. Wilson, March 1972.
[5] Polezhaev, Y. V., and Yurevich, F. B., "Heat Protection," *Energy*, edited by A. V. Lykova, Energiya, Moscow, Russia, 1976 (in Russian).

ACKNOWLEDGMENTS

The activity on hypersonic reentry began at Commissariat à l'Energie Atomique et aux Énergies Alternatives (CEA) in the 1970s thanks to Henri Berthoumieu, who gave impetus to a field that had a low level of scientific development in France. With Patrick Gallais, who was as ignorant of the subject as I was, we put our enthusiasm and our knowledge of physics to reviewing engineering models used in the field. For the thermal protection system (TPS), which interested us the most, Bernard Leroy developed the numerical part. Subsequently, this activity has benefited from the contributions of many people from CEA, Office National d'Etudes et Recherches Aérospatiales (ONERA), Aérospatiale (now EADS Astrium), and other agencies or industrial laboratories. I want to thank them without being able to cite all.

At maturity, this activity received a boost from younger people who were just as enthusiastic as us, such as Corinne Canton-Desmeuzes and Thanh-Ha Nguyen-Bui. I want to highlight the work of Bruno Dubroca who, besides his talent in numerical analysis, renewed in their entirety certain physical models used in this text.

My thanks also go to the university researchers whose assistance was important, especially Patrick Le Tallec, Raymond Brun, Pierre Charrier, and Gérard Vignoles.

NOMENCLATURE

a	sound velocity, $m \cdot s^{-1}$
	or radiative constant, $J \cdot m^{-3} \cdot K^{-4}$
a_0	Bohr radius, m
A_{ij}	Einstein coefficient for transition i $\rightarrow$ j (induced emission), s^{-1}
b	collision distance, m
B	temperature exponent in Arrhenius law
B_{ij}	Einstein coefficient for transition $i \rightarrow j$ (stimulated emission, absorption), $s \cdot sr \cdot kg^{-1}$
B_ν	Planck function, $J \cdot sr^{-1} \cdot m^{-2}$
$\mathcal{B}'$	mass flow injection
c	mass fraction
	or light speed, $m \cdot s^{-1}$
$\tilde{c}$	elemental fraction
C	pre-exponential factor, $m^3 \cdot mol^{-1} \cdot s^{-1} \cdot K^{-B}$
C_A	drag coefficient (vehicle axis)
C_p	mass heat coefficient at constant pressure, $J \cdot kg^{-1} \cdot K^{-1}$
$\mathcal{C}_p$	molar heat coefficient at constant pressure, $J \cdot mol^{-1} \cdot K^{-1}$
C_V	mass heat coefficient at constant volume, $J \cdot kg^{-1} \cdot K^{-1}$
$\mathcal{C}_V$	molar heat coefficient at constant volume, $J \cdot mol^{-1} \cdot K^{-1}$
$\mathcal{C}_H$	Stanton number
$\mathcal{C}_M$	mass transfer coefficient
D	multicomponent diffusion coefficient, $m^2 \cdot s^{-1}$
$\mathbf{D}_R$	Eddington tensor, s
$\mathcal{D}$	binary diffusion coefficient, $m^2 \cdot s^{-1}$
$\mathcal{D}^T$	thermal diffusion coefficient, $kg \cdot m^{-1} \cdot s^{-1}$
$\mathcal{D}$a	Damköhler number,
e	mass internal energy, $J \cdot kg^{-1}$
ε	molar internal energy, $J \cdot mol^{-1}$
$\mathbf{E}$	electrical field, $V \cdot m^{-1}$
E	mass total energy, $J \cdot kg^{-1}$
	or atomic or molecular energy level, J

E_0	ionization energy, J
E_R	volume radiative energy, $J \cdot m^{-3}$
$E_{R\nu}$	volume spectral radiative energy, $J \cdot s \cdot m^{-3}$
f	velocity function distribution, $m^{-1} \cdot s$
F	collisional term, $m^2 \cdot kg^{-\frac{1}{2}}$
$\mathbf{F}$	volume force, $N \cdot m^{-3}$
	or energy flux, $W \cdot m^{-2}$
g	gravity acceleration, $m \cdot s^{-1}$
	or conversion rate
G	impact mass conversion
	or mass free enthalpy (Gibbs function), $J \cdot kg^{-1}$
$\mathcal{G}$	molar free enthalpy (Gibbs function), $J \cdot m^{-3}$
	or Gaunt factor
h	Planck constant, $J \cdot s$
	or mass enthalpy, $J \cdot kg^{-1}$
	or geopotential altitude, m
h_0	atmospheric scale factor, m
h_a	adiabatic wall enthalpy, $J \cdot mol^{-1}$
$\mathcal{H}$	molar enthalpy, $J \cdot mol^{-1}$
I_ν	spectral radiative luminance (intensity), $J \cdot sr^{-1} \cdot m^{-2}$
$\mathbf{j}$	electric current density, $A \cdot m^{-2}$
$\mathbf{J}$	mass flux density, $kg \cdot m^{-2} \cdot s^{-1}$
$\mathcal{J}$	molar flux density, $mol \cdot m^{-2} \cdot s^{-1}$
$\mathcal{J}_\nu$	radiative source term, $J \cdot m^{-3} \cdot sr^{-1}$
$\tilde{\mathbf{J}}$	elemental flux, $kg \cdot m^{-2} \cdot s^{-1}$
k	Boltzmann constant, $J \cdot K^{-1}$
	or gaseous reaction constant, $\left(m^3 \cdot mol^{-1}\right)^{\left(\Sigma_i \nu_{ij} - 1\right)} \cdot s^{-1}$
	or heterogeneous reaction constant, $m^{-2} \cdot s^{-1} \cdot Pa^{\left(1 - \Sigma_i \mu_{ij}\right)}$
	or kinetic turbulent energy, $m^2 \cdot s^{-2}$
$\mathbf{k}$	propagation unit vector
k^T	multicomponent thermal diffusion number
$\mathbf{K}$	permeability, m^2
K_p	thermodynamic equilibrium constant
l	characteristic length, m
l_D	Debye length, m
$\mathcal{L}$	Lewis–Semenov number
m	mass, kg
$\mathcal{M}$	molar mass, $kg \cdot mol^{-1}$
	or radiative strength, $C^2 \cdot m^2$
$\overline{\mathcal{M}}$	mean molar mass, $kg \cdot mol^{-1}$

$\mathcal{M}a$	Mach number
$\dot{m}$	mass flow (equivalent to $\lVert\mathbf{J}\rVert$), $kg \cdot m^{-2} \cdot s^{-1}$
n	particle numerical density, m^{-3}
n_A	number of elements
n_e	number of gaseous species
n_r	number of reactions
$\mathbf{n}$	unit vector perpendicular to fluid–solid interface, toward the fluid
$\mathcal{N}$	Avogadro's number, mol^{-1}
N_s	site number by surface unit, m^{-2}
p_0	reference pressure, Pa
p	pressure, Pa
$\bar{p}$	equilibrium vapor pressure, Pa
$\mathbf{P}$	pressure tensor, Pa
$\mathcal{P}$	Prandtl number
$\mathcal{P}_e$	Peclet number
$P_\nu(\mathbf{\Omega}' \rightarrow \mathbf{\Omega})$	phase function
q_e	electron charge, C
$\dot{\mathbf{q}}$	heat flux, $W \cdot m^{-2}$
Q	volume density of charges, $C \cdot m^{-3}$
Q^*	apparent heat of ablation, $J \cdot kg^{-1}$
r	radius, m
	or recovery factor
	or efficiency
R	reflectivity
	or stagnation point radius, m
R_a	arithmetic roughness, m
$\mathcal{R}e$	Reynolds number
$\mathcal{R}$	gas universal constant, $J \cdot mol^{-1} \cdot K^{-1}$
S	steric factor
	or surface, m^2
	or source term, $J \cdot m^{-3} \cdot s^{-1}$
	or mass entropy, $J \cdot kg^{-1} \cdot K^{-1}$
S_c	Schmidt number
S_ν	volume surface, m^{-1}
	or spectral source term, $J \cdot m^{-2} \cdot sr^{-1}$
$\mathcal{S}$	source term, $m^{-3} \cdot s^{-1}$
	or molar entropy, $J \cdot mol^{-1} \cdot K^{-s}$
	or differential collisional cross-section, m^2
$\mathcal{S}_{tot}$	total collisional cross-section, m^5
t	time, s
T	temperature, K

T_0	reference temperature, K
$\mathcal{T}$	viscous stress tensor, Pa
	or relaxation term, $J \cdot mol^{-1} \cdot s^{-1}$
T_u	turbulence intensity
v	velocity, $m \cdot s^{-1}$
V	molecular interaction potential, J
$\mathcal{V}$	molar volume, $mol \cdot m^{-3}$
$\mathbf{V}$	fluid velocity, $m \cdot s^{-1}$
$\mathbf{V}_D$	diffusion velocity, $m \cdot s^{-1}$
W	power, W
W_e	Weber number
x	volumic molar fraction
$\mathbf{x}$	space coordinates, m
Z	electrical charge number
Z^{rot}	vibrational relaxation rate

Greek Letters

α	thermal diffusivity, $m^2 \cdot s^{-1}$
α^{T}	binary thermal diffusivity coefficient
β	ballistic coefficient, $kg \cdot m^{-2}$
γ	reactional efficiency (or sticking probability)
	or collisional rate, $m^{-3} \cdot s^{-1}$
	or reentry path angle, degree
δ	Kröneker symbol
	or boundary layer thickness, m
Δe^{vib}	vibrational energy lost during a chemical reaction, $J \cdot kg^{-1}$
Δe^e	ionization energy, $J \cdot kg^{-1}$
Δh	apparent ablation heat energy, $J \cdot kg^{-1}$
Δh_f^0	enthalpy of formation at normal pressure, $J \cdot kg^{-1}$
ϵ	emissivity
	or open porosity
	or turbulent kinetic energy dissipation rate, $m^2 \cdot s^{-3}$
ϵ_0	vacuum permittivity, $C^2 \cdot N^{-1} \cdot m^{-2}$
ϵ_M	turbulent mass diffusion coefficient, $m^2 \cdot s^{-1}$
ϵ_H	turbulent thermal mass diffusion coefficient, $m^2 \cdot s^{-1}$
η	volume viscosity, $Pa \cdot s$
κ	Von Karman constant
κ_ν	spectral absorption (extinction) coefficient, m^{-1}
λ	thermal conductivity, $W \cdot m^{-1} \cdot K^{-1}$
	or periodicity length of system, m

λ^r	reactive thermal conductivity, $W \cdot m^{-1} \cdot K^{-1}$
Λ	screening parameter,
$\boldsymbol{\Lambda}$	conductivity tensor, $W \cdot m^{-1} \cdot K^{-1}$
μ	dynamical viscosity, $Pa \cdot s$
ν	frequency, s^{-1}
$\dot{\omega}$	mass production rate by unit volume, $kg \cdot m^{-3} \cdot s^{-1}$
	or mass production rate by unit surface, $kg \cdot m^{-2} \cdot s^{-1}$
Ω	stoichiometric matrix
	or unit direction vector
$\Omega_{ij}^{(l,s)}$	collision integral for the couple of species(i, j), $m^3 \cdot s^{-1}$
ϖ	$\dot{\omega}$ vector
ρ	volume mass, $kg \cdot m^{-3}$
$\bar{\rho}$	apparent volume mass, $kg \cdot m^{-3}$
σ	collisional cross-section, m^2
	or Stefan–Boltzmann constant, $W \cdot m^{-2} \cdot K^{-4}$
	or electrical conductivity, $S \cdot m^{-1}$
	or surface strain, $N \cdot m^{-1}$
θ	angle, rad
	or momentum thickness, m
	or fraction of occupied sites
	or characteristic vibrational temperature, K
$\mathcal{T}$	volume reaction rate, $mol \cdot m^{-3} \cdot s^{-1}$
	or surface reaction rate, $mol \cdot m^{-2} \cdot s^{-1}$
	or shear, Pa
$\mathcal{T}_\nu$	optical depth
$\mathcal{V}$	molar concentration (molarity), $mole \cdot m^{-3}$
Φ	volume collision numerical flux, $m^{-3} \cdot s^{-1}$
	or surface collision numerical flux, $m^{-2} \cdot s^{-1}$
χ	Eddington factor
ζ	chemical reaction progression rate

Subscripts or Superscripts

a	relative to stagnation point or to ablation
A	activation
app	apparent value
b	backward
chim	chemical
coll	collision
cond	relative to conduction
conv	relative to convection
D	relative to dissociation

e	boundary layer or liquid layer external value
	or relative to electron
eq	equivalent
	or equilibrium
f	forward
g	gas
	or formation
I	independent
	or relative to ion
int	internal
l	liquid
	or laminar
p	relative to wall
	or relative to pyrolysis
R	relative to resin, radiation, or roughness
reac	relative to chemical reaction
red	reduced
rot	rotation
s	relative to surface, solid, or scattering
SD	relative to "hard spheres" potential
t	turbulent
T	relative to temperature
tot	total
trans	translation
v	relative to virgin material
	or relative to frequency
vib	relative to vibration
0	thermodynamic value at reference pressure p_0
2	relative to quantities behind a shock
∞	at infinity
$*$	nondimensional value
$+$	relative to inner layer of boundary layer

Functions and Operators

$erf(\bullet)$	error function
$E_n(\bullet)$	exponential integral of rank n
Id	unit tensor
$\mathrm{Ker}(\bullet)$	matrix kernel
$\min(\bullet)$, $\max\{\bullet)$	minimum and maximum functions
$\mathrm{rank}\{\bullet)$	matrix rank
$\mathcal{W}(\bullet)$	Lambert function (or omega function)
δ	Kronecker symbol

Δ	Laplace operator
∇	nabla operator
∇^t	transpose nabla operator
$\otimes$	tensor product operator
$\bullet''$	turbulent fluctuation
$\|\|\bullet\|\|$	vector modulus
$< \bullet >$	moment of function
$\bar{\bullet}$	time average
$\tilde{\bullet}$	Reynolds average, $\tilde{\bullet} = \dfrac{\overline{\rho \bullet}}{\bar{\rho}}$
$\zeta(\bullet)$	Riemann zeta function

Constants

CODATA reference values*

a	$\dfrac{8\pi^5 k^4}{15h^3c^3} = 7.5659143 \times 10^{-16} \mathrm{J \cdot m^{-3} \cdot K^{-4}}$
a_0	$h^2\epsilon_0/\pi m_e q_e^2 = 0.52917720859 \times 10^{-10}$ m
c	$2.99792458 \times 10^8 \mathrm{\ m \cdot s^{-1}}$
h	$6.62606896 \times 10^{-34} \mathrm{\ J \cdot s}$
k	$1.3806504 \times 10^{-23} \mathrm{\ J \cdot K^{-1}}$
m_e	$9.10938215 \times 10^{-31}$ kg
$\mathcal{N}$	$6.02214179 \times 10^{23} \mathrm{\ mol^{-1}}$
p_0	1 atm $= 1.013250 \times 10^5$ Pa
q_e	$1.602176487 \times 10^{-19}$ C
$\mathcal{R}$	$\mathcal{N}k = 8.314472 \mathrm{\ J \cdot mol^{-1} \cdot K^{-1}}$
T_0	300 K (JANAF: $T_0 = 297.15$ K, HITRAN: $T_0 = 296$ K ...)
ϵ_0	$8.854187817\ldots \times 10^{-12} \mathrm{\ C^2 \cdot N^{-1} \cdot m^{-2}}$
σ	$\dfrac{ac}{4} = 5.670400 \times 10^{-8} \mathrm{W \cdot m^{-2} \cdot K^{-4}}$

Acronyms Used in Text, Figures, and References

AAS	American Astronomical Society
ACM	Association for Computing Machinery
AEDC	Arnold Engineering Development Center
AFML	Air Force Materials Laboratory
AFOSR	Air Force Office of Scientific Research
AGARD	Advisory Group for Aeronautical Research and Development

* National Institute of Science and Technology, "Reference on Constants, Units and Uncertainty," 2006, http://physics.nist.gov/cuu/Constants/ [retrieved 24 April 2008].

AHF	Aerodynamic Heating Facility
AHS	American Helicopter Society
AIAA	American Institute of Aeronautics and Astronautics
AIChE	American Institute of Chemical Engineers
ANZIAM	Australian and New Zealand Industrial and Applied Mathematics
ARAD	Analog Resistance Ablation Detector
ARD	Advanced Reentry Demonstrator
ASCE	American Society of Civil Engineers
ASEE	American Society for Engineering Education
ASME	American Society of Mechanical Engineers
ASTM	American Society for Testing and Materials
AVT	Applied Vehicle Technology
BET	Brunauer, Emett, Taylor
BHWT	Boeing Hypersonic Wind Tunnel
CEA	Commissariat à l'Energie Atomique et aux Énergies Alternatives (French Alternative Energies and Atomic Energy Commission)
	NASA Chemical Equilibrium with Applications
CERMICS	Centre d'Enseignement et de Recherches en Mathématiques, Informatique et Calcul Scientifique (Center for Teaching and Research in Mathematics, Computer Science and Scientific Calculation)
CFL	Courant, Friedrich, Levy
CIRA	Centro Italiano Ricerche Aerospaziali (Italian Aerospace Research Center)
CODATA	Committee on Data for Science and Technology
COTA	Capteur Optique de Thermique et d'Ablation (Ablation and Thermics Optical Sensor)
CPAT	Centre de Physique des Plasmas et de leurs (Center of Plasma Physics and Applications of Toulouse) Applications de Toulouse
CRC	Chemical Rubber Company
CVD	chemical vapor deposition
CVI	chemical vapor infiltration
DET	dust erosion tunnel
DLR	Deutschen Zentrum für Luft- und Raumfahrt (German Aerospace Center)
DNA	Defense Nuclear Agency
DNS	direct numerical simulation
DSC	differential scanning calorimetry
EDLS	entry, descent, and landing system

ENSAE	Ecole Nationale Superieure de l'Aeronautique et de l'Espace (National High School of Aeronautics and Space)
ERV	Elementary Representative Volume
ESTEC	European Space and Technology Centre
ETI	Effects Technology Incorporated
GCL	Geometric Conservation Law
GRAM	Global Reference Atmospheric Model
HACA	hydrogen abstraction C_2H_2 addition
HEAT	High Enthalpy Aerothermal Test
ICBM	intercontinental ballistic missile
IEEE	Institute of Electrical and Electronics Engineers
IES	Industrial Engineering Society
IHF	Interaction Heating Facility
INRIA	Institut de Recherche en Informatique et en Automatique (National Institute for Research in Computer Science and Control)
IR	infrared
IRDT	Inflatable Reentry and Descent Technology
IREV	Inflatable Reentry Experimental Vehicle
IWC	ice water content
KMSI	Knowledge Management Solutions Incorporated
LES	large eddy simulation
LWC	liquid water content
MER	Mars Exploration Rover
MiRKa	Mikro Rückkehrkapsel (Micro Reentry Capsule)
MOLA	Mars Orbiter Laser Altimeter
MSL	Mars Science Laboratory
NASA	National Aeronautics and Space Administration
NMR	nuclear magnetic resonance
NSWC	Naval Surface Warfare Center
NTRS	NASA Technical Report Server
ODF	opacity distribution function
ONERA	Office National d'Etudes et Recherches Aérospatiales (French Aerospace Lab)
PAN	polyacrylonitrile
PANT	Passive Nosetip Technology
PICA	phenolic-impregnated carbon ablator
PMMA	polymethylmetacrylate
PTF	Panel Test Facility
PTFE	polytetrafluoroethylene
QSS	quasi steady-state
RHS	right-hand side

RKR	Rydberg, Klein, Rees
RTO/AVT	Research and Technology Organization, Applied Vehicle Technology (OTAN)
SAE	Society of Automotive Engineers
SAI	Science Applications Incorporated
SAMSO	Space and Missile Systems Organization
SCEBD	Self-Consistent Effective Binary Diffusion
SFT	Société Francaise des Thermiciens (French Heat Engineering Association)
SIAM	Society for Industrial and Applied Mathematics
SLA	super lightweight ablator
SoRI	Southern Research Institute
SPA	surface-protected ablator
SQP	sequential quadratic programming
SSH	Schwartz, Stretton, Herzfeld
STS	Space Transportation System
TPS	thermal protection system
TPSX	Thermal Protection Systems Expert
TWCP	tape-wrapped carbon phenolic
VDC	vibration–dissociation coupling
VDVC	vibration–dissociation–vibration coupling
UV	ultraviolet
WGS	World Geodetic System

Chapter 1 Thermal Protection System Conception

- Calculate the trajectory of reentry
- Learn classes of materials
- Identify physical problems

1.1 Planetary Reentry

This chapter deals with the use of thermal protection for reentry probes or ballistic vehicles. The problems are similar to those of nozzles for solid rocket boosters. Traditionally, the role of experience is important, from the design of material on the bench of the chemist to the full-scale test in a ground test or in flight. This chapter will focus on terrestrial reentry, but the information can be applied to different atmospheres.

1.1.1 Planetary Atmospheres

The composition of the atmosphere is, of course, an important factor conditioning the reentry. Table 1.1 provides some information concerning the atmospheres of planets (as well as the Titan satellite) in which probes or ballistic objects were used.

In addition to composition, we must determine the profiles of pressure and temperature with altitude. This knowledge is based on two approaches:

- *A statistical description based on measurements:* This applies to the Earth's atmosphere, with the models in *U.S. Standard Atmosphere 1966* (US66) [1] giving typical atmospheres for different latitudes and different seasons. This type of model has been updated by NASA [2]: the Global Reference Atmospheric Model (GRAM) database has been extended to the atmospheres of Mars and Venus [3]. The data are relatively scarce for these planets, which makes this kind of model rather imprecise. These measurements are made from nonreentering probes, such as those made

Table 1.1 Characteristics of Various Planetary Atmospheres*

Planet	Composition (Vol. Fraction)	Minor Species (<0.5%)	Mean Ground Pressure (10^5 Pa)
Earth	$N_2 = 78.1\%$ $O_2 = 21.0\%$ $Ar = 0.9\%$	CO_2	1.01
Mars	$CO_2 = 95\%$ $N_2 = 2.7\%$ $Ar = 1.6\%$	O_2, H_2O	0.064
Venus	$CO_2 = 96.5\%$ $N_2 = 3.4\%$	Ar, SO_2	90
Jupiter	$H_2 = 86\%$ $He = 14\%$	CH_4, N_2	—
Titan	$N_2 = 94\%$ $CH_4 = 5\%$	Ar, H_2	1.5

*The values are average values for low latitudes. In the case of Earth, H_2O is present in a variable fraction, 0 to 1%.

on Titan by Voyager-1. From these indirect measurements [refraction of radio waves, infrared spectroscopy, and ultraviolet (UV) absorption], atmospheres were built and fairly well-confirmed by measurements in situ [4].

- *A description based on a numerical climate model for Mars [5]:* This is made possible by the absence of water (oceans, clouds), whose presence makes Earth's calculations very complex.

Various design studies will also require knowledge of the dispersion of each parameter, including winds (horizontal and vertical components). An example for the Martian atmosphere is given in Fig. 1.1.

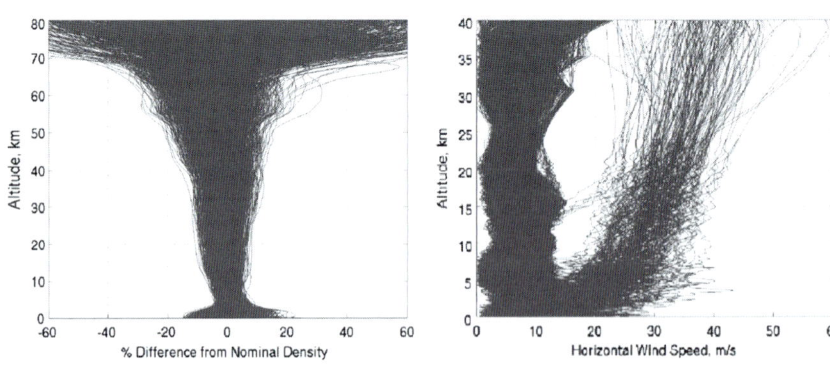

Fig. 1.1 Scattering of Mars atmospheric parameters (6).

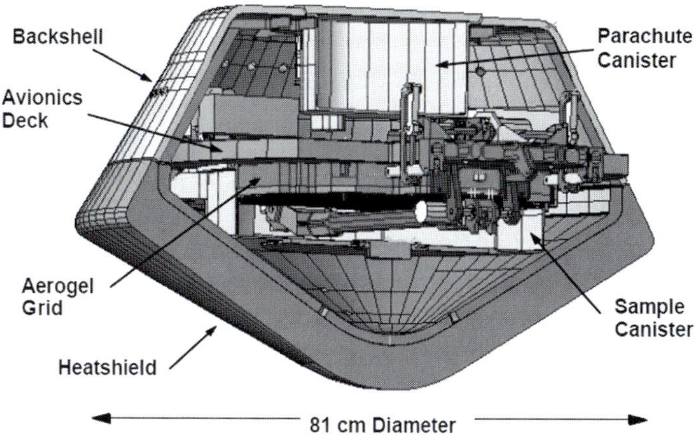

Fig. 1.2 Stardust.

The calculation of the trajectory will also require knowledge of the gravity field. In this area, the standard is now the WGS84 (World Geodetic System), whose description is given in Gallais [7].

1.1.2 The Reentry Phase

The reentering object {entry, descent, and landing system (EDLS); see examples in Figs. 1.2 and 1.3 [8]}, consisting of the lander and ablative thermal protection system (TPS), is separated from its carrier with a

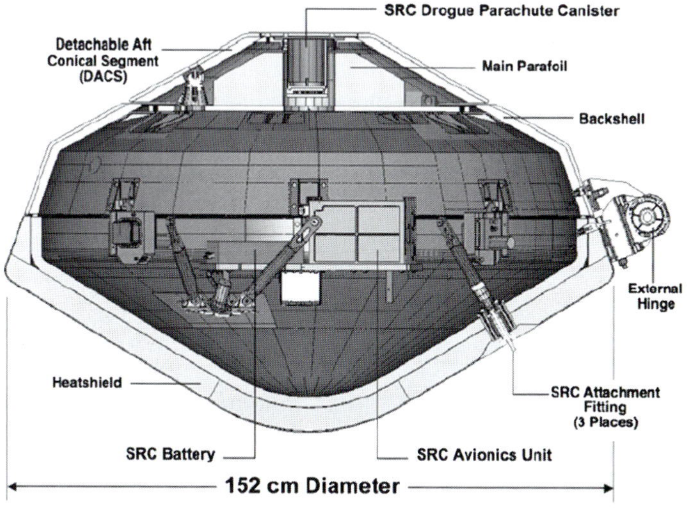

Fig. 1.3 Genesis.

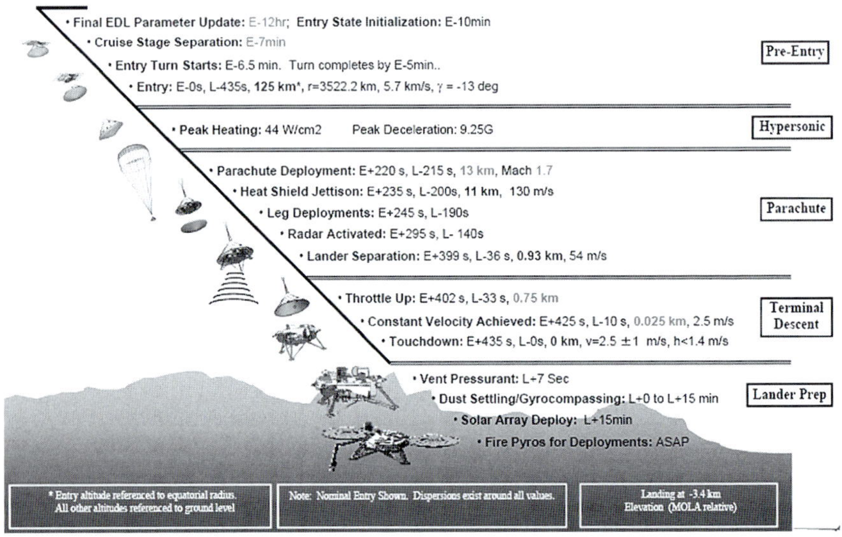

Fig. 1.4 Reentry synoptic example: Netlander (6).

low-speed rotation, which ensures a fixed direction in Galilean frame of reference. The angle of reentry (the angle between the momentum and the local horizontal) is chosen to be as low as possible. The ballistic phase lasts from a few tens of minutes to several hours.

The entry phase is divided into several stages (Fig. 1.4):

- The slowdown to transonic phase, during which all the phenomena of heating occur.
- The deep stall, usually divided into two phases—transonic pilot chute of small diameter, also functioning as a stabilizer, and main parachute of large diameter, which reduces the speed at a few tens of meters per second.
- During the parachute phase, the front portion of the thermal protection is dropped and the rear portion remains attached to the parachute, both of which are released.
- Finally, a landing system (speed less than 10 $m \cdot s^{-1}$) is used, consisting of thrusters and articulated arms or airbags.

1.2 Orders of Magnitude

Reentry velocities vary greatly, depending on the application (Fig. 1.5), but always are very high: a few thousand meters per second, or even tens of thousands for a direct rendezvous on a giant planet (47.4 $km \cdot s^{-1}$ in the case of Galileo on Jupiter).

This section will address issues of reentry and thermal protection of the object. Some calculations of orders of magnitude will provide enlightenment

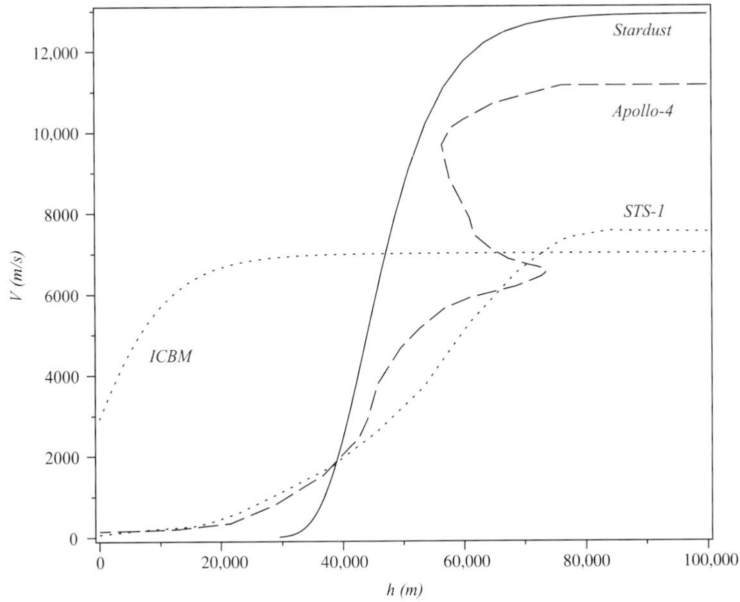

Fig. 1.5 Typical reentries.

to the discussion. First, note that in supersonics, the reentry vehicle is preceded by a detached shock wave (Fig. 1.6); however, this is not true for sharp bodies, which are discarded because of the high heat flux levels associated with them (see Sec. 4.5). The effects associated with transport coefficients (diffusion, viscosity, conductivity) are confined to a very thin

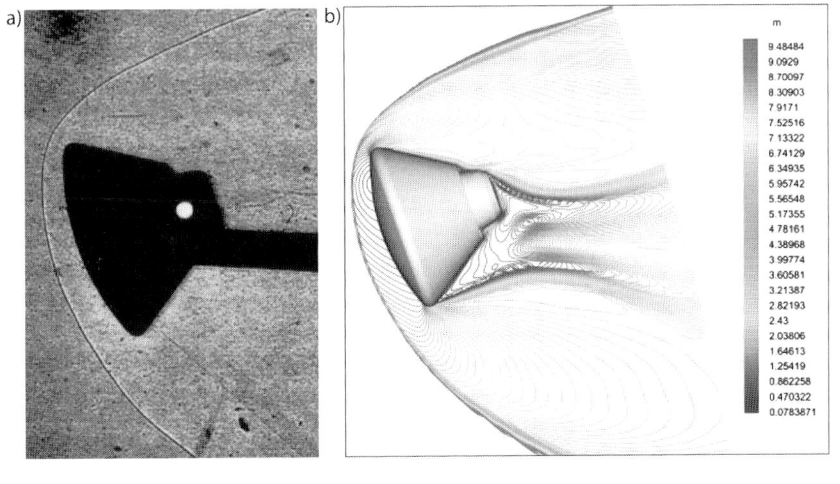

Fig. 1.6 a) ARD mock-up in wind tunnel ($\mathcal{M}a = 4$), and b) corresponding numerical calculation (isomach lines) (9).

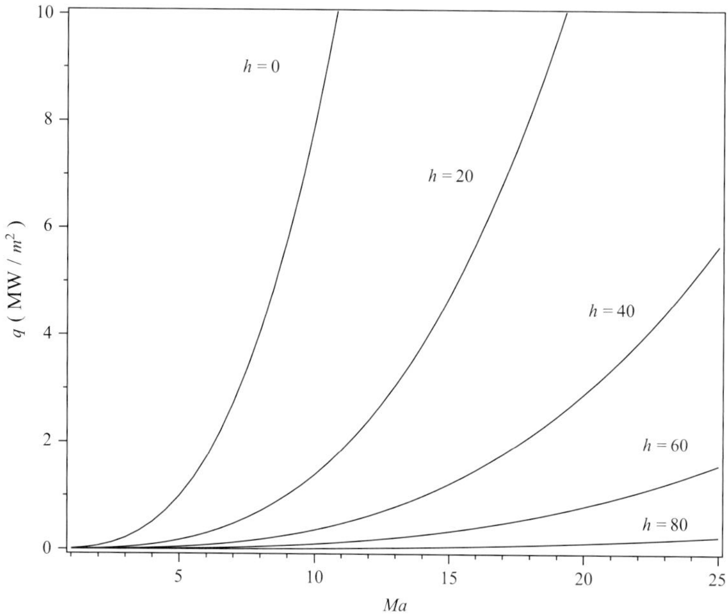

Fig. 1.7 Mach number influence on terrestrial reentry fluxes for varying altitudes (in kilometers), $R = 1$ m.

boundary layer (a few micrometers to some millimeters), laminar or turbulent, with the exception of detached areas, especially the back of the object. The calculations of orders of magnitude that will be made do not relate to objects with a stable angle of attack, such as Apollo ($\simeq$25 to 30 deg) or the Advanced Reentry Demonstrator (ARD). This effect, caused by offsetting of the center of gravity, gives a lift that is directly managed by controlling the roll using gas jets. This explains why the early reentry includes a resource (the object regains altitude, Fig. 1.5).

The difficulties of the design problem in terms of material and experimental simulation can be illustrated as follows. Figure 1.7 gives the stagnation point heat flux on a cold and inert wall, depending on the Mach number $\mathcal{M}a = V/a$, where a is the speed of sound,

$$a = \left(\gamma \frac{\mathcal{R}\mathcal{T}}{\mathcal{M}}\right)^{\frac{1}{2}}$$

($\gamma = 1.4$, $\overline{\mathcal{M}} = 28.8 \cdot 10^{-3}$ kg $\cdot$ mole^{-1} for air). This flow is calculated from a correlation derived from systematic parameterization (see Sec. 4.5.3), assuming an isothermal atmosphere,

$$\dot{q} = \alpha \rho^{\frac{1}{2}} R^{-\frac{1}{2}} V^3 \tag{1.1}$$

The coefficient α depends on the given atmosphere:

- $\alpha = 1.83 \times 10^{-4}\,\mathrm{kg}^{-1/2} \cdot \mathrm{m}^{-1}$ for air [10]
- $\alpha = 1.35 \times 10^{-4}\,\mathrm{kg}^{-1/2} \cdot \mathrm{m}^{-1}$ for Mars atmosphere [11]

R is the radius of the object, and ρ is the density of air at the given altitude [1]. You can also roughly estimate this value from elementary considerations. Suppose the atmosphere is in hydrostatic equilibrium (which is almost always true) and isothermal (which is a rough approximation). The hydrostatic equilibrium is written as

$$\frac{\mathrm{d}p}{\mathrm{d}h} = -g\rho \tag{1.2}$$

where h is the altitude and $g = 9.806\ \mathrm{m \cdot s^{-2}}$, the terrestrial gravity ($g = 3.6\ \mathrm{m \cdot s^{-2}}$ on Mars) is assumed constant. If we substitute p's value from the equation of the state of the air,

$$p = \rho \frac{\mathcal{R}\mathcal{T}}{\overline{\mathcal{M}}}$$

we get

$$\frac{\mathrm{d}\rho}{\mathrm{d}h} = -\frac{\rho}{h_0} \tag{1.3}$$

The solution to Eq. (1.3) is

$$\rho = \rho_0 e^{-\frac{h}{h_0}}$$

where

$$h_0 = \frac{\overline{\mathcal{M}}g}{\mathcal{R}\mathcal{T}}$$

is the scale factor. Various values of this parameter are given in Table 1.2.

Figure 1.7 measures the drastic effects of altitude and Mach number. You can evaluate the effects of such quantities by calculating the radiative equilibrium that would be obtained for a reusable thermal protection (by $\dot{q} = \epsilon\sigma T^4$), the conduction flux not exceeding a few percent in this

Table 1.2 Atmospheric Scale Factor (7, 12)

Planet	h_0 (km)
Earth	7
Mars	7.6
Venus	5.3
Titan	38

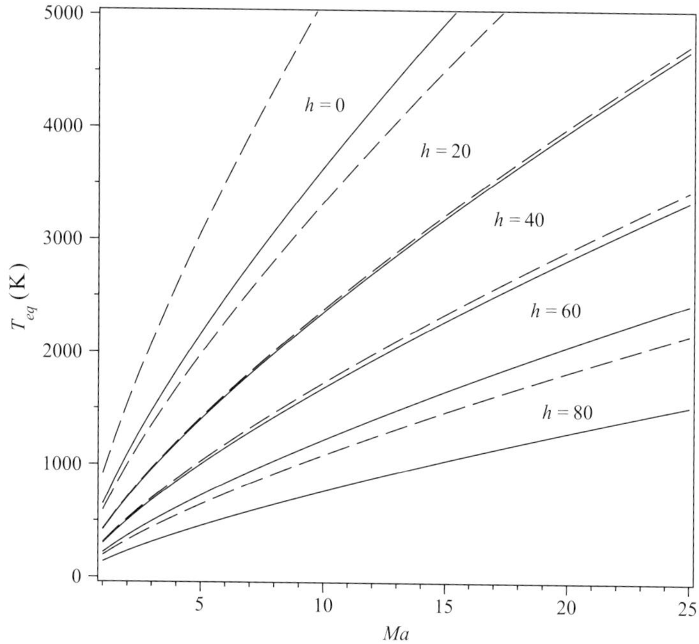

Fig. 1.8 Mach number influence on radiative equilibrium temperature for varying altitudes (in km), $R = 1$ m, metal ($\epsilon = 0.2$) and ceramic composite ($\epsilon = 0.8$).

configuration. Figure 1.8 shows that the values prohibit any solution for low altitudes. Indeed, metals do not exceed 1500 K and are penalized by a low emissivity. Ceramic composites do not exceed about 2000 K in an oxidizing atmosphere. This explains why we try to slow down at very high altitudes when possible. This is the condition for being able to use reusable systems such as the space shuttle. Braking is enhanced by a high incidence of the plane. Another method is to deploy a large structure [Inflatable Re-entry and Descent Technology (IRDT) or Inflatable Re-entry Vehicle Experiment (IRVE)] [13, 14] (see Fig. 1.9). This method has not yet passed the testing stage, however.

The influence of the radius R also can be observed: the heat flux increases with decreasing radius ("small is difficult"). That is the conclusion of work being done on the draft European Space Shuttle Hermes, which is smaller than the U.S. shuttle.

It should be noted that, historically, the first heatshields were of the heat sink type, such as copper or beryllium (e.g., the first Mercury flights, the flight of the Fire II Apollo technology demonstrator), because of the good conductivity of these metals. In this type of solution, the amount of energy stored in the system is very important. This method was quickly abandoned because of the weight and thermomechanical problems.

It is possible to evaluate the time history of flux from trajectory computations on an isothermal atmosphere and assuming constant drag and a straight path. (This is called the Allen trajectory [7].) The input data is the ballistic coefficient

$$\beta = \frac{m}{S_{ref} C_x}$$

mass m divided by

$$S_{ref} C_x = \frac{2F_x}{\rho_\infty V_\infty^2}$$

where F_x is the projection of aerodynamic force on the velocity vector, and the slope γ, of course, relative to the horizontal local return ($h \simeq 120$ km). Figure 1.10 shows that:

- Flux first increases, V_∞ remaining approximately constant and ρ_∞ increasing quickly.
- Then flux decreases while ρ_∞ continues to increase; V_∞^3 rapidly decreases in the deceleration phase.

This same type of approximation method provides access to the total energy received. If time and space are integrated into the local flux, we obtain the energy transferred to the object, which is very small as compared to the kinetic energy change of the object (<10% [15]), almost all the energy

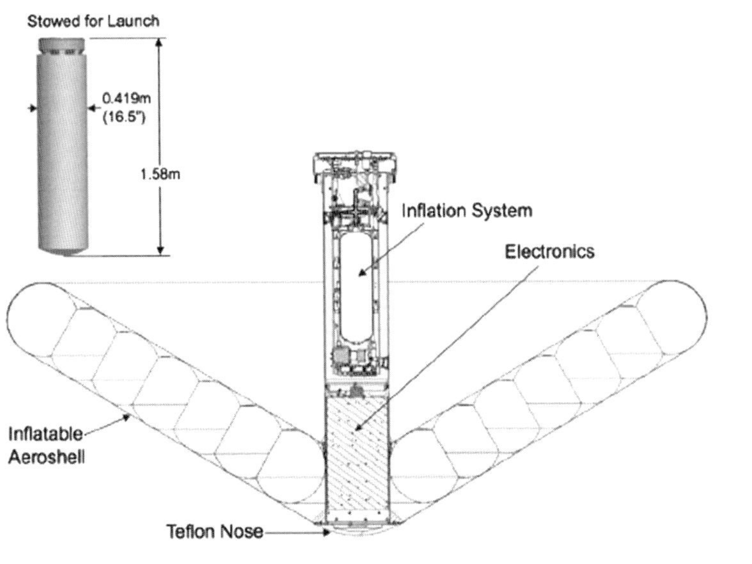

Fig. 1.9 Inflatable reentry vehicle experiment.

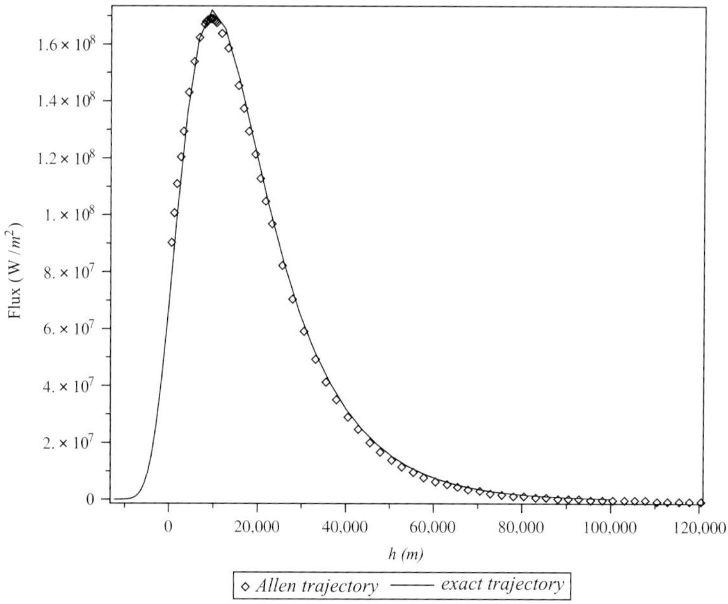

Fig. 1.10 Stagnation point fluxes; $\beta = 200$ kg · m^{-2}, $\gamma =$ 20 deg, R = 0.1 m.

being used to heat the atmosphere. Assuming in the first approximation that the portion of total energy transferred to the vehicle is constant, the energy received is proportional to V_∞^2 for all trajectories for which the final velocity is small as compared to the initial one.

Note in Table 1.3 that entry path angles, counted from the local horizontal, are small. This configuration allows for slowing down in low-density regions of the atmosphere, thus minimizing maximum fluxes. The much higher entry path angle of the Huygens probe (see Table 1.2) is due to the low gravity of Titan, which has the lowest vertical density gradient. This causes the same effect as a lower reentry angle in an atmosphere with a higher gradient. Another higher entry path angle (Pioneer North Probe) results from constraints on trajectories: the two other small spacecraft of this mission have lower entry path angles.

The geometries of reentering objects are extremely simple, at least the windward part: sphere, cone (possibly), and torus (possibly). Some recent studies have been concerned with optimizing shape problems [26, 27], but they have not been followed successfully in the engineering domain. Geometries are provided in Table 1.3. Figures 1.11 and 1.12 show photographs of two of these objects after their return. The similarity among the various geometries is due to technical concerns (stability, maximum heat flux, etc.), as well as the desire to minimize development costs by incorporating existing results. For example, MER and Phoenix have identical geometries, which

Table 1.3 Geometry of Vehicles and Reentry Conditions (8, 16–25)*,†

Reentry Vehicle	Planet, Flight Date	Radius (mm)	Cone Angle (Deg)	Torus Radius (mm)	Diameter (m)	Mass (kg)	Velocity ($m \cdot s^{-1}$)	Angle (Deg)
Apollo 4‡	Earth, 1967	4694	—	196	3.912	5425	11,140	−7.07
Venera-9§	Venus, 1975	1000	—	—	1	1560	10,700	−20.5
Viking	Mars, 1976	880	70	—	3.54	980	4420	−17
Pioneer 13	Venus, 1978							
– Large probe		360	45	—	1.42	316.5	11,540	−32.4
– North probe		190	45	—	0.78	91	11,670	−68.7
Galileo	Jupiter, 1995	222	44.9	—	1.264	335	47,400	−8.5
MiRKa	Earth, 1997	1000	—	—	1	154.4	7620	−2.51
Pathfinder	Mars, 1997	663.8	70	66.3	2.65	584	7260	−14.06
ARD	Earth, 1998	3360	—	140	2.80	2800	7451.7	−2.6
MER A	Mars, 2004	660	70	66	2.65	827	5400	−11.49
MER B	Mars, 2004	660	70	66	2.65	832	5500	−11.47
Genesis	Earth, 2004	442	60	34	1.52	205.6	10,770	−9
Huygens	Titan, 2006	1250	60	25	2.70	319	6100	−65
Stardust	Earth, 2006	220.2	59.5	20	0.811	45.2	12,799	−8.21
Phoenix	Mars, 2008	660	70	66	2.65	602	5900	−13
Hayabusa (MUSES-C)	Earth, 2010	202	45	—	0.404	16.27	11,300	−13.3

*The angles are counted back from the local horizontal. The values are approximate and vary from one source to another.
†The arbitrary reentry altitude is 120 km for the Earth or Mars, 1270 km for Titan because of the large scale factor, and 450 km above the isobar $p = 1$ bar for Jupiter.
‡Apollo-4 (unmanned) was the fastest of the Apollo flights.
§RussianSpaceWeb.com, "Venera-9/10," http://www.russianspaceweb.com/venera75.html [retrieved 24 Nov. 2008].

are close to those of the Pathfinder. Geometries found in these three families of cone angles are as follows:

- 8- to 10-deg angles for low-drag ballistic vehicles.
- 45-deg angle, in which the flow is supersonic on the conical part.
- 60- and 70-deg angles, in which the flow is subsonic until the rear of the conical part, at zero incidence; a small angle of attack is sufficient to make supersonic the leeward meridian.

The reentry environment is usually very severe, and the mass of thermal protection is a significant part of the mission, up to 50% in the case of Galileo [23]. The materials used are diverse, each tailored to the specific mission. They are described in Sec. 1.3.

We note in passing that very few planetary probes were monitored; those monitored include Viking, Galileo, and Pathfinder. In contrast, a number of tests dedicated to knowledge of phenomena have taken place in the terrestrial atmosphere, such as RAM C-III [28], Fire II [29], Apollo 2 [30], Re-entry

Fig. 1.11 ARD capsule after reentry (courtesy of European Space Agency).

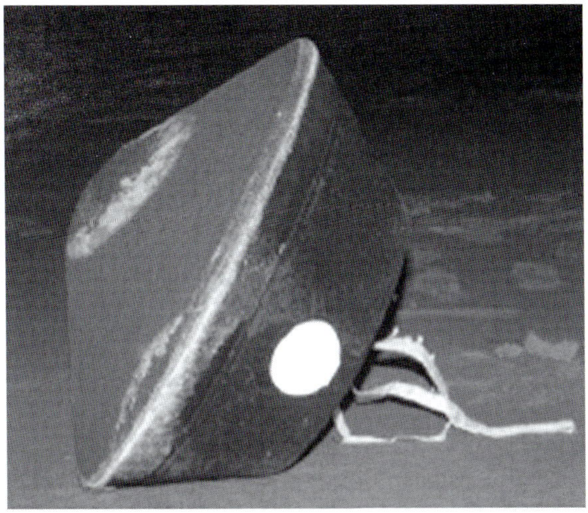

Fig. 1.12 Stardust capsule after reentry.

F [31], MiRKa [24], and ARD [25]. Note that, beyond the difficulty and cost of flight test measurements, wall ablation measurements are very difficult:

- They are limited in number and their best location is often known a posteriori.
- They are often intrusive (examples are given in Sec. 12.4).
- Measurements in the flow, challenging on an inert wall, are much more difficult here.

Moreover, characterization of the atmosphere is difficult. Traditional means such as balloons only work up to an altitude of approximately 30 km. For all these reasons, this type of test is a validation of a project much more than a scientific validation of calculations.

1.3 Major Classes of Materials for Thermal Protection Systems

The temperatures mentioned in the previous section seriously limit the materials possible for a TPS because they require high melting (or sublimation, or decomposition) temperatures. This is achieved only by having a solid, strong chemical covalent bond (as is found in silica, boron, and carbon) or a metallo-covalent bond (d-block elements: titanium, tungsten, tantalum, rhenium, and so on). In addition, lightness is an important issue in reducing the energy costs of propulsion. So in practice, the choice is reduced to the use of silica, carbon, boron (chemically similar to carbon but not used because of its cost), or possibly carbides, nitrides, or refractory oxides (SiC, B_4C, Si_3N_4, Al_2O_3, and so on).

Progressing further in the specification, we can see that the materials listed are performing well in terms of enthalpy of sublimation. (It is highly endothermic, so there is a good consumption of energy improving the protection.) However, enduring mechanical shear stresses is a problem: a polycrystalline ceramic generally is not suitable because it is too fragile. Indeed, as for traditional ceramics, its toughness is related to the average grain size and constitutive defects in the material. Because of this, it is very difficult to produce high quality ceramics. We then turn to the concept of a composite material, whose benefit is that we can significantly increase the energy for material failure by cleverly combining fiber and matrix. Section 1.3.1 briefly reviews the various types of fibers and binders (matrix) that can be employed. More detailed information can be found in [32], for example.

1.3.1 Composite Materials

The number of feasible materials is limited by usable fibers and binders, but multiplied by the possible architectures. The material can be machined from a block or directly deposited on the structure of the vehicle or on a mandrel by draping (Fig. 1.17 later in the chapter) or another technique.

Fibers are made of the following:

- Silica (SiO_2).
- Carbon (C), obtained by carbonization of a synthetic fiber such as rayon (cellulose), polyacrylonitrile (PAN), or pitch.
- Fibers such as boron fiber (on core W) or alumina fibers (Nextel fiber, made with 70% Al_2O_3, 28% SiO_2, and 2% B_2O_3) are seldom used because of their cost or difficulties of manufacturing. Others, such as silicon carbide (SiC) fiber, are used only for reusable thermal protection.
- Nylon (polyamide) was used in the first ablative materials [33], such as in the leeward part of Galileo [20].

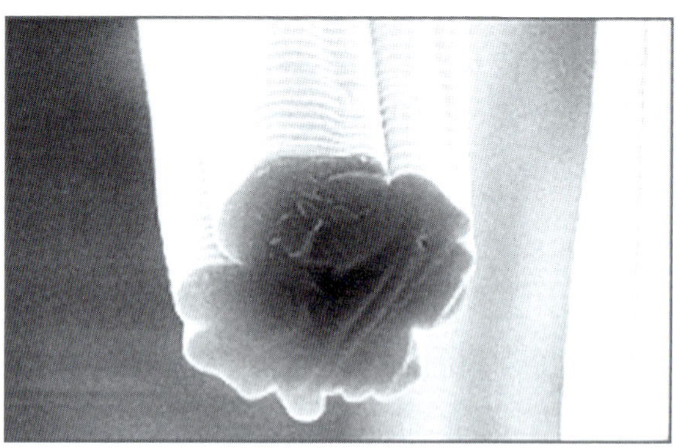

Fig. 1.13 Cellulosic fiber (micrometric scale); diameter 7 to 10 μm.

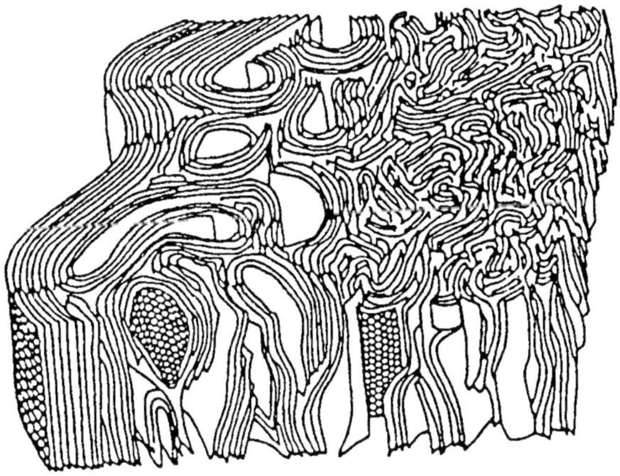

Fig. 1.14 Schematic structure of an ex-PAN carbon fiber (nanometric scale) (34). It shows the folding of graphite planes.

These fibers may be short, randomly arranged by mixing, or long and arranged in regular arrays of yarns with a few hundred to several thousand fibers, each having a diameter on the order of ten micrometers (Fig. 1.13). These fibers are structured; for example, carbon fibers (Fig. 1.14) have the graphite planes generally parallel to the fiber. This leads to highly anisotropic mechanical or thermal properties: heat conduction in graphite is about two orders of magnitude larger in the plane between planes. The quality of their structure gives them remarkable mechanical properties; such a fiber is about five times more resistant in traction than a steel fiber of the same diameter.

Depending on the application, the means used for producing materials may be:

- Weaving in a 2-D or 3-D form of fabric: twill, canvas (or taffeta), or satin (Fig. 1.15)
- Twining (rarely)
- Knitting (Fig. 1.16)
- Coiling (used mainly for producing reservoirs)
- Draping (a stack of tissues, sometimes with a change of direction to promote isotropy in terms of mechanical properties); among the materials produced in this way, note the carbon-resin tape-wrapped carbon phenolic (TWCP) made from strips of fabric cut to 45 deg with respect to the axes of wires and connected by sewing (Fig. 1.17); this angle (bias) was chosen due to the need to have no wire parallel to the surface (and thus able to be pulled)

Nonwoven material can also be developed by, for example, using an assembly of parallel yarns in which a third direction is created by pulling some of them with a needle. The idea of creating a third direction is found in the binding of tissues by sewing.

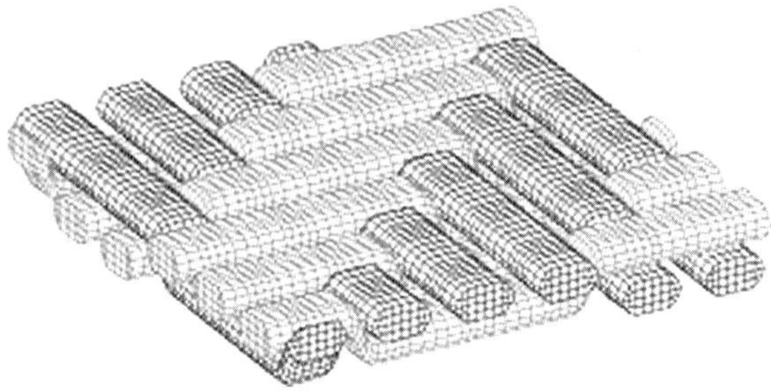

Fig. 1.15 TWCP fabric (millimetric scale).

The binder (matrix) can be of different types:

- Resin (epoxy, phenolic, silicone) for low-temperature and ablative materials.
- A refractory material (resistant to high temperature), such as SiO_2, C, SiC, or Ti, obtained by infiltration in liquid or the vapor phase of the preform; that is to say, the fibers are held in place by any method. (This is often a technological problem that is difficult to solve.)

The implementation of the matrix or "infiltration" is the subject of sometimes very complex processes. Some examples include [36]

- The pitch matrix of carbon–carbon (C–C) materials requires a pressure of several hundred bars, followed by carbonization at over 2800 K.
- The process of gas impregnation (chemical vapor infiltration or CVI) makes it difficult to optimize the quality of the carbon deposit.

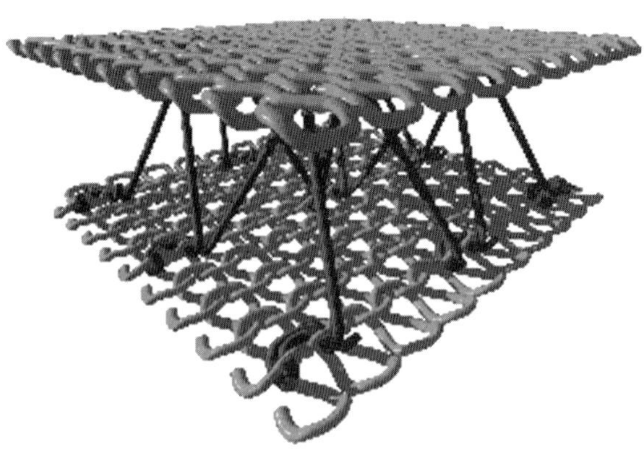

Fig. 1.16 Double surface with spacer obtained by knitting (courtesy of ETECT Society).

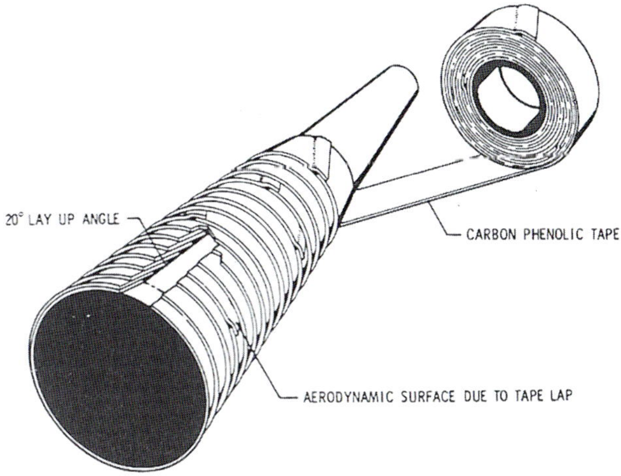

Fig. 1.17 Manufacturing of a TWCP carbon-resin heatshield material (35).

Quite often, an adjuvant is included in the matrix for a particular property, for example:

- Carbon powders, silica beads, or whiskers (microfibers) in resins to improve mechanical properties.
- Nanofibers to improve the fiber–matrix interface.
- Antioxidants for carbon matrix pieces; for example, a boron-based additive will create B_2O_3 oxide, which is liquid after formation, clogging the cracks in which it was created.

A final element of the design is the fiber–matrix interface, whose adhesive properties play a role in the mechanical properties of the composite. This interface can consist of real chemical bridges. The physical composition of the material forming the matrix can be significantly influenced by the presence of fiber in the vicinity thereof (e.g., graphite structure in the case of carbonaceous materials).

1.3.2 Other Materials

1.3.2.1 Foams

Carbon foams have been used to insulate against high temperatures; one example used for thermal protection is the C–C dual-layer type in which the heatshield is decomposed into two parts:

- The ablative part, for example, a thin layer of C–C.

- The insulator, such as a carbon foam able to retain its physical integrity and some mechanical properties up to 3000 K. This type of material has a very low density, typically less than 100 $kg \cdot m^{-3}$. An example of this type of material is shown in Fig. 1.18.

1.3.2.2 Ghost Materials

Also worth mentioning is the use of the following homogeneous materials for ablation tests at low temperature [38]:

- Ice (H_2O, CO_2)
- Camphor ($C_{10}H_{16}O$, developed formula 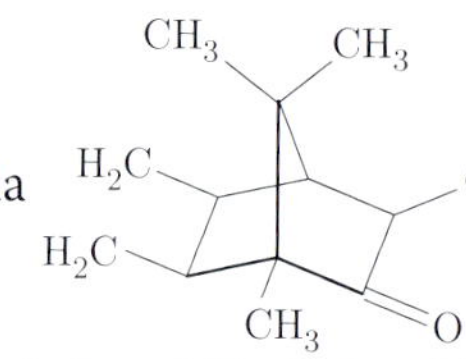 for which an ablation model is described in Rafinejad and Derbidge [39])
- Paraffin (alkans C_nH_{2n+2}, $n > 20$)
- Ethyl carbamate (urethane) 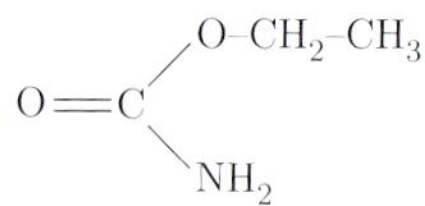
- Polymethylmetacrylate (PMMA, polymer 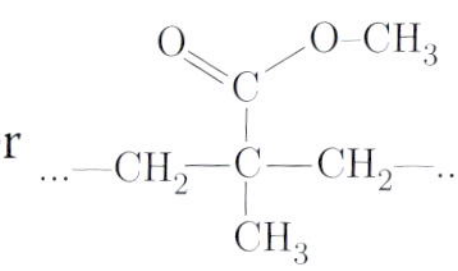 Plexiglas, Lucite)

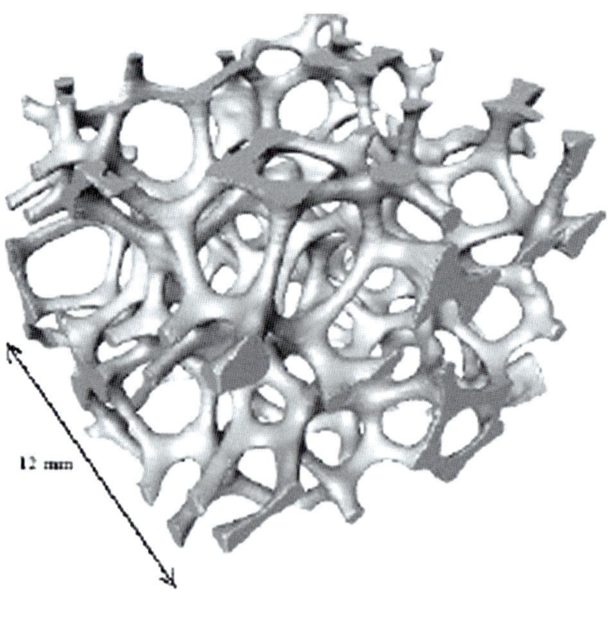

Fig. 1.18 Carbon foam (37).

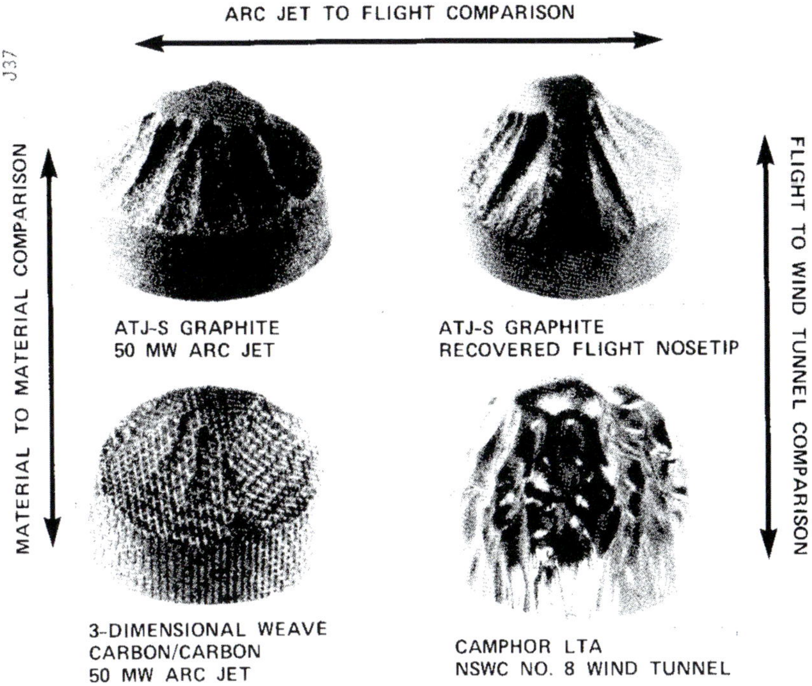

Fig. 1.19 Representativeness of hot or cold wind tunnel tests (40).

- Polycarbonate (polymer 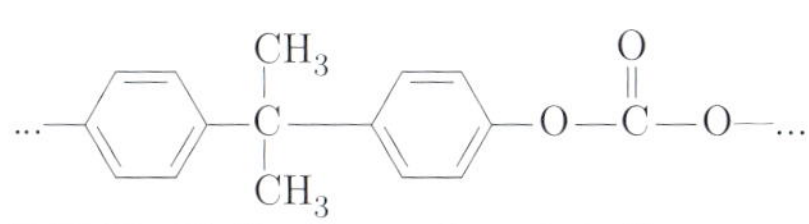 Lexan)
- Polytetrafluoroethylene (PTFE, Teflon; see Sec. 8.1.1)

These tests on alternative materials can be extremely successful if there is good certainty as to their representativeness. Figure 1.19 shows the quality of this type of approach, which may lead to misinterpretations when the phenomenology is not fully understood.

1.3.3 Some Examples

The following are some examples of materials used in different situations:

- The carbon–carbon composites (C–C) are used as ablative nozzles in powder propellants, in the noses of ballistic reentry vehicles, and in the Genesis probe (for terrestrial reentry). They are also used for the nose and leading edges of the space shuttle due to their excellent mechanical properties at high temperature; in this case they are covered by a ceramic (nonablative).
- Silica composites and resin (S–R) have been used for reentry in various forms: long fibers and phenolic resin for ballistic reentry; short fibers,

epoxy resin, and micro-balloons of phenolic resin in the case of Apollo (Avcoat) [41]; and short fibers and phenolic resin in the case of Huygens (material AQ60) or ARD (material Aleastrasil) [17].

- The tape-wrapped carbon phenolic (TWCP) carbon resin is used for nozzles of solid rocket boosters and for reentry probes including Pioneer (Venus) and Galileo (Jupiter). Note that the areas adjacent to the stagnation point, which are not feasible for winding fabrics like TWCP, are made by draping. Other types of fiber architectures were also used, such as the phenolic impregnated carbon ablator (PICA [42]) for the Stardust spacecraft (Earth reentry).
- The C–SiC composite has been used in the MiRKa mission (Micro Reentry Capsule) [43]. It was the outer part of an architecture composed of a pyrolyzable material separated from the C–SiC by a vented spacer. With the conditions encountered during this reentry, ablation was negligible.
- Mild missions or less-exposed parts of the body (rear) used materials with cork powder, silicon resin, and phenolic resin microspheres such as the super lightweight ablator (SLA) for Mars missions: Viking, Pathfinder [44], and MER (Mars Exploration Rover). Huygens used a silicone resin with silica spheres [45]. Materials with cork powder and phenolic resin have been studied since the Apollo design [46].

Finally, we mention that the first examples of ablative composite materials used nylon fiber (polyhexamethylene adipamide, $C_{12}H_{22}N_2O_2$) with a phenolic resin [47]. Moreover, PTFE has been studied (see Sec. 8.1.1) and tested in flight and plasma tests [48]. It was used as an ablative material in scientific flight tests, including RAM C III [49] and IRVE [13].

1.3.4 Use by Type of Mission

Table 1.4 summarizes the materials used and the severity of the corresponding mission determined through various parameters, of course dependent on the shape, mass, atmosphere, and other parameters of the mission.

Maximum pressure is an indicator of severity because it determines the thickness of the boundary layer, so the exchanges, directly and through the laminar-turbulent transition, as well as possible roughness affect the turbulent flow. This is shown in Fig. 1.20.

Maximum flux (calculated for an inert wall at normal temperature) depends not only on the pressure, but also on the total enthalpy. This quantity gives an indication of the thermomechanical constraints experienced by the material, which limits its use. The shear stress of the boundary layer (10^2 to 10^3 Pa) is negligible as compared to the thermal stress in the material (10^6 to 10^7 Pa), but plays a role when a liquid layer is present at the surface.

The thickness ablated is roughly proportional to the integrated flux (see Sec. 5.9). This is related to ablation mechanisms that will be discussed in Chapter 5, and the fact that elevated values of ablation velocity are associated

Table 1.4 Materials and Severity of Missions*,† (11, 17, 18, 20, 23, 24, 42, 50–55)

Material	ρ (kg . m^{-3})	Reentry Vehicle	Maximum Pressure (Pa)	Maximum Flux (MW . m^{-2})	Radiative Part (%)	Integrated Flux (MJ . m^{-2})
PICA	$\simeq$250	Stardust	3×10^4	11	11	360
		MSL‡	3.7×10^4	1.97	$\simeq 0$	54.8
SLA 561-V	260	Viking	5×10^3	0.21	$\simeq 0$	11
		Pathfinder	2×10^4	1.1	5	—
		MER B	1×10^4	0.5	$\simeq$0	45
		Phœnix	—	0.44	$\simeq$0	—
AQ60	280	Huygens	1×10^4	1.0	80	30
Prosial 1000	540–600	Huygens	—	0.025	$\simeq$0	—
Avcoat 5026-39G	530	Apollo 4	8×10^4	5.3	$\simeq$35	300
TWCP	1450	Pioneer§	7×10^5	72	47	117
		Galileo¶	7×10^5	350	$\simeq$50	2000
		Hayabusa	0.53×10^5	11.1	$\simeq$17	212
Aleastrasil	1650	ARD	0.22×10^6	1.2	—	—
C–SiC	3200	MiRKa	1.8×10^4	1.2	$\simeq$0	120
C–C	1800	Genesis	1×10^5	7	4	170

*Figure values shown are rough cold wall; they can be slightly different from one source to another.
†National Aeronautiics Space Administration, TPSX Material Properties Database, http://tpsx.arc.nasa.gov [retrieved 30 April 2009].
‡Edquist, K. T., Dyakonov, A. A., Wright, M. J., and Tang, C. Y., "Aerothermodynamic Design of the Mars Science Laboratory Heatshielc," AIAA Paper 2007-1206, 45th AIAA Aerospace Sciences Meeting and Exhibit, Reno, Nevada, Jan. 2007.
§Values for Galileo are blocked hot wall values.
¶Values for Pioneer are those of the early "North Probe."

with a quasi-stationary regime. The temperature rise at the beginning of the mission and decrease in the end produce only low ablation.

The part of the radiation in the incident total flux varies greatly from one mission to the next. In general, the radiation heat flux is low at low speeds but increases much faster with speed, as shown in Fig. 1.21 and discussed in Sec. 10.2.2. We saw earlier that the convective flow was $R^{-1}V_{\infty}^{3}$, which makes it the dominant mode up to high speeds. There is a counter-example with Huygens, which is related to the formation of a chemical species (CN) particularly active in the UV and formed behind the shock, so it is in a particularly hot region. It is a particular case associated with the composition of the atmosphere of Titan (Table 1.1).

Materials are often classified according to their density. This is an obvious parameter classification because, all else being equal, the ablation is inversely proportional to this parameter. This quantity also determines the homogeneity of composite materials and hence the state of surface ablation, the thermomechanical properties, and so on. However, relationships between these quantities are complex, and the quality of the material cannot be reduced to its density only.

Note that these materials are often a family, whose exact composition varies during its development. Thus, the material used for Apollo was Avcoat 5026-39G, the various SLA probes used SLA-561-V, and so on.

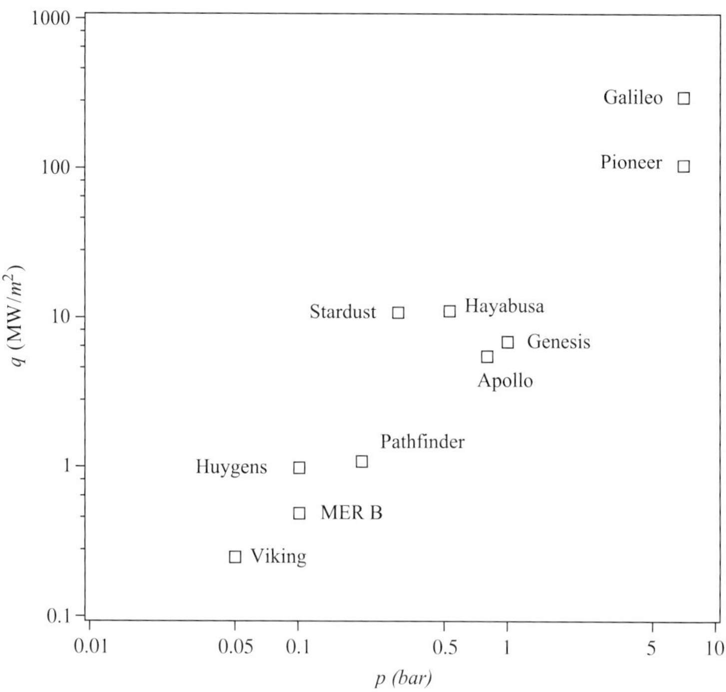

Fig. 1.20 Pressure–heat flux relationship.

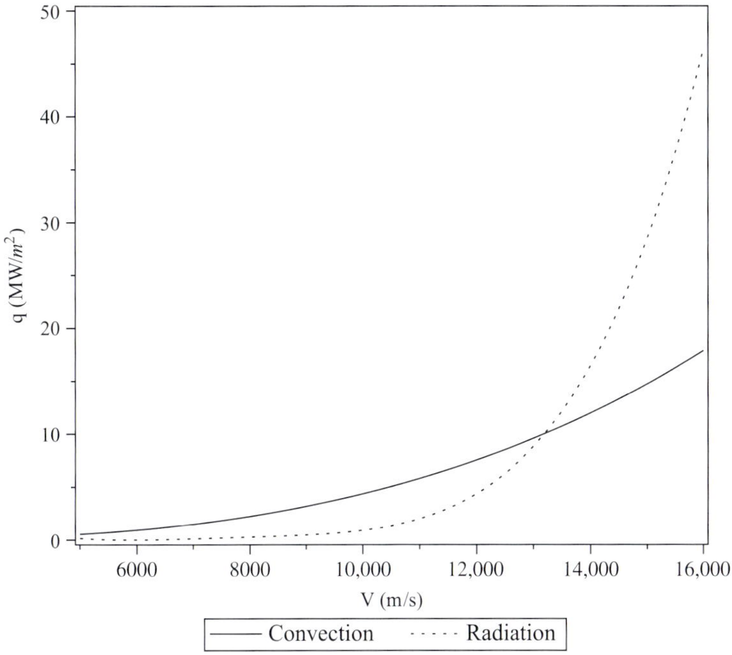

Fig. 1.21 Comparison of convective and radiative heat fluxes; Earth atmosphere, $h = 60$ km, $R = 0.5$ m.

Most ablative materials perform an ablative function, provide thermal insulation, and serve a structural function. However, there are counterexamples such as the following:

- The C–C dual layer contains ablative highly conductive carbon completed by an insulator (foam) of low-density carbon fiber for thermal protection.
- The surface protected ablator (SPA) consists of a pyrolyzable material covered with Si–C. Between the two materials, channels permit gas to escape. This is a complex architecture in which the external part seems not to play a significant role in ablation in the unique test made.

Care must be taken to choose an appropriate material because thermal protection can be an important part of the vehicle's mass—up to half in the case of Galileo. It is roughly proportional to the integrated flux [23], the ablation speed being roughly proportional to heat flux.

1.4 Physical Problems

A very diverse group of problems is encountered in the thermal protection materials. We will discuss some of them in this section for their representativeness, not only from an application point of view, but also because

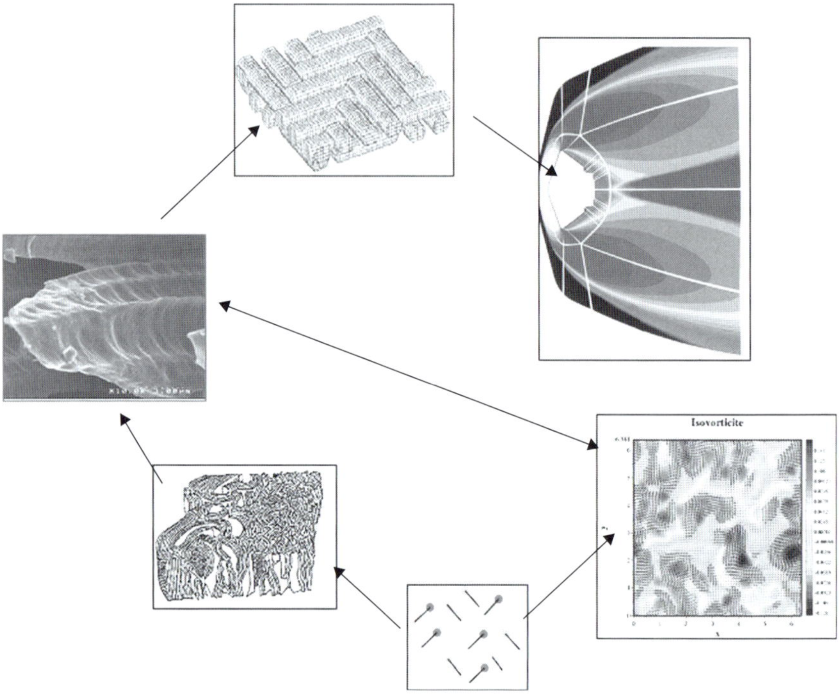

Fig. 1.22 Problem scales and interactions.

they involve a number of physical (or physicochemical) and numerical modeling techniques. These problems are present at very different scales (see Fig. 1.22):

- At the atomic level in homogeneous and heterogeneous reactive processes
- At the flow level, including very small scales (up to 10 μm) in turbulence
- At the level of fiber and its surroundings (microrugosity)
- At the cloth level (macrorugosity)
- At the level of the vehicle

There are many possible couplings (see Fig. 1.22) at both the physical and engineering levels. In some cases this coupling may be the source of problems at new scales (Chapter 6).

1.4.1 Material–Flow Interaction

The calculation of the external flow is a major initial difficulty. This problem is the subject of numerous books and will not be discussed here. We will address the multispecies aspect, particularly strong in this area, due to gas–material reactions. Specific aspects of transport (multispecies diffusion, calculation of transport coefficients) are discussed in Chapter 2.

1.4.1.1 Heterogeneous Surface Reactions

The first major class of problems concerns the interaction between material and flow. This includes the heterogeneous chemical reactions of the fluid medium with the wall. These reactions may be:

- *Mass balanced at steady state:* This is a problem encountered in reusable thermal protection. The temperature in the flow is sufficiently high ($\simeq$2000 K) to dissociate the molecular oxygen. A number of materials, such as metals, have the property of absorbing oxygen atoms and recombining them into O_2. This reaction is exothermic, and the corresponding energy is transferred to the wall; this is called a catalytic reaction (heterogeneous catalysis of recombination of oxygen). These reactions are coupled with the ablation reactions described next.
- *Unbalanced even at steady state:* In this case, part of the material providing the thermal protection reacts with the fluid molecules (oxygen, nitrogen) or changes phase (solid to liquid or solid to gas, but also gas to solid or gas to liquid). Indeed, if there is a disappearance of solid material this is not always true for a particular reaction or a particular location of the object (cold zone). This reaction is called ablation, and it is produced from temperatures of several hundred Kelvin for low-temperature ablators used in a "cold" wind tunnel to 4500 K for C–C material in terrestrial reentry. It should be noted that in this type of phenomenon, the material and its gaseous environment drives the temperature much more than heat flux alone. This is because there is a "runaway" reaction from a certain temperature (depending on the material and the atmosphere). We will improperly call this value "ablation temperature." In reality, ablation has a rather rapid variation rate around this temperature that results from the balance of energy phenomenon.

Note that in systems with ablation there are many catalytic reactions (Chapter 5).

The modeling of heterogeneous reactions is itself a difficult problem for a material such as carbon, which has a multitude of physical and chemical forms in the environment of interest. These phenomena are discussed in Chapters 3 and 8.

In some cases, the species present in the material or formed during heterogeneous reactions leads to the presence at the surface of a viscous liquid layer that migrates downstream. An important example of this is materials containing silica. This problem is addressed in Chapter 9.

1.4.1.2 Roughness

The material that undergoes ablation is inhomogeneous, with multiple characteristic scales, as described in Sec. 1.3. It therefore creates a differential ablation at the microscopic (fiber) or millimetric (scale of the yarn or cloth) level. You can also see the creation of a roughness at millimeter scale (scallop) on homogeneous materials. Roughness can drastically change the exchanges of mass, momentum, and energy between the material and the fluid in some cases:

- Directly, when the flow is turbulent
- Indirectly, by inducing a laminar–turbulent transition because of the locally created vortex

As we have said, the effects can be very important. For example, the wall heat flux can be increased to 300% (experimental result [56]). These phenomena are discussed in Sec. 7.1.

1.4.2 Heterogeneous Reactions in Porous Media

The phenomena mentioned here are for exchanges of mass and energy in porous media. The mass exchange induced by heterogeneous reactions leads to an evolution of the porosity of the medium. The porosity either increases, in the case of pyrolysis of the resins used in ablative materials, or decreases, in the case of carbon deposition (CVI). This phenomenon is also observed for the pyrolysis gases in the region near the surface of the material.

This variation in porosity results in a change of transport coefficients associated with permeability and thermal conductivity due to real or apparent radiative transfer (important in porous media above 1500 K). It also causes a variation in the specific surface area (open area gas per unit volume of material) that determines the amount of mass deposited or removed per unit volume and time.

Problems will be raised again at different scales:

- Microscopic to calculate the internal topology of the material and transport properties of the medium
- Macroscopic (homogenization) to calculate the phenomenon at a higher scale

It is easily conceivable that, in addition, the experimental problems themselves lead to new problems, for example:

- Reconstructing the internal topology of the material from a 3-D image obtained by tomography, consisting of volume elements (voxels) associated with shades of "grays" associated with local density
- Reconstructing by inverse method based on global experience transport quantity inaccessible to direct measurement

These problems are addressed in Chapters 8 and 12.

1.4.3 Radiation

When the velocity is very important or when the flow contains some very active species (CN, for example), the radiation must be taken into account.

Depending on circumstances, we may or may not assume the problem is decoupled from the fluid flow except through the wall terms. Solving the problem of radiation is cumbersome if one solves the kinetic equation of radiative transfer, particularly if coupled to the flow. How far can we simplify this problem? This is discussed in Chapter 10, using the moments method and closure by using the H-theorem (maximization of radiative entropy) for a gas of photons. That chapter also discusses the effect of gas injection on radiative wall heat flux.

The problem of radiation also arises in porous materials, especially at the very high temperatures encountered in ablation. The solution to the problem is not unique, and depends on the mean optical free path as compared to a length characteristic of macroscopic heat transfer. According to the ratio of these quantities, the problem must be treated as a problem of radiative transfer or led to the knowledge of an equivalent thermal conductivity. In this latter case, the problem of calculating this equivalent conductivity from the geometry of the medium and its thermo-optics properties is discussed in Sec. 10.4.

1.4.4 Erosion and Thermomechanical Effects

In some atmospheres, solid or liquid particles in suspension are present in clouds. Crossing these clouds at high speed leads to a difficult problem with considerable theoretical challenges: the effect of flow on the particle, of particle on the flow, the impact and creation of ejectas, and the effects of these ejectas. All of these very complex problems are discussed in Chapter 11, which shows there are still many unresolved issues.

Many publications speak for removing material in solid form under thermomechanical effects. If this phenomenon exists and is visible on the test films, the mechanisms that are at the origin are not clearly justified. In some cases this phenomenon is simply given as an ad hoc explanation for discrepancies between theory and experiments. This phenomenon is not studied in this book.

1.4.5 Approximative Methods and Experiments

You can now see the enormity of the physical problems involved in the phenomena of reentry. It is obvious that approximate methods are needed for understanding such a system. Such methods are developed in Chapter 4. They still constitute the basis for many computer codes of engineering.

For similar reasons, experience remains an important base of knowledge. Chapter 12 focuses on experiments and instrumentation specific to the field, particularly plasma jet experiments and their limitations. A few examples in that chapter will also show the utilization of this type of test and how we can characterize a material's ablation and thermal properties.

References

[1] U.S. Committee on Extension to the Standard Atmosphere, *U.S. Standard Atmosphere* 1966, U.S. Government Printing Office, Washington, DC, 1966.

[2] American Institute of Aeronautics and Astronautics, "Guide to Reference and Standard Atmosphere Models," AIAA Standards Series G-003B, Jan. 2004.

[3] Leslie, F. W., and Justus, C. G., "Earth Global Reference Atmospheric Model (GRAM99): Short Course," *Earth Global Reference Atmospheric Model Workshop and Short Course*, Denver, Colorado, Aug. 2007.

[4] Yelle, R. V., Strobell, D. F., Lellouch, E., and Gautier, D., "The Yelle Titan Atmosphere Engineering Models," ESA SP-1177, *Huygens: Science, Payload and Mission, Proceedings of an ESA Conference*, edited by Wilson, A., ESA Publications, Noordwijk, The Netherlands, 1997, p. 243.

[5] Lewis, S. R., Collins, M., Read, P. L., Forget, F., Hourdin, F., Fournier, R., Hourdin, C., Talagrand, O., and Huot, J.-P., "A Climate Database for Mars," *Journal of Geophysical Research, Planets*, Vol. 104, No. E10, 1999, 24177–24194.

[6] Prince, J. L., Grover, M. R., Desai, P. N., and Queen, E. M., "2007 Mars Phoenix Entry and Landing Simulation and Modeling Analysis," *NASA Technical Document Server*, Document ID 20080013471 [retrieved 24 August 2008].

[7] Gallais, P., *Atmospheric Re-Entry Vehicle Mechanics*, Springer-Verlag, Berlin, 2007.

[8] Lyons, D. T., and Desai, P. N., "Adventures in Parallel Processing: Entry, Descent and Landing Simulation for the Genesis and Stardust Missions," AAS Report 05-267, *AIAA/AAS Astrodynamics Specialist Conference*, Lake Tahoe, California, Aug. 2005.

[9] Muylaert, J., Kordulla, W., Marraffa, L., Schwane, R., Spel, M., Welpot, L., and Wong, H., "Aerothermodynamic Analysis of Space-Vehicle Phenomena," *ESA Bulletin*, Vol. 105, 2001, pp. 69–79.

[10] Fay, J. A., and Riddell, F. R., "Theory of Stagnation Point Heat Transfer in Dissociated Air," *Journal of the Aerospace Sciences*, Vol. 25, No. 2, 1958, pp. 83–86.

[11] Laub, B., and Curry, D., "Tutorial on Ablative TPS," *NASA Ames Conference Center, Second International Planetary Probe Workshop*, Moffett Field, CA, August 2004.

[12] Smith, A. J., Gogel, T., and Vandevelde, P., "Plasma Radiation Database PARADE, Final Report," Fluid Gravity Engineering Ltd Report TR28/96, 1996.

[13] Hughes, S. J., Dillman, R. A., Starr, B. R., Stephan, R. A., Lindell, M. C., and McNeil Cheatwood, F., "Inflatable Re-entry Vehicle Experiment (IRVE) Design Overview," AIAA Paper No. 2005–1636, *18th Decelerator Systems Technology Conference and Seminar*, Munich, Germany, May 2005.

[14] Marraffa, L., Kassing, D., Baglioni, P., Wilde, D., Walther, S., Pitchkhadze, K., and Finchenko, V., "Inflatable Reentry Technologies: Flight Demonstration and Future Prospects," *ESA Bulletin*, Vol. 103, 2000, pp. 78–85.

[15] Laub, B., Wright, M. J., and Venkatapathy, E., "Thermal Protection Systems Technology (TPS) Design and the Relationship to Atmospheric Entry Environments," *Sixth International Planetary Probe Workshop, Short Course on Extreme Environments Technologies*, Atlanta, Georgia, June 2008.

[16] Ball, A. J., Garry, J. R. C., Lorenz, R. D., and Kerzhanovich, V. V. *Planetary Landers and Entry Probes*, Cambridge University Press, Cambridge, UK, 2007.

[17] Bouilly, J.-M., Dariol, L., and Leleu, F., "Ablative Thermal Protections for Atmospheric Entry. An Overview of Past Missions and Needs for Future Programmes," ESA SP-631, *5th European Workshop on Thermal Protection Systems and Hot Structures*, edited by Fletcher, K. ESTEC, Noordwijk, The Netherlands, 2006.

[18] Braun, R. D., and Manning, R. M., "Mars Exploration Entry, Descent and Landing Challenges," *IEEE Transactions on Automatic Control Conference, Paper No. 0076*, Toronto, Canada, Aug. 2005.

[19] Cassell, A. M., Allen, G. A., Winter, M., Grinstead, J. H., Antimisiaris, M. E., Albers, J., and Jennsikens, P., "Hayabusa Reentry: Trajectory Analysis and

Observation Mission Design," *International Planetary Probe Workshop 8*, Portsmouth, Virginia, June 2011.

[20] Davies, C., and Arcadi, M., *Planetary Mission Entry Vehicles, Quick Reference Guide, Version 3.0*, NASA SP-2006-3401, 2006.

[21] Desai, P. M., and Qualls, G. D., "Stardust Entry Reconstruction," *Journal of Spacecraft and Rockets*, Vol. 47, No. 5, 2010.

[22] Hillje, E. R., "Entry Aerodynamics at Lunar Return Conditions Obtained from the Flight of Apollo 4 (AS-501)," NASA Report TN D-5399, Oct. 1969.

[23] Laub, B., and Venkatapathy, E., "Thermal Protection Systems Technology and Facility Needs for Demanding Future Planetary Missions," *Second International Workshop on Planetary Probe Atmospheric Entry and Descent Trajectory Analysis and Science*, Lisbon, Portugal, Oct. 2003.

[24] Müller-Eigner, R., Koppenwallner, G., and Fritsche, B., "Pressure and Heat Flux Measurement with RAFLEX II During MIRKA Reentry," *Proceedings of the Third European Symposium on Aerothermodynamics for Space Vehicles*, ESTEC, Noordwijk, The Netherlands, Nov. 1998.

[25] Tran, P., Paulat, J.-C., and Boukhobza, P., "Re-entry Flight Experiments Lessons Learned: The Atmospheric Reentry Demonstrator ARD," in *Flight Experiments for Hypersonics Vehicle Development*, AVT/VKI Lecture Series RTO-EN-AVT-130, Paper 10, June 2007.

[26] Theisinger, J. E., Braun, R. D., and Clark, I. G., "Aerothermodynamic Shape Optimization of Hypersonic Entry Aeroshells," AIAA Paper 2010–9200, *Thirteenth AIAA/ISSMO Multidisciplinary Analysis Optimization Conference*, Fort Worth, Texas, Sept. 2010.

[27] Brown, J. L., Garcia, J. A., and Kinney, D. J., "An Asymmetric Capsule Geometry Study for CEV," AIAA Paper 2007–704, *45th AIAA Aerospace Sciences Meeting and Exhibit*, Reno, Nevada, Jan. 2007.

[28] Weaver, W. L., and Bowen, J. T., "Entry Trajectory, Entry Environment, and Analysis of Spacecraft Motion for the RAM C-III Flight Experiment," NASA Report TM X-2562, June 1972.

[29] Cauchon, D. L., "Radiative Heating Results from the Fire II Flight Experiment at a Reentry Velocity of 11.4 Kilometers per Second," NASA Report TM X-1402, July 1967.

[30] Wright, M. J., Prabhu, D. K., and Martinez, E. R., "Analysis of Afterbody Heating Rates on the Apollo Command Modules, Part 1: AS-202," AIAA Paper 2004–2456, *37th AIAA Thermo-physics Conference*, Portland, Oregon, June 2004.

[31] Wright, M. J., and Zoby, E. V., "Flight Measurements of Boundary-Layer Transition on a 5° Half-Angle Cone at Freestream Mach Number of 20 (Reentry F)," NASA Report TM X-2253, May 1971.

[32] Matthews, F. L., and Rawlings, R. D., *Composite Materials: Engineering and Science*, Chapman & Hall, London, 1999.

[33] Maahs, H. G., "Ablation Performance of Glasslike Carbons, Pyrolytic Graphites, and Artificial Graphite in the Stagnation Pressure Range 0.035 to 15 Atmospheres," NASA Technical Note TN D-7005, Dec. 1970.

[34] Johnson, J. W., and Thorne, D. J., "Effect of Internal Polymer Flaws on Strength of Carbon Fibres Prepared from an Acrylic Precursor," *Carbon*, Vol. 7, No. 6, 1969, pp. 659–660.

[35] Kryvoruka, J. K., and Bramlette, T. T., "Effect of Ablation-Induced Roll Torques on Re-entry Vehicles," *Journal of Spacecraft and Rockets*, Vol. **14**, No. 6, 1977, pp. 370–375.

[36] Savage, G., *Carbon-Carbon Composites*, Chapman and Hall, London, 1993.

[37] Doermann, D., and Sacadura, J.-F., "Heat Transfer in Open Cell Foam Insulation," *Journal of Heat Transfer*, Vol. **118**, No. 1, 1996, pp. 88–93.

[38] Larson, H. K., and Mateer, G. G., "Cross-Hatching. A Coupling of Gas Dynamics with the Ablation Process," AIAA Paper 68–670, 1968.

[39] Rafinejad, D., and Derbidge, C., "Computer User's Manual: Erosion Shape (EROS) Computer Code," Defense Technical Information Center, Document ADA026613, Passive Nosetip Technology (PANT) Program, Acurex/Aerotherm Interim Report, Vol. 17, Dec. 1974.
[40] Koyabashi, W. S., and Saperstein, J. L., "Low-Temperature Ablator Test for Shape-Stable Nosetip Applications on Manoeuvering Re-entry Vehicles," *Thermophysics of Atmospheric Entry*, Vol. 82, Progress in Astronautics and Aeronautics, AIAA, New York, 1982, pp. 148–176
[41] Pavlovsky, J. E., and Leger, L. G., "Apollo Experience Report. Thermal Protection Subsystem," NASA Report TN D-7564, Jan. 1974.
[42] Tran, H. K., Johnson, C. E., Rasky, D. J., Hui, F. C. L., Hsu, M.-T., Chen, T., Chen, Y. K., Paragas, D., and Koyabachi, L., "Phenolic Impregnated Carbon Ablators (PICA) as Thermal Protection Systems for Discovery Missions," NASA Technical Memorandum 110440, April 1997.
[43] Hald, H., and Ulmann, T., "Reentry Flight and Ground Testing Experiences with Hot Structures of C/C-SiC Material," AIAA Paper 2003-1667, *44th AIAA/ASME/ASCE/AHS Structures, Structural Dynamics, and Materials Conference*, Norfolk, Virginia, April 2003.
[44] Willcockson, W. H., "Mars Pathfinder Heatshield Design and Flight Experience," *Journal of Spacecraft and Rockets*, Vol. 36, No. 3, 1999.
[45] Labaste, V., and Mignot, Y., "EADS Launch Vehicles' Main Achievements for Planetary Exploration Missions," ESA SP-521, *Proceedings of 4th European Workshop "Hot Structures and Thermal Protection Systems for Space Vehicles,"* Palermo, Italy, Nov. 2002, European Space Agency, Paris, 2003, p. 99.
[46] Graves, R. A., Jr., and Walton, T. E., Jr., "Free-Flight Test Results on the Performance of Cork as a Thermal Protection Material," NASA Technical Report TN D-2438, Sept. 1964.
[47] Pike, R. W., Farmer, R. C., and Balhoff, J. F., "Coupled Radiating Shock Layers with Finite Rate Chemistry Effects," Final Report Part II, Reacting Fluid Laboratory, Department of Chemical Engineering, Louisiana State University, June 1975.
[48] Winters, C. W., Witte, W. G., Rashis, B., and Hopko, N., "A Free-Flight Investigation of Ablation of a Blunt Body to a Mach Number of 13.1," NASA Report TN D-2354, July 1954.
[49] Dunn, M. G., and Kang, S.-W., "Theoretical and Experimental Studies of Reentry Plasmas," NASA Technical Report CR-2232, April 1973.
[50] Bouilly, J.-M., "Thermal Protection System of the Huygens Probe During Titan Entry: Flight Preparation and Lessons Learned," ESA SP-631, *4th International Symposium on Atmospheric Reentry Vehicles and Systems*, Arcachon, France, May 2005, European Space Agency, Paris, 2006.
[51] Gupta, R. N., "Aerothermodynamics Analysis of Stardust Sample Return Capsule with Coupled Radiation and Ablation," AIAA Paper 99-0227, *37th Aerospace Science Meeting*, Reno, Nevada, Jan. 1999.
[52] Jenniskens, P., Wercinski, P., Olejniczack, J., Raiche, G., Kontinos, D., Allen, G., Desai, P. N., Revelle, D., Hatton, J., Baker, R. L., Russell, R. W., Taylor, M., and Rietmeijer, F., "Preparing for Hyperspeed MAC: An Observing Campaign to Monitor the Entry of the Genesis Sample Return Capsule," *Earth, Moon and Planets*, Vol. **95**, 2004, pp. 339–360.
[53] Mazoué, F., and Marraffa, L. "Flowfield/Radiation Coupling Analysis for Huygens Probe Entry into Titan Atmosphere," AIAA Paper 2005-5392, *38th AIAA Thermophysics Conference*, Toronto, Ontario, June 2005.
[54] Mazoué, F., "Huygens Probe Entry into Titan Atmosphere. Heat Fluxes Studies. Part 2: Heat Fluxes Around Huygens Probe," ESA Technical Report, Feb. 2004.
[55] Schaefer, J. W., and Dahm, T. J., "Studies of Nozzle Ablative Material Performance for Large Solid Boosters," NASA Technical Report CR-72080, Dec. 1966.
[56] Wool, M. R., "Summary of Experimental and Analytical Results," Interim Report: Passive Nosetip Technology (PANT) Program, Vol. X, SAMSO-TR-74-86, Jan. 1975.

Chapter 2 Conservation Laws for a Multispecies Gaseous Medium

- Review mathematics of conservation laws
- Apply the conservation laws
- Describe multispecies environments
- Approximating diffused media
- Calculation of transport coefficients
- Write conductivity expressions

2.1 Introduction

2.1.1 Microscopic and Macroscopic Aspects

The conservation laws will be named later Navier–Stokes equations even this is not the history of fluid mechanics where the equations applicable to a moving fluid were first established for an incompressible medium. In this context, the Navier–Stokes equations referred to the conservation of mass and momentum, but not to the energy equation.

The modern version of the conservation laws comes from the works of Chapman, Enskog, and Burnett, who established these laws from the microscopic level, described by the Boltzmann equation, written for the species i

$$\frac{\partial f_i}{\partial t} + \mathbf{V} \cdot \nabla f_i = \sum_j \mathbf{Q}(f_i, f_j) \tag{2.1}$$

This kinetic equation describes the evolution of the probability density of velocity f_i and energy states corresponding to internal degrees of freedom of the molecule or molecular ion. $Q(f_i, f_j)$ is the collision kernel, a quadratic term describing complex molecular interactions. This approach has the merit of allowing an explanation of all the terms involved in the transport at macroscopic level, which are, in the traditional approach, subject to a first-gradient closure between the flux and the gradient of a quantity. For example, Fick's law connects the diffusion flux to the gradient of a mass fraction

$$\mathbf{J}_i = -\rho \mathcal{D}_i \nabla c_i \tag{2.2}$$

The diffusion coefficient $\mathcal{D}_i$ has to be measured in the traditional approach.

The integration of the Boltzmann equation in velocity and energy spaces leads to macroscopic laws sometimes called Enskog equations. However, these equations are not closed because they reveal unknown quantities—heat flux vector and pressure tensor—which are collision-dependent. These quantities are obtained by a method of small perturbations of the dimensionless equation provided by Chapman and Enskog [1–3]. Solving these equations rigorously is possible only in the case of certain potentials (hard sphere or Maxwellian potentials). The solution to the zero-order leads to the Euler equations, in which transport properties are absent. The corresponding distribution of velocities and energies, however, is sufficient to study reactive collisions, both homogeneous and heterogeneous (see Chapter 3). This distribution is the Maxwell–Boltzmann equilibrium (given here for a single temperature)

$$f_i = \frac{\alpha_i m_i^3}{h^3 Q_{\mathrm{trans}}} \exp\left(-\frac{m_i v_i^2}{2kT}\right) \tag{2.3}$$

α_i is the fraction of molecules in the state i, of energy E_i counted from the fundamental state, described by Boltzmann statistics [4]

$$\alpha_i = \frac{g_i}{Q_{\mathrm{int}}} \exp\left(-\frac{E_i}{kT}\right) \tag{2.4}$$

Translational and internal partition functions are given by the following expressions:

$$\begin{cases} Q_{\mathrm{trans}}(T) &= \left(\frac{2\pi m_i kT}{h^2}\right)^{\frac{3}{2}} \\ Q_{\mathrm{int}}(T) &= \sum_j g_j \exp\left(-\frac{E_j}{kT}\right) \end{cases} \tag{2.5}$$

Partition functions play a fundamental role in thermodynamics. They can also be used for the approximation of the thermodynamic quantities (see Appendix B).

The modifications at the first order of the distribution function, corresponding to the Navier–Stokes equations [3], are not taken into account (e.g., for modeling chemical reactions [Chapter 3]). Moreover, interaction with a wall substantially alters the distribution over a distance of some mean free paths (Knudsen layer). This effect is ignored in this book. Note, however, that some chemical kinetics data come from the interpretation of experiments where this effect is taken into account, inducing a problem of consistency.

In severe cases, such as with shock waves, the internal degrees of freedom of the molecules are excited and their distribution is no longer described by a Boltzmann distribution, or, to a lesser level of complexity, it is described by a Boltzmann distribution different from that described by the translational temperature T of a Maxwellian distribution. Section 2.6 describes this type of environment and physical approximations allowing a macroscopic calculation at a reasonable cost, although still high. Note that the various studies on the subject do not lead to a unique physical model.

2.1.2 Intermolecular Potentials

2.1.2.1 Atoms or Molecules

The calculation of transport coefficients requires knowledge of the interaction potential $V(r)$ (we limit ourselves to spherically symmetric potentials) between two particles, molecules, atoms, ions, or electrons. Various types of potentials are used:

1. Potentials with limited physical contents to push far beyond the analytical calculations. In this category are:
   - The hard sphere, representing an interaction of billiard balls
   - The "variable hard sphere" potential, an adaptation of precedent in which the radius of interaction depends on the collision energy, a modification introduced for better physical representation by Bird for direct Monte Carlo resolution of the Boltzmann equation [5]
   - Potentials to account approximately for the rotation: a rough sphere, taking account of the tangential impulsion at the point of contact, a loaded hard sphere, and spherical geometry with cylinder ends [2]
   - The Maxwell potential $V = a/r^4$ of small physical representativeness but authorizing calculation of an exact macroscopic solution [2]
2. More physical potentials include:
   - Lennard-Jones potential

     $$V = 4\epsilon\left[\left(\frac{\sigma}{r}\right)^{12} - \left(\frac{\sigma}{r}\right)^{6}\right]$$

     which is the basis of calculations in the field and therefore widely used. The parameters of this potential are the collision cross-section σ (a distance of interaction for which $V = 0$, typically a fraction of a nanometer) and ϵ/k values present in the reduced temperature $T^* = kT/\epsilon$, representing the intensity of potential in the equilibrium of attractive and repulsive forces. This potential is valid at low temperatures (typically $T < 5000$ K), the repulsive part corresponding to high energies not being described properly. Cross-sections of collision are such that $\epsilon_{ik} \simeq (\epsilon_{ii}\epsilon_{kk})^{1/2}$ and $\sigma_{ik} \simeq (\sigma_{ii} + \sigma_{kk})/2$ [3].

- At the other end of the physical spectrum, one can mention the potential of diatomic molecules based on spectroscopic properties such as the Hulburt–Hirschfelder potential [6, 7] or potentials given by the RKR method (Rydberg–Klein–Rees) [8] to obtain a potential in numeric form, each point being the image of a portion of the optical spectrum.

2.1.2.2 Charged Species

The low mass of electrons makes them highly mobile. The electrons can be viewed like a set of charged particles, ions and electrons, creating a null macroscopic electric field and electric current in the absence of an external source (e.g., a conductive wall). The ions move through this environment, disrupting the local potential created by electrons. The electric field of the ion is shielded by the electrons at a characteristic distance called the Debye length. This quantity is defined by the following expression, where the medium is defined by two temperatures, T_e for electrons and T for the translational motion of ions, molecules, or atoms (see Sec. 2.6)

$$l_D = \left[\frac{q_e^2\left(\frac{n_e}{T_e} + \sum_i Z_i^2 \frac{n_i}{T}\right)}{\epsilon_0 k}\right]^{\frac{1}{2}} \tag{2.6}$$

where Z_i is the charge number of ion i.

In the case of a medium (see Sec. 2.4.2) with one ion having an unique charge, we get

$$l_D = \left[\frac{n_e q_e^2}{\epsilon_0 k}\left(T^{-1} + T_e^{-1}\right)\right]^{\frac{1}{2}} \tag{2.7}$$

The macroscopic electric current is null. This phenomenon can be seen as a movement of ions accompanied by an electron cloud. This analogy is the origin of the notion of ambipolar diffusion.

The potentials used are derived from an approximate calculation of the Poisson equation, with sources using either the Debye potential (when considering an ion surrounded by electrons) or the Debye–Hückel (when taking into account the influence other ions).

Unlike previous potentials, the potentials of Debye or Debye–Hückel explicitly depend on the rate of attendance of charged particles. The dependence of associated transport properties on temperature is markedly different than for neutral particles. This observation has implications on approximations used in an environment where neutral and charged particles coexist.

Potential can be expressed in two ways:

- On a continuous form

$$V(r) = -\frac{Zq_e}{r}\exp\left(-\frac{r}{l_D}\right) \tag{2.8}$$

- On a discontinuous form

$$\begin{cases} V(r) = -\dfrac{Zq_e}{r} & \text{if } r < l_D \\ V(r) = 0 & \text{if } r > l_D \end{cases} \tag{2.9}$$

The numerical results on the collision integrals related to these two assumptions are not significantly different.

2.1.3 Collision Integrals

2.1.3.1 Neutral Particles

This section concerns interactions in which at least one particle is neutral; therefore, the interaction potential in binary interaction is short range, significantly lower than the mean free path.

The passage to the macroscopic scale involves a number of average quantities obtained by summing over all possible interactions, defined by the distance of collision b (smaller distance between the initial trajectories of the particles) and the relative velocity of impact v_{ij} (in the center of mass). These quantities are the collision integrals, defined in the case of elastic collisions by [3]

$$\Omega_{ij}^{(l,s)}(T) = \left(\frac{kT}{2\pi m_{ij}}\right)^{\frac{1}{2}} \int_0^\infty \gamma_{ij}^{2s+3} e^{-\gamma_{ij}^2} \mathcal{S}_{ij}^l \mathrm{d}\gamma_{ij} \tag{2.10}$$

with

$$\begin{cases} \gamma_{ij} = \left(\dfrac{m_{ij}}{kT}\right)^{\frac{1}{2}} \dfrac{v_{ij}}{2} \\ m_{ij} = \frac{m_i m_j}{m_i + m_j} \\ \chi_{ij} = \pi - 2b \displaystyle\int_{r_m}^\infty \frac{dr}{r^2\left[1 - \frac{b^2}{r^2} - \frac{V(r)}{\frac{1}{2}m_{ij}v_{ij}^2}\right]^{\frac{1}{2}}} \end{cases} \tag{2.11}$$

χ_{ij} is the deflection angle, m_{ij} is the reduced mass, r is the distance between particles, and r_m is the minimum approach distance. We also use the differential cross-section of collision-dependent potential V via the deflection angle χ_{ij}

$$\mathcal{S}_{ij}^{l}(v_{ij}) = 2\pi \int_0^{\infty} (1 - \cos^l \chi_{ij}) \, \mathrm{bdb} \tag{2.12}$$

In the case of inelastic collisions, that is, with change in internal degrees of freedom, these definitions continue to be used but describe the internal state of each chemical species. Thus the molecule 1 will pass from state i to state m, the molecule 2 from j to n, and so on. The differential cross-section is then $\mathcal{S}_{i\to m}^{j\to n}(g \to g', \chi)$.

The collision integral is often written using the following operator:

$$\langle \bullet \rangle = \left(\frac{kT}{2\pi m_{ij}}\right)^{\frac{1}{2}} \int_0^{\infty} (\bullet) \, \gamma_{ij}^3 e^{-\gamma_{ij}^2} \mathcal{S}_{ij}^{l} \, \mathrm{d}\gamma_{ij} \tag{2.13}$$

Table 2.1 shows that μ_i and λ_i use the same collision integrals. We therefore would expect to find relationships between these two quantities. To the contrary, any relationship of any of these two quantities with $\mathcal{D}_{ii}$ can be obtained only in the case of a particular potential, the hard sphere being one of them.

The collision integrals are analytic in the case of a hard sphere potential

$$\left[\Omega_{ij}^{(l,s)}\right]_{HS} = \left(\frac{kT}{2\pi m_{ij}}\right)^{\frac{1}{2}} \frac{(s+1)!}{2} \left[1 - \frac{1 + (-1)^l}{2(1+l)}\right] \pi\sigma_{ij}^2 \tag{2.14}$$

where σ_{ij} is the interaction distance.

The hard sphere integrals are used to define dimensionless collision integrals

$$\Omega_{ij}^{(l,s)^*} = \frac{\Omega_{ij}^{(l,s)}}{\left[\Omega_{ij}^{(l,s)}\right]_{HS}} \tag{2.15}$$

Thus, the quantities $\Omega_{ij}^{(l,s)^*}$ are of order unity and vary slowly with T (or T^*). They are tabulated for air and some carbon compounds [3, 6, 9–15].

Table 2.1 Dependence of Transport Properties on Collision Integrals*

Property	Species Alone	Mixture
Viscosity	$\Omega^{(2,2)}$	$\Omega^{(2,2)}$, $\Omega^{(1,1)}$
Conductivity	$\Omega^{(2,2)}$	$\Omega^{(2,2)}$, $\Omega^{(1,s)}$, $s = 1, 2, 3$
Diffusion	$\Omega^{(1,1)}$	$\Omega^{(1,1)}$
Thermal diffusion	—	$\Omega^{(2,2)}$, $\Omega^{(1,s)}$, $s = 1, 2, 3$

*In the case of the property of a mixture, the terms $\Omega^{(1,3)}$ are generally neglected.

Subsequently, we also use quantities directly related to these values

$$\begin{cases} A_{ij}^{*} = \dfrac{\Omega_{ij}^{(2,2)^{*}}}{\Omega_{ij}^{(1,1)^{*}}} \\ B_{ij}^{*} = \dfrac{5\Omega_{ij}^{(1,2)^{*}} - 4\Omega_{ij}^{(1,3)^{*}}}{\Omega_{ij}^{(1,1)^{*}}} \\ C_{ij}^{*} = \dfrac{\Omega_{ij}^{(1,2)^{*}}}{\Omega_{ij}^{(1,1)^{*}}} \end{cases} \tag{2.16}$$

These quantities are of order unity for neutral particles.

The recurrence relation to compute C_{ij}^{*} makes it a dependent variable. This relationship is as follows [3]:

$$T\frac{\partial\Omega_{ij}^{(l,s)}}{\partial T} + \left(s + \frac{3}{2}\right) = \Omega_{ij}^{(l,s+1)} \tag{2.17}$$

After some calculations, we get

$$C_{ij}^{*} = \frac{1}{3}\left[\frac{\partial\left(\ell n\,\Omega_{ij}^{(1,1)}\right)}{\partial(\ell n T)} + \frac{5}{2}\right] \tag{2.18}$$

The collision integrals are known with an accuracy of about 10 to 20%, slightly less when ab initio calculations were made, usually for atom–atom or atom–atomic ion couples. The lack of precision is higher for some interactions that are difficult to study experimentally, such as for the electron with species existing only at high temperatures [16]. Precision on C^{*}, quantity close to unity, is particularly damaging to the calculation of thermal diffusion coefficients, which vanish for $C^{*} = 6/5$.

2.1.3.2 Charged Particles

In the case of charged particles, atomic ions, molecular ions, or electrons, the calculation of collision integrals uses the effective Coulomb potential defined by Eqs. (2.8) or (2.9). The calculation leads to the following result [17]:

$$\Omega_{ij}^{(l,s)} = \left(\frac{2\pi k T_e}{m_{ij}}\right)^{\frac{1}{2}}\frac{(s-1)!}{2}\left(\frac{Z_i Z_j q_e^2}{kT}\right)^2 \mathcal{S}^l(\alpha) \tag{2.19}$$

with

$$\mathcal{S}^l(\alpha) \simeq \int_0^{\alpha}\left[1 - \left(\frac{4x^2-1}{4x^2+1}\right)^l\right] x\mathrm{d}x \simeq \frac{l}{4}\,\ell n\,(4\alpha^2) \tag{2.20}$$

where

$$\alpha = \frac{8\pi\epsilon_0 k T_e l_D}{Z_i Z_j q_e^2} \tag{2.21}$$

2.2 Conservation Laws

2.2.1 The Fluid Medium

We will not repeat here the theory of Chapman and Enskog [2, 3]. This section provides the main results for a multispecies environment, compressible, weakly or not ionized, and in which the internal degrees of freedom of molecules (vibration, rotation) are not excited. This therefore relates to moderate temperature environments, typically less than 10,000 K. The case of weakly ionized media (typically less than a few percent in volume of charged particles, electrons, or ions) is also within this framework. Indeed, in this type of medium, the macroscopic electric field is weak and long-range collective effects in the medium are absent. As such, the term "plasma," which is often used in aerothermal literature, is not really justified.

The variables used to describe the environment are:

1. Pressure p and temperature T (basic thermodynamic variables).
2. For the attendance rate for each species, one of the following:
   - The particle number by unit volume n_i
   - The molar concentration $\mathcal{V}_i$, such that $\sum_i \mathcal{V}_i = p/\mathcal{R}T$
   - The partial density ρ_i, such that $\sum_i \rho_i = \rho$
   - The mass fraction $c_i = \rho_i/\rho$, such that $\sum_i c_i = 1$
   - The volume fraction x_i , such that $\sum_i x_i = 1$
   - The partial pressure $p_i = px_i$

   These quantities are equivalent and linked by the following relations:

$$c_i = \frac{\mathcal{M}_i}{\mathcal{M}} x_i \qquad \mathcal{V}_i = \frac{p_i}{\mathcal{R}T} \qquad n_i = \frac{\mathcal{N} p_i}{\mathcal{R}T} = \mathcal{N}\mathcal{V}_i$$

3. The molar mass $\mathcal{M}_i$ and mean molar mass

$$\overline{\mathcal{M}} = \sum_i x_i \mathcal{M}_i = \left[\sum_i \frac{c_i}{\mathcal{M}_i}\right]^{-1} \tag{2.22}$$

4. The volume electric charge density of species i, $Q_i = \mathcal{V}_i \mathcal{N} Z_i q_e$, where Z_i is the charge number of species i ($Z = -1$ for the electron, $Z = N$ for an ion of charge N), and the volume total charge density $Q = \sum_i Q_i$.
5. The overall velocity (barycentric velocity) $\mathbf{V}$ and the total mass flow $\mathbf{J} = \rho\mathbf{V}$.
6. The diffusion velocity $\mathbf{V}_{D_i} = \mathbf{V}_i - \mathbf{V}$, the difference between the average velocity of the species i and overall velocity together, and the

corresponding mass flow $\mathbf{J}_i = \rho_i \mathbf{V}_{D_i}$. As a consequence of the previous definition, the sum of diffusion mass flows is zero, $\sum_i \mathbf{J}_i = 0$.

7. The mass formation rate of the species i, per unit volume and time $\dot{\omega}_i$, is such that $\sum_i \dot{\omega}_i = 0$ by conservation of mass elements.
8. The viscous stress tensor

$$\mathcal{T} = \mu\left(\nabla \mathbf{V} + \nabla^t \mathbf{V} - \frac{2}{3}\nabla \cdot \mathbf{V}\,\mathbf{Id}\right) + \eta \nabla \cdot \mathbf{V}\,\mathbf{Id} \tag{2.23}$$

where μ is the dynamic viscosity and η bulk viscosity, assuming Stokes' hypothesis $\eta + 2/3\mu = 0$.

9. The internal energy $e = \sum_i e_i$, the total mass energy $E = e + (V^2/2)$, and the mass enthalpy $h_i = e_i + (p_i/\rho_i)$ and so, via the state equation, $h = \sum_i h_i = e + p/\rho$.
10. The heat flux

$$\dot{\mathbf{q}} = -\lambda \nabla T + \sum_i \rho c_i h_i\, \mathbf{V}_{D_i} + RT \sum_i \sum_k \frac{c_i \mathcal{D}_k^T}{\mathcal{M}_i \mathcal{D}_{ik}} \times (\mathbf{V}_{D_i} - \mathbf{V}_{D_k}) + \dot{\mathbf{q}}_{\mathrm{rad}} \tag{2.24}$$

where λ is the thermal conductivity, $\mathcal{D}_{ij}$ is the binary diffusion coefficient for the pair of species (i, j), and $\mathcal{D}_i^T$ is the thermal diffusion coefficient that, despite its name, is not a diffusion coefficient. It would have been more logical to define a coefficient diffusion $\mathcal{D}_i^T/\rho$, which would have the additional advantage of highlighting the dependence in $1/p$ of this quantity, similar to other diffusion coefficients.

The radiative heat flux, highly dependent on velocity, is discussed in Chapter 10.

The term corresponding to thermal diffusion can be expressed in a simpler form [18]

$$\dot{\mathbf{q}} = -\lambda \nabla T + \sum_i \rho c_i h_i \mathbf{V}_{D_i} + p \sum_i k_i^T \mathbf{V}_{D_i} + \dot{\mathbf{q}}_{\mathrm{rad}} \tag{2.25}$$

The coefficient of multicomponent thermal diffusion k_i^T is a dimensionless number.

Conservation equations for the n_e species are written as follows:

- Mass conservation equation for species i ($n_e - 1$ independent equations)

$$\frac{\partial \rho_i}{\partial t} + \nabla \cdot [\rho_i(\mathbf{V} + \mathbf{V}_{D_i}] = \dot{\omega}_i \tag{2.26}$$

The mass production rate per unit volume $\dot{\omega}_i$ is expressed in Sec. 3.1.1.

- By summing on i, the total mass conservation equation

$$\frac{\partial \rho}{\partial t} + \nabla \cdot (\rho \mathbf{V}) = 0 \tag{2.27}$$

- The momentum conservation equation

$$\frac{\partial (\rho \mathbf{V})}{\partial t} + \nabla \cdot (\rho \mathbf{V} \otimes \mathbf{V} + p\, \mathbf{Id} - \mathcal{T}) = Q\mathbf{E} \tag{2.28}$$

  where $\mathbf{E}$ is the electric field
- The energy conservation equation

$$\frac{\partial (\rho \mathbf{E})}{\partial t} + \nabla \cdot (\rho \mathbf{V} h + \dot{\mathbf{q}} - \mathcal{T}\ \mathbf{V}) = \mathbf{j} \cdot \mathbf{E} \tag{2.29}$$

  where $\mathbf{j}$ is the electric current density defined in Eq. (2.30)
- The equation giving the current density is written

$$\mathbf{j} = \sum_i Q_i \mathbf{V}_{D_i} \tag{2.30}$$

- The relationship closing the electrical system $f(\mathbf{j}, \mathbf{E}) = 0$, which differs depending on the problems addressed (see Secs. 2.4 and 2.5.5).

Finally, the system of conservation laws must be closed by an equation of state. This is obtained very simply by assuming that each species behaves as an ideal gas (a realistic assumption in our case), which implies that the mixture behaves in the same way

$$\frac{p_i}{\rho_i} = \frac{\mathcal{R}T}{\mathcal{M}_i} \Rightarrow \frac{p}{\rho} = \frac{\mathcal{R}T}{\overline{\mathcal{M}}}$$

2.2.2 Boundary Conditions

In the most general case, the solid surface is the boundary that exchanges mass, momentum, and energy. It is also a place where gases and energy are created (or disappear) through the physical and chemical reactions likely to take place there. In reality, the physical boundary is more complicated, and the mathematical one described here is defined in the sense of asymptotic description (see Chapter 6 and Sec. 8.1.2). This boundary moves with the recession velocity $\mathbf{v}_a$ in a fixed reference frame. The coordinates are n in the fixed frame and y in the moving frame. Derivatives expressed in these two variables are related by the following relations:

$$\begin{cases} \left.\dfrac{\partial \bullet}{\partial t}\right|_y = \left.\dfrac{\partial \bullet}{\partial t}\right|_n - \mathbf{v}_a \dfrac{\partial \bullet}{\partial y} \\ \dfrac{\partial \bullet}{\partial y} = \dfrac{\partial \bullet}{\partial n} \end{cases} \tag{2.31}$$

2.2.2.1 Energy Exchanges

If $\mathbf{J}_E$ is the flow of energy E across the surface, the conservation equation of this quantity is written in fixed axes

$$\frac{\partial(\rho E)}{\partial t} + \frac{\partial \mathbf{J}_E}{\partial n} = 0 \tag{2.32}$$

This equation is written, given Eq. (2.31), in the axes related to the wall,

$$\frac{\partial(\rho E)}{\partial t} + \frac{\partial}{\partial y}(\mathbf{J}_E - \rho \mathbf{v}_a E) = 0 \tag{2.33}$$

The energy fluxes are:

- In the fluid

$$\mathbf{J}_{E_g} = \rho_g \mathbf{V}_g \left(e_g + \frac{p}{\rho} + \frac{V_g^2}{2} \right) - \rho \mathbf{v}_a e_g - \mathcal{T}\,\mathbf{V}_g + \dot{\mathbf{q}}_g \tag{2.34}$$

- In the solid

$$\mathbf{J}_{E_s} = \dot{\mathbf{q}}_s - (1 - \epsilon_s)\rho_s \mathbf{v}_a e_s - \epsilon_s \rho_{g_p} v_a e_{g_p} \tag{2.35}$$

where ϵ_s is the fraction of the area occupied by the porosity, which in some cases can be equated to the porosity ϵ itself (e.g., if the porosity is formed by cylinders perpendicular to the surface).

2.2.2.2 Momentum Exchanges

In the systems we study, exchanges of momentum are generally reduced to their simplest expression, namely:

- For the normal derivative (zero flux), a boundary layer–type equation

$$\frac{\partial p}{\partial y} = 0 \tag{2.36}$$

- For the velocity derivative parallel to the surface, we impose the continuity of the shear. We may indeed have a nonzero velocity in the case of a porous medium or a liquid layer. This speed is still very low. This does not, however, impose $u = 0$ because this condition, reasonable for a boundary layer, may be inappropriate at the interface of a medium such as a porous solid like tape-wrapped carbon phenolic (TWCP) or phenolic-impregnated carbon ablator (PICA).

2.2.2.3 Mass Exchanges

Mass exchanges are handled by the continuity of mass flow rates of each species

$$(\mathbf{J}_{m_i} - \rho_i \mathbf{V}_g) \cdot \mathbf{n} = \dot{m}_i \tag{2.37}$$

The calculation of mass-ablation rates $\dot{m}_i$, and possibly pyrolysis, is treated in Sec. 3.2 and Chapters 5 and 9.

2.3 Diffusion in Neutral Medium

The system described in Sec. 2.2 is not closed because we must express the diffusion rates. These quantities are the solutions of the following linear system of equations of rank $n_e - 1$, called the Stefan–Maxwell system. This system is demonstrated in the classic work of Hirschfelder et al. [3]

$$\sum_{k \neq i} \frac{x_i x_k}{\rho \mathcal{D}_{ik}} \left(\frac{\mathbf{J}_k + \mathcal{D}_k^T \nabla(\ell n\, T)}{c_k} - \frac{\mathbf{J}_i + \mathcal{D}_i^T \nabla(\ell n\, T)}{c_i} \right) = \mathbf{d}_i \tag{2.38}$$

where $\mathbf{d}_i$ is the "engine" of the phenomenon of diffusion

$$\mathbf{d}_i = \nabla x_i + (x_i - c_i) \nabla(\ell n\, p) \tag{2.39}$$

We could use a different, more coherent version [18, 19]

$$\sum_{k \neq i} \frac{x_i x_k}{\rho \mathcal{D}_{ik}} \left(\frac{\mathbf{J}_k}{c_k} - \frac{\mathbf{J}_i}{c_i} \right) = \mathbf{d}'_i \tag{2.40}$$

where $\mathbf{d}'_i$ is containing thermal diffusion

$$\mathbf{d}'_i = \nabla x_i + (x_i - c_i) \nabla(\ell n\, p) + k_i^T \nabla(\ell n\, T) \tag{2.41}$$

k_i^T is the multicomponent thermal diffusion coefficient, related to the coefficients of "ordinary" (binary) thermal diffusion by a system similar to Eq. (2.38). Here we adopt the sign convention chosen in Chapman and Cowling [2], and Hirschfelder et al. [3]

$$k_i^T = \sum_{j \neq i} \frac{x_i x_j}{\rho \mathcal{D}_{ij}} \left(\frac{\mathcal{D}_j^T}{c_j} - \frac{\mathcal{D}_i^T}{c_i} \right) \tag{2.42}$$

This quantity can be positive (diffusion to the cold medium) or negative (downward hot).

The formal resolution of the system in Eq. (2.38) gives

$$\mathbf{J}_i = -\frac{\rho}{\overline{\mathcal{M}}^2} \sum_{k \neq i} \mathcal{M}_i \mathcal{M}_k D_{ik} \mathbf{d}_k - \mathcal{D}_i^T \nabla(\ell n\, T) \tag{2.43}$$

One can see the terms appearing due to the gradients of:

- Mass fraction.
- Temperature (Soret effect), negligible in a boundary layer [20], but of some importance in chemical nonequilibrium detached flows [21].
- Pressure (Dufour effect), negligible except in some cases such as expansion of a gas mixture containing species of very different molecular masses [22]. The shock does not justify the inclusion of this term, the Navier–Stokes equations being wrong in such a situation when $\mathcal{M}a > 2$ [23].

Some properties can be given:

- Summation on Eq. (2.42) produces $\sum_i k_i^T = 0$.
- Equations (2.39) and (2.42) imply that $\sum_i \mathrm{d}_i = 0$ (also true for $\sum_i \mathrm{d}'_i = 0$).
- In the same way, Eq. (2.43) implies that $\sum_i \mathcal{D}_i^T = 0$. Indeed, $\sum_i \mathbf{J}_i = 0$ should be verified for any $\nabla(\ell n\, T)$, ∇x_i, and $\nabla(\ell n\, p)$.
- The multispecies coefficients D_{ik} are defined by certain linear combinations. Traditionally one imposes that $D_{12} = -\mathcal{D}_{12}$ in the binary case, inducing n_e constraints. This leads to an asymmetric and nonsingular matrix, such that $D_{ii} = 0$ [18]. The multispecies diffusion coefficients are a solution of the system [3]

$$\sum_{j \neq i} \frac{1}{\mathcal{D}_{ij}} \left[x_i(\mathcal{M}_h D_{jh} - \mathcal{M}_k D_{jk}) - x_j(\mathcal{M}_h D_{ih} - \mathcal{M}_k D_{ik}) \right]$$
$$= \overline{\mathcal{M}}(\delta_{ih} - \delta_{ik}) \quad i, j, h, k = 1, n_e \tag{2.44}$$

The formal solution of this system is written [3]

$$D_{ij} = \frac{\overline{\mathcal{M}}}{\mathcal{M}_j} \frac{K^{ji} - K^{ii}}{K} \tag{2.45}$$

K is the determinant of the matrix defined by the following elements:

$$\begin{cases} K_{ii} = 0 \\ K_{ij} = \dfrac{x_i}{\mathcal{D}_{ij}} + \dfrac{\mathcal{M}_j}{\mathcal{M}_i} \dfrac{1}{\overline{\mathcal{D}}_i} \quad j \neq i \end{cases} \tag{2.46}$$

K^{ij} is the minor of K (line i, column j) and $\overline{\mathcal{D}}_i$ is a quantity that is subsequently defined by

$$\overline{\mathcal{D}}_i = \left(\sum_{k \neq i} \frac{x_k}{\mathcal{D}_{ik}} \right)^{-1} \tag{2.47}$$

Moreover, the multicomponent diffusion coefficients satisfy the following constraint [3]:

$$\sum_i (\mathcal{M}_i \mathcal{M}_j D_{ij} - \mathcal{M}_i \mathcal{M}_k D_{ik}) = 0 \quad \forall j \neq k, \quad k = 1, n_e \tag{2.48}$$

that is to say, $1/2\, n_e(n_e - 1)$ constraints. In total, the matrix D_{ij} includes $1/2\, n_e(n_e - 1)$ independent terms.

The resolution of Eq. (2.42) is analogous to the Stefan–Maxwell system. Indeed, Eqs. (2.40) and (2.42) are identical up to sign and accept the same constraint ($\sum_i \mathbf{J}_i = 0$ and $\sum_i \mathcal{D}_i^T = 0$). The solution is therefore analogous to Eq. (2.43)

$$\mathcal{D}_i^T = \frac{\rho}{\mathcal{M}^2} \sum_k \mathcal{M}_i \mathcal{M}_k D_{ik} k_k^T \tag{2.49}$$

We also use another quantity: the binary thermal diffusion factor α_{ij}^T, which is connected to the multicomponent thermal diffusion coefficient by

$$k_i^T = \sum_j x_i x_j \alpha_{ij}^T \tag{2.50}$$

This antisymmetric matrix is calculable from the collision integrals (Sec. 2.5.4.1). Its elements are dimensionless, independent of pressure but dependent on mass fractions (in addition to temperature), which makes them costly to calculate.

2.3.1 Binary Diffusion

In the case of a binary system in which we retain only the diffusion concentration gradient, the Stefan–Maxwell equation [Eq. (2.38)] is solved immediately. Its solution leads to the following equation (to be compared to more general forms appearing in Sec. 2.3.2):

$$\mathbf{J}_i = \rho_i \mathbf{V}_{D_i} = -\rho c_i \mathcal{D}_i \nabla x_i, \quad i = 1, 2 \tag{2.51}$$

with

$$\mathcal{D}_i = (1 - c_i)\overline{\mathcal{D}}_i \tag{2.52}$$

where $\overline{\mathcal{D}}_i$ is given by Eq. (2.47). Moreover, there is a statistical relationship between the binary diffusion coefficient and molar mass (see Sec. 2.3.3)

$$\mathcal{D}_{ij} \simeq \left(\frac{\mathcal{M}_{\mathrm{ref}}^2}{\mathcal{M}_i \mathcal{M}_j}\right)^{\alpha} \mathcal{D}_{\mathrm{ref}} \tag{2.53}$$

where $\mathcal{D}_{\mathrm{ref}}$ is a diffusion coefficient for a reference species and α a constant of about 0.5.

This property has been exploited in environments in which the molecules can be split into two groups according to molar masses. This is the case for air, with O, N on one side and O_2, N_2, NO on the other. This type of approach is used in Fay and Riddell [24]. The presence of heavier molecules in the environment resulting from the ablation boundary limits this approach.

2.3.2 Diagonal Approximations

In this section we are interested only in approximating the first term of Eq. (2.43). The first type of method is to diagonalize the transfer matrix of $\mathbf{d}_i$ to $\mathbf{J}_i$ to obtain a law of the following type:

$$\mathbf{J}_i = -\rho \mathcal{D}_i^{\mathrm{eq}} \mathbf{d}_i' \tag{2.54}$$

where $\mathcal{D}_i^{\mathrm{eq}}$ is an "equivalent" diffusion coefficient to be defined.

In some cases, we seek a law of Fick type based on the gradient of mass fraction

$$\mathbf{J}_i = -\rho \mathcal{D}_i^{\mathrm{eq}} \nabla c_i \tag{2.55}$$

The Stefan–Maxwell equation involves ∇x_i, so we will have to spend from x_i to c_i and vice versa. The gradients of these quantities are related by the following relations:

$$\begin{cases} \nabla x_i = \dfrac{\overline{\mathcal{M}}}{\mathcal{M}_i} \nabla c_i - \dfrac{\overline{\mathcal{M}}^2 c_i}{\mathcal{M}_i} \sum_j \dfrac{\nabla c_j}{\mathcal{M}_j} \\ \nabla c_i = \dfrac{\mathcal{M}_i}{\overline{\mathcal{M}}} \nabla x_i - \dfrac{\mathcal{M}_i x_i}{\overline{\mathcal{M}}^2} \sum_j \mathcal{M}_j \nabla x_j \end{cases} \tag{2.56}$$

2.3.2.1 Blottner Approximation

Blottner showed that one could write the diffusion gradients of mass fractions as follows (see Appendix A.2.1):

$$\mathbf{J}_i = -\rho \mathcal{D}_i \nabla c_i + \rho \sum_{j \neq i} b_{ij} \nabla c_j \tag{2.57}$$

with

$$\begin{cases} \mathcal{D}_i = (1 - x_i)\overline{\mathcal{D}}_i \\ b_{ij} = \mathcal{D}_i - \dfrac{\mathcal{M}_i}{\mathcal{M}_j} D_{ij} - \left(1 - \dfrac{\mathcal{M}_i}{\mathcal{M}_j}\right) \sum_k D_{ik} c_k \end{cases} \tag{2.58}$$

where $\overline{\mathcal{D}}_i$ is given by Eq. (2.47).

It is possible to retain only the diagonal terms in a first approximation in the general case [21], even if this expression was first used to express the flow of minor species.

Note that this expression does not satisfy $\sum_i \mathrm{J}_i = 0$. We can solve this problem by renormalization (Sec. 2.3.2.4).

2.3.2.2 Effective Binary Diffusion

The Stefan–Maxwell system [Eq. (2.38)] is written in terms of diffusion velocities

$$\sum_{k \neq i} \frac{x_i x_k}{\mathcal{D}_{ik}} (\mathbf{V}_{D_k} - \mathbf{V}_{D_i}) = \mathbf{d}'_i \tag{2.59}$$

The assumption of effective binary diffusion is to write that

$$\forall k \neq i \quad \mathbf{V}_{D_k} = \overline{\mathbf{V}}_D$$

It gives the expression of $\mathbf{V}_{D_i}$ [25]

$$\mathbf{V}_{D_i} = \overline{\mathbf{V}}_D - \frac{\overline{\mathcal{D}}_i}{x_i} \mathbf{d}'_i \tag{2.60}$$

where $\overline{\mathcal{D}}_i$ is given by Eq. (2.47).

$\overline{\mathbf{V}}_D$ will be drawn from the constraint of zero total flux written as follows:

$$\rho_i \mathbf{V}_{D_i} + \sum_{j \neq i} \rho_j \mathbf{V}_{D_j} = 0 \tag{2.61}$$

that is,

$$\rho_i \mathbf{V}_{D_i} + (\rho - \rho_i) \overline{\mathbf{V}}_D = 0 \tag{2.62}$$

The average velocity derived from this expression and introduced in Eq. (2.60) gives the final expression of the flow

$$\mathbf{J}_i = \rho_i \mathbf{V}_{D_i} = -\rho c_i (1 - c_i) \overline{\mathcal{D}}_i \mathbf{d}'_i \tag{2.63}$$

This form generalizes Eq. (2.51). It is close to that derived from Eq. (2.58), leading to possible confusion.

This method is used in the CHEMKIN package [26].

Note that this approximation is inconsistent in the sense that the hypothesis "all $\mathbf{V}_{D_k}$ are equal except for $\mathbf{V}_{D_i}$" implies that they are all equal. That is why the sum of fluxes calculated in this way is not zero. However, we can give a theoretical justification and an extension to this approximation [18]. Let $\mathbf{V}_{D_i}$ be the diffusion velocity. We will find its expression in a form similar to Eq. (2.63)

$$\mathbf{V}_{D_i} = -(1 - c_i) \overline{\mathcal{D}}_i \mathbf{b}_i \tag{2.64}$$

where $\mathbf{b}_i$ is an unknown function. Introducing this term into the Stefan–Maxwell equation [Eq. (2.38)], provides

$$-\sum_{k\neq i}\left[\frac{x_i(1-c_k)\overline{\mathcal{D}}_k}{\mathcal{D}_{ik}}\mathbf{b}_k-\frac{x_k(1-c_i)\overline{\mathcal{D}}_i}{\mathcal{D}_{ik}}\mathbf{b}_i\right]=\mathbf{d}'_i \tag{2.65}$$

or, in matrix form

$$(\mathbf{Id}-\mathbf{A})\mathbf{b}=\mathbf{d}' \tag{2.66}$$

where $\mathbf{A}$ is the matrix with element

$$A_{ik}=c_i\delta_{ik}+\frac{x_i(1-c_k)\overline{\mathcal{D}}_k}{\mathcal{D}_{ik}}(1-\delta_{ik}) \tag{2.67}$$

This system can be inverted and formally developed

$$\mathbf{b}=(\mathbf{Id}-\mathbf{A})^{-1}\mathbf{d}'=(\mathbf{Id}+\mathbf{A}+\mathbf{A}^2+\cdots)\,\mathbf{d}' \tag{2.68}$$

At zeroth order, we find a particular solution that is the effective binary diffusion approximation.

2.3.2.3 Weighted Methods

This section presents a class of methods for finding the approximations outlined in the previous section. They are known as Self-Consistent Effective Binary Diffusion (SCEBD) [27]. We start from the Stefan–Maxwell equation, written as

$$\sum_{k\neq i}\alpha_{ik}(\mathbf{V}_{D_k}-\mathbf{V}_{D_i})=\mathbf{d}'_i \tag{2.69}$$

with $\alpha_{ik}=\frac{x_ix_k}{\mathcal{D}_{ik}}$.

This equation will be rewritten as follows:

$$\mathbf{V}_i=\frac{1}{\alpha_i}\sum_{k\neq i}\alpha_{ik}\mathbf{V}_{D_k}-\frac{\mathbf{d}'_i}{\alpha_i} \tag{2.70}$$

with $\alpha_i=\sum_{k\neq i}\alpha_{ik}$.

The resolution of the system will restrict the general solution to the following form:

$$\alpha_{ik}=w_iw_k \tag{2.71}$$

α_i is therefore expressed as follows:

$$\alpha_i=w_i(w-w_i) \tag{2.72}$$

with $w=\sum_k w_k$.

Equation (2.70) becomes

$$\mathbf{V}_{D_i} = \overline{\mathbf{V}}_D - \frac{\mathbf{d}_i'}{ww_i} \tag{2.73}$$

where $\overline{\mathbf{V}}_D$ is an average velocity weighted by w_i

$$\overline{\mathbf{V}}_D = \frac{1}{w}\sum_k w_k \mathbf{V}_k \tag{2.74}$$

Equation (2.73) can be rewritten in a form identical to Eq. (2.60) by showing an equivalent diffusion coefficient

$$\mathcal{D}_i = \left(1 - \frac{w_i}{w}\right)\overline{\mathcal{D}}_i \tag{2.75}$$

We found as a special case

- Blottner method $w_i = x_i$
- Effective binary diffusion $w_i = c_i$

However, the form of the expression of diffusion coefficients [Eq. (2.101)] suggested an expression of w_i in the form

$$w_i = x_i \mathcal{M}_i^{\frac{1}{2}} \quad \text{or} \quad w_i = c_i \mathcal{M}_i^{-\frac{1}{2}}$$

Note that it is also possible to use the method of separation of variables described in Sec. 2.3.3 and take a weight

$$w_i = \frac{x_i F_i}{\mathcal{D}_{\text{ref}}^{\frac{1}{2}}} \text{(method valid for neutral media only)}$$

2.3.2.4 Renormalization

In general, if $\mathbf{J}_i^\dagger$ is a particular solution of Eq. (2.38) (in fact, in our case is an approximate solution), $\mathbf{J}_i^\dagger + \rho_i\mathbf{C}$ is also. We can identify the constant $\mathbf{C}$, writing the nullity of diffusion flux $\sum_i \mathbf{J}_i = 0$. By replacing $\mathbf{J}_i$ by its value $\mathbf{J}_i^\dagger + \rho_i\mathbf{C}$, we get

$$\mathbf{C} = -\frac{1}{\rho}\sum_j \mathbf{J}_j^\dagger \tag{2.76}$$

The normalized flux is then

$$\mathbf{J}_i = \mathbf{J}_i^\dagger - c_i\sum_j \mathbf{J}_j^\dagger \tag{2.77}$$

2.3.2.5 Constant Lewis–Semenov or Schmidt Numbers

In this approximation, using the gradient of mass fraction [Eq. (2.2)], diagonalization is obtained very simply:

- By defining a priori a number of Lewis–Semenov $\mathcal{L}_{ij} = \frac{\lambda}{\rho C_p \mathcal{D}_{ij}} = \mathcal{L}$ constant. The diffusion coefficient is then given by

$$\mathcal{D}_i = \mathcal{D} = \frac{\lambda}{\rho C_p \mathcal{L}} \tag{2.78}$$

  $\mathcal{L}$ is, in fact, variable, ranging between 1 and 1.4 for air in the range 300–6000 K (see Fig. 4.1 in Chapter 4), with a domain of much greater variation in the presence of a light species (hydrogen) or ablation products (0.6 to 2.2 in the latter case [28]).
- Equivalently, a Schmidt number Sc constant

$$\mathcal{D}_i = \mathcal{D} = \frac{\mu}{\rho\, Sc} \tag{2.79}$$

  Sc is between 0.6 and 1 for air in the range 300–6000 K (Fig. 4.1).

2.3.3 Splitting

Let a property of binary diffusion coefficients, justified in Sec. 2.5.1, be

$$\mathcal{D}_{ik} \simeq (\mathcal{D}_{ii}\mathcal{D}_{kk})^{\frac{1}{2}} \tag{2.80}$$

This property allows splitting of variables

$$\mathcal{D}_{ij} = \frac{\mathcal{D}_{\text{ref}}}{F_i F_j} \tag{2.81}$$

where $\mathcal{D}_{\text{ref}}$ is any of the $\mathcal{D}_{ij}$. The F_i in this case are precalculated. Their dependence on temperature is low (Fig. 2.1) and sometimes neglected, particularly when using Lennard–Jones potentials leading to very small variations in temperature. The coefficients F_i can be approximated by the following correlation, if one takes $\mathcal{D}_{\text{ref}} = \mathcal{D}_{O_2-O_2}$ [29]

$$F_i = \left(\frac{\mathcal{M}_i}{\mathcal{M}_{O_2}}\right)^{0.461} \tag{2.82}$$

In practice, an approximation by least squares will give accurate results. Error will be evaluated through the mean absolute error

$$\frac{1}{n_e(n_e-1)} \sum_{i=1}^{n_e} \sum_{j=i}^{n_e} \left(\mathcal{D}_{ij} - \frac{D_{\text{ref}}}{F_i F_j} \right) \tag{2.83}$$

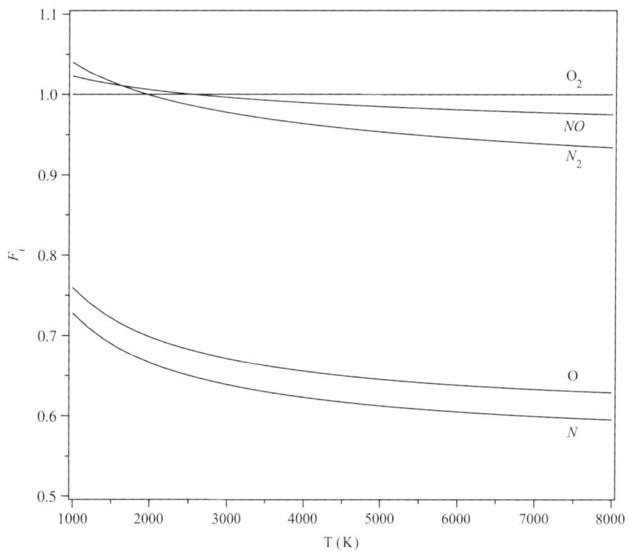

Fig. 2.1 Approximation of binary diffusion coefficients; air, five species, $p = p_0$. Collision integrals given by (10) and (16).

In the case treated in [29] with a Lennard–Jones potential, the error on $\mathcal{D}_{ij}$ is low:

- In the absence of light species (H, H_2, or He), the mean error is approximately 1% and the error always less than 5%.
- In the presence of light species (for mixtures of species of very different molecular weights), the average error is 5% and the error is always less than 10%.

In the cases treated, the coefficients of F_i are weakly dependent on temperature.

This particular form of writing binary diffusion coefficients allows an analytical solution of the system in Eq. (2.38) when the Dufour effect is neglected (Appendix A)

$$\mathbf{J}_i + \mathcal{D}_i^T \nabla(\ell n\ T) = -\frac{\rho \mathcal{D}_{\text{ref}}}{\mu_1}\left(\frac{\mu_2}{\overline{\mathcal{M}}}\nabla Z_i - \frac{Z_i - c_i}{\overline{\mathcal{M}}}\nabla c_i\right) \tag{2.84}$$

with

$$\left\{\begin{array}{lcl} Z_i & = & \dfrac{\overline{\mathcal{M}} c_i}{F_i \mu_2} \\ \mu_1 & = & \displaystyle\sum_k \dfrac{\overline{\mathcal{M}} c_k F_k}{\mathcal{M}_k} \\ \mu_2 & = & \displaystyle\sum_k \dfrac{\overline{\mathcal{M}} c_k}{F_k} \end{array}\right. \tag{2.85}$$

Note that this method is illustrated here using the example of air at thermodynamic equilibrium. The problem of calculating F_i is solved by a nonlinear method (sequential quadratic programming, SQP). Results, shown in Fig. 2.1, show little variation of F_i with temperature. The average absolute error is 2% at most when we take into account the evolution with temperature, and less than 8% using average values over the interval. Even so, the inaccuracy of the approximation is the same order of magnitude as (or less than) the knowledge of integrals collision [16].

Please note that this splitting assumes that all diffusion coefficients have the same temperature dependence. This is verified for the neutral species but absolutely not for charged species. This approximation is not usable in that case.

2.3.4 Accuracy of the Various Methods

This section will compare the different approaches to approximate the Stefan–Maxwell system. We will use the example of air at thermodynamic equilibrium, such that we will use a thread to present numerical results of all the approximations of transport properties. This example permits us to reduce the number of variables because the results are a function only of thermodynamic variables. In the case of diffusion, we will further restrict the problem by assuming constant pressure (which is a good approximation near the stagnation point, for example) and we restrict ourselves to diffusion by concentration gradient. We can write

$$\nabla x_i = \frac{\partial x_i}{\partial T} \nabla T \tag{2.86}$$

We can also write a similar equation for c_i. The calculation of partial derivatives $\partial x_i / \partial T$ is outlined in Sec. 3.1.5.2.

The diffusion fluxes are expressed as a linear combination of concentrations [Eq. (2.43)]. It is therefore possible, replacing x_i or c_i by its above expression, to express these fluxes in terms of ∇T and, even better, to express these relations as scalar flux because the species gradient and thermal gradient are linear. For example, the general expression in Eq. (2.43) is written here

$$\mathbf{J}_i = \nabla T \frac{\rho}{\overline{\mathcal{M}}^2} \sum_{j \neq i} \mathcal{M}_i \mathcal{M}_j D_{ij} \frac{\partial x_i}{\partial T} \tag{2.87}$$

Note first that the $\mathbf{J}_i$ are collinear and second that the relative error on the $\mathbf{J}_i$ is the same as on $\mathbf{J}_i / \|\nabla T\|$; therefore, we found a way to synthetically evaluate the approximation errors. However, note that this method has a problem with the fact that $\partial x_i / \partial T$ vanishes in two instances:

- At low temperature when the composition varies slowly, although this did not affect the analysis.

- Because the x_i passes through a local extrema, the flux vanishes and changes sign. The exact value and approximation do not vanish at exactly the same temperature, which causes a singularity on the value of relative error. This fact has no special significance, and the analysis of error will be circumvented by using the standard error

$$Error = \frac{\sum_i ||\mathbf{J}_i^{\text{exact}} - \mathbf{J}_i^{\text{app}}||}{\sum_i ||\mathbf{J}_i^{\text{exact}}||} \tag{2.88}$$

where $\mathbf{J}_i^{\text{exact}}$ and $\mathbf{J}_i^{\text{app}}$ are the exact and approximate values, respectively.

Comparison of various methods (Fig. 2.2) drew the following comments:

- In general, the approximations degrade with temperature, as does the heterogeneity of the medium.
- Unsurprisingly, the separation of variables gives good results, whether or not the dependence of F_i with T is taken into account.
- The very simple method $\mathcal{L} = C^{st}$ gives poor results throughout the temperature range. Note that this method is used here with the approximation of Mason and Saxena (Sec. 2.5.3.3) for conductivity because we computed an approximate calculation. The exact expression does not give better results.

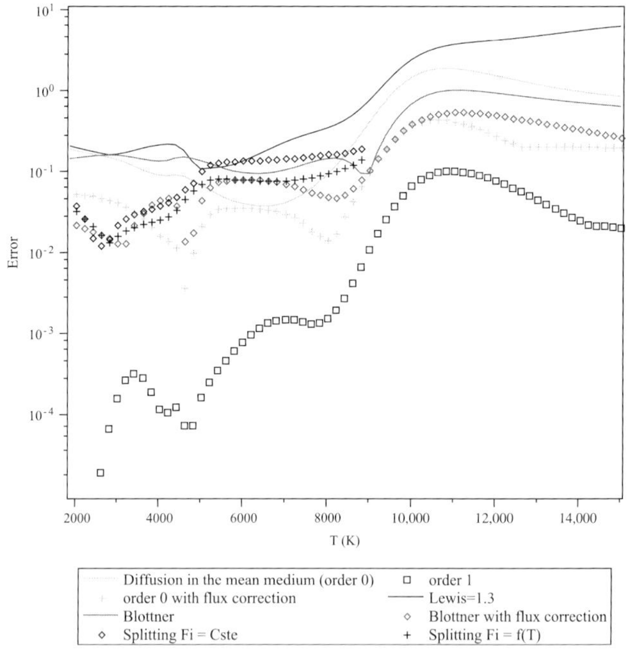

Fig. 2.2 Approximation errors of the diffusion matrix. Methods with corrected fluxes were tested on a vehicle at reentry and give excellent results (30).

- The Blottner method gives good results at low temperature; it is weaker at temperatures higher than 3000 K.
- The effective binary diffusion method at order 0 leads to results similar to the Blottner method.
- The SCEBD method (result not shown to avoid overloading the figure) also gives very similar results.
- The normalization corrections provide an important gain, especially in the case of the effective binary diffusion method, for which we obtain an accuracy of about 2 to 3% over a great part of the domain for zeroth order, much less for order 1. The gain is more limited for the Blottner approximation.

2.3.5 Numerical Calculation

The multicomponent diffusion coefficient (or multinary diffusion coefficient) D_{ik} can be calculated numerically. However, several obstacles arise: the system (of dimension n_e but of rank $n_e - 1$) can be ill-conditioned if some species are in trace amounts. The method used should take into account this specificity. Studies have compared various methods [31]. They conclude the greater efficiency of iterative methods. Among these, the effective binary diffusion of Sec. 2.3.2.2 provides the basis for an accurate method [32]. Equation (2.68) can be rewritten in the following form, using the Horner scheme

$$\mathbf{J}_i = -\rho(1 - c_i)\overline{\mathcal{D}}_i(\mathbf{Id} + \mathbf{A}(\mathbf{Id} + \mathbf{A}(\cdots)))\,\mathbf{d}'_i \tag{2.89}$$

After renormalization, the error, as defined by Eq. (2.88), is decreased by one to three orders of magnitude per iteration (Fig. 2.3). Recall that the order 0 is analytical and generally widely sufficient [30].

2.4 Diffusion in Weakly Charged Media

2.4.1 Zero Current Closure

The Stefan–Maxwell equation [Eq. (2.38)] is written for neutral media as its formal solution, Eq. (2.43). In the general case a new term appears in Eq. (2.39), linked to external forces. Here the electric field generated by charged particles applies a force per unit volume $\mathbf{F}_i = Q_i\mathbf{E}$ [3].

$$\mathbf{d}^\bullet_i = \mathbf{d}'_i - \frac{1}{p}(Q_i - c_iQ)\mathbf{E} \tag{2.90}$$

where $\mathbf{d}'_i$ is given by Eq. (2.4).

The generally accepted hypothesis in hypersonic flow is a zero electrical current density, which implies a macroscopic charge density of zero in a stationary case. Indeed, the equation of charge conservation is easily obtained by summing Eq. (2.26); premultiplied by $\frac{\mathcal{N} Z_i q_e}{\mathcal{M}_i}$, we get

$$\frac{\partial Q}{\partial t} + \nabla \cdot (Q\mathbf{V}) = -\nabla \mathbf{j} \tag{2.91}$$

In stationary case and with the assumption of zero diffusion current, we get

$$\nabla \cdot (Q\mathbf{V}) = 0 \tag{2.92}$$

The charge density being zero upstream "at infinity," it is zero everywhere in the flow. This may be defective in the immediate vicinity of the wall, which may have a region of current generated or absorbed by it.

The Stefan–Maxwell equation can therefore be written as a linear system whose unknowns are diffusion rates and the electric field

$$\sum_{k \neq i} \frac{x_i x_k}{\mathcal{D}_{ik}} (\mathbf{V}_{D_k} - \mathbf{V}_{D_i}) + \frac{Q_i}{p} \mathbf{E} = \mathbf{d}'_i \tag{2.93}$$

taking into account that the electric field modifies the solution moderately but significantly, [i.e., that the gap brought about in the solution (Fig. 2.4) is larger than the error of approximation of a good method]. (See Sec. 2.3.4 for definitions of error and the values of these for various methods.)

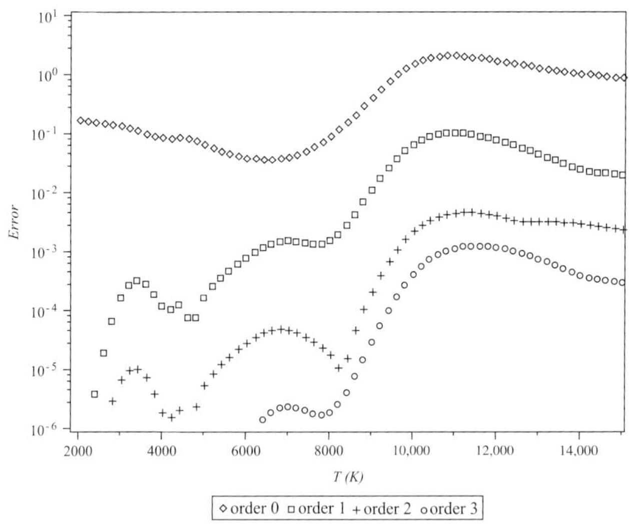

Fig. 2.3 Errors of numerical approximation of the diffusion matrix.

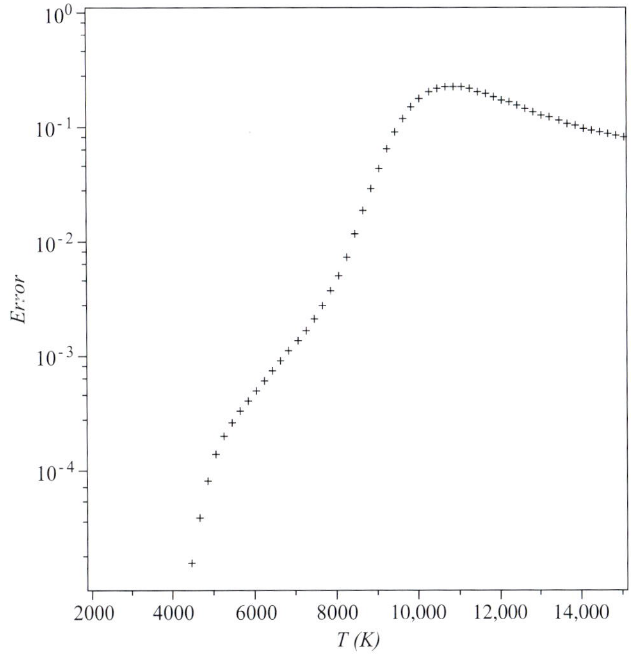

Fig. 2.4 Error on the diffusion for a charged medium. Error on the mass flux when neglecting the electric field.

2.4.2 Ambipolar Approximation

2.4.2.1 General Case

We use a diagonal approximation, that is, where the diffusion flux is expressed using Fick's law (Sec. 2.3.2). Indeed, there is no demonstration of what follows in the general case. The expression of the diffusion flux is

$$\mathbf{J}_i = -\rho_i \mathcal{D}_i \left(\mathbf{d}'_i - \frac{Q_i}{p} \mathbf{E} \right) \tag{2.94}$$

The zero current is expressed by

$$\sum_i \mathbf{V}_{D_i} Q_i = 0 \tag{2.95}$$

We deduce the expression of the field

$$\mathbf{E} = \frac{p}{\sum_i \mathcal{D}_i Q_i^2} \sum_i \mathcal{D}_i Q_i \mathbf{d}'_i \tag{2.96}$$

This calculation can be done with a low-level approximation, for example, by the effective binary diffusion method at zero order. But precision reasserts itself in the calculation of fluxes: the accuracy of the results is substantially the same as that obtained in the case of a neutral medium (Fig. 2.2).

2.4.2.2 A Single Ion: The Classical Approximation

The expression in Eq. (2.96) can be simplified by an asymptotic analysis on the parameter $\epsilon = M_e^{1/2}$ [23]. Indeed, the diffusion coefficients involving the electron are of order ϵ^{-1} from Eq. (2.101) later in this chapter. This implies that $\overline{\mathcal{D}}_e$ and $\mathcal{D}_e$ are also of order ϵ^{-1}. Moreover, Q_e is of order ϵ^{-2} and ρ_e of order ϵ^2. Terms for the electron are dominant in the numerator and denominator of the expression in Eq. (2.96), and we have

$$\mathbf{E} \simeq \frac{p}{Q_e} \mathbf{d}'_e \tag{2.97}$$

We can further simplify this expression, k_e^T being low (Sec. 2.5.4.1) and the influence of the pressure gradient generally negligible

$$\mathbf{E} \simeq \frac{p}{Q_e} \nabla x_e \tag{2.98}$$

In the classic case of a medium comprising a single ion (neutral N, ion I, and electron e), we have $x_e = x_I = \frac{1-x_N}{2}$, so $\nabla x_I = \nabla x_e$. Moreover $Q_I = -Q_e$. It follows

$$\begin{cases} \mathbf{J}_N = -\rho_N \mathcal{D}_N \nabla x_N & \forall N \\ \mathbf{J}_I = -\rho_I \mathcal{D}_A \nabla x_I \\ \mathbf{J}_e = 0 \end{cases} \tag{2.99}$$

with

$$\mathcal{D}_A \simeq 2\mathcal{D}_I \tag{2.100}$$

We find the classical doubling of ion flux: the apparent or "ambipolar" diffusion coefficient is doubled and the electron flux is negligible. Note also that the expression for neutral flux is unchanged.

The validity of this approximation can be questioned in the general case. Comparing with the mass flows and without electric field, from the previous calculations, we find that doubling the flow ion is not verified in general (see Fig. 2.5).

2.5 Calculation of Transport Coefficients

2.5.1 Diffusion Coefficients

The binary diffusion coefficient, as the coefficient of thermal diffusion, and other transport coefficients (conductivity and viscosity) are solutions

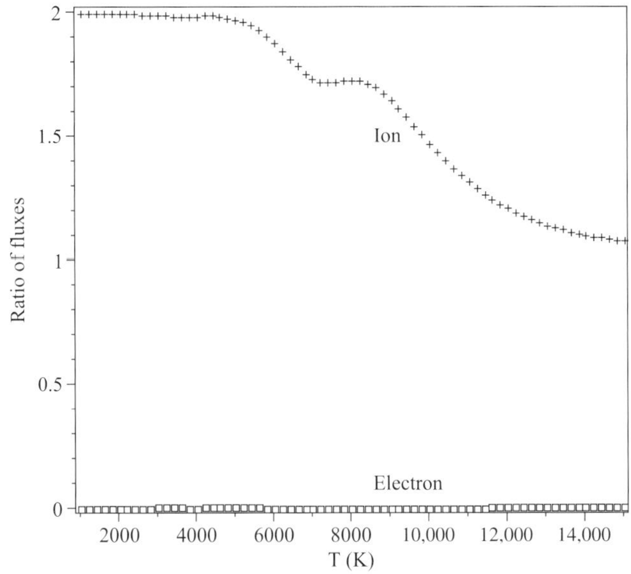

Fig. 2.5 Ambipolar diffusion. Ratio of ion and electron mass fluxes with and without consideration of electric field (exact calculation).

of linear systems involving collision integrals between molecules [2, 3, 31]

$$\mathcal{D}_{ik} = \frac{\alpha_1 \left(T^3 \frac{\mathcal{M}_i + \mathcal{M}_k}{2\mathcal{M}_i \mathcal{M}_k} \right)^{\frac{1}{2}}}{p \sigma_{ik}^2 \Omega_{ik}^{(1.1)^*} (T_{ik}^*)} \tag{2.101}$$

with

$$\alpha_1 = \frac{3}{8} \left(\frac{\mathcal{N} k^3}{\pi} \right)^{\frac{1}{2}} = 8.422885 \times 10^{-24} \ \mathrm{J}^{\frac{3}{2}} \cdot \mathrm{K}^{-\frac{3}{2}} \cdot \mathrm{mol}^{-\frac{1}{2}} \tag{2.102}$$

The expression in Eq. (2.101) highlights the general dependence in $\frac{T^{\frac{3}{2}}}{p}$ (within the variation of $\Omega^{(1,1)^*}$ with T^*, low). It also gives an explanation of the property in Eq. (2.85). Indeed, from Eq. (2.101)

$$\frac{\mathcal{D}_{ik}^2}{\mathcal{D}_{ii} \mathcal{D}_{kk}} = \frac{1}{2} \left[\frac{(\mathcal{M}_i + \mathcal{M}_k)^2}{\mathcal{M}_i \mathcal{M}_k} \right]^{\frac{1}{2}} \times \frac{\Omega_{ii}^{(1,1)*} \Omega_{kk}^{(1,1)^*}}{\Omega_{ik}^{(1,1)*2}} \times \left(\frac{\sigma_{ii} \sigma_{kk}}{\sigma_{ik}^2} \right)^2 \tag{2.103}$$

Each of the terms contained in this expression is near unity when the molecular weights and cross-sections do not differ by more than 10–30%. In practice, only the presence of hydrogen or helium in an environment can lead to significant errors on the calculated diffusion coefficients. This same expression is used to measure the very low temperature dependence through the second term.

Generally, one method to approximate numerically $\mathcal{D}_{ik}$ is an expression like $\mathcal{D}_{ik} = \frac{f(T)}{p}$

$$\mathcal{D}_{ik} = \frac{1}{p}\exp\left[\sum_{l=0}^{N} \alpha_{ikl}(\ln T)^l\right] \tag{2.104}$$

The order N is generally taken to be 2 or 3 [26, 34].

2.5.2 Viscosity

2.5.2.1 Viscosity of a Species

The viscosity of a species i is given by an expression of the same type as in Eq. (2.101)

$$\mu_i = \alpha_2 \frac{(\mathcal{M}_i T)^{\frac{1}{2}}}{\sigma_{ii}^2 \Omega_{ii}^{(2,2)^*}} \tag{2.105}$$

with

$$\alpha_2 = \frac{5}{16}\left(\frac{k}{\pi \mathcal{N}}\right)^{\frac{1}{2}} = 8.44195 \times 10^{-25}\ \mathrm{J}^{\frac{1}{2}} \cdot \mathrm{K}^{-\frac{1}{2}} \cdot \mathrm{mol}^{\frac{1}{2}} \tag{2.106}$$

For the numerical approximation of μ, expressions similar to that used for the diffusion coefficient are used [Eq. (2.104)], of course without the term $1/p$ [26, 34].

2.5.2.2 Viscosity of a Mixture: The Exact Value

The exact expression for the viscosity of a mixture is obtained by solving a linear system. The formal solution is [3]

$$\mu = -\frac{\begin{vmatrix} H_{11} & H_{12} & \dots & H_{1n_e} & x_1 \\ H_{21} & H_{22} & \dots & H_{1n_e} & x_2 \\ \dots & \dots & \dots & \dots & \dots \\ H_{n_e1} & H_{n_e2} & \dots & H_{n_en_e} & x_{n_e} \\ x_1 & x_2 & \dots & x_{n_e} & 0 \end{vmatrix}}{\begin{vmatrix} H_{11} & H_{12} & \dots & H_{1n_e} \\ H_{21} & H_{22} & \dots & H_{1n_e} \\ \dots & \dots & \dots & \dots \\ H_{n_e1} & H_{n_e2} & \dots & H_{n_en_e} \end{vmatrix}} \tag{2.107}$$

with

$$\begin{cases} H_{ii} = \frac{x_i^2}{\mu_i} + \sum_{k \neq i} \frac{2x_i x_k}{\mathcal{M}_i + \mathcal{M}_k} \Delta_{ik} \left(1 + \frac{3}{5}\frac{\mathcal{M}_k}{\mathcal{M}_i} A_{ik}^*\right) \\ H_{ij} = -\frac{2x_i x_j}{\mathcal{M}_i + \mathcal{M}_j} \Delta_{ij} \left(1 - \frac{3}{5} A_{ij}^*\right), \quad j \neq i \\ \Delta_{ij} = \frac{\mathcal{R}T}{p\mathcal{D}_{ij}} \end{cases} \tag{2.108}$$

Figure 2.6 gives the viscosity of air at thermodynamic equilibrium. We see an increase with temperature related to the individual variation of viscosities with this quantity [Eq. (2.105)]. The various species present at medium temperature having viscosities of the same order of magnitude, the total viscosity is insensitive to the composition, and thus the pressure. All the foregoing is false at high temperatures with the emergence of charged species, low viscosity because of their strong interactions, and, in the case of the electron, the low mass. The viscosity of the mixture is decreasing and pressure-sensitive [35].

2.5.2.3 Approximations for the Viscosities of Mixed Gases

Various approximations for the viscosity of a gas mixture are available in the literature. They are based on approximations of development on the exact expression [Eq. (2.107)], noting that the nondiagonal elements are small as compared to the diagonal ones. This development is written as [36]

$$\mu = \sum_i \frac{x_i^2}{H_{ii}} - \sum_i \sum_{j \neq i} \frac{x_i x_j H_{ij}}{H_{ii} H_{jj}} + \sum_i \sum_{j \neq i} \sum_{k \neq i} \frac{x_k x_k H_{ij} H_{ik}}{H_{ii} H_{jj} H_{kk}} - \cdots \tag{2.109}$$

- A first approximation (order 0) was found by Buddenberg and Wilke [3], taking $A_{ij}^* = 5/3$ and thereby eliminating all nondiagonal terms in the equation. Note that this quantity is close to 1.1 for realistic potential collisions between molecular species [3], but may be significantly different for interactions involving an atom or a charged species. Thus

$$\mu \simeq \sum_i \frac{x_i^2}{H_{ii}} = \sum_i \frac{\mu_i}{1 + \sum_{j \neq i} \frac{x_j}{x_i} \psi_{ij}} \tag{2.110}$$

with

$$\psi_{ij} = \alpha \frac{\mu_i}{\mathcal{M}_i} \Delta_{ij} \tag{2.111}$$

The constant α (equal to 2 strictly speaking) has been numerically adjusted to give the best approximation, obtained with $\alpha = 1.385$ for selected gases.

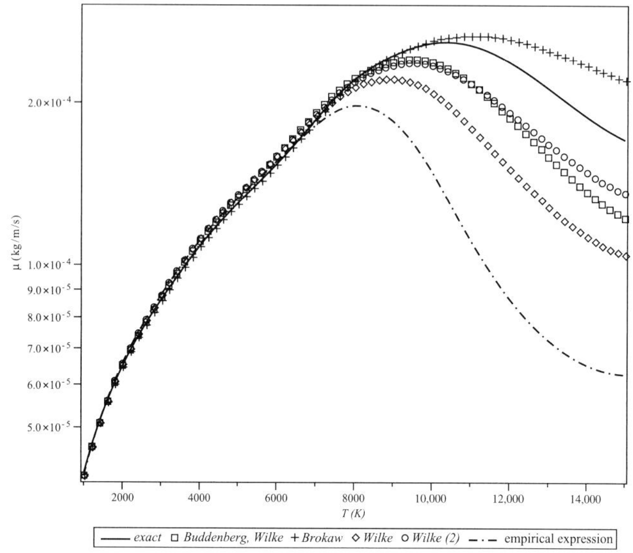

Fig. 2.6 Viscosity of air in thermodynamic equilibrium at normal pressure.

When we compare this approximation to the exact value in the case of air in thermodynamic equilibrium (Fig. 2.6) there is a reasonable agreement for medium temperatures, which can also be improved by taking $\alpha = 1.45$. We also see a significant disagreement at high temperatures for the reason cited earlier. Note that this disagreement, corresponding to a high volume fraction of electrons, is not important. Indeed, the constitutive equations described earlier in this chapter (Sec. 2.2) are themselves not very good in this case.

- Brokaw suggests another approximation for ψ_{ij} [37] based on an approximation of the development determinants of Eq. (2.107)

$$\mu \simeq \sum_i \frac{x_i^2}{H_{ii}} \frac{1}{1 + \sum_{j \neq i} \dfrac{H_{ij} x_j}{H_{jj} x_i}} \tag{2.112}$$

This expression is exact up to the second order.
We will rewrite this expression in an equivalent form

$$\mu \simeq \sum_i \frac{x_i^2}{b_i} \frac{1 + \sum_{j \neq i} \dfrac{H_{ij}}{H_{jj}}}{1 + \sum_{j \neq i} \dfrac{H_{ij} x_j}{H_{jj} x_i}} \tag{2.113}$$

with

$$b_i = H_{ii} + \sum_{j \neq i} H_{ij} \tag{2.114}$$

If we retain only the main part, we get

$$\mu \simeq \sum_i \frac{x_i^2}{b_i} = \frac{\mu_i}{1 + \sum_{j \neq i} \frac{x_j}{x_i} \psi_{ij}} \tag{2.115}$$

with

$$\psi_{ij} = \frac{6}{5} A_{ij}^* \frac{\mu_i}{\mathcal{M}_i} \Delta_{ij} \tag{2.116}$$

Δ_{ij} is given by Eq. (2.108).

For a realistic intermolecular potential of Lennard-Jones type, the quantity $(6/5)A_{ij}^*$ is close to 1.35 [3]. This justifies the value found numerically by Buddenberg and Wilke. The values calculated by this approximation give errors by excess of up to 8% [37]. This is confirmed by Fig. 2.6 for mean temperatures. For the same reasons already stated, the approximation gives poor results for partially ionized gases.

μ_i can be expressed as a function of self-diffusion coefficient $\mathcal{D}_{ii}$ by [3]

$$\frac{\mathcal{M}_i}{\mu_i \Delta_{ii}} = \frac{6}{5} A_{ii}^* \tag{2.117}$$

Substituting into Eq. (2.116), we get

$$\psi_{ij} = \frac{\mathcal{D}_{ii}}{\mathcal{D}_{ij}} \frac{A_{ij}^*}{A_{ii}^*} \tag{2.118}$$

We see that $\psi_{ii} = 1$. In this case the expression in Eq. (2.110) can be written in an equivalent form

$$\mu \simeq \sum_i \frac{x_i \mu_i}{\sum_{j \neq i} x_j \psi_{ij}} \tag{2.119}$$

Note that the $\psi_{ij}(T)$ can be approximated beforehand to reduce computation time.

- The expression in Eq. (2.118) is similar to that credited to Wilke [38]. This is expressed as a function of μ_i. The expression proposed by Wilke uses an approximation of $\mathcal{D}_{ij}$ based on a hard sphere potential

$$\mathcal{D}_{ij} \simeq \frac{\left(8\dfrac{\mathcal{M}_i+\mathcal{M}_j}{\mathcal{M}_i\mathcal{M}_j}\right)^{\frac{1}{2}}}{\left[\left(\dfrac{1}{\mathcal{M}_i\mathcal{D}_{ii}^2}\right)^{\frac{1}{4}}+\left(\dfrac{1}{\mathcal{M}_j\mathcal{D}_{jj}^2}\right)^{\frac{1}{4}}\right]^2} \tag{2.120}$$

Moreover, we assume that all $A_{ij}*$ are equal. Substituting μ_i for $\mathcal{D}_{ij}$, we get

$$\psi_{ij} \simeq \frac{\left[1+\left(\dfrac{\mu_i}{\mu_j}\right)^{\frac{1}{2}}\left(\dfrac{\mathcal{M}_j}{\mathcal{M}_i}\right)^{\frac{1}{4}}\right]^2}{\left[8\left(1+\dfrac{\mathcal{M}_i}{\mathcal{M}_j}\right)\right]^{\frac{1}{2}}} \tag{2.121}$$

Note that this expression is particularly simple because it does not involve cross-terms $\mathcal{D}_{ij}$. The approximation gives good results (error <2% if there is not too large a difference between molecular weights of various constituents, and less than 5% in the presence of hydrogen). These values are valid only for the examples chosen in this publication. Indeed, if we look at the results provided by expression from an ionized medium (Fig. 2.6), we see that the results are not very good.

If we use the expression in Eq. (2.80) instead of in Eq. (2.120), and make the same assumption on the A^*_{ij}, we obtain an even simpler expression of equivalent precision (Fig. 2.6)

$$\psi_{ij} = \left(\frac{\mathcal{M}_j\mu_i}{\mathcal{M}_i\mu_j}\right)^{\frac{1}{2}} \tag{2.122}$$

We can approximate Eq. (2.113), noting that $H_{ij}/H_{ii} \ll 1$, and consequently the terms behind the sum signs, are small compared to 1. It makes a development of the numerator and denominator at first order and is introduced into the first equality $\frac{H_{ii}}{b_i} = \frac{H_{jj}}{b_j}$

$$\mu \simeq \sum_i \frac{x_i^2}{b_i} \frac{1-\sum_{j\neq i}\dfrac{H_{ij}x_j}{H_{jj}x_i}}{1-\sum_{j\neq i}\dfrac{H_{ij}}{H_{jj}}} \tag{2.123}$$

By postponing the approximation given by Eq. (2.115), in which the denominator is neglected, then equivalent to Eq. (2.110), we get

$$\mu \simeq \sum_i \mu_i \frac{1-\sum_{j\neq i}\psi_{ij}}{1-\sum_{j\neq i}\dfrac{x_j}{x_i}\psi_{ij}} \tag{2.124}$$

with

$$\psi_{ij} = \frac{H_{ij}}{x_i x_j}\mu_i \tag{2.125}$$

The accuracy of this approximation is comparable to that of Wilke in the absence of hydrogen, a little better in the presence of light species [39].

- If we develop the first order equation [Eq. (2.124)], we get

$$\mu_i \simeq \sum_i \mu_i (1 + b_{ij}) x_j \tag{2.126}$$

with

$$b_{ij} = \sum_{j \neq i} \frac{H_{ij}}{x_i x_j}\left(\frac{\mu_i}{x_i} - \frac{\mu_j}{x_j}\right) \tag{2.127}$$

The quantity b_{ij} depends only weakly on composition, in particular the term $H_{ij}/x_i x_j$ does not depend on it [Eq. (2.107)]. We can replace the expression with its numerical value calculated for a judiciously chosen reference composition. This type of approximation [10, 37] leads to errors in the order of 0.1%.

- Note that, in the case of a gas having only electrically neutral particles, the approximations described can be rewritten using the method of separation of variables described in Sec. 2.3.3 [29].
- There is a relationship that seems empirically derived that gives good results up to medium temperatures

$$\mu = \left(\sum_i \frac{c_i}{\mu_i}\right)^{-\frac{1}{2}} \tag{2.128}$$

These good results (Fig. 2.6) are at least partly due to the fact that the values μ_i of various atomic or molecular species are of the same order of magnitude (maximum difference of about 30% in the example of air).

2.5.3 Conductivity

2.5.3.1 Conductivity of a Species

The expression of conductivity has a form very similar to the viscosity for a monatomic gas (without degree of internal freedom, $\mathcal{C}_{V_i} = \mathcal{C}_{V_i}^{\text{trans}} = \frac{3\mathcal{R}}{2}$)

$$\lambda_i = \frac{5}{2}\frac{\mathcal{C}_{V_i}}{\mathcal{M}_i}\mu_i = \frac{5\alpha_2}{2}\frac{(\mathcal{M}_i T)^{\frac{1}{2}}}{\sigma_{ii}^2 \Omega_{ii}^{(2,2)^*}}\frac{\mathcal{C}_{V_i}^{\text{trans}}}{\mathcal{M}_i} \tag{2.129}$$

In the general case of polyatomic species, Eucken proposed a modification to this expression, taking into account internal degrees of freedom

$$\frac{\lambda_i \mathcal{M}_i}{\mu_i} = f_i^{\text{trans}} \mathcal{C}_{V_i}^{\text{trans}} + f_i^{\text{int}} \mathcal{C}_{V_i}^{\text{int}} \tag{2.130}$$

with $f_i^{\text{trans}} = \frac{5}{2}$ to recover the previous expression as a particular case.

We rewrite Eq. (2.130) as follows:

$$\frac{\lambda_i \mathcal{M}_i}{\mu_i} = \frac{5}{2} \mathcal{C}_{V_i}^{\text{trans}} + f_i^{\text{int}} \left(\mathcal{C}_{V_i} - \frac{3\mathcal{R}}{2} \right) \tag{2.131}$$

where $f_i^{\text{int}} = 1$ corresponds to the Eucken correlation. In this case, Eq. (2.131) reduces to

$$\frac{\lambda_i \mathcal{M}_i}{\mu_i} = \mathcal{C}_{V_i} + \frac{9\mathcal{R}}{4} \tag{2.132}$$

Other expressions have been proposed, for example, assuming that the internal energy of translational energy diffuses as the species itself, characterized by the coefficient $\mathcal{D}_{ii}$. Furthermore, we assume that the internal energy is described by a Boltzmann distribution at translational temperature T. The f_i^{int} is then the inverse of the Schmidt number [36, 40]

$$f_i^{\text{int}} = \frac{\rho_i \mathcal{D}_{ii}}{\mu_i} = \frac{6}{5} A_{ii}^* \tag{2.133}$$

This expression, often called the modified Eucken correlation, generally gives excessive values whereas the Eucken correlation gives default values [39]. Indeed, the internal degrees of freedom relax more slowly than translational and their real diffusion coefficient is less than $\mathcal{D}_{ii}$ ($\mathcal{D}_{\text{int}}/\mathcal{D}_{ii}$ is the order of 0.4 to 1 depending on the species and temperature [41]).

In the case of a molecular medium, this quantity is approximately constant $f_i^{\text{int}} \simeq 0.35$ [3]. Numerically, the expression is then

$$\frac{\lambda_i \mathcal{M}_i}{\mu_i} \simeq 1.35 \mathcal{C}_{V_i} + 1.73\mathcal{R} \tag{2.134}$$

The most general expression will involve all degrees of freedom of the molecule: translation, rotation, and internal (vibrational and electronic levels)

$$\frac{\lambda_i \mathcal{M}_i}{\mu_i} = f_i^{\text{trans}} \mathcal{C}_{V_i}^{\text{trans}} + f_i^{\text{int}} \mathcal{C}_{V_i}^{\text{int}} \tag{2.135}$$

This form is justified by a kinetic analysis (Wang Chang and Uhlenbeck, Mason and Monchik) [40], involving a quantity quantifying translation–rotation of the molecule through a number of effective collisions (vibrational relaxation rate), Z^{rot} or $\tau_{\mathrm{rot}} = \mu / pZ^{\mathrm{rot}}$, exchange characteristic time).

Unlike translation–rotation, the translation–vibration exchanges are not taken into account. Moreover, the diffusion coefficients of energy are com bined with the mass diffusion coefficients $\mathcal{D}_{ii}$ because of ignorance of cross-sections involved in the calculation. Various expressions are available. We give two of them here:

1. The classical expression of Mason and Monchick

$$\begin{cases} f_i^{\mathrm{trans}} & = \dfrac{5}{2}\left[1 - \dfrac{5}{6Z_i^{\mathrm{rot}}}\left(1 - \dfrac{2}{5}\dfrac{\rho_i \mathcal{D}_{ii}}{\mu_i}\right)\dfrac{\mathcal{C}_{V_i}^{\mathrm{int}}}{\mathcal{R}}\right] \\ f_i^{\mathrm{int}} & = \dfrac{\rho_i \mathcal{D}_{ii}}{\mu_i}\left[1 + \dfrac{5}{4Z_i^{\mathrm{rot}}}\left(1 - \dfrac{2}{5}\dfrac{\rho_i \mathcal{D}_{ii}}{\mu_i}\right)\right] \end{cases} \tag{2.136}$$

Note that the modified Eucken approximation is recovered in the limit of high Z_i^{rot}. By replacing $\rho_i \mathcal{D}_{ii}/\mu_i$ with its numerical approximation, we get

$$\frac{\lambda_i \mathcal{M}_i}{\mu_i} = 1.35\mathcal{C}_{Vi} + 1.73\mathcal{R} - \frac{\mathcal{C}_{V_i}^{\mathrm{rot}}}{Z_i^{\mathrm{rot}}\mathcal{C}_{V_i}} \tag{2.137}$$

2. An expression proposed by Warnatz [42] used in CHEMKIN [43]

$$\begin{cases} f_i^{\mathrm{trans}} & = \dfrac{5}{2}\left(1 - \dfrac{2}{\pi}\dfrac{\mathcal{C}_{V_i}^{\mathrm{rot}}}{\mathcal{C}_{V_i}^{\mathrm{trans}}}\dfrac{A_i}{B_i}\right) \\ f_i^{\mathrm{rot}} & = \dfrac{\rho \mathcal{D}_{ii}}{\mu_i}\left(1 + \dfrac{2}{\pi}\dfrac{A_i}{B_i}\right) \\ f_i^{\mathrm{int}} & = \dfrac{\rho \mathcal{D}_{ii}}{\mu_i} \\ A_i & = \dfrac{5}{2} - \dfrac{\rho \mathcal{D}_{ii}}{\mu_i} \\ B_i & = Z_i^{\mathrm{rot}} + \dfrac{2}{\pi}\left(\dfrac{5}{3}\dfrac{\mathcal{C}_{V_i}^{\mathrm{rot}}}{\mathcal{R}} + \dfrac{\rho \mathcal{D}_{ii}}{\mu_i}\right) \\ Z_i^{\mathrm{rot}} & = Z_i^{\mathrm{rot}}(T_{\mathrm{ref}})\dfrac{F(T)}{F(T_{\mathrm{ref}})} \end{cases} \tag{2.138}$$

$Z_i^{\mathrm{rot}}(T_{\mathrm{ref}})$ is actually the average of Z_{ij}^{rot}, given for each species i that collide with the species J. The temperature dependence of Z_i^{rot} in the case of a Lennard–Jones potential for similar molecules is given by expression [44, 45]

$$F(T^*) = 1 + \frac{\pi^{\frac{3}{2}}}{2} T^{*\frac{1}{2}} + \left(\frac{\pi^2}{4} + 2\right) T^* + \pi^{\frac{3}{2}} T^{*\frac{3}{2}} \tag{2.139}$$

In the case of another potential, we can define a new reduced temperature T^* from the definition of ϵ/k given in Sec. 2.1.2.1.

Expressions used for the numerical approximation of λ are similar to those used for the coefficient diffusion [Eq. (2.104)], of course without the term $1/p$ [26, 34].

2.5.3.2 Conductivity of a Mixture: Exact Expression for the Translational Conductivity

The translational conductivity of a mixture is given by [3]

$$\lambda^{\mathrm{trans}} = \lambda' - \frac{\mathcal{R}}{2} \sum_i \sum_{j \neq i} x_i x_j \Delta_{ij} \left(\frac{\mathcal{D}_i^T}{x_i \mathcal{M}_i} - \frac{\mathcal{D}_j^T}{x_j \mathcal{M}_j} \right)^2 \tag{2.140}$$

Δ_{ij} is given by Eq. (2.108).

The term λ' is obtained by solving a linear system. Formally, the solution is the quotient of two determinants whose numerator is of dimension $2n_e+1$ [3]. The most common simplifying assumption is to make $\frac{6}{5} C_{ij}^* = 1$, which corresponds to neglecting the thermal diffusion. For a realistic potential interaction between molecules, this value is around 1.1. Because of this simplification, the dimension numerator is reduced to $n_e + 1$ and the second term of Eq. (2.140) is zero. The expression of translational conductivity is then [10, 46]

$$\lambda^{\mathrm{trans}} = 4 \frac{\begin{vmatrix} L_{11} & L_{12} & \ldots & L_{1n_e} & x_1 \\ L_{21} & L_{22} & \ldots & L_{1n_e} & x_2 \\ \ldots & \ldots & \ldots & \ldots & \ldots \\ L_{n_e1} & L_{n_e2} & \ldots & L_{n_en_e} & x_{n_e} \\ x_1 & x_2 & \ldots & x_{n_e} & 0 \end{vmatrix}}{\begin{vmatrix} L_{11} & L_{12} & \ldots & L_{1n_e} \\ L_{21} & L_{22} & \ldots & L_{1n_e} \\ \ldots & \ldots & \ldots & \ldots \\ L_{n_e1} & L_{n_e2} & \ldots & L_{n_en_e} \end{vmatrix}} \tag{2.141}$$

with

$$\begin{cases} L_{ii} = -\dfrac{4x_i^2}{\lambda_i} - \dfrac{16}{25\mathcal{R}} \sum_{k \neq i} \dfrac{x_i x_k \left(\dfrac{15}{2}\mathcal{M}_i^2 + \dfrac{25}{4}\mathcal{M}_k^2 - 3\mathcal{M}_k^2 B_{ik}^* + 4\mathcal{M}_i\mathcal{M}_k A_{ik}^* \right)}{(\mathcal{M}_i + \mathcal{M}_k)^2} \Delta_{ik} \\ L_{ij} = \dfrac{16}{25\mathcal{R}} \dfrac{x_i x_j \mathcal{M}_i \mathcal{M}_j \left(\dfrac{55}{4} - 3B_{ij}^* - 4A_{ij}^* \right)}{(\mathcal{M}_i + \mathcal{M}_j)^2} \Delta_{ij} \quad j \neq i \end{cases} \tag{2.142}$$

The calculation in the case of air at local thermodynamic equilibrium (Fig. 2.7) shows that the approximation is excellent for the first term of Eq. (2.140), except at high temperature. The second term is almost completely negligible everywhere, except at high temperatures. In the latter case, errors are partly compensated for and the approximation remains accurate within approximately 1%.

At low temperature (Fig. 2.8), translational and internal conductivities are of the same order of magnitude. At high temperature (Fig. 2.9), the conductivity of the electron becomes dominant.

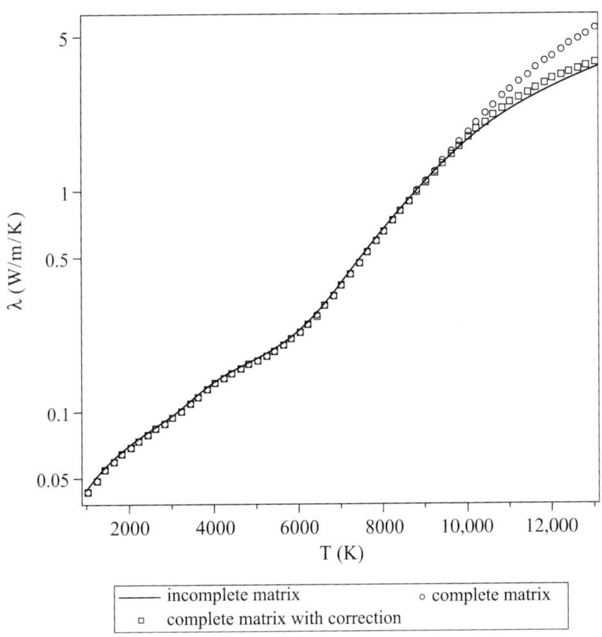

Fig. 2.7 Translational conductivity of air at thermodynamic equilibrium ($p = p_0$). First level of approximation.

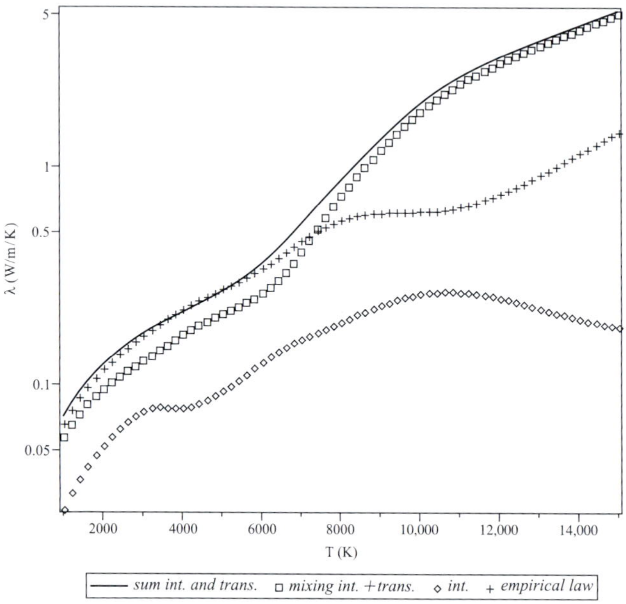

Fig. 2.8 Internal and translational conductivities of air at thermodynamic equilibrium ($p = p_0$).

2.5.3.3 Approximations of the Translational Conductivity of a Mixture

Approximations used for determining viscosity (Sec. 2.5.2.3) are usable here. They lead to an expression of the following type:

$$\lambda^{\text{trans}} \simeq \sum_i \frac{\lambda_i^{\text{trans}}}{1 + \sum_{j \neq i} \psi_{ij} \dfrac{x_j}{x_i}} \tag{2.143}$$

Subsequently, we noted

$$\lambda_i^{(1)} = \frac{\lambda_i^{\text{trans}}}{1 + \sum_{j \neq i} \psi_{ij} \dfrac{x_j}{x_i}} \tag{2.144}$$

We will thus find:

1. An approximation obtained by a development analogous to that of the determinants of Buddenberg and Wilke, based on a similar development

to Eq. (2.109)

$$\lambda^{\text{trans}} = -4\sum_i \frac{x_i^2}{L_{ii}} + 4\sum_i \sum_{j\neq i} \frac{x_i x_j L_{ij}}{L_{ii}L_{jj}} - 4\sum_i \sum_{j\neq i} \sum_{k\neq i} \frac{x_k x_k L_{ij} L_{ik}}{L_{ii}L_{jj}L_{kk}} - \cdots \quad (2.145)$$

At order 0 (neglecting the terms L_{ij} and $j \neq i$), we get

$$\psi_{ij} = \alpha \frac{4\lambda_i x_i x_j \Delta_{ij}}{25\mathcal{R}} \frac{\left(\frac{15}{2}\mathcal{M}_i^2 + \frac{25}{4}\mathcal{M}_j^2 - 3\mathcal{M}_j^2 B_{ij}^* + 4\mathcal{M}_i\mathcal{M}_j A_{ij}^*\right)}{(\mathcal{M}_i + \mathcal{M}_j)^2} \quad (2.146)$$

The constant α is adapted to the case treated. For the case of air, we took $\alpha = 1.3$. This approximation works very well at low temperature (Fig. 2.8), but more poorly when the electronic conductivity becomes important (Fig. 2.9). As for viscosity, recall that the general equations described in Sec. 2.2 are not valid for a highly ionized medium, which limits the importance of differences recorded here.

2. Mason and Saxena [47] established an approximation for conductivity similar to that for viscosity by Wilke [Eq. (2.121)]. The cross-terms $\mathcal{D}_{ij}$ are

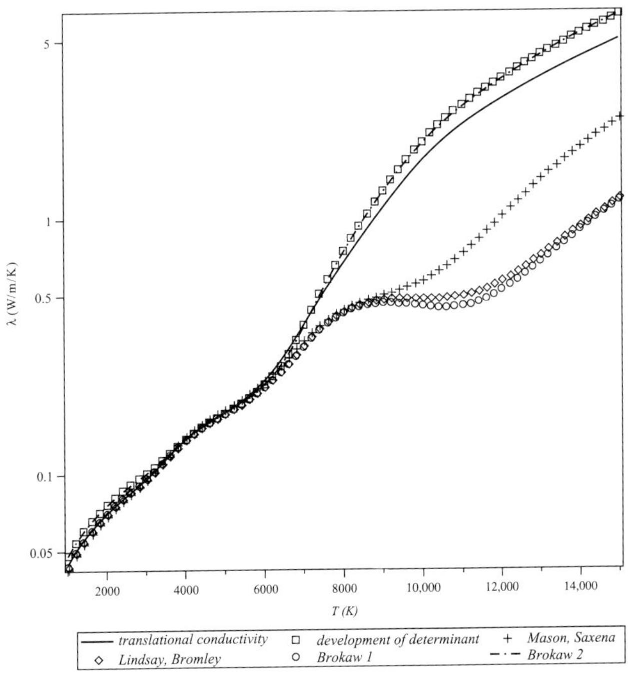

Fig. 2.9 Translational conductivity of air at thermodynamic equilibrium ($p = p_0$). Second level of approximation.

replaced by self-diffusion constants $\mathcal{D}_{ii}$ using Eq. (2.120). Moreover, the term $f(\mathcal{M}_i, \mathcal{M}_j, A^*_{ij}, B^*_{ij})$ contained in Eq. (2.146) is replaced by a constant adjusted to best approach the gaseous medium studied

$$\psi_{ij} = \frac{1.065}{\sqrt{8}} \frac{\left[1 + \left(\frac{\lambda_i^{\text{trans}}}{\lambda_j^{\text{trans}}}\right)^{\frac{1}{2}} \left(\frac{\mathcal{M}_j}{\mathcal{M}_i}\right)^{\frac{1}{4}}\right]^2}{\left(1 + \frac{\mathcal{M}_i}{\mathcal{M}_j}\right)^{\frac{1}{2}}} \tag{2.147}$$

For example, in the absence of hydrogen, the error is very low: <3% on average, <10% at maximum. (With hydrogen, the error is 4% and 15%, respectively.) In our example, the results are very good at low temperature (Fig. 2.8), but bad in an ionized medium (Fig. 2.9).

3. An approximation due to Lindsay and Bromley, issued from the development of the first order determinants [37] ψ_{ij} [given by Eq. (2.116)]

$$\frac{\phi_{ij}}{\psi_{ij}} = 1 + \left(\frac{\mathcal{M}_i - \mathcal{M}_j}{\mathcal{M}_i + \mathcal{M}_j}\right)^2 \left(\frac{15}{4A^*_{ij}} - 1\right) \times \left[1 + \left(\frac{12B^*_{ij} + 5}{30 - 8A^*_{ij}}\right) \frac{\mathcal{M}_j}{\mathcal{M}_i - \mathcal{M}_j}\right] \tag{2.148}$$

The error, calculated for binary gases H_2–CO_2, is less than 3%. This approximation, not particularly good in our case (see Fig. 2.9), is mentioned to introduce the next approximation.

4. At higher order, we find an expression similar to Eq. (2.124)

$$\lambda^{\text{trans}} \simeq \sum_i x_i \lambda_i^{(1)} \frac{1 + \sum_{j \neq i} a_{ij} \lambda_j^{(1)}}{x_i + \sum_{j \neq i} a_{ij} \lambda_i^{(1)} x_j} \tag{2.149}$$

with

$$a_{ij} = \frac{8}{25\mathcal{R}} \frac{\mathcal{M}_i \mathcal{M}_j}{(\mathcal{M}_i + \mathcal{M}_j)^2} \left(\frac{55}{8} - \frac{3B^*_{ij}}{2} - 2A^*_{ij}\right) \Delta_{ij} \tag{2.150}$$

$\lambda_i^{(1)}$ is any lower-order approximations, defined by Eq. (2.144).

If one uses the quantity $\lambda_i^{(1)}$ defined by Eq. (2.144), the expression of Lindsay and Bromley, this expression is accurate to a few per thousand in a binary mixture H_2–CO_2 [47]. In general, it significantly improves results when the approximation of order 1 is already good. In our case, the results for the ionized media are not significantly improved (Fig. 2.9, curve "Brokaw 1"). The term is also used with any approximation of order 0, such as the first in the list (Buddenberg and Wilke type). The conclusion is the same as previously (Fig. 2.9, curve "Brokaw 2").

2.5.3.4 Approximations of the Internal Conductivity of a Mixture

The most general case is of inextricable complexity. We will refer to [1] for the binary case. It is generally used for the internal conductivity expression [Eq. (2.141)], assuming a diffusive exchange between degrees of freedom in translation and internal [47]. For approximating the exact expression, we can use the expression already encountered

$$\lambda^{\text{int}} \simeq \sum_i \frac{x_i \lambda_i^{\text{int}}}{x_i + \sum_{j \neq i} \psi_{ij} x_j} \tag{2.151}$$

- With an approximation provided by Yun, Weissman, and Mason [49] giving reasonable results in all the temperature ranges (Fig. 2.8)

$$\psi_{ij} = \frac{\mathcal{D}_{ii}}{\mathcal{D}_{ij}} x_j \tag{2.152}$$

- With an approximation provided by Mason and Saxena [47], using the same procedure as that used for λ^{trans}

$$\psi_{ij} = \frac{1}{\sqrt{8}} \frac{\left[1 + \left(\dfrac{\lambda_i^{\text{trans}}}{\lambda_j^{\text{trans}}}\right)^{\frac{1}{2}} \left(\dfrac{\mathcal{M}_i}{\mathcal{M}_j}\right)^{\frac{1}{4}}\right]^2}{\left(1 + \dfrac{\mathcal{M}_i}{\mathcal{M}_j}\right)^{\frac{1}{2}}} \tag{2.153}$$

The results in our case (Fig. 2.8) are slightly worse than before.

2.5.3.5 Approximations of the Total Conductivity of a Mixture

Note that if we compare Eqs. (2.147) and (2.153) the ψ_{ij} are identical without the factor 1.065. Assuming this term equal to 1 and summing these two equations, we obtain directly the conductivity of the mixture

$$\lambda \simeq \sum_i \frac{x_i \lambda_i}{x_i + \sum_{j \neq i} \psi_{ij} x_j} \tag{2.154}$$

The λ_i here are the total conductivities of each species. The simplicity of this approach explains its popularity.

Also found in the literature is an expression that seems totally empirical

$$\lambda = \frac{1}{2} \left[\sum_i x_i \lambda_i + \left(\sum_i x_i \lambda_i^{-1} \right)^{-1} \right] \tag{2.155}$$

This expression may be related to the relation in Eq. (2.128) for viscosity. The remarks that can be done on this expression are also the same: the expression gives a reasonable approximation up to medium temperatures in the case of the air because the λ_i are the same order of magnitude (less than 50% deviation between different values).

2.5.4 Thermal Diffusion

The diffusion gradient of temperature is generally considered much less than that due to the concentration gradient. We can quantify the difference in the example of air in thermodynamic equilibrium, as described in Sec. 2.3.4. The flux ratio is given by $[(||k_i^T \nabla \ell n(T)||)/||\nabla x_i||]$ [see Eq. (2.41)]. In the example described, this term is

$$\frac{k_i^T}{T\,\frac{\partial x_i}{\partial T}}$$

Its maximum (regardless of species) is approximately order 0.2 (except in the vicinity of singularities $\partial x_i/\partial T = 0$). It is therefore of little influence in our example. Despite the decoupling between composition and temperature in nonequilibrium chemical environments, it seems that the Soret effect remains low. That is the conclusion of macroscopic calculations [21]: a maximum deviation of 5% on the wall heat flux and 15% on the composition.

Note that, due to the presence of the term $6C_{ij}^* - 5$ in Eq. (2.156) and the fact that C_{ij}^* is the order of unity, the accuracy of calculation of diffusion constants is small.

2.5.4.1 Binary Thermal Diffusion Coefficients

The matrix element α_{ij}^T used in the calculation of multicomponent thermal diffusion numbers [Eq. (2.50)] is given by [3]

$$\alpha_{ij}^T = (6C_{ij}^* - 5)\frac{x_i S_{ij} - x_j S_{ji}}{x_i^2 P_{ij} + x_i x_j Q_{ij} + x_j^2 P_{ji}} \tag{2.156}$$

with

$$\begin{cases} S_{ij} = \left(\frac{\mathcal{M}_i+\mathcal{M}_j}{2\mathcal{M}_j}\right)^{\frac{3}{2}} - \frac{15}{4A_{ij}^*}\left(\frac{\mathcal{M}_j-\mathcal{M}_i}{2\mathcal{M}_i}\right) - 1 \\ P_{ij} = 6\left(\frac{\mathcal{M}_i+\mathcal{M}_j}{2\mathcal{M}_j}\right)^{\frac{1}{2}}\left[1 + \frac{4A_{ij}^*}{15} - \frac{1}{12}\left(\frac{12B_{ij}^*}{5}+1\right)\frac{\mathcal{M}_i}{\mathcal{M}_j} + \frac{1}{2}\frac{(\mathcal{M}_i-\mathcal{M}_j)^2}{\mathcal{M}_i\mathcal{M}_j}\right] \\ Q_{ij} = 12\left[1 + \frac{2A_{ij}^*}{15}\frac{(\mathcal{M}_i+\mathcal{M}_j)}{(\mathcal{M}_i\mathcal{M}_j)^{\frac{1}{2}}} - \frac{1}{12}\left(\frac{12B_{ij}^*}{5}+1\right) - \frac{5}{32A_{ij}^*}\left(\frac{12B_{ij}^*}{5}-5\right)\frac{(\mathcal{M}_i-\mathcal{M}_j)^2}{\mathcal{M}_i\mathcal{M}_j}\right] \end{cases} \tag{2.157}$$

All these quantities depend weakly on T^* (or T). It is therefore possible to approximate them by a polynomial. The matrix α_{ij}^T is antisymmetric ($Q_{ij} = Q_{ij}$).

We can significantly simplify the expression for the α_{ij} [Eq. (2.156)] in the special case $[(\mathcal{M}_i - \mathcal{M}_j)/\mathcal{M}_i + \mathcal{M}_j] < \epsilon$. This is the case of an isotopic mixture, or any mixture in which $\epsilon < 0.15$ [3]. This equation becomes

$$\alpha_{ij}^T = \frac{15}{2}\frac{(2A_{ij}^* + 5)(6C_{ij}^* - 5)}{A_{ij}^*(16A_{ij}^* - 12B_{ij}^* + 55)}\frac{\mathcal{M}_j - \mathcal{M}_i}{\mathcal{M}_i + \mathcal{M}_j} \tag{2.158}$$

2.5.4.2 Multicomponent Thermal Diffusion Numbers

We can calculate k_i^T from the following expression [2]:

$$k_i^T = \frac{T}{5p}\sum_j \frac{6C_{ij}^* - 5}{\mathcal{D}_{ij}}\frac{x_i\mathcal{M}_i\lambda_{p_j} - x_j\mathcal{M}_j\lambda_{p_i}}{\mathcal{M}_i + \mathcal{M}_j} \tag{2.159}$$

where the λ_{p_i} are "partial" thermal conductivities such that $\lambda = \sum_i \lambda_{pi}$ [2], which are solutions of the following linear system:

$$\lambda_{p_i}\left\{\frac{x_i}{\lambda_i} + \sum_{j\neq i}\frac{\Delta_{ij}x_j}{5\mathcal{R}(\mathcal{M}_i+\mathcal{M}_j)^2}\left[6\mathcal{M}_i^2 + (5 - 4B_{ij}^*)\mathcal{M}_j^2 + 8\mathcal{M}_i\mathcal{M}_jA_{ij}^*\right]\right\}$$
$$-x_i\left[1 + \sum_{j\neq i}\frac{\Delta_{ij}\mathcal{M}_i\mathcal{M}_j}{5\mathcal{R}(\mathcal{M}_i+\mathcal{M}_j)^2}\left(11 - 4B_{ij}^* - 8A_{ij}^*\right)\lambda_{p_j}\right] = 0 \tag{2.160}$$

Δ_{ij} is given by Eq. (2.108).

This system is equivalent to Eq. (2.141), which has already been given approximate solutions. The solution of the system in Eq. (2.160) can be approximated by the following expression [50]:

$$\lambda_{p_i} = \lambda_i\left(1 + \frac{1}{x_i}\sum_{j\neq i}x_j\psi_{ij}\right)^{-1} \tag{2.161}$$

where ψ_{ij} is the Mason and Saxena approximation given in Sec. 2.5.3.3. This approximation is shown in Fig. 2.10. The coefficients necessary to calculate this approximation being necessary for the conductivity, the cost of this method is low and the precision relatively good as compared to other methods of calculating thermal diffusion.

2.5.4.3 Approximation of Thermal Diffusion Coefficients

The exact expression of the thermal diffusion coefficient $\mathcal{D}_i^T$ involves a ratio of determinant of order $2n_e + 1$, irreducible [3]. The $\mathcal{D}_i^T$ are positive or negative (Fig. 2.11).

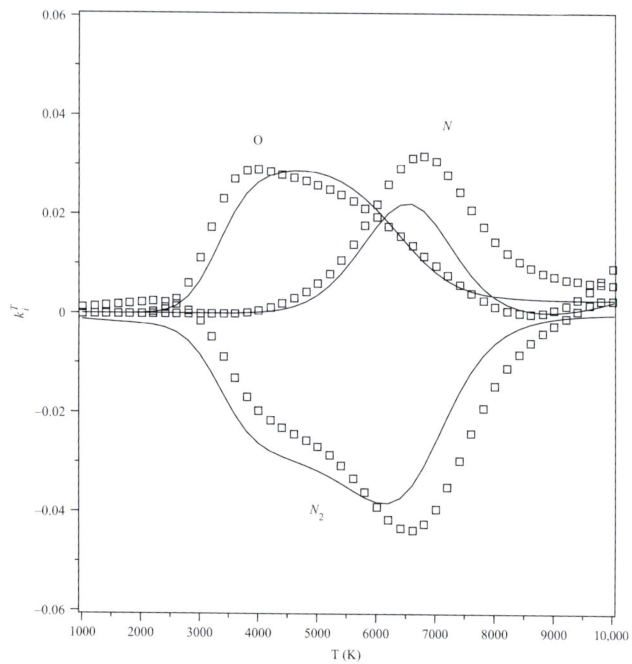

Fig. 2.10 Multicomponent thermal diffusion numbers of N_2, O, and N (continuous curves).

A simple approximation of the thermal diffusion has the same origin as the approximation described by splitting in Sec. 2.3.3 [51]. It is written (see Appendix A.3)

$$\mathcal{D}_i^T = \frac{0.5\rho\mathcal{D}_{\text{ref}}\mu_2(Z_i - c_i)}{\mu_1\overline{\mathcal{M}}} \tag{2.162}$$

Despite the lack of rigor in the derivation of this approximation, its accuracy is quite reasonable. In air, at temperatures where this ratio is not as low as O_2 or NO, the error is on the order of 10 to 20% (Fig. 2.11). Recall that this type of approach is not usable in an ionized medium.

2.5.4.4 Influence of Internal Degrees of Freedom on the Thermal Diffusion

The internal degrees of freedom can be taken into account by an additive term in α_{ij} [52, 53]

$$\alpha_{ij}^{\text{int}} = \frac{T}{5p\mathcal{D}_{ij}}\left[\frac{(6\tilde{C}_{ji} - 5)\lambda_j^{\text{int}}}{x_j} - \frac{(6\tilde{C}_{ij} - 5)\lambda_i^{\text{int}}}{x_i}\right] \tag{2.163}$$

Internal conductivities λ_i^{int} are defined in Sec. 2.5.3.1. The collision integrals $\tilde{C}_{ij} \neq \tilde{C}_{ji}$ involving rotational exchanges are computable with simple potentials (loaded hard sphere) capable of taking into account the angular momentum of the colliding molecules. It is generally assumed that $\tilde{C}_{ij} = \tilde{C}_{ii}$. These quantities are then expressed in terms of characteristic number of collisions of i to return to equilibrium Z_i^{rot}, these quantities being calculated by this model but also measurable. One can use the approximate relation [52]

$$\tilde{C}_{ii} - 5 \simeq \frac{0.8}{Z_i^{\text{rot}}} \tag{2.164}$$

The contribution of α_{ij}^{int} to the total value may reach 30% in absolute value for hydrogen [53]. This quantity can be negative when the translational component is positive, reducing k_i^T. In the example air in equilibrium thermodynamics, the internal component does not exceed 1% of the total, certainly lower than the precision available for the α_{ij}. This term can legitimately be neglected.

2.5.5 Electrical Conductivity

We addressed the problem of electric field and current in Sec. 2.4 in an electrically unbounded medium in which we have set the current to zero.

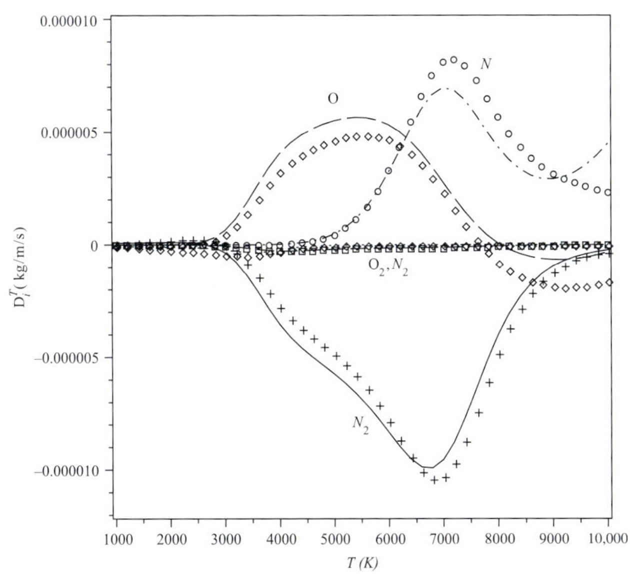

Fig. 2.11 Approximations of $\mathcal{D}_i^T$ for various species of air at thermodynamic equilibrium ($p = p_0$).

However, there are examples where this assumption is not verified. This is the case for plasma jet generators (see Chapter 12) in which the gas heating is provided by the electrons of an electric arc. In this type of installation, the gas is heated to temperatures of around 20,000 K and is significantly ionized. The magnetic field is usually small enough to be neglected, and the electrical conductivity σ is scalar

$$j = \sigma \mathbf{E} \tag{2.165}$$

Equation (2.38) can be simplified in our case because of the low mass of the electron

$$\left\| \frac{J_e}{c_e} \right\| >> \left\| \frac{J_i}{c_i} \right\| \quad \forall i \tag{2.166}$$

Neglecting the thermal diffusion gradient, we get

$$-\sum_{k \neq e} \frac{x_e x_k}{\rho \mathcal{D}_{ek}} \frac{\mathbf{J}_e}{c_e} = \mathbf{d}_e \tag{2.167}$$

The right term in this expression is given by Eq. (2.90), in which we will retain only the terms related to the movement of charged species in the electric field, whose effects are predominant. Assume further that the overall

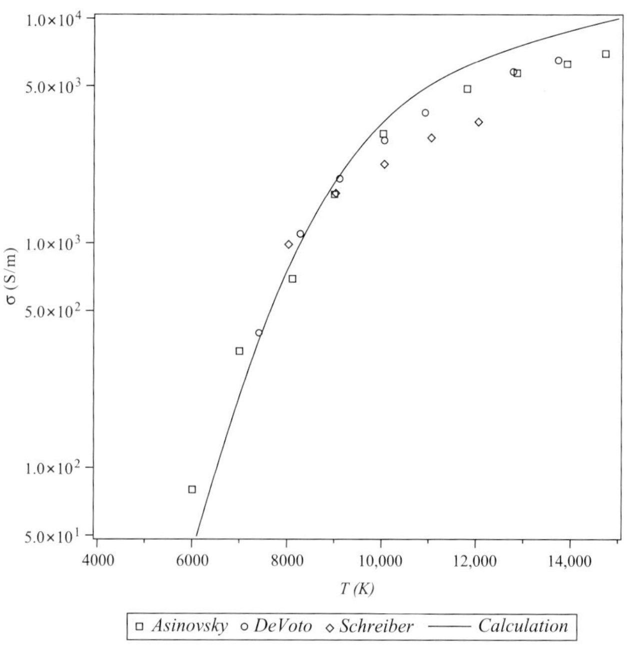

Fig. 2.12 Electrical conductivity of air at thermodynamic equilibrium $p = p_0$; comparison with experiments.

charge is zero (see Sec. 2.4).

$$\mathbf{j} = -\frac{Q_e E}{p} \tag{2.168}$$

Hence the expression of electrical conductivity

$$\sigma = \frac{Q_e^2 \overline{\mathcal{D}}_e}{p_e} \tag{2.169}$$

where $\overline{\mathcal{D}}_e$ is given by Eq. (2.47).

The accuracy of this expression for coefficients' binary distribution using the development to the order 1 of Sonin polynomials is about 10 to 20% depending on the degree of ionization [54], which is consistent with the precision of the data of the collision integrals. One might add that the higher order that is sometimes used needs knowledge of the quantities $\Omega^{(l,s)}$ defined by Eq. (2.10) with $l = 1$ and $s \in [1, 5]$, uncommon in databases. The comparison of Eq. (2.169) with measurements (Fig. 2.12) is satisfactory.

2.6 Medium in Thermodynamic Nonequilibrium

2.6.1 Description of Environment

Environments previously studied involve a number of collisions between molecules, atoms, and electrons sufficient to achieve thermodynamic equilibrium, that is, a microscopic description of the various degrees of freedom (translation, rotation, vibration, and electronic levels) by the Maxwell–Boltzmann statistics defining a unique temperature T. This is not true immediately behind the shock and can span a greater or lesser distance depending on the value of the density, so the number of collisions between the constituent particles of the medium differs. The consequences of this phenomenon can be important in two cases:

- The medium is not dense enough, then the consequences of such effects will be felt up the wall.
- The nonequilibrium zone plays a particular role (e.g., in the case of radiation).

The analysis is therefore based on the collision frequency, as well as the nature of energy exchange in a shock:

- Translational energy only (elastic collision)
- Translational and rotational energies
- Translational, rotational, and vibrational energies

This analysis will be completed by the characteristic relaxation times τ^{trans}, τ^{rot}, and τ^{vib}, and the characteristic time of chemical reaction τ^{reac}

and compared with the characteristic time of the flow τ. In general we have

$$\tau^{\text{vib}} \gg \tau^{\text{rot}} \gg \tau^{\text{trans}} \tag{2.170}$$

Moreover, the characteristic time associated with a chemical reaction is generally greater than that of the vibration, with few instances of the same order of magnitude (e.g., the dissociation of oxygen at high temperature).

The characteristic time of flow (relative to the macroscopic gradients) may be:

- *Large as compared to any other value:* Local thermodynamic equilibrium is usually checked when the pressure is high.
- *Large as compared to* τ^{vib} *but the same order of magnitude as* τ^{reac}*:* It is the chemical nonequilibrium, generally obtained at lower pressures.
- *Of the same order of magnitude as* τ^{vib}*:* It is the vibrational nonequilibrium, present in areas behind shock or heavy relaxation zones.

The lower values of the characteristic time of the flow correspond to situations of rarefied flows or near these and have no relevance in the problems of ablation. They are, in fact, the domain of low heat fluxes.

This analysis shows the extreme complexity of the problem because the situations encountered vary over time (variation of p_∞), in space (at a given time), and even with the gas. Complete modeling of such a system is based on the detailed study of populations of each energy state of each molecule. This study is reduced to a few simple cases, for example, downstream of a shock or in a 1-D isentropic expansion to serve for calculating reference [55, 56]. We will investigate the various assumptions to describe this type of environment through a system simple enough to calculate a flow with a reasonable duration.

- The internal energies involved are assumed small as compared to the relative kinetic energy in the collision.
- We have seen that τ^{rot} was greater than τ^{trans} (except perhaps in the case of hydrogen), and both are still small as compared to other characteristic times. We therefore assume $T_{\text{rot}} = T$. (T here is the translational temperature of heavy particles.)
- The effectiveness of the exchange of energy between free electrons and electronic levels of molecules, atoms, or ions induces the electron temperature molecules T_e to be equal to the translational temperature of electrons.
- Electronic exchanges by molecular shock are negligible because of the energy gap between the ground and the first excited level (except maybe for NO).
- Vibration–translation exchange (denoted VT) during a shock, important at high temperatures, induces little modification of the internal state of the molecule (mono-quantum transition). Furthermore, the vibration–vibration exchange (denoted VV) mainly leads to a simple single-quantum

vibrational exchange (resonant transitions). This leads to a distribution of vibrational energy remains of Boltzmann type [57], authorizing the definition of a temperature of vibration, different for each molecule, denoted by T_i^{vib}. The characteristic times depend on the collision partner [58], and we note τ_{ij}^{VT} and τ_{ij}^{VV}, the relaxation times for the species i during its interaction with species j.

- It should also simplify the problem to get a usable system, and this can be done in at least two ways:
  1. We assume the harmonic oscillator, to express terms related to VV collisions [59].
  2. We assume that the energy exchange in VV collisions are zero. In reality, given the assumption of resonant collision made previously, these exchanges are either zero (collisions between similar molecules) or low. This assumption eliminates VV collision terms in the equations and will be adopted later.

This gives a description of the environment with temperatures T, T_e, and T_i^{vib}. Some authors [60] choose to simplify this description for situations closer to thermodynamic equilibrium in making $T_i^{\text{vib}} = T_e$, retaining only the preponderant part of the nonequilibrium. Figure 2.15 later in this chapter gives an idea of the consequences of this approximation.

2.6.2 Thermodynamic Properties

In the general case, the thermodynamic properties of molecules and molecular ions depend on T, T_e, and T_i^{vib}. The expression of these values is difficult to write and is available only in some cases [61]. In general, one assumes the various forms of freedom of the molecule are independent, and thus neglects vibration–rotation interaction. This allows us to write

$$e_i(T, T_i^{\text{vib}}, T_e) \simeq e_i^e(T_e) + e_i^{\text{vib}}(T_i^{\text{vib}}) + e_i^{\text{rot}}(T) + e^{\text{trans}}(T) \tag{2.171}$$

This writing can significantly simplify the calculation of energy for each term using the approximations described in Appendix B.

Figure 2.13 compares the exact values for N_2, O_2, and NO with:

- The calculation made in the previous approximation. The results are good for O_2 and NO, molecules for which rovibrational characteristics are very similar for all electronic levels. The results are poor for N_2 because of differences on these characteristics between the ground state and first excited level.
- The calculation made with the assumption of a harmonic oscillator. The anharmonicity of the potential leads to the considerable differences at high temperatures. These differences have to be considered in the range of existence of the species; for example, at 10,000 K, O_2 is totally dissociated.

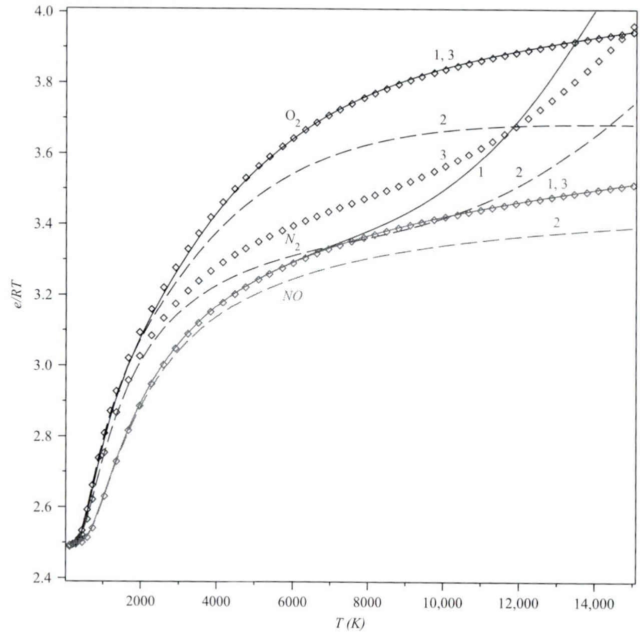

Fig. 2.13 Reduced molar internal energies; (1) exact values; (2) harmonic oscillator approximation; (3) independent modes hypothesis.

This remark is intended to show that this approximation is not necessarily as poor as suggested by the figure.

The vibration–rotation coupling term ignored in the previous approach is not entirely negligible at high temperature. The choice of the reference temperature with which it is assessed (T_V or T) does not induce much difference [62].

2.6.3 Transport Properties

Mass diffusion coefficients and the dynamic viscosity of ions and neutral species are not sensitive at the first order to internal degrees of freedom. They will therefore depend on the single variable T [1, 36], as they were expressed in previous sections (exact or approximate values). These same coefficients for electrons are obviously expressed in terms of T_e. The approximate ambipolar diffusion coefficient [Eq. (2.100)] now reads [66]

$$\mathcal{D}_A \simeq \left(1 + \frac{T_e}{T}\right) \mathcal{D}_I \tag{2.172}$$

The bulk viscosity shows the delay of the degree of freedom in rotation relative to the translation and rotation in a sudden change in volume. If we limit ourselves to the interaction VT, this quantity is expressed as follows [1, 3, 36]:

$$\eta = \frac{\mathcal{R}p}{C_V^2} \sum_i C_{V_i}^{\text{int}} \tau^{VT} \tag{2.173}$$

The characteristic time τ_i^{VT}s given, assuming the harmonic oscillator hypothesis, by

$$\tau_i^{VT} = \left(\sum_j \frac{x_j}{\tau_{ij}^{VT}} \right)^{-1} \tag{2.174}$$

Effects related to the bulk viscosity are generally low or negligible, even under pressure changes as severe as a shock wave at low Mach number, typically $\mathcal{M}a < 2$ [64]. Beyond this value, equations describing a continuous medium are no longer valid [23].

The τ_{ij}^{VT}, at least necessary to calculate the relaxation terms [Eq. (2.180)], can be evaluated from an experimental correlation derived by Millikan and White and widely cited [58, 60], partially justified theoretically [65], valid for $T < 8000$ K

$$\ell n \left(\frac{p_j \tau_{ij}^{VT}}{p_0} \right) \simeq 0.0367 \mathcal{M}_{ij}^{\frac{1}{2}} \theta_i^{\frac{4}{3}} (T^{-\frac{1}{3}} - 0.0844 \mathcal{M}_{ij}^{\frac{1}{4}}) - 18.452 \tag{2.175}$$

where $\mathcal{M}_{ij}$ is the reduced molar mass of the couple ij [expression identical to Eq. (2.11)] and $\theta_i = \frac{E_{V_i}}{k}$ is the characteristic vibrational temperature. This expression has been established in the following area:

$$\begin{cases} 1.75 \times 10^{-3}\ \text{kg} \cdot \text{mol}^{-1} < \mathcal{M}_{ij} < 127 \times 10^{-3}\ \text{kg} \cdot \text{mol}^{-1} \\ 310\ \text{K} < \theta_i < 3395\ \text{K} \\ 280\ \text{K} < T < 8000\ \text{K} \end{cases} \tag{2.176}$$

The experimental dispersion may reach 50%. In addition, for high temperature $T > 5000$ K, this expression gives too low values, the assumption of little change in internal energy as compared to the total energy involved in the collision being not well verified.

Park [58] proposes to correct τ_{ij}^{VT} from the following reasoning. At high energy, a collision will cause a vibrational exchange. Moreover, the repulsive part of the interaction potential dominates. The time feature will be close to the average time between collisions for a collision between hard spheres [2]

$$\tau_{ij}^{VT}\Big|_{HS} p_j = \frac{1}{4\mathcal{N}\sigma_{ij}^2} \left(\frac{\mathcal{M}_{ij} \mathcal{R} T}{\pi} \right)^{\frac{1}{2}} \tag{2.177}$$

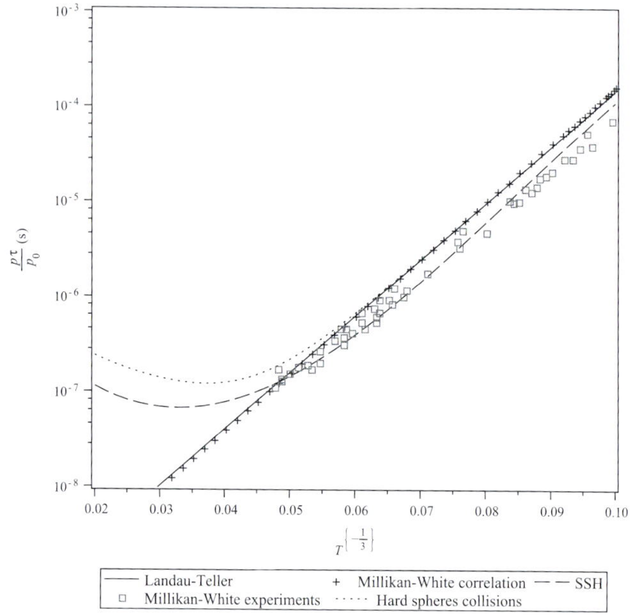

Fig. 2.14 VT relaxation time for the collision O_2–O_2.

More elaborate calculations of interaction (SSH theory, for Schwartz, Stretton, and Herzfeld) have been made to calculate the characteristic time for large interaction energies [65].

Figure 2.14 compares various approximations with measurements made in a temperature range that is limited as compared to applications of interest. τ_{ij}^{VT} can be approximated numerically by a numerical expression of the following type [65]:

$$\ell n\left(\frac{p_j \tau_{ij}^{VT}}{p_0}\right) = \sum_{k=1}^{4} \alpha_{ijk} T^{\frac{k-1}{3}} \tag{2.178}$$

The data are fairly well known in the case of pure air [58, 60, 65], but much less so for couples involving certain products of ablation.

2.6.4 Conservation Laws

The new conservation equations are as follows [60, 163, 327]:

1. The equation of mass conservation [Eq. (2.26)] is formally unchanged; the production term $\dot{\omega}_i(T, T_j^{\text{vib}})$ has to be redefined (see Sec. 3.1.6).
2. The equation of conservation of momentum [Eq. (2.28)] is also unchanged (neglecting pressure vibrational relaxation [1]).

3. Each degree of freedom should be examined separately. Then a conservation equation of vibrational energy of molecules or molecular ions follows

$$\frac{\partial(\rho_i e_i^{\text{vib}})}{\partial t} + \nabla \cdot (\rho_i e_i^{\text{vib}} \mathbf{V}) + \nabla \cdot \dot{\mathbf{q}}_i^{\text{vib}} = S_i^{VT} + S_i^e + S_i^{\text{reac}} \tag{2.179}$$

where

- S_i^{VT} is the amount of vibrational energy created by the exchanges VT. This term is expressed by the Landau–Teller law assuming a simple harmonic oscillator

$$S_i^{VT} = \frac{\rho_i \left[e_i^{\text{vib}}(T) - e_i^{\text{vib}} \right]}{\tau_i^{VT}} \tag{2.180}$$

- The term S_i^e is related to the exchange of internal energy and vibration

$$S_i^e = \frac{\rho_i \left[e_i^{\text{vib}}(T_e) - e_i^{\text{vib}} \right]}{\tau_i^e} \tag{2.181}$$

- The term S_i^{reac} is related to the fact that the species is created by a chemical reaction causing a loss of vibrational energy Δe_i^{vib}. Noting that the mean state obtained after chemical reaction is the same as when

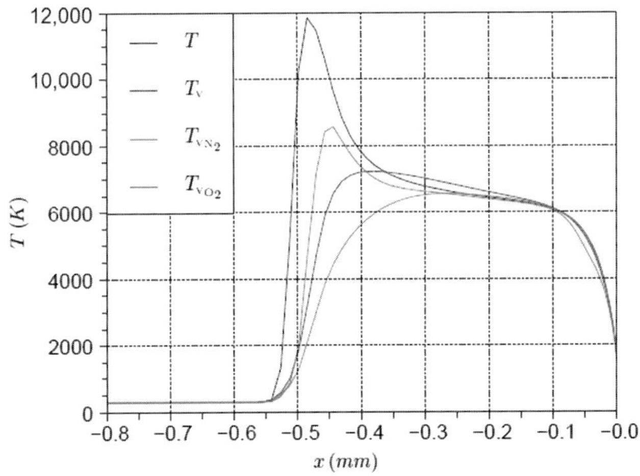

Fig. 2.15 Temperatures behind a shock in air ($\mathcal{M}a = 13.35$, $P_\infty = 664$ Pa, $T_\infty = 293$ K). Comparison of models with one or more temperatures of vibration (67).

producing the chemical reaction (especially true at thermodynamic equilibrium), then

$$S_i^{\mathrm{reac}} = \dot{\omega}_i \Delta e_i^{\mathrm{vib}} \tag{2.182}$$

The valuation of $\Delta e_i^{\mathrm{vib}}$ is treated in Sec. 3.1.6.

- $\dot{\mathbf{q}}_i^{\mathrm{vib}}$ is the vibrational heat flux given by

$$\dot{\mathbf{q}}_i^{\mathrm{vib}} = -\lambda_i^{\mathrm{vib}} \nabla \cdot T_i^{\mathrm{vib}} \tag{2.183}$$

Vibrational thermal conductivity $\lambda_i^{\mathrm{vib}}(T_i^{\mathrm{vib}})$ is given in Sec. 2.5.3.1. Note, however, that exchanges $V–T$, important in our case, are not taken into account in assessing this conductivity. The radiative term corresponding to vibrational transitions is negligible.

4. Similarly, there is also an equation for the conservation of electronic energy $e^e = \sum_i c_i e_i^e$

$$\frac{\partial(\rho e^e)}{\partial t} + \nabla \cdot [(\rho e^e + p_e)\mathbf{V}] + \nabla \cdot \dot{\mathbf{q}}^e = -\sum_i S_i^e + S_e^{\mathrm{coll}} - \sum_i \dot{\omega}_e \Delta e_i^e + Q\,\mathbf{E} \cdot \mathbf{V} \tag{2.184}$$

where

- p_e is the electronic pressure.
- The electronic heat flux $\dot{\mathbf{q}}^e$ is given by

$$\dot{\mathbf{q}}^e = -\lambda_e \nabla T_e + \sum_i \rho c_i e_i^e (T_e) \mathbf{V}_{D_i} + \dot{\mathbf{q}}_{\mathrm{rad}}^e \tag{2.185}$$

Electronic thermal conductivity $\lambda_e(T_e)$ corresponding to elastic collisions between charged particles is calculated by the expression in Eq. (2.129). In this equation we use the radiation term $\dot{\mathbf{q}}_{\mathrm{rad}}^e$ or radiative electronic transitions (Sec. 10.1.2)

- S_e^{coll} is the production of energy by elastic exchange between electrons and neutral atoms or molecules on one hand, and between electrons and ions on the other. Assume that almost all collisions are elastic, a small part being sufficient to allow the equilibrium that led to the definition of T_e (see Sec. 2.6.1).

This quantity is given by the following expression:

$$S_e^{\mathrm{coll}} = 3\rho_e \mathcal{R}(T - T_e) \sum_i \frac{\Phi_{ei}}{\mathcal{M}_i} \tag{2.186}$$

The number of collisions is given by the following expression

[Eq. (3.5), in which $\mathcal{M}_e \ll \mathcal{M}_i$]

$$\Phi_{ei} = \mathcal{N}^2 \, p_e p_i \sigma_{ej} \left[\frac{8\pi}{\mathcal{M}_e (\mathcal{R}T)^3} \right]^{\frac{1}{2}} \tag{2.187}$$

The given cross-sections for electron-neutral interactions σ_{ei} can be found in [9, 10].

For electron–ion interactions, the cross-section is [66, 58]

$$\sigma_{ei} = \frac{8\pi}{27\epsilon_0^2} \frac{q_e^4}{(kT_e)^2} \ell n \left[\frac{9\epsilon_0^3 (kT_e)^4}{q_e^6 p_e} \right] \tag{2.188}$$

- The term $-\dot{\omega}_e \Delta e_i^e$ represents the amount of energy lost by ionization of the electron leading to the ionization of the species i. The term Δe_i^e is the ionization energy that is ordinarily taken in the ground state of the molecule, which increases the corresponding term.
- The last term, corresponding to the work of forces associated with charged particles, is usually ignored (assuming $Q = 0$, Sec. 2.4.1).

5. The conservation equation of total energy [Eq. (2.29)] is formally unchanged, the heat flux being written here

$$\dot{\mathbf{q}} = -\lambda^{\text{trans}} \nabla T + \sum_i \dot{\mathbf{q}}_i^{vib} - \lambda_e \nabla T_e + \sum_i \rho c_i h_i \mathbf{V}_{D_i} + p \sum_i k_i^T \mathbf{V}_{D_i} + \dot{\mathbf{q}}_{\text{rad}} \tag{2.189}$$

2.6.5 Example

The nonequilibrium is seen in Fig. 2.15, which represents the effect of a shock wave in air. The degrees of freedom in the vibration of nitrogen and moreover of oxygen are slowly populated given the characteristic time of flow, causing a delay of chemical reactions from the thermodynamic equilibrium composition. The energy conservation imposes a total value of the translational temperature behind the shock greater than its value at thermodynamic equilibrium. One can see in this figure the difference between distinct vibrational temperatures and a unique temperature equal to an electronic one, in this case $T_i^{\text{vib}} = T_e$.

Ablation phenomena occur at rather high pressures, so the environment generally will be returned to thermodynamic equilibrium at wall. These phenomena are therefore important in cases where the radiation gas is a significant part of energy exchange, that is to say, at very high speeds or with the presence in the upstream environment of species that can lead to the formation of a particularly active species such as CN.

References

[1] Brun, R., *Introduction to Reactive Gas Dynamics*, Oxford University Press, Oxford, UK, 2009.

[2] Chapman, S., and Cowling, T. G., *The Mathematical Theory of Non-Uniform Gases*, Cambridge University Press, New York, 1991.

[3] Hirschfelder, J. O., Curtiss, C. F., and Bird, R. B., *Molecular Theory of Gases and Liquids*, John Wiley & Sons, New York, 1966.

[4] Anderson, J. D., *Hypersonic and High Temperature Gas Dynamics*, McGraw-Hill, New York, 1989.

[5] Bird, G. A., *Molecular Gas Dynamics and the Direct Simulation of Gas Flows*, Engineering Science Series, Vol. 42, Oxford University Press, Oxford, UK, 1994.

[6] Biolsi, L., Fenton, J., and Owenson, B., "Transport Properties for a Mixture of the Ablation Products C, C_2, and C_3," *Thermophysics of Atmospheric Entry*, edited by Nixon D., Vol. 82, Progress in Astronautics and Aeronautics, AIAA, New York, 1982, pp. 17–36.

[7] Rainwater, J. C., Holland, P. M., and Biolsi, L., "Numerical Calculation of Gaseous Transport Properties from the Hulburt-Hirschfelder Potential with Applications to Planetary Entry Thermal Protection," *Thermophysics of Atmospheric Entry*, edited by Nixon D., Vol. 82, Progress in Astronautics and Aeronautics, AIAA, New York, 1982, pp. 3–16.

[8] Stone, A. J., *The Theory of Intermolecular Forces*, Oxford University Press, Oxford, UK, 2002.

[9] Bacri, J., and Raffanel, S., "Calculation of Transport Coefficients of Air Plasmas," *Plasma Chemistry and Plasma Processing*, Vol. 9, No. 1, 1989, pp. 133–154.

[10] Gupta, R. N., Yos, J. M., Thomson, R. A., and Lee, K., "Review of Reaction Rates and Thermodynamic and Transport Properties for an 11-Species Air Model for Chemical and Thermal Nonequilibrium Calculations to 30,000 K," NASA Technical Report RP 1232, 1990.

[11] Park, C., Jaffe, R. L., and Partridge, H., "Chemical-Kinetic Parameters of Hyperbolic Earth Entry," *Journal of Thermophysics and Heat Transfer*, Vol. 15, No. 1, 2001, pp. 76–90.

[12] Wright, M. J., Hwang, H. H., and Schwenke, D. W., "Recommended Collision Integrals for Transport Properties Computations, Part 2: Mars and Venus Entries," *AIAA Journal*, Vol. 45, No. 1, 2007, p. 281.

[13] Wright, M. J., Bose, D., and Olejnicsak, J., "The Impact of Flowfield-Radiation Coupling on Aeroheating for Titan Aerocapture," AIAA Paper 2004-0484, 42nd AIAA Aerospace Sciences Meeting and Exhibit, Reno, Nevada, Jan. 2004.

[14] Yos, J., "Revised Transport Properties for High Temperature Air and Its Components," AVCO Technical Report Z220, Nov. 1967.

[15] Yun, K. S., and Mason, E. A., "Collision Integrals for the Transport Properties of Dissociating Air at High Temperatures," *The Physics of Fluids*, Vol. 43, No. 12, 2005, pp. 2558–2564.

[16] Wright, M. J., Bose, D., Palmer, G. E., and Levin, E., "Recommended Collision Integrals for Transport Properties Computations, Part 1: Air Species," *AIAA Journal*, Vol. 5, No. 4, 1962, pp. 380–386.

[17] Finson, M. L., Pirri, A. N., Nebolsine, P. E., Simons, G. A., and Wu, P. K. S., "Advanced Reentry Aeromechanics. Final Report," Physical Sciences Incorporated Report AD-A052 744, Jan. 1978.

[18] Copeland, D. A., "On the Solution of Stefan-Maxwell Equations," AIAA Paper 2003-851, 41st Aerospace Science Meeting and Exhibit, Reno, Nevada, Jan. 2003.

[19] Koyabashi, N., and Yamamoto, I., "Comparison of Expressions of Thermal Diffusion Phenomena in Multi-component Mixture," *Journal of Nuclear Science and Technology*, Vol. 34, No. 1, 1997, pp. 73–79.

[20] Moore, J. A., "Chemical Equilibrium to Viscous Flows," Ph.D. Thesis, State University of New York at Buffalo, May 1967.

[21] Desmeuzes, C., Duffa, G., and Dubroca, B., "Different Levels of Modeling for Diffusion Phenomena in Neutral and Ionized Mixtures," *Journal of Thermophysics and Heat Transfer*, Vol. 11, No. 1, pp. 36–44.
[22] Rothe, D. E., "Electron Beam Studies of the Diffusive Separation of Helium-Argon Mixtures," *The Physics of Fluids*, Vol. 9, 1966, pp. 1643–1658.
[23] Braeunig, J.-P., Bourgat, J.-F., Charrier, P., Coatanea, F., Dubroca, B., Duffa, G., Le Tallec, P., Mieussens, L., and Perthame, B., "Comparison of Discrete Velocity and Random Particle Methods for Rarefied Reentry Flows," AIAA Paper 2002-5180, 11th AIAA/AAAF Congress, Space Planes and Hypersonic Systems and Technologies, Orléans, France, Sept. 2002.
[24] Fay, J. A., and Riddell, F. R., "Theory of Stagnation Point Heat Transfer in Dissociated Air," *Journal of the Aerospace Sciences*, Vol. 25, No. 2, 1958, pp. 83–86.
[25] Bird, R. B., Stewart, W. E., and Lightfoot, E. N., *Transport Phenomena*, 2nd ed., John Wiley & Sons, New York, 2007.
[26] Kee, R. J., Dixon-Lewis, G., Warnatz, J., Coltrin, M. E., Miller, J. A., and Moffat, H. K., "CHEMKIN III: A Fortran Computer Code Package for the Evaluation of Gas-Phase, Multicomponent Transport Properties," Sandia National Laboratories Report SAND86-8246B, March 1998.
[27] Ramshaw, J. D., and Chang, C. H., "Friction-Weighted Self-Consistent Effective Binary Diffusion," *Journal of Nonequilibrium Thermodynamics*, Vol. 21, No. 3, 1996, pp. 223–232.
[28] Scala, S. M., and Gilbert, L. M., "Sublimation of Graphite at Hypersonic Speeds," *AIAA Journal*, Vol. 3, No. 9, 1965.
[29] Bartlett, E. P., Kendall, R. M., and Rindal, R. A., "An Analysis of the Coupled Chemically Reacting Boundary Layer and Charring Ablator," Part IV, A Unified Approximation for Mixture Transport Properties for Multicomponent Boundary-Layer Applications, NASA Report CR-1063, June 1968.
[30] Sutton, K., and Gnoffo, P. A., "Multi-Component Diffusion with Application to Computational Aerothermodynamics," AIAA Paper 1998-2575, 7th AIAA/ASME Joint Thermophysics and Heat Transfer Conference, Albuquerque, New Mexico, June 1998.
[31] Ern, A., and Giovangigli, V., "Multicomponent Transport Algorithms," in *Physico-Chemical Models for High-Enthalpy and Plasma Flows*, Lecture Series 2002-07, edited by Fletcher, D. et al., Von Karman Institute for Fluid Dynamics, Rhode Saint Genese, Belgium, 2002.
[32] Oran, E. S., and Boris, J. P., "Detailed Modeling of Combustion System," *Progress in Energy and Combustion Science*, Vol. 7, 1981, pp. 1–72.
[33] Ramshaw, J. D., and Chang, C. H., "Ambipolar Diffusion in Multicomponent Plasmas," *Plasma Chemistry and Plasma Processing*, Vol. 11, No. 3, 1991, pp. 395–402.
[34] McBride, B. J., Gordon, S., and Reno, M. A., "Coefficients for Calculating Thermodynamics and Transport Properties of Individual Species," NASA Technical Memorandum 4513, 1993.
[35] Bruno, C., and Barbato, M., "Aerothermochemistry for Hypersonic Technology," Lecture Series 1995-04, Von Karman Institute for Fluid Dynamics, Rhode Saint Genese, Belgium, April 1995.
[36] Ferziger, J. H., and Kaper, H. G., *Mathematical Theory of Transport Processes in Gases*, North Holland Publishing Amsterdam, 1972.
[37] Brokaw, R. S., "Approximate Formulas for the Viscosity and Thermal Conductivity of Gas Mixtures," *Journal of Chemical Physics*, Vol. 29, No. 2, 1958, pp. 391–397.
[38] Wilke, C. R., "A Viscosity Equation for Gas Mixtures," *Journal of Chemical Physics*, Vol. 18, No. 4, 1950, pp. 517–519.
[39] Reid, R. C., Prausnitz, J. M., and Sherwood, T. K., *The Properties of Gases and Liquids*, McGraw-Hill, New York, 1977.
[40] Hirschfelder, J. O., "Heat Conductivity in Polyatomic or Electronically Excited Gases," *Journal of Chemical Physics*, Vol. 26, No. 2, 1957, pp. 282–285.

[41] Ahtye, W. F., "Thermal Conductivity in Vibrationally Excited Gases," *Journal of Chemical Physics*, Vol. 57, 1972, pp. 5542–5555.
[42] De Filippis, F., Del Vecchio, A., Martucci, A., Trifoni, E., Marraffa, L., Savino, R., and Paterna, D., "70 MW Plasma Wind Tunnel Upgrades for ESA Aurora TPS Testing," 4th International Planetary Probe Workshop, Pasadena, California, June 2006.
[43] Chen, Y.-K., Milos, F. S., Reda, D. C., and Stewart, D. A., "Graphite Ablation and Thermal Response Simulation Under Arc-Jet Flow Conditions," AIAA Paper 2003-4042, 36th AIAA Thermophysics Conference, Orlando, Florida, June 2003.
[44] Brau, C. C., and Jonkman, R. M., "Classical Theory of Rotational Relaxation in Diatomic Gases," *Journal of Chemical Physics*, Vol. 52, No. 2, 1970, pp. 477–484.
[45] Parker, J. G., "Rotational and Vibrational Relaxation in Diatomic Gases," *The Physics of Fluids*, Vol. 2, No. 4, 1959, pp. 449–462.
[46] Muckenfuss, C., and Curtiss, C. F., "Thermal Conductivity of Multicomponent Gas Mixtures," *Journal of Chemical Physics*, Vol. 29, No. 6, 1958, pp. 1273–1277.
[47] Mason, E. A., and Saxena, S. C.,"Approximate Formula for the Thermal Conductivity of Gas Mixtures," *The Physics of Fluids*, Vol. 1, No. 5, 1958, pp. 361–369.
[48] Mohammadi, B., and Pironneau, O., *Analysis of the $k-\epsilon$ Turbulence Model*, John Wiley and Sons, New York, 1993.
[49] Yun, K. S., Weissman, S., and Mason, E. A., "High-Temperature Transport Properties of Dissociating Nitrogen and Dissociating Oxygen," *The Physics of Fluids*, Vol. 5, No. 6, 1962, pp. 672–678.
[50] Oran, E. S., and Boris, J. P., *Numerical Simulation of Reactive Flow*, 2nd ed., Cambridge University Press, Cambridge, UK, 2001.
[51] Kendall, R. M., Rindal, R. A., and Bartlett, E. P., "Thermochemical Ablation," AIAA Paper 65-642, 1965.
[52] Monchick, L., Yun, K. S., and Mason, E. A., "Formal Kinetic Theory of Transport Phenomena in Polyatomic Gas Mixtures," *Journal of Chemical Physics*, Vol. 39, No. 3, 1963, pp. 654–669.
[53] Yamakawa, H., Koyabashi, N., Enokida, Y., and Yamamoto, I., "Thermal Diffusion Factor for Diatomic Gas Mixture in Multicomponent System," *Journal of Nuclear Science and Technology*, Vol. 37, No. 4, 2000, pp. 397–404.
[54] Schweitzer, S., and Mitchner, M., "Electrical Conductivity of Partially Ionized Gases," AIAA Paper 1965-1030, 6th Symposium on Engineering Aspects of Magnetohydrodynamics, Pittsburgh, Pennsylvania, April 1965.
[55] Capitelli, M., Colonna, G., and Armenise, I., "State to State Electron and Vibrational Kinetics in One Dimensional Nozzle and Boundary Layer Flows," *Physico-Chemical Models for High-Enthalpy and Plasma Flows*, edited by Fletcher D. et al. Lecture Series 2002-07, Von Karman Institute for Fluid Dynamics, Rhode Saint Genese, Belgium, 2002.
[56] Johnston, C. O., "Nonequilibrium Shock-Layer Radiative Heating for Earth and Titan Entry," Ph.D. Dissertation, Virginia Polytechnic Institute and State University, Blacksburg, Virginia, Nov. 2006.
[57] Brown, J. L., Garcia, J. A., and Kinney, D. J., "An Asymmetric Capsule Geometry Study for CEV," AIAA Paper 2007-704, 45th AIAA Aerospace Sciences Meeting and Exhibit, Reno, Nevada, Jan 2007.
[58] Park, C., *Nonequilibrium Hypersonic Aerothermodynamics*, John Wiley & Sons, New York, 1990.
[59] Thivet, F., and Laboudigue, B., *Introduction à la modélisation des écoulements hypersoniques*, Ecole de printemps de mécanique des fluides numérique, Aussois, France, 1991.
[60] Gnoffo, P. A., Gupta, R. N., and Shinn, J. L., "Conservation Equations and Physical Models for Hypersonic Air Flows in Thermal and Chemical Nonequilibrium," NASA Technical Paper 2867, Feb. 1989.
[61] Jaffe, R. L., "The Calculation of High-Temperature Equilibrium and Nonequilibrium Specific Heat Data for N_2, O_2 and NO," AIAA Paper 1987-1633, 22nd Thermophysics Conference, Honolulu, Hawaii, June 1987.

[62] Babou, Y., Rivière, Ph., Perrin, M.-Y., and Soufiani, A., "High-Temperature and Nonequilibrium Partition Function and Thermodynamic Data of Diatomic Molecules," *International Journal of Thermophysics*, Vol. 30, 2009, pp. 416–438.

[63] Conrad, J. R., and Schunk, R. W. "Diffusion and Heat Flow Equations with Allowance for Large Temperature Differences Between Interacting Species," *Journal of Geophysical Research*, Vol. 84, No. A3, 1979, pp. 811–822.

[64] Billet, G., Giovangigli, V., and de Gassowski, G., "Impact of Volume Viscosity on a Shock/Hydrogen Bubble Interaction," *Combustion Theory and Modelling*, Vol. 12, 2008, pp. 221–248.

[65] Thivet, F., "Modélisation et calcul d'écoulements hypersoniques en déséquilibre chimique et thermodynamique," Ph.D. thesis, Ecole Centrale de Paris, No. 39, Sept. 1992.

[66] Huba, J. D., "NRL Plasma Formulary," Naval Research Laboratory Report NRL/PU/6790-07-500, Revision 2007.

[67] Lopez, B., "Développement d'un code d'aérothermodynamique pour la simulation des écoulements hypersoniques hors équilibre thermochimique. Applications au écoulements d'arcjets et de rentrée atmosphérique," Ph.D. thesis, Université d'Orléans, Orléans, France, March 2010.

Chapter 3 Elementary Chemical Reactions Modeling

- Find the form of reaction constants
- Solve problems related to local thermodynamic equilibrium
- Identify the phases of heterogeneous reactions
- Understand the relationship between homogeneous and heterogeneous reactions

Two types of modeling problems will be discussed in this chapter: Homogeneous chemical reactions in gas phase, in particular at local thermodynamic equilibrium with a variable elemental composition, and heterogeneous chemical reactions and surface phase changes. In general we ignore the effects of internal degrees of freedom of the molecule, which are particularly important behind the shock; however, these effects largely disappear between the shock wave and the wall. The higher the pressure, the more quickly this occurs. For now we are interested in parietal phenomena whose importance increases with pressure. This does not mean that internal degrees of freedom are not acting in the gas–wall interaction.

The beginning of the first part of this chapter will consist solely of recalls, a problem treated by many books.

3.1 Gaseous Reactions

3.1.1 The Medium in Thermodynamic Equilibrium

The n_r gas phase chemical reactions are described by the formal system

$$\sum_{i=1}^{n_e} \nu'_{ij} A_i \rightleftharpoons \sum_{i=1}^{n_e} \nu''_{ij} A_i \quad j = 1, n_r \tag{3.1}$$

where A_i are the n_e chemical species and the ν_{ij} are stoichiometric coefficients (molecule i involved in the reaction j). We introduce the stoichiometric matrix $\mathbf{\Omega}$ defined by

$$\Omega_{ij} = \nu''_{ij} - \nu'_{ij} \tag{3.2}$$

This matrix is sometimes defined by the inverse relationship.

In a homogeneous medium, the molar concentrations $\mathcal{V}_i$ obey the law of "mass action"

$$\frac{\mathrm{d}\mathcal{V}_i}{\mathrm{d}t} = \frac{\dot{\omega}_i(\mathcal{V}_k, T)}{\mathcal{M}_i} = \sum_{j=1}^{n_r} \Omega_{ij}\tau_j \tag{3.3}$$

with

$$\tau_j = k_{f_j} \prod_{i=1}^{n_e} \mathcal{V}_i^{\nu'_{ij}} - k_{b_j} \prod_{i=1}^{n_e} \mathcal{V}_i^{\nu''_{ij}} \tag{3.4}$$

τ_j is the amount of moles formed in the reaction j per unit volume and time, and $\dot{\omega}_i$ is the mass of the species i formed per unit volume and time.

The kinetics theory of gases [1] allows us to find the form of the reaction constants, proportional to the following:

- The collision frequency between molecules i and j

$$\Phi = \mathcal{N}^2 \mathcal{V}_i \mathcal{V}_j \sigma_{ij} \left(\frac{\mathcal{M}_i + \mathcal{M}_j}{\mathcal{M}_i \mathcal{M}_j} 8\pi \mathcal{R} T \right)^{\frac{1}{2}} \tag{3.5}$$

- The effectiveness of the collision, as measured by the cross-section. This may be
  - With a threshold in energy: the constant γ measuring the effectiveness will be of Arrhenius type

$$\gamma = \alpha f(T) \exp\left(-\frac{E_A}{RT} \right) \tag{3.6}$$

  - The expression of $f(T)$ depends on the potential of collision used [1].
  - For any possible collision speed: this is the case when a long-range potential exists, for example, in the case of ion–molecule collisions. In this case, γ will be constant and the reaction called non-Arrhenius.

This gives a law of the form (with or without exponential term)

$$k_{f/b} = A T^{\frac{1}{2}} f(T) \exp\left(-\frac{E_A}{\mathcal{R} T} \right) \tag{3.7}$$

To reasonably reconstruct the experimental results, one uses a heuristic expression

$$k_{f/b} = AT^B \exp\left(-\frac{E_A}{\mathcal{R}T}\right) \tag{3.8}$$

This expression takes into account various effects such as the vibrational excitation of one of the reactants or the variation of cross-sections with temperature.

The thermodynamics [1] defines the relations between k_f and k_b.

- The quantities k_f and k_b are related by the thermodynamic equilibrium constant pressure constant (Appendix B)

$$K_{p_j} = \frac{k_{f_j}}{k_{b_j}} \left(\frac{\mathcal{R}T}{p_0}\right)^{\Delta\Omega_j} \tag{3.9}$$

  where $\Delta\Omega_j = \sum_i \Omega_{ij}$ is the change of stoichiometry in the reaction j.
- Activation energies are related to the enthalpy change of reaction at the reference pressure p_0

$$E_{A_f} - E_{A_b} = \Delta\mathcal{H}^0 \tag{3.10}$$

- Pre-exponential factors are related to the entropy change of reaction

$$\ell n\left(\frac{A_f}{A_b}\right) = \frac{\Delta S^0}{\mathcal{R}} \tag{3.11}$$

Quantities K_p, $\mathcal{H}^0$, and S^0 are functions of temperature, given in thermodynamic data tables [2]. There are many sources for k_f and k_b [1, 3–8]. Sometimes only one of these values exists, the other being recalculated from the relation in Eq. (3.9). The compatibility between all these values is, in principle, taken into account in the data available. Appendix B.1.2 provides an efficient approximation to ensure that compatibility.

3.1.2 Local Thermodynamic Equilibrium

To illustrate the model, we take the case of Earth atmosphere. The simplest chemical system has 2 elements (N, O), 5 species (O_2, N_2, O, N, NO), and 17 reactions (Table 3.1), counting all possible catalysts. These catalysts are all species present in the environment and involved in the chemical reaction by absorbing the internal energy of the reaction products, usually in excited states. Production rates for the same overall reaction but involving

Table 3.1 Chemical Reactions for Air*

$N_2 + M \rightleftharpoons 2N + M$
$O_2 + M \rightleftharpoons 2O + M$
$N_2 + O \rightleftharpoons NO + N$
$NO + O \rightleftharpoons O_2 + N$
$NO + M \rightleftharpoons N + O + M$

*Species noted M are any species N_2, O_2, N, O, or NO.

different catalysts differ only by a multiplicative constant giving effectiveness. Indeed, interaction leading to chemical reaction (giving the value of E_A) is the same regardless of the catalyst.

At higher temperatures, ionized species such as NO^+ and e^- (the electron) appear, followed by N_2^+, O_2^+, N^+, and O^+. An example of the existence domain of these species is given in Fig. 3.1.

3.1.3 Problems Related to Local Thermodynamic Equilibrium

In the case where the characteristic times of each gas phase reaction are lower than other characteristic times of the medium (advection, diffusion), the composition of the medium can be described by thermodynamic characteristics of the species and the elemental composition. Therefore, the problem in this case is to build an equation of state (whose graphical representation for a given elemental composition is called a *Mollier diagram*). Given commonly used numerical methods, the thermodynamic quantities used are the density and internal energy (or enthalpy). Because we are in a fluid medium that mixes the gas from "infinity upstream" and species from the wall, created by ablation or pyrolysis, elemental composition varies from one point to another. The equation of state building is therefore in the form $g = f(\rho,\ e,\ \tilde{c}_i)$, where g is the desired quantity (temperature, pressure, etc.) and $\tilde{c}_i$ the mass fraction of element i (O, N, C, H, etc.). The size of the problem is very important: If N_A is the number of elements, then the number of independent variables is $N_A + 1$ ($n_A - 1$ independent mass fractions, plus any pair of thermodynamic variables). Given the number of variables, some methods used in aerodynamics where basic fractions are uniform, such as approximation [3] or numerical tables, are prohibited by cost (time or memory size). It will be necessary to either calculate the equation of state "online" and find cost-effective methods of numerical computation or reduce the size of the problem.

In the case of direct calculation, we know that a very effective method of resolution is the minimization of energy (free enthalpy) at constant temperature and pressure

$$\mathcal{V}\mathcal{G} = \sum_i \mathcal{V}_i(\mathcal{H}_i - T\mathcal{S}_i) \tag{3.12}$$

that is,

$$\frac{\mathcal{G}}{RT} = \sum_i x_i \left[\frac{\mathcal{H}_i^0}{\mathcal{R}T} - \frac{\mathcal{S}_i^0}{\mathcal{R}} + \ell n \left(\frac{px_i}{p_0} \right) \right] \tag{3.13}$$

For each species, the molar enthalpy $\mathcal{H}_i^0$ and molar entropy $\mathcal{S}_i^0$ related to the reference pressure p_0 are given by the databases (see Appendix B).

This minimization should be performed under the constraint of respecting the elemental composition of the system

$$\sum_i n_{ki}\mathcal{V}_i = C^{ste} \quad k = 1, n_A \tag{3.14}$$

where n_{ki} is the number of elements k in species i.

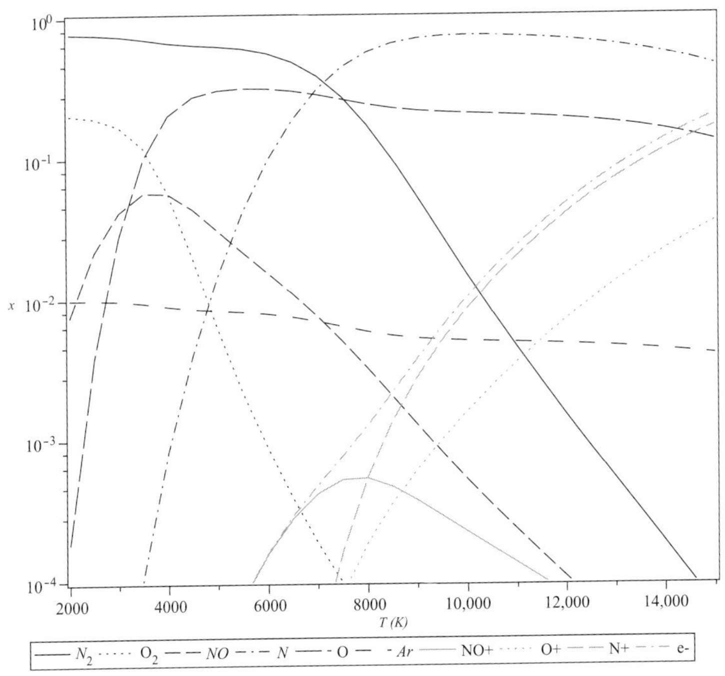

Fig. 3.1 Volumic composition of air at normal pressure (calculated from the kinetic data of Dunn & Kang (9)).

The quantities contained in Eq. (3.12)—temperature, pressure, or volume fractions—are not known a priori in a global calculation, which implies the use of nested loops calculation. The following methods did not have this drawback, which is particularly sensitive in computing time.

3.1.4 Direct Resolution of the Equilibrium

3.1.4.1 Minimum Reaction System

Before tackling the problem, we will establish the minimum system of reactions conducting to thermodynamic equilibrium. Indeed, the result is independent of the number of reactions, provided that the proposed system is complete.

This system can be derived from the matrix **n** of element n_{kj}. The system size is $n_e - n_A$. This can be seen from the matrix $\mathbf{E} = \mathbf{nn}^t$. Line $j \in [n_e - n_A + 1, n_e]$ corresponds to the species j. It can be obtained from species $[1, j-1]$ by linear combination. In general, the reaction system corresponds to the null eigenvalues of **E**. Take, for example, the reactions of Table 3.2.

The value of the matrix **E** is

$$\mathbf{E} = \begin{vmatrix} 4 & 0 & 2 & 0 & 2 \\ 0 & 4 & 0 & 2 & 2 \\ 2 & 0 & 1 & 0 & 1 \\ 0 & 2 & 0 & 1 & 1 \\ 2 & 2 & 1 & 1 & 2 \end{vmatrix} \tag{3.15}$$

One can see that

- Line 3 is half of line 1.
- Line 4 is half of line 2.
- Line 5 is half the sum of lines 1 and 2.

The minimum number of pseudo-chemical reactions is then $n_e - n_A$ [the rank of the matrix is $rank$ $(\mathbf{E}) = 3$]. We deduce the chemical system

Table 3.2 Transfer Matrix Species-Elements (n_{kj})

n_{kj}	N	O
N_2	2	0
O_2	0	2
N	1	0
O	0	1
NO	1	1

Table 3.3 One Possible Minimal Reactions System for Air (Five Species)

$N_2 \rightleftharpoons 2N$
$O_2 \rightleftharpoons 2O$
$N_2 + O_2 \rightleftharpoons 2NO$

minimum of Table 3.3. Note that if, in **n** we had put N and O first in the list, the third chemical equation would have been written

$$\mathrm{N} + \mathrm{O} \Leftrightarrow \mathrm{NO}$$

This illustrates the fact that the minimal system is not unique.

In general, we can write the general system as follows. The eigenvectors of E corresponding to null eigenvalues are

$$\begin{vmatrix} 0 \\ 1 \\ 0 \\ -2 \\ 0 \end{vmatrix} \quad \begin{vmatrix} 1 \\ 0 \\ -2 \\ 0 \\ 0 \end{vmatrix} \quad \begin{vmatrix} 0 \\ 0 \\ -1 \\ -1 \\ 1 \end{vmatrix} \tag{3.16}$$

These eigenvectors give the stoichiometry of the minimum reaction system written in conventional form $\sum_j \Omega_{ij} A_j = 0$. For example, the first eigenvector corresponds to $1 \cdot \mathrm{O}_2 - 2 \cdot \mathrm{O} = 0$ (to be understood as $\mathrm{O}_2 \rightleftharpoons 2\mathrm{O}$). We notice that while looking at the third eigenvector the linear solver used for analysis (Maple©) as privileged one solution, the second mentioned previously.

3.1.4.2 Resolution of Thermodynamic Equilibrium

Solving conservation equations of species [Eq. (3.3)] is relatively expensive because of very different orders of magnitude of reaction constants. But we are interested only in the stationary solution: the reaction coefficients k_f and k_b have no importance in and of themselves; only their relationship plays a role. We will take this observation into account and renormalize the time characteristics of the system (proportional to $1/k_f$ and $1/k_b$, and equal to those values in the case of reactions of first order) as follows:

- If $k_f > k_b$, that is, $K_p/(\mathcal{R}T)^{\Delta\Omega} > 1$, then $k_f = 1$ and $k_b = (\mathcal{R}T)^{\Delta\Omega}/K_p$.
- If $k_f < k_b$, that is, $K_p/(\mathcal{R}T)^{\Delta\Omega} < 1$, then $k_b = 1$ and $k_f = K_p/(\mathcal{R}T)^{\Delta\Omega}$.

This operation has the effect of bringing closer the characteristic times corresponding to each reaction and, consequently, accelerating the calculation of the steady state. It is a method that minimizes the domain of values:

- If $K_p/(\mathcal{R}T)^{\Delta\Omega} > 1$, the interval is $[(\mathcal{R}T)^{\Delta\Omega}/K_p,\ 1]$.
- If $K_p/(\mathcal{R}T)^{\Delta\Omega} < 1$, the interval is $[K_p/(\mathcal{R}T)^{\Delta\Omega},\ 1]$.

To the equations of conservation of the species contained in the pseudo-reactions defined here, we will add the constraint to satisfy a given value for $\rho\epsilon$. This constraint will be written in a relaxed form

$$\frac{\partial(\rho e)}{\partial t} = \frac{1}{\tau}(\rho_0 e_0 - \rho e) \tag{3.17}$$

In this equation $\rho_0 e_0$ is the constraint. The relaxation time τ is arbitrary. It will be chosen to be the same order of magnitude as that of the pseudo-chemical reactions. By expanding, we get

$$\frac{\partial}{\partial t}\left(\sum_i \rho c_i e_i\right) = \sum_i e_i \frac{\partial(\rho c_i)}{\partial t} + \sum_i \rho c_i \left.\frac{\partial e_i}{\partial T}\right|_\rho \frac{\partial T}{\partial t} \tag{3.18}$$

Given the thermodynamic relation $C_{V_i} = (\partial e_i/\partial T)\big|_\rho$, we obtain the temperature equation, to solve with the system

$$\frac{\partial T}{\partial t}\left(\sum_i \rho c_i C_{V_i}\right) = \frac{1}{\tau}(\rho_0 \epsilon_0 - \rho\epsilon) - \sum_i \epsilon_i \frac{\partial(\rho c_i)}{\partial t} \tag{3.19}$$

Other quantities are calculated by the same method. This is done in Appendix C.

The computational cost of such a procedure is much more important than the use of an equation of state. The number of pseudo-reactions is lower than the actual number of possible reactions. However, the cost is much lower than that of a nonequilibrium chemical calculation (assuming solved problems having a very low characteristic time, in this case by using an implicit solver).

3.1.5 Transport Properties

For the transport properties in the problems outlined in the previous section, it is not practicable to give an approximation of these properties according to ρ, e, and $\tilde{c}_k$. One has to know all the values characterizing the environment (particularly c_i) and apply the approximations given in Chapter 2.

For aerodynamics (constant $\tilde{c}_k$ values of the medium), many sources provide the transport properties of air [10–14].

3.1.5.1 Total Thermal Conductivity

Using an equation of state does not permit direct knowledge of volume fractions at all points. This constitutes a challenge to calculate the terms related to diffusion in the flux of energy [Eq. (2.24)]. However, the resolution of the Maxwell–Stefan equation allows one to write flows as

$J_i = f(\nabla x_i, \nabla p, \nabla T)$ [Eq. (2.43)]. We can also write

$$\nabla x_i = \frac{\partial x_i}{\partial T}\nabla T + \frac{\partial x_i}{\partial p}\nabla p + \sum_k \frac{\partial x_i}{\partial \tilde{c}_k}\nabla \tilde{c}_k \tag{3.20}$$

In our problems, we can usually neglect the pressure gradient, but we cannot a priori eliminate the elementary fraction gradient. Therefore, the result can be understood only within the framework of a medium at fixed elemental composition.

Equation (2.25) will be written

$$\dot{q} = -\lambda \nabla T - \lambda^r \nabla T \tag{3.21}$$

with

$$\lambda^r \nabla T = -\sum_{i=1}^{n_e} \mathbf{J}_i^c \left(h_i + \frac{\mathcal{R} T k_i^T}{\overline{\mathcal{M}} c_i} \right) \tag{3.22}$$

where $\mathbf{J}_i^c$ is the diffusive flux relative to the concentration gradient.

The term λ^r, improperly called "reactive conductivity" because it is a term of energy transfer by mass diffusion, is calculable and can be included in the true conductivity. The single term corresponding to the diffusion gradient of concentration can represent up to half of the heat transfer by conduction [15]. The term corresponding to the true thermal diffusion is generally negligible. Indeed, the gradient of mass fraction is more important in an environment at thermodynamic equilibrium, thermal gradient, because the reactions are faster in an environment far from chemical equilibrium. Subsequently we will neglect the true thermal diffusion equation, and Eq. (3.22) will simplify to

$$\lambda^r \nabla T = -\sum_{i=1}^{n_e} \mathbf{J}_i^c h_i \tag{3.23}$$

The diffusion flux is calculated from its exact expression [Eq. (2.43)], from which comes

$$\lambda^r = -\frac{\rho}{\overline{\mathcal{M}}^2} \sum_i h_i \sum_{j \neq i} \mathcal{M}_i \mathcal{M}_j D_{ij} \frac{\partial x_j}{\partial T} \tag{3.24}$$

We may also use any approximation of diffusive flux given in Chapter 2 (Sec. 2.3).

We must now express the relationship between gradients of species and temperature. For this, we start from the definition of the equilibrium constant in Eq. (B.8)

$$\Delta\Omega_j \, \ell n \, p + \sum_i \Omega_{ij} \, \ell n \, x_i = \ell n \, K_{p_j} \quad j = 1, n_r \tag{3.25}$$

Deriving with respect to temperature we get

$$\sum_i \Omega_{ij} \frac{\partial(\ell n\, x_i)}{\partial T} = \frac{\partial(\ell n\, K_{p_j})}{\partial T} \quad j = 1, n_r \tag{3.26}$$

The second part of the equation is obtained from a database, for example using the expression from Eq. (B.10).

The system is not closed, and the stoichiometric matrix is not invertible. The n_A missing equations will be obtained from the constraint of conservation of elements [Eq. (4.15)]. This will be written as follows:

$$\sum_i \left(n_{ki} - \tilde{c}_k \frac{\mathcal{M}_i}{\mathcal{M}_k} \right) x_i = 0 \quad k = 1, n_A \tag{3.27}$$

Differentiating with respect to T, it becomes (remember that here $\tilde{c}_k$ is assumed constant)

$$\sum_i \left(n_{ki} - \tilde{c}_k \frac{\mathcal{M}_i}{\mathcal{M}_k} \right) x_i \frac{\partial \ell n\,(x_i)}{\partial T} = 0 \quad k = 1, n_A \tag{3.28}$$

The system is still not closed because there are only $n_A - 1$ independent relationships. We add the equation $\sum_i x_i = 1$, which gives by differentiation

$$\sum_i x_i \frac{\partial\, \ell n\, x_i}{\partial T} = 0 \tag{3.29}$$

The systems in Eqs. (3.26) and (3.28) constitute a linear system in $(\partial(\ell n\, x_i)/\partial T)$

$$\begin{vmatrix} \Omega_{1,1} & \dots & \Omega_{1,n_e} \\ \dots & \dots & \dots \\ \Omega_{n_r,1} & \dots & \Omega_{n_r,n_e} \\ K_{1,1} & \dots & K_{1,n_e} \\ \dots & \dots & \dots \\ K_{n_A-1,1} & \dots & K_{n_A-1,n_e} \\ x_1 & \dots & x_{n_e} \end{vmatrix} \begin{vmatrix} \dfrac{\partial \ell n\,(x_1)}{\partial T} \\ \dots \\ \dots \\ \dots \\ \dots \\ \dfrac{\partial \ell n\,(x_{n_e})}{\partial T} \end{vmatrix} = \begin{vmatrix} \dfrac{\partial(\ell n\, K_{p_1})}{\partial T} \\ \dots \\ \dfrac{\partial(\ell n\, K_{p_{n_r}})}{\partial T} \\ 0 \\ \dots \\ 0 \end{vmatrix} \tag{3.30}$$

with

$$K_{ki} = \left(n_{ki} - \tilde{c}_k \frac{\mathcal{M}_i}{\mathcal{M}_k} \right) x_i \tag{3.31}$$

In the particular case of a single reaction (e.g., the decomposition of dioxygen), the resolution gives

$$\lambda^r = \rho \mathcal{D}_{O,O_2} \frac{x_{O_2}(1 - x_{O_2})}{1 + x_{O_2}} (h_O - h_{O_2}) \frac{\partial\, \ell n\, K_p}{\partial T} \tag{3.32}$$

We can write this equation in a different form by using the Van't Hoff law

$$\frac{\partial (\ell n K_p)}{\partial T} = \frac{\Delta \mathcal{H}}{\mathcal{R} T^2} \tag{3.33}$$

It becomes

$$\lambda^r = \frac{2 \mathcal{M}_O \rho \mathcal{D}_{O,O_2}}{\mathcal{R} T^2} \frac{x_{O_2}(1 - x_{O_2})}{1 + x_{O_2}} (h_O - h_{O_2})^2 \tag{3.34}$$

λ^r is positive and vanishes in the absence of reaction ($x_{O_2} = 0$ or $x_{O_2} = 1$). We find this type of curve in the general case (Fig. 3.2). (Many experimental values exist at low temperature [$T < 1000$ K] [16], few in the domain that interests us [17].) The various peaks correspond to reactions:

- The first peak corresponds to the dissociation of dioxygen and the formation of NO.
- The second corresponds to the dissociation of N_2.
- The third corresponds to ionization.

In our case, the reactive conductivity is of the same order of magnitude as the true conductivity.

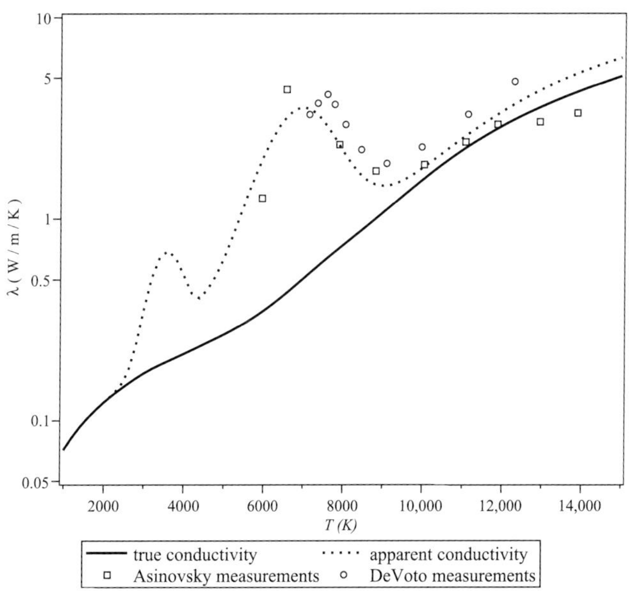

Fig. 3.2 Reactive conductivity of air at thermodynamic equilibrium, $p = p_0$.

3.1.5.2 Remarks on the Flux Species

Note that the total fluxes of species are not independent. First, in the $n_I = n_e - n_A$ independent reactions defined previously (e.g., in Table 3.3), we can find n_I species appearing only one time in both reactions. These will be placed high on the list to simplify the notation and will be referred to as "independent" species. In the example chosen, it will be N, O, and NO.

The total molar flow $J_i = (\rho c_i \mathbf{V} + \mathbf{J}_i^c/\mathcal{M}_i)$ is such that the thermodynamic equilibrium is verified at any point. Molar fluxes must comply with the various reactions and associated stoichiometry. Otherwise, we would be dealing with a reaction–diffusion problem. Moreover, the relative composition, determined by independent species, should be verified. There are thus two types of constraints:

- The constraint of respect of the given composition is written using the independent species fluxes

$$\mathcal{J}_i = \mathcal{J}_0 \quad \forall i = 1, n_I \tag{3.35}$$

$\mathcal{J}_0$ is an arbitrary value. In the example, we have $\mathcal{J}_{N_2} = \mathcal{J}_{O_2} = \mathcal{J}_{NO} = \mathcal{J}_0$.

- The thermodynamics equilibrium constraint is written, through "dependent" species fluxes,

$$\mathcal{J}_i + \sum_{k=1}^{n_I} m_{ki} \mathcal{J}_k = 0 \quad i = n_I + 1,\ n_e \tag{3.36}$$

with $m_{ki} = (\Omega_{ki}/\Omega_{kk})$.

In our example, the fluxes of O and N will thus be imposed

$$\begin{cases} \mathcal{J}_O = 2\mathcal{J}_{O2} + \mathcal{J}_{NO} \\ \mathcal{J}_N = 2\mathcal{J}_{N_2} + \mathcal{J}_{NO} \end{cases} \tag{3.37}$$

This method is used by Brokaw [18, 19] in the special case where convection is absent. The diffusion flux by the concentration gradient $\mathcal{J}_i^c$ will be written by decomposing the parts corresponding to dependent and independent species and introducing the molar enthalpy $\mathcal{H}_i = h_i \mathcal{M}_i$

$$\sum_{i=1}^{n_e} \mathbf{J}_i^c h_i = \sum_{i=1}^{n_I} \mathcal{J}_i^c \mathcal{H}_i + \sum_{i=n_I+1}^{n_e} \mathcal{J}_i^c \mathcal{H}_i \tag{3.38}$$

By inserting Eq. (3.36) in the second term on the right-hand side, and rearranging the expression, we get

$$\sum_{i=1}^{n_e} \mathbf{J}_i^c h_i = \sum_{i=1}^{n_e} \frac{\mathcal{J}_i \Delta \mathcal{H}_i}{\Omega_{ii}} \tag{3.39}$$

where $\Delta\mathcal{H}_i$ is the molar enthalpy for reaction i

$$\Delta\mathcal{H}_i = \sum_{k=1}^{n_e} \Omega_{ik}\mathcal{H}_k \quad i = 1, n_I \tag{3.40}$$

We will now link the enthalpies of reaction to the gradients of volume fractions. To do this, start from the relationship definition of the equilibrium constant

$$\ell n\, K_{p_j} = \sum_{k=n_I+1}^{n_e} \Omega_{jk}\, \ell n\, p_k - \ell n\, p_j \quad j = 1,\ n_I \tag{3.41}$$

for example,

$$\ell n\, K_{p_j} = \sum_{k=n_I+1}^{n_e} \Omega_{jk}\, \ell n\, x_k - \ell n\, x_j + \left(\sum_{k=n_I+1}^{n_e} \Omega_{jk} - 1\right) \ell n\, p \quad j = 1, n_I \tag{3.42}$$

By taking the spatial gradient of this expression and neglecting ∇p and $\nabla \tilde{c}_k$, we get

$$\frac{\partial\, \ell n\, K_{p_j}}{\partial T}\nabla T = \sum_{k=n_I+1}^{n_e} \frac{\Omega_{jk}}{x_k}\nabla x_k - \frac{\nabla x_j}{x_j} \quad j = 1, n_I \tag{3.43}$$

The first term is connected to the reaction enthalpy by the Van't Hoff law

$$\frac{\Delta\mathcal{H}_j}{\mathcal{R}T^2}\nabla T = \sum_{k=n_I+1}^{n_e} \frac{\Omega_{jk}}{x_k}\nabla x_k - \frac{\nabla x_j}{x_j} \quad j = 1, n_I \tag{3.44}$$

With the law of Stefan–Maxwell [Eq. (2.38)], we can express the gradients of mass fractions

$$\begin{cases} \nabla x_k = \sum_{i=1}^{n_I} \Delta_{ik}\left(x_k\mathcal{J}_i^c - x_i\mathcal{J}_k^c\right) + \sum_{j=n_I+1}^{n_e} \Delta_{ik}\left(x_k\mathcal{J}_i^c - x_i\mathcal{J}_k^c\right) & k = 1, n_I \\ \nabla x_j = \sum_{i=1}^{n_I} \Delta_{ik}\left(x_k\mathcal{J}_i^c - x_i\mathcal{J}_k^c\right) + \sum_{j=n_I+1}^{n_e} \Delta_{il}\left(x_1\mathcal{J}_i^c - x_i\mathcal{J}_l^c\right) & l = n_I + 1, n_e \end{cases} \tag{3.45}$$

where $\Delta_{ik} = (\mathcal{R}T/\mathcal{D}_{ik}p)$ [Eq. (2.108)].

Substituting these expressions in Eq. (3.44) and taking into account Eq. (3.36), we get

$$\frac{\Delta\mathcal{H}_j}{\mathcal{R}T^2}\nabla T = \sum_{i=1}^{n_e} R_{ik}\mathcal{J}_i^c \tag{3.46}$$

with

$$R_{ij} = R_{ji} = \sum_{k=1}^{n_e-1} \sum_{l=k+1}^{n_e} \Delta_{kl} x_k x_l \left(\frac{\Omega_{ik}}{x_k} - \frac{\Omega_{il}}{x_l} \right) \times \left(\frac{\Omega_{jk}}{x_k} - \frac{\Omega_{jl}}{x_l} \right) \quad i,\ j = 1,\ n_I \tag{3.47}$$

Hence the expression of the reactive conductivity, written as a formal solution of the linear system in Eq. (3.47)

$$\lambda^r = \frac{1}{\mathcal{R}T^2} \frac{\begin{vmatrix} 0 & \Delta\mathcal{H}_1 & \Delta\mathcal{H}_2 & \dots & \Delta\mathcal{H}_{n_I} \\ \Delta\mathcal{H}_1 & R_{11} & R_{12} & \dots & R_{1n_r} \\ \Delta\mathcal{H}_2 & R_{21} & R_{22} & \dots & R_{1n_r} \\ \dots & \dots & \dots & \dots & \dots \\ \Delta\mathcal{H}_{n_I} & R_{n_r1} & R_{n_I} & \dots & R_{n_rn_r} \end{vmatrix}}{\begin{vmatrix} R_{11} & R_{12} & \dots & R_{1n_I} \\ R_{21} & R_{22} & \dots & R_{1n_I} \\ \dots & \dots & \dots & \dots \\ R_{n_r1} & R_{n_r2} & \dots & R_{n_In_I} \end{vmatrix}} \tag{3.48}$$

The results of this method are identical to those of the first method (Fig. 3.2). In fact, it is possible to show that in the case of a zero total flux, which is the assumption made by Brokaw, solving Eqs. (3.35) and (3.36) leads to $\tilde{\mathbf{J}}_k = 0$. This respects the elemental concentration and shows that Eq. (3.48) is more general than supposed by its demonstration.

Note that this method cannot be extended to environments in which there is a gradient of elementary fractions.

3.1.6 Medium Out of Equilibrium

In the case of a medium in thermodynamic nonequilibrium described by temperatures of translation–rotation T, electronic T_e, and vibrational T_i^{vib} for the molecule or molecular ion i (see Sec. 2.6), we must redefine reaction constants and the effect on the vibrational state of molecules formed. Indeed, chemical reactions lead preferentially to excited states. In most cases we will study the dissociation of diatomic molecules like O_2, N_2, or H_2, sometimes triatomic molecules like CO_2, but this approach may also apply to the exchange reactions of the type $AB + C \rightarrow A + BC$.

The modeling of these phenomena is based on the work of Treanor and Marrone [20, 21] and the vibration–dissociation coupling (VDC) and vibration–dissociation–vibration–coupling (VDVC) models.

3.1.6.1 Effects of Chemical Reactions on Vibrational Nonequilibrium

Chemical reactions occur preferentially on the highest vibrational levels, at least those whose energy exceeds the dissociation energy. The average mass energy lost by dissociating molecules is

$$\Delta e^{vib} = \alpha \frac{\mathcal{R} T_D}{\mathcal{M}} \tag{3.49}$$

Temperatures of dissociation T_D are given by thermodynamic tables (e.g., $T_D = 59{,}000$ K for O_2, 113,000 K for N_2, and 52,800 K for H_2). The constant α varies with species and temperature T^{vib} between 0.1 and 0.8 [22]. The harmonic oscillator model gives a value of about 0.5, often used for calculations in fluid mechanics. Note that the quantity Δe^{vib} is the same order of magnitude as e^{vib} itself or superior to it (see values in Fig. 2.13).

The coupling effect is also felt through the modification of the composition. An increase in light species causes a return to equilibrium by decreasing τ^{VT} (Eq. 2.180): Atom–molecule interactions are more efficient than molecule–molecule interactions.

3.1.6.2 Effect of Vibrational Nonequilibrium on Reactions

With probability a priori (of Arrhenius type) dissociation of the molecule [20], we show that the dissociation constant is expressed as follows:

$$k_f'(T, T^{vib}) = k_f(T) \frac{Q_{vib}(T) Q_{vib}(T_F)}{Q_{vib}(T^{vib}) Q_{vib}(-T_U)} \tag{3.50}$$

with

$$T_F^{-1} = T^{vib^{-1}} - T^{-1} - T_U^{-1} \tag{3.51}$$

k_f is the reaction constant at thermodynamics equilibrium, Q the vibrational partition function, and $-T_U$ represents the vibrational temperature at which the products are formed from the reaction. The negative sign shows simply that a majority of molecules formed are in a high energy level, higher than the maximum value possible in the case of equilibrium described by a Boltzmann equation. One can use an experimental approximate value $T_U = (T_A/3)$, where $T_A = (E_A/k)$ is the activation temperature of the reaction.

Park [7] has proposed a simpler expression that has an experimental origin

$$k_f'(T, T^{vib}) = k_f T^{\alpha} T^{vib^{(1-\alpha)}} \tag{3.52}$$

The original value $\alpha = 0.5$ was amended later to be $\alpha = 0.7$, a value closer to microscopic calculations [23].

3.1.6.3 Atom Ionization

In the case of atoms, the nonequilibrium appears again on the higher levels (Fig. 3.3). In this case, populations of these levels are correctly described by the Saha law [Eq. (10.26)] using the temperature T_e. The case

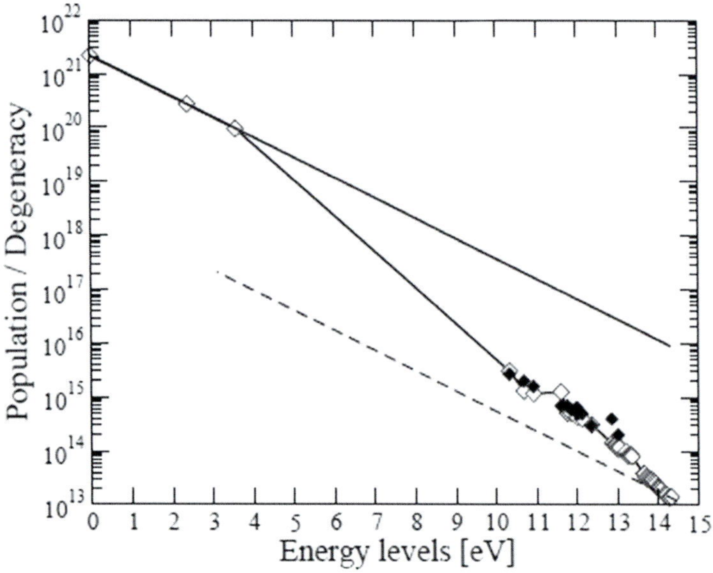

Fig. 3.3 Populations of electronic levels of atomic nitrogen. Low levels follow Boltzmann statistics (solid line) and weakly bounded levels follow Saha's law (dashed curve).

in Fig. 3.3, obtained from a collisional-radiative model, corresponds to a terrestrial reentry (Fire II) [24].

3.2 Heterogeneous Reactions

Heterogeneous reactions occur in different phases, in practice between a gas and a solid or a liquid. Indeed, reactions between two condensed phases are always occurring through the gas from one or the other of the reactants.

3.2.1 Basic Mechanisms

The chemical interaction between the gas molecule and the wall is made on a specific location of the latter called the site, denoted *Site*. The material is characterized by the number of sites per unit area N_S and the proportion of them unoccupied θ or occupied by the species A (often reduced to a single atom) θ_A. There are various types of interactions, of which we cite the major types here:

- Adsorption (fixation of A on a site)

$$Site + A \rightarrow Site - A$$

- Desorption (the inverse phenomenon)

$$Site - A \rightarrow Site + A$$

- Eley–Rideal mechanism

$$Site - A + B \leftrightharpoons Site + AB$$

- Langmuir–Hinshelwood mechanism involving two neighboring sites

$$Site - A + Site - B \leftrightharpoons 2\ Site + AB$$

Also note that the molecules or atoms attached to the surface can move on it by surface diffusion. This mechanism will not intervene directly on the quantities selected to describe the phenomena (N_s and θ), but it is obvious that the phenomenon of diffusion, by spreading occupied sites on the surface, will alter the effectiveness of the Langmuir–Hinshelwood mechanism. We assume also that a molecule (or atom) adsorbed on a site does not change its environment, especially the properties of neighboring sites. This last item is not really true in the case of ablation or deposition on carbon, as demonstrated by molecular dynamics.

3.2.2 Reaction Kinetics

The reaction kinetics are derived from microscopic analysis showing the rate of efficient phenomena per unit area and time. We analyze each of these mechanisms in the following sections.

3.2.2.1 Adsorption

In the case of the collision of a gas molecule with the wall, we have

$$\frac{\mathrm{d}}{\mathrm{d}t}(N_s\theta) = -\frac{\mathrm{d}}{\mathrm{d}t}(N_s\theta_A) = -\Phi_A\,\gamma_A\,\theta \tag{3.53}$$

This equation simply says that the number of molecules "stuck" per unit area and time is proportional to

1. Φ_A, the number of collisions per unit area and time of the molecule A with the wall, so, assuming a Maxwellian velocity distribution

$$\Phi_A = \frac{Np_A}{(2\pi\mathcal{M}_A\mathcal{R}T)^{\frac{1}{2}}} \tag{3.54}$$

Maxwellian distribution is a good approximation of the exact distribution (Chapman–Enskog distribution [25, 26]) when the flow is continuous. We neglect the effects of the Knudsen layer.

2. The probability θ to have a free site. In fact, the sticking probability is superior to θ. This is likely related to the potential attractiveness of the sites occupied. This is particularly true at low temperature. This effect will be ignored here.
3. The effectiveness of the collision γ_A (or sticking probability), constant or given by a term of Arrhenius type representing the effect of the potential barrier between the average gas molecule and the surface of the material [7]

$$\gamma_A = \alpha f(T) \exp\left(-\frac{E_A}{\mathcal{R}T}\right) \tag{3.55}$$

In this case the reaction coefficient is defined by

$$k_f = \frac{\Phi_A \gamma_A}{p_A} \tag{3.56}$$

and Eq. (3.53) becomes

$$\frac{\mathrm{d}}{\mathrm{d}t}(N_s \theta) = -k_f p_A \theta \tag{3.57}$$

In this writing, the reaction coefficient k_f becomes data, for example, of experimental origin.

3.2.2.2 Desorption

In the case of desorption (Eyring's law), given the probability that the thermal agitation brings sufficient energy to break the bond [27]

$$\frac{\mathrm{d}}{\mathrm{d}t}(N_s \theta) = -\frac{\mathrm{d}}{\mathrm{d}t}(N_s \theta_A) = N_s \frac{\mathcal{R}T}{h\mathcal{N}} \gamma_A \theta_A = k_f \theta_A \tag{3.58}$$

where the term $(\frac{\mathcal{R}T}{h\mathcal{N}})\gamma_A$ represents the frequency of thermal desorption of a site and $N_s \theta_A$ the number of sites occupied by the element that interests us. The reaction coefficient is

$$k_f = N_s \frac{\mathcal{R}T}{h\mathcal{N}} \gamma_A \tag{3.59}$$

γ_A being described by an Arrhenius law.

3.2.2.3 Eley-Rideal Mechanism

For this mechanism involving an atom or molecule in the gas phase, we have

$$\begin{aligned} \frac{1}{2}\frac{\mathrm{d}}{\mathrm{d}t}(N_s \theta) &= -\frac{\mathrm{d}}{\mathrm{d}t}(N_s \theta_A) = -\frac{\mathrm{d}}{\mathrm{d}t}(N_s \theta_B) = N_s \Phi_{AB}^S \theta_A \theta_B - \theta^2 \gamma_b \Phi_{AB} \\ &= k_f \theta_A \theta_B - k_b \theta^2 p_{AB} \end{aligned} \tag{3.60}$$

with

$$\begin{cases} k_f = N_s \Phi^S_{AB} \\ k_b = \dfrac{\gamma_b \Phi_{AB}}{p_{AB}} \end{cases} \tag{3.61}$$

where Φ^S_{AB} is the formation frequency of a molecule from atoms (or molecules) attached to two neighboring sites.

3.2.2.4 Langmuir-Hinshelwood Mechanism

The molecule B, interacting with an occupied site, forms a molecule that is carried away (and the reverse phenomenon, for example, the dissociative adsorption of O_2)

$$\frac{\mathrm{d}}{\mathrm{d}t}(N_s \theta) = -\frac{\mathrm{d}}{\mathrm{d}t}(N_s \theta_A) = \theta_A \gamma_f \Phi_B - \theta \gamma_b \Phi_{AB} = k_f \theta_A p_B - k_b \theta p_{AB} \tag{3.62}$$

γ_f is the reaction probability for direct interaction. This type of interaction must overcome a potential barrier: this term is of Arrhenius type. For a backward reaction the probability γ_b also includes a potential barrier.

The reaction constants are therefore of the form

$$\begin{cases} k_f = \dfrac{\gamma_b \Phi_B}{p_B} \\ k_b = \dfrac{\gamma_b \Phi_{AB}}{p_{AB}} \end{cases} \tag{3.63}$$

3.2.2.5 Other Mechanisms

Note the possibility for a molecule to bind to two neighboring sites [24]. This type of mechanism is advanced when an intermediate of the reaction can explain experimental observations [28]. It may perhaps explain schemes referring to sites of a different nature [29].

In addition, surface diffusion, which tends to distribute evenly occupied sites, can alter rate reactions through the Langmuir–Hinshelwood mechanism by decreasing the probability that two neighboring sites are occupied.

3.2.3 Sublimation

The phase change can be described by a Knudsen–Langmuir mechanism [30]

$$\dot{m}_i = \alpha_i \left(\frac{\mathcal{M}_i}{2\pi \mathcal{R} T} \right)^{\frac{1}{2}} (p_{eq_i} - p_i) \tag{3.64}$$

The kinetic approach of sublimation is based on the idea that one can replace the solid with a gas at the equilibrium vapor pressure p_{eq_i} [30].

Indeed, $(\mathcal{M}_i/2\pi\mathcal{R}T)^{1/2}p_i$ is the incident mass flux. Having $\dot{m}_i = 0$ at thermodynamic equilibrium implies that the flux emitted by the wall is equal to $(\mathcal{M}_i/2\pi\mathcal{R}T)^{1/2}p_{eq_i}$. The sticking coefficient α_i can be estimated by a semi-quantum approach [7] as $\alpha_i \simeq (\theta_R/T)^{1/2}$, where θ_R is the rotational constant of the molecule; α_i is significantly less than unity.

Another way to look at the problem is to consider that these mechanisms are similar to those involving active sites. These reactions are assumed not to change the number of free sites $N_S\theta$ (no change in geometry at the nanoscale).

One assumes that the phase change requires adjacent sites to create molecules C_n (Langmuir–Hinshelwood mechanism). The mass flows will be given by

$$\dot{m}_i = \frac{\mathcal{M}_i\theta^i}{\mathcal{N}}(k_{f_i} - k_{b_i}p_i) \tag{3.65}$$

with

$$\begin{cases} k_{f_i} = k_{b_i}p_{eq_i} \\ k_{b_i} = \dfrac{\mathcal{N}\alpha_i}{(2\pi\mathcal{M}_i\mathcal{R}T)^{\frac{1}{2}}} \end{cases} \tag{3.66}$$

These last two relations verify, of course, $p_{eq_i} = K_{p_i} = k_{fC_i}/k_{bC_i}$

Note that these two approaches are equivalent only when $\theta = 1$. In the case where $\theta \neq 1$, it is conceivable that the presence of sites occupied by an atom disturbs sublimation, which cannot occur on this site, or at least we do not know that could describe this be a phase change in such a configuration. Moreover, the sublimation is difficult to describe in such physical behavior because it destroys the graphitic cycle. We will see in applications that this problem does not arise (Sec. 5.7).

This new script seems artificial (the number of sites here is an artifact). It has the advantage, however, of possibly leading to a general formalism that we use later.

3.2.4 General Equations

A very general way to write this type of heterogeneous reaction is given in Dubroca et al. [31]. We consider the site as an element and the free or occupied site as a species. The reaction is written in all cases (homogeneous or heterogeneous)

For $k = 1, n_r$

$$\sum_{i=1}^{n_e} \nu'_{ik}A_i + \sum_{j=1}^{n_s} \mu'_{jk}\mathrm{Site}_j \rightleftharpoons \sum_{i=1}^{n_e} \nu''_{ik}A_i + \sum_{j=1}^{n_s} \mu''_{jk}\mathrm{Site}_j \tag{3.67}$$

Conservation of sites

$$\frac{\mathrm{d}}{\mathrm{d}t}(N_S\theta_j) = \frac{\mathcal{N}\dot{\omega}_j}{\mathcal{M}_j} \tag{3.68}$$

and

$$\begin{cases} \sum_j \theta_j = 1 \\ \sum_j \dot{\omega}_j = 0 \end{cases} \tag{3.69}$$

Production rates are given by

$$\dot{\omega}_j = -\mathcal{M}_j \sum_{k=1}^{n_r} \dot{\omega}_{jk}\tau_k \tag{3.70}$$

$$\mathcal{N}\tau_k = k_{f_k}\prod_{i=1}^{n_e} p_i^{\nu'_{ik}} \prod_{l=1}^{n_s} \theta_l^{\mu'_{lk}} - k_{b_k}\prod_{i=1}^{n_e} p_i^{\nu''_{ik}} \prod_{l=1}^{n_s} \theta_l^{\mu''_{lk}} \tag{3.71}$$

The ablation mass rate is given by the equation

$$\dot{m}_C = \mathcal{M}_C \sum_{j=1}^{n_r} \frac{\tau_j}{\mathcal{M}_j} \tag{3.72}$$

We can obviously neglect the gain or loss of mass due to catalytic reactions.

3.2.5 Heterogeneous Catalysis

Note that the reactions of heterogeneous catalysis, such as the recombination reaction of atomic oxygen involved in the problems of reusable thermal protection, are a special case in which the mass flow of wall material is zero at steady state. However, experimental results are often interpreted in a more restrictive way. Let the catalytic reaction of recombination (e.g., recombination of atomic oxygen, often studied for reusable thermal protection) be $O + O \rightleftharpoons O_2$. The mass flux of O to the wall is simply given as

$$\dot{m}_O = \frac{1}{\mathcal{N}}\left(\gamma_{O_2}\mathcal{M}_{O_2}\Phi_{O_2} - \gamma_O\mathcal{M}_O\Phi_O\right) \tag{3.73}$$

The relatively low temperature of the material, typically less than 2000 K, makes possible the assumption that the dissociation of oxygen is negligible ($\gamma_{O_2} = 0$). Assuming Maxwellian fluxes, the phenomenon is then fully described by the sticking probability γ

- $\gamma_O = 0$ corresponds to a noncatalytic wall.
- $\gamma_O = 1$ corresponds to a perfectly catalytic wall.

In certain situations this type of model can lead to significant errors, typically 15% in some points and much more near the connection between two materials with very different properties [32].

3.3 Relationship Between Homogeneous and Heterogeneous Reactions

The problem of data compatibility here is analogous to the problem in homogeneous systems, taking into account in the conception of databases. But the equivalent of these databases for heterogeneous reactions does not exist.

Let $\mathbf{\Omega}$ be the stoichiometric matrix in which one has written first the gaseous species, and after that the sites, counted as one species. Let $\mathbf{\Omega}_S$ on the submatrix be relative to sites only.

We will try to find a linear combination of lines that lead to $\mathbf{\Omega}_S = 0$ (which refers only to homogeneous reactions). Let l be the row vector canceling $\mathbf{\Omega}_S$

$$l\Omega = |\Omega_E \quad 0| \tag{3.74}$$

where $\mathbf{\Omega}_E$ is the new submatrix for the gas species. This equation implies that

$$l\,\Omega_S = 0 \tag{3.75}$$

As a result, we deduce that $l \in Ker(\Omega_S)$.

The case of ablation reactions induces a problem because these reactions do not respect the conservation of gaseous species. It overcomes this by arbitrarily adding the missing species in the reaction, so also in the ad hoc column of the matrix. One can see an example of this method in Table 5.1 in Chapter 5.

3.3.1 Relationship Between Reaction Constants

Let us write the relationship that gives the steady state occupancy of sites for the reaction k

$$\prod_{i=1}^{n_e} p_i^{-\Omega_{ik}} \prod_{j=1}^{n_s} \theta_j^{-\Omega_{jk}} = K_k \tag{3.76}$$

If we take the logarithm of this expression, we get

$$\sum_{i=1}^{n_e} \Omega_{ik} \ell n\, p_i + \sum_{j=1}^{n_s} \Omega_{jk} \ell n\, \theta_j = \ell n\, K_k \tag{3.77}$$

Multiply by one of the vectors l_k of dimension n_r

$$\sum_{k=1}^{n_l} l_k \left(-\sum_{i-1}^{n_e} \Omega_{ik} \ell n\, p_i - \sum_{j=1}^{n_s} \Omega_{jk} \ell n\, \theta_j \right) = \sum_{k=1}^{n_r} l_k \ell n\, K_k \tag{3.78}$$

However, according to the preceding section

$$\sum_{k=1}^{n_l} l_k \Omega_{ik} = \begin{cases} 0 & \text{for a site} \\ \alpha_i & \text{for one gaseous species} \end{cases} \tag{3.79}$$

Equation (3.78) is thus simplified

$$-\sum_{i=1}^{n_e} \alpha_i \ell n\, p_i = \sum_{k=1}^{n_r} l_k \ell n\, K_k \tag{3.80}$$

or

$$\prod_{i=1}^{n_e} p_i^{-\alpha_i} = \prod_{m=1}^{n_r} K_k^{l_k} \tag{3.81}$$

The left term is the equilibrium constant K_p (F_k) for the gas phase reaction F_k. It becomes

$$\forall\, k = 1, n_l \qquad \prod_{m=1}^{n_r} K_p^{l_k} = K_p(F_k) \tag{3.82}$$

This system assesses the consistency of data. An example is developed in Sec. 5.4.

3.3.2 Stationarity of the Heterogeneous System

Stationarity will be achieved when the following equation is true:

$$\frac{\mathrm{d}(N_S \theta_i)}{\mathrm{d}t} = f(\theta_i, p_j, T) = 0 \tag{3.83}$$

This is a nonlinear system for θ_i whose resolution will typically be made numerically by the solution of an unsteady case. It is mentioned here for the issues discussed in the following sections.

3.3.2.1 Langmuir Isotherm

We can give a classic case of stationarity: the adsorption–desorption system defined by

$$Site + A \leftrightharpoons Site - A$$

In this case we write

$$\frac{d(N_S\theta)}{dt} = k_f p_A \theta + k_b(1-\theta) = 0 \tag{3.84}$$

The immediate solution is $\theta = (1 + K_p p_A)^{-1}$ where $K_p = (k_j/k_b)$. This law describes the (adsorption) Langmuir isotherm. This result can also be obtained by thermodynamic analysis that can also handle configurations not considered here (multilayer absorption, interactions between adsorbed molecules, etc.).

3.3.2.2 Catalytic Recombination

Another example allows us to show that in special cases, the reactions may show an "order," that is, a dependency to any power of a partial pressure. Falling into this category are results from experiments on catalytic recombination of oxygen [33], including the following example:

1. Consider the system shown in Table 3.4, characterized by adsorption and desorption of species A, and recombination of this species by a Langmuir–Hinshelwood mechanism, assumed irreversible ($k_{b_2} \ll k_{f_2}$).
   Stationarity of the system will be given by the following equation:

$$k_{f_1} p_A (1-\theta_A) - k_{b_1}\theta_A + k_{f_2}\theta_A^2 = 0 \tag{3.85}$$

   The production rate of A_2 will be proportional to θ_A^2.
   Two limiting cases have an analytical solution:
   - T is large and the coverage of sites is low (see Fig. 5.5), $\theta_A \simeq 0$. The resolution of Eq. (3.85) gives

$$\theta_A \simeq \frac{K_p p_A}{1 + K_p p_A} \tag{3.86}$$

     where $K_p = (k_{f_1}/k_{b_1})$. This implies that $K_p p_A \ll 1$, so $\theta_A \simeq K_p p_A$ and the production rate is proportional to p_A^2.
   - T is small and the coverage of sites is large (see Fig. 5.5), $\theta_A \simeq 1$. The approximate solution to order 1 gives

$$\theta_A \simeq \frac{K_p p_A + 1 + \left(\frac{k_{f_2}}{k_{b_1}}\right)}{K_p p_A + 1 + 2\left(\frac{k_{f_2}}{k_{b_1}}\right)} \tag{3.87}$$

Table 3.4 Example of Catalytic Recombination (Langmuir–Hinshelwood Recombination)

Reaction 1	$Site + A \rightleftharpoons Site - A$
Reaction 2	$2\ Site - A \rightarrow 2\ Site + A_2$

Table 3.5 Second Example of Catalytic Recombination (Eley–Rideal Recombination)

Reaction 1	$Site + A \rightleftharpoons Site - A$
Reaction 2	$Site - A + A \rightarrow Site + A_2$

$\theta \simeq 1$ force $K_p p_A \gg 1$. In this case, the recombination rate is independent of the partial pressure of species A.

2. If we assume a recombination by the Eley–Rideal mechanism is also irreversible (Table 3.5), the stationarity gives (K_p has the same definition as previously)

$$\theta_A = \frac{K_p p_A}{\left(K_p - \left(\frac{k_{f_2}}{k_{b_1}}\right)\right) p_A + K_p} \tag{3.88}$$

- At low temperature, $\theta_A \simeq 1$ (see Fig. 5.5). This means that the second term in the denominator of Eq. (3.88) is negligible. We have

$$\theta_A \simeq \frac{K_p}{K_p - \left(\frac{k_{f_2}}{k_{b_1}}\right)} \tag{3.89}$$

  and the formation rate of A_2 the first order (proportional to p_A).
- At high temperature, $\theta_A \simeq 0$ and it is not possible to define an apparent order for the reaction.

This kind of law is found in the literature, for example for the flux of carbon oxidation [34]

$$\dot{m} = A \exp\left(-\frac{T_A}{T}\right) p_{O_2}^{\frac{1}{2}} \tag{3.90}$$

This is an interpretation of experimental results, the phenomenon discussed in Chapter 5 being unable to be reduced to a simple law.

Another example of this kind of simplification is given in Sec. 8.2.1.6 on treating carbon deposition by pyrolysis gas.

3.3.3 Reduction of Chemical System

The number of species and reactions can be very high, especially in the presence of hydrogen issued from reactions of pyrolysis. This problem has been particularly studied in the field of combustion, where it is particularly relevant because of the important number of species present in the medium. Methods of reducing the system have been developed, but are rarely used in aerothermodynamics. An example is given in Goussis et al.

[35]. These methods are based on a partition of fast and slow reactions (this is valid for the forward reaction and the backward reaction), the latter then being assumed almost instantaneous. From a mathematical point of view, this results in the study of the manifolds of the differential system $dc_i/dt = \dot{\omega}_i$. There are, however, problems associated with these methods:

- Partition between the reactions is not necessarily the same for all regions of the area studied or may vary with time.
- This causes a problem for the calculation of transport properties of the medium.

Note also another method: "repro-modeling." This method is actually a parameterization of source terms of equations, obtained by detailed calculations carried out independently.

References

[1] Burcat, A., Dixon-Lewis, G., Frenklach, M., Gardiner, W. C, Jr., Hanson, R., Salimian, S., Troe, J., Warnatz, J. and Zellner, R., "Combustion Chemistry," edited by Gardiner W. C., Jr., Springer-Verlag, New York, 1984, pp. 127–172.
[2] Chase, M. W., Jr., Davies, C. A., Downey, J. R., Jr., Frurip, D. J., MacDonald, R. A. and Syverud, A. N., "JANAF Thermochemical Tables, Third Edition," *Journal of Physical and Chemical Reference Data*, Vol. 14, Suppl. p1 & 2, 1985.
[3] Anderson, J. D., *Hypersonic and High Temperature Gas Dynamics*, McGraw-Hill, New York, 1989.
[4] Gökçen, T., "N2–CH4–Ar Chemical Kinetic Model for Simulation of Atmospheric Entry to Titan," AIAA Paper 2004–2469, June 2004.
[5] Park, C., "Review of Chemical Kinetics Problems of Future NASA Missions. I: Earth Entries," *Journal of Thermophysics and Heat Transfer*, Vol. 7, No. 3, 1993, pp. 385–398.
[6] Park, C., Howe, J. T., Jaffe, R. L. and Candler, G. V., "Review of Chemical Kinetics Problems of Future NASA Missions. II: Mars Entries," *Journal of Thermophysics and Heat Transfer*, Vol. 8, No. 1, 1994, pp. 9–23.
[7] Park, C., *Nonequilibrium Hypersonic Aerothermodynamics*, John Wiley & Sons, New York, 1990.
[8] Surzhikov, S. T., "Convective and Radiative Heating of MSRO, Predicted by Different Kinetic Models," *Proceedings of the 2nd International Workshop on Radiation of High Temperature Gases in Atmospheric Entry, Rome, Italy*, ESA, Noordwijk, The Netherlands, September 2006.
[9] Gnoffo, P. A., Gupta, R. N. and Shinn, J. L., "Conservation Equations and Physical Models for Hypersonic Air Flows in Thermal and Chemical Nonequilibrium," NASA Technical Paper 2867, Feb. 1989.
[10] Armaly, B. F. and Sutton, K., "Viscosity of Multicomponent Partially Ionized Gas Mixtures," *Thermophysics of Atmospheric Entry*, edited by Horton, T. E., Vol. 82, Progress in Astronautics and Aeronautics, AIAA, New York, 1982, pp. 53–67.
[11] Bacri, J. and Raffanel, S., "Calculation of Transport Coefficients of Air Plasmas," *Plasma Chemistry and Plasma Processing*, Vol. 9, No. 1, 1989, pp. 133–154.
[12] Gupta, R. N., Yos, J. M., Thomson, R. A. and Lee, K., "Review of Reaction Rates and Thermodynamic and Transport Properties for an 11-Species Air Model for Chemical and Thermal Nonequilibrium Calculations to 30,000 K," NASA Technical Report RP 1232, 1990.
[13] Hansen, C. F., "Approximations for the Thermodynamic and Transport Properties of High-Temperature Air," NASA Technical Report R-50, 1960.

[14] Yos, J., "Revised Transport Properties for High Temperature Air and Its Components," AVCO Technical Report Z220, Nov. 1967.
[15] Dunn, G. J. and Eagar, T. W., "Calculation of Electrical and Thermal Conductivities of Metallurgical Plasmas," Welding Research Council (WRC) Bulletin No. 357, Sept. 1990.
[16] Lemmon, E. W. and Jacobsen, R. T., "Viscosity and Thermal Conductivity Equations for Nitrogen, Oxygen, Argon and Air," *International Journal of Thermophysics*, Vol. 25, No. 1, 2004, pp. 21–69.
[17] Sarma, G. S. R., "Physico-Chemical Modeling in High-Enthalpy and Plasma Flows: General Introduction," Von Karman Institute for Fluid Dynamics Lecture Series 2002–07, June 2002.
[18] Brokaw, R. S., "Thermal Conductivity of Gas Mixtures in Chemical Equilibrium. II," *Journal of Chemical Physics*, Vol. 32, No. 4, 1960, pp. 1005–1006.
[19] Butler, J. N. and Brokaw, R. S., "Thermal Conductivity of Gas Mixtures in Chemical Equilibrium," *Journal of Chemical Physics*, Vol. 26, No. 6, 1957, pp. 1636–1643.
[20] Marrone, P. V. and Treanor, C. E., "Chemical Relaxation with Preferential Dissociation from Excited Vibrational Levels," *Physics of Fluids*, Vol. 6, No. 9, 1963, pp. 1215–1221.
[21] Treanor, C. E. and Marrone, P. V., "Effect of Dissociation on the Rate of Vibrational Relaxation," *Physics of Fluids*, Vol. 5, No. 9, 1962, pp. 1022–1026.
[22] Brun, R., Introduction to Reactive Gas Dynamics, Oxford University Press, Oxford, UK, 2009.
[23] Sharma, S. P., Huo, W. M. and Park, C., "The Rate Parameters for Coupled Vibration- Dissociation in a Generalized SSH Approximation," AIAA Paper 1988–2714, *Thermophysics, Plasmadynamics and Lasers Conference*, San Antonio, Texas, June 1988.
[24] Panesi, N., Magin, T., Bourdon, A., Bultel, A., Chazot, O. and Babou, Y., "Collisional-Radiative Modeling in Flow Simulations," NATO Research and Technology Organization Report RTO-EN-AVT-162, Sept. 2009.
[25] Chapman, S. and Cowling, T. G., *The Mathematical Theory of Non-Uniform Gases*, Cambridge University Press, New York, 1991.
[26] Hirschfelder, J. O., Curtiss, C. F. and Bird, R. B., *Molecular Theory of Gases and Liquids*, John Wiley & Sons, New York, 1966.
[27] Atkins, P. W., Physical Chemistry, 6th ed., W. H. Freeman & Co., New York, 2000.
[28] Koenig, P. C., Squires, R. G. and Laurendeau, N. M., "Evidence for Two-Site Model of Char Gasification by Carbon Dioxide," *Carbon*, Vol. 23, No. 5, 1985, pp. 531–536.
[29] Nagle, J. and Strickland-Constable, R. F., "Oxydation of Carbon between 1000–2000°C," *Proceedings of Fifth Conference on Carbon*, Pergamon Press, Oxford, UK, 1962, pp. 154–164.
[30] Schrage, R. W. *A Theoretical Study of Interphase Mass Transfer*, Columbia University Press, New York, 1953.
[31] Dubroca, B., Duffa, G. and Leroy, B., "High Temperature Mass and Heat Transfer Fluid-Solid Coupling," AIAA Paper 2002–5180, *11th Meeting of AIAA/AAAF Space Planes and Hypersonic Systems and Technologies*, Orléans, France, Sept. 2002.
[32] Canton-Desmeuzes, C., Couzi, J., Frayssinet, O. and Goyhénèche, S., "Behaviour of Two Catalycity Models on Surface Presenting Material Discontinuities," *4th International Symposium on Atmospheric Reentry Vehicles and Systems*, Arcachon, France, Mar. 2005.
[33] Balat-Pichelin, M., "Physico-Chemical Models for High Enthalpy and Plasma Flow Modelling. Applications: Experimental Investigation of Surface Reactions," *Von Karman Institute for Fluid Dynamics Lecture Series 2002-07*, edited by Fletcher D., Charbonnier J.-M., Sarma G. S. R. and Magin T., Von Karman Institute for Fluid Dynamics, Rhode Saint Genese, Belgium, June 2002.

[34] Metzger, J. W., Engels, M. J. and Diaconis, N. J., "Oxydation and Sublimation of Graphite in Simulated Re-Entry Environments," *AIAA Journal*, Vol. 5, No. 3, 1967, pp. 451–460.
[35] Goussis, D. A., Lam, S. H. and Gnoffo, P. A., "Reduced and Simplified Chemical Kinetics for Air Dissociation Using Computational Singular Perturbation," AIAA Paper 90–0644, *28th Aerospace Sciences Meeting*, Reno, Nevada, Jan. 1990.

Chapter 4 Approximate Methods

- Solve mass and heat transfer in the boundary layer
- Describe the effect of wall injection
- Calculate the quantities at the stagnation point of an axisymmetric body

4.1 Introduction

The analytical methods for solving mass and heat transfer in the boundary layer were developed in the 1960s, when the numerical calculation possibilities where restricted. They are now of great interest for understanding the physical phenomena and calculating orders of magnitude. In addition, most ablation phenomena computer codes developed in the 20th century still widely use all or part of these approximations. Some examples of these methods can be found in D.B. Spalding's book, *Convective Mass Transfer: An Introduction.*

4.2 Reactive Laminar Boundary Layers

4.2.1 Approximation of Heat Flux

By neglecting the diffusion by temperature and pressure gradients and retaining only the dominant terms (gradients normal to the wall along the coordinate y, positive toward the fluid), the heat flux is written

$$\dot{q} = -\lambda \frac{\partial T}{\partial y} + \sum_i \rho_i V_{i_y} h_i \tag{4.1}$$

The diffusion rate can be approximated (Sec. 2.3.2.5) by using an expression using a single diffusion coefficient

$$\rho_i V_{i_y} \simeq -\rho \mathcal{D} \frac{\partial c_i}{\partial y} \tag{4.2}$$

We can then rewrite the flux [Eq. (4.1)] as follows:

$$\dot{q} = -\lambda \frac{\partial T}{\partial y} - \rho \mathcal{D} \sum_j h_j \frac{\partial c_j}{\partial y} \tag{4.3}$$

Moreover, the total enthalpy is defined by $h = \sum_i c_i h_i$. Differentiating this expression, it becomes

$$\mathrm{d}h = C_p \mathrm{d}T + \sum_j h_j \mathrm{d}c_j \tag{4.4}$$

In this expression, the "frozen" specific heat at constant pressure is defined by

$$C_p = \sum_i c_i C_{p_i}, \quad C_{p_i} = \frac{\partial h_i}{\partial T}$$

Hence the new expression for the heat flux

$$\dot{q} = -\frac{\lambda}{C_p} \left[\left(\frac{\partial h}{\partial y} - \sum_j h_j \frac{\partial c_i}{\partial y} \right) + \frac{\rho \mathcal{D} C_p}{\lambda} \sum_j h_j \frac{\partial c_j}{\partial y} \right] \tag{4.5}$$

We can express $\mathcal{D}$ from a given Lewis–Semenov number

$$\mathcal{L} = \frac{\lambda}{\rho C_p \mathcal{D}} \tag{4.6}$$

Substituting into Eq. (4.5), it becomes

$$\dot{q} = -\frac{\lambda}{C_p} \left[\frac{\partial h}{\partial y} - (\mathcal{L} - 1) \sum_j h_j \frac{\partial c_j}{\partial y} \right] \tag{4.7}$$

The Lewis number is close to 1, ranging between 0.8 and 1.4 for air, for example (Fig. 4.1), in the temperature range of interest. A good approximation of heat flux is then given by

$$\dot{q} = -\frac{\lambda}{C_p} \frac{\partial h}{\partial y} \tag{4.8}$$

We find an expression identical to that obtained in nonreactive gas. In this case, chemical reactions influence the heat flux only through the transport properties and boundary conditions on the surface of the material.

This analysis extends easily to the turbulent boundary layer when using a simple model of turbulent viscosity:

- The mass diffusion coefficient $\mathcal{D}$ is replaced by $\mathcal{D} + \in_M$, where $\in_M$ represents the turbulent mass "diffusion."

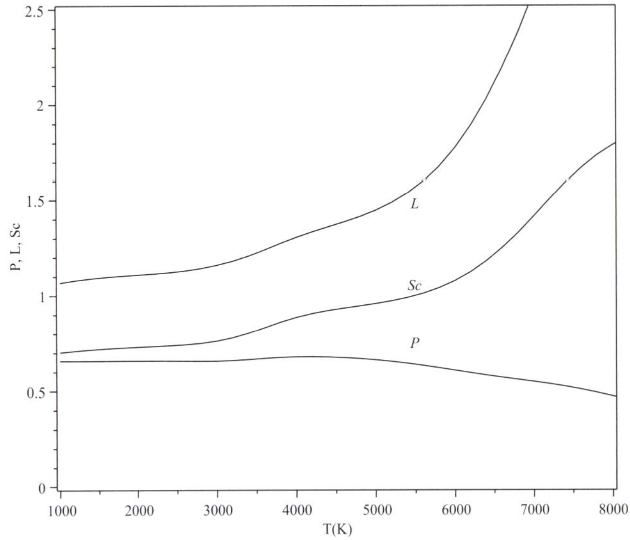

Fig. 4.1 Prandtl, Lewis, and Schmidt numbers for air at thermodynamic equilibrium, $p = p_0$; diffusion coefficient: $\mathcal{D}_{O_2,N_2}$.

- The thermal diffusion coefficient $\frac{\lambda}{C_p}$ is replaced by $\frac{\lambda}{C_p} + \epsilon_H$, where $\in_H$ represents the "diffusion" of turbulent energy.

4.2.2 Analogy Between Mass and Heat Transfer

4.2.2.1 A Simplification of the Energy Equation

For a 2-D boundary layer (longitudinal coordinate s) in steady regime, the equation of total enthalpy $h_T = h + (V^2/2)$ with $\vec{V} = (u, v)$ reads [1] (2-D plane, no term related to the acceleration of the external flow)

$$\rho u \frac{\partial h_T}{\partial s} + \rho v \frac{\partial h_T}{\partial y} = \frac{\partial}{\partial y}\left[\frac{\mu}{\mathcal{P}}\frac{\partial h_T}{\partial y} + \mu\left(1 - \frac{1}{\mathcal{P}}\right)\frac{\partial}{\partial y}\left(\frac{u^2}{2}\right) + \left(1 - \frac{1}{\mathcal{L}}\right)\sum_j \rho\mathcal{D}\frac{\partial c_j}{\partial y}\right] \tag{4.9}$$

The Prandtl number $\mathcal{P} = \frac{\mu C_p}{\lambda}$ is also close to unity, between 0.5 and 0.7 for the air in our temperature range (Fig. 4.1).

We obtain an approximation of Eq. (4.9) by writing $\mathcal{P} = \mathcal{L} = 1$

$$\rho u \frac{\partial h_T}{\partial s} + \rho v \frac{\partial h_T}{\partial y} = \frac{\partial}{\partial y}\left(\mu \frac{\partial h_T}{\partial y}\right) \tag{4.10}$$

4.2.2.2 Conservation of Elements

The equation for conservation of species in the boundary layer approximation is written

$$\frac{\partial}{\partial s}(\rho u c_i) + \frac{\partial}{\partial y}\left(\rho v c_i - \rho \mathcal{D}\frac{\partial c_i}{\partial y}\right) = \dot{\omega}_i \tag{4.11}$$

This equation can be used to write a conservation equation of elements. This is done using the Shvab–Zeldovich transformation

$$\tilde{c}_k = \mathcal{M}_k \sum_i n_{ki}\frac{c_i}{\mathcal{M}_i} \qquad k = 1, n_A \tag{4.12}$$

where n_{kj} is the number of element k in species j, and n_A is the number of elements in the system. This system includes $n_A - 1$ independent relationships because $\sum_k \tilde{c}_k = \sum_i c_i = 1$. The production rate of elements is written

$$\dot{\omega}_k = \mathcal{M}_k \sum_j \frac{n_{kj}}{\mathcal{M}_j} = 0 \tag{4.13}$$

Multiplying Eq. (4.11) by $\frac{\mathcal{M}_k}{\mathcal{M}_j} n_{kj}$ and summing over j there is, given the relationship of global conservation $\frac{\partial(\rho u)}{\partial s} + \frac{\partial(\rho v)}{\partial y} = 0$,

$$\rho u \frac{\partial \tilde{c}_k}{\partial s} + \rho v \frac{\partial \tilde{c}_k}{\partial y} = \frac{\partial}{\partial y}\left(\rho \mathcal{D}\frac{\partial \tilde{c}_k}{\partial y}\right) \tag{4.14}$$

4.2.2.3 Analogy Between Elemental Fractions and Total Enthalpy

The Schmidt number $Sc = \mathcal{LP}$ is unity, hence $\mu = \rho\mathcal{D}$, so Eqs. (4.10) and (4.14) governing h_T and $\tilde{c}_k$ have the same form. These two quantities are then linearly dependent

$$\frac{\tilde{c}_k - \tilde{c}_{k_p}}{\tilde{c}_{k_e} - \tilde{c}_{k_p}} = \frac{h_T - h_{T_p}}{h_{T_e} - h_{T_p}} \tag{4.15}$$

where e and p correspond to the values outside the boundary layer and at the wall, respectively. Note that $\rho_p v_p^2 \ll h_p$, implying that $h_{T_p} \simeq h_p$.

4.2.3 Other Analogies

4.2.3.1 Analogy with the Equation of Conservation of Momentum

The equation for longitudinal momentum in the boundary layer is written

$$\rho u \frac{\partial u}{\partial s} + \rho v \frac{\partial u}{\partial y} = -\frac{\partial p_e}{\partial s} + \frac{\partial}{\partial y}\left(\frac{\partial u}{\partial y}\right) \tag{4.16}$$

where p_e is the Eulerian external pressure, constant in the the boundary layer. Comparing this equation for a slowly varying boundary layer (neglecting $\frac{\partial p_e}{\partial s}$ compared to the dissipative term) with Eq. (4.14), we see a new analogy. Then we generalize the already written

$$\frac{\tilde{c}_k - \tilde{c}_{k_p}}{\tilde{c}_{k_e} - \tilde{c}_{k_p}} = \frac{h_T - h_{T_p}}{h_{T_e} - h_{T_p}} = \frac{u}{u_e} \tag{4.17}$$

4.2.3.2 Reynolds Analogy

Define the Stanton number C_H by

$$\dot{q} = -\rho_e u_e C_H (h_a - h_p) \tag{4.18}$$

Also define the shear $\tau_p = (\mu \frac{\partial u}{\partial y})|_p$. Differentiating Eq. (4.17), it becomes

$$\frac{\left.\frac{\partial u}{\partial y}\right|_p}{u_e} = \frac{\left.\frac{\partial h}{\partial y}\right|_p}{h_{T_e} - h_{T_p}} \tag{4.19}$$

By introducing the heat flux [Eq. (4.8)] in this expression

$$\mathcal{P} C_H = \frac{\tau_p}{\rho_e u_e^2} \tag{4.20}$$

with the approximation $\mathcal{P} = 1$, we find $C_H = \tau_p / \rho_e u_e^2$, the equation called the Reynolds analogy, widely used in the literature. It must be said that this equation is invalidated by the following:

- An important wall injection
- The presence of roughness

4.3 Injection (Blowing or Blocking) Coefficient

The injection of gas at the wall causes a strong thickening of the boundary layer (Fig. 4.2). This results in reduced convective exchanges, which we will discuss in this section.

4.3.1 Definition of Blowing Coefficient

The conservation of elemental mass at wall implies

$$\rho_p v_p c_{i_p} + J_{i_p} = \dot{m} c_{i_s} + \dot{\omega}_i \tag{4.21}$$

where $\dot{\omega}_i$ is the surface mass formation for element k issued from ablation and contained in the material with the fraction c_{i_s}, J_{i_p} is the diffusion flux,

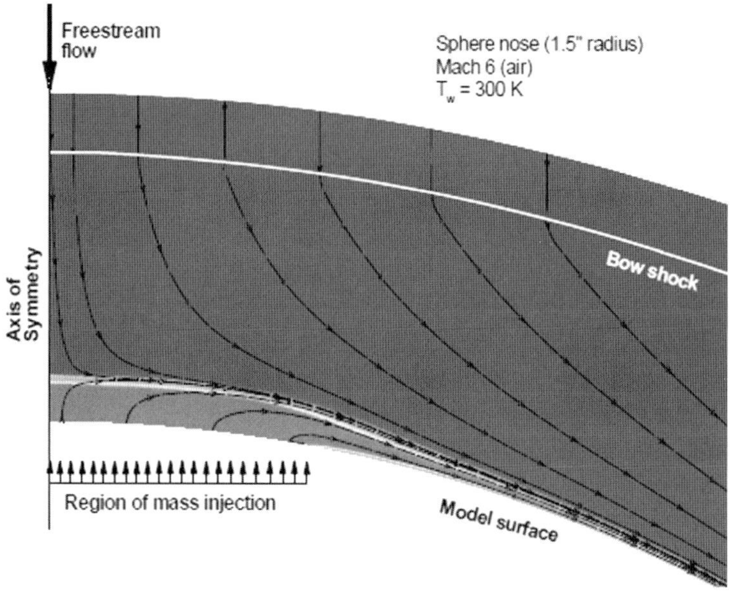

Fig. 4.2 Influence of blowing on heat flux (2).

and $\dot{m}$ is the ablation mass flux. Conservation of mass at solid–fluid interface is written in a frame in axis moving with the surface

$$\dot{m} = \rho_p v_p \tag{4.22}$$

By performing a weighted sum of the system in Eq. (4.21) in order to make the basic mass fractions appear [Eq. (4.12)], it follows

$$\tilde{J}_k = -\dot{m}(\tilde{c}_k - \tilde{c}_{k_p}) \tag{4.23}$$

Applying the same method to the definition of the relationship of mass transfer coefficient [Eq. (4.41)], we can write

$$\tilde{J}_k = -\rho_e u_e C_M(\tilde{c}_{k_e} - \tilde{c}_{k_p}) \tag{4.24}$$

By identification

$$(\mathcal{B}' + 1)\tilde{c}_{k_p} = \mathcal{B}'\tilde{c}_{k_s} + \tilde{c}_{k_e} \tag{4.25}$$

In this last equation we introduced the dimensionless parameter representing the injection at wall (or the wall deposition) also called *blowing factor* $\mathcal{B}'$ defined by

$$\mathcal{B}' = \frac{\dot{m}}{\rho_e u_e C_M} \tag{4.26}$$

These quantities are available from a model describing the ablation by chemical reactions and phase changes (Chapter 3). The relation in Eq. (4.25) connects the basic fractions to the wall and outside the boundary layer through $\mathcal{B}'$. Values outside the boundary layer are known (they are those of the atmosphere at upstream infinity). These relations allow us to know the wall values.

For example, given the expressions obtained in the case of air on a carbon wall ($\tilde{c}_{O_e} \simeq 0.19$, $\tilde{c}_{N_e} \simeq 0.80$, $\tilde{c}_{C_s} = 1$)

$$\frac{\tilde{c}_{O_e}}{\tilde{c}_{O_p}} = \frac{\tilde{c}_{N_e}}{\tilde{c}_{N_p}} = \mathcal{B}' + 1 \tag{4.27}$$

$$\tilde{c}_{C_p} = \frac{\mathcal{B}'}{\mathcal{B}' + 1} \tag{4.28}$$

4.3.2 Generalization to the Case with Pyrolysis

In the case where a gas from within the material (e.g., pyrolysis) with the rate $\dot{m}_g$ overlaps with the outgassing due to ablation to give a total mass heat flux $\dot{m} + \dot{m}_g$, the previous expressions generalize without difficulty. Equation (4.25) becomes

$$(\mathcal{B}' + \mathcal{B}'_g + 1)\tilde{c}_{k_p} = \mathcal{B}'\tilde{c}_{k_s} + \mathcal{B}'_g\tilde{c}_{k_g} + \tilde{c}_{k_e} \tag{4.29}$$

with

$$\mathcal{B}'_g = \frac{\dot{m}_g}{\rho_e u_e C_M} \tag{4.30}$$

$\mathcal{B}'$ is defined by $\mathcal{B}' = \dot{m}/\rho_e u_e C_M$ and $\rho_p v_p = \dot{m} + \dot{m}_g$.

The hypothesis of a stationary problem permits us to find a relationship between $\mathcal{B}'$ and $\mathcal{B}'_g$ (Sec. 4.7).

4.4 The Couette Problem Analogy

The Couette problem describes the viscous flow between two plates separated by δ. (By analogy, this quantity represents the boundary layer thickness.) The relative velocity of these two plates is u_e. (By analogy, this quantity is the velocity outside the boundary layer.) Moreover, the transport properties are assumed to be independent of temperature. The problem studied is stationary. This very simple example helps one to understand some expressions that will be used later in this chapter.

4.4.1 Adiabatic Wall Enthalpy and Stanton Number

The equation of conservation of momentum is written

$$\frac{\mathrm{d}}{\mathrm{d}y}\left(\mu\frac{\mathrm{d}u}{\mathrm{d}y}\right) = 0 \tag{4.31}$$

This equation has the following solution:

$$\frac{u}{u_e} = \frac{y}{\delta} \tag{4.32}$$

The energy equation is written

$$\frac{\mathrm{d}}{\mathrm{d}y}\left(\lambda\frac{\mathrm{d}T}{\mathrm{d}y} - \mu u\frac{\mathrm{d}u}{\mathrm{d}y}\right) = 0 \tag{4.33}$$

The solution is a second order polynomial in y

$$T = -\mathcal{P}\frac{u_e^2}{2C_p}\left(\frac{y}{\delta}\right)^2 + \left(T_e + \mathcal{P}\frac{u_e^2}{2C_p} - T_p\right)\frac{y}{\delta} + T_p \tag{4.34}$$

where $\mathcal{P}$ is the Prantdl number

$$\mathcal{P} = \frac{\mu C_p}{\lambda} \tag{4.35}$$

From this last expression we get the wall heat flux

$$\dot{q} = -\lambda\left.\frac{\mathrm{d}T}{\mathrm{d}y}\right|_{y=0} = -\frac{\lambda}{\delta}\left(T_e + \mathcal{P}\frac{u_e^2}{2C_p} - T_p\right) \tag{4.36}$$

So, by introducing the enthalpy $h = C_pT$

$$\dot{q} = -\frac{\lambda}{\delta C_p}\left(h_e + \mathcal{P}\frac{u_e^2}{2} - h_p\right) \tag{4.37}$$

Viscous dissipation is taken into account by replacing the enthalpy h_e by the adiabatic wall enthalpy

$$h_a = h_e + \mathcal{P}\frac{u_e^2}{2} \tag{4.38}$$

One can generalize this equation to the boundary layer by introducing the wall "recovery factor" r estimated as $r = f(\mathcal{P})$ from experimental correlations:

- For a Couette flow: $r = \mathcal{P}$
- For a laminar boundary layer: $r \simeq \mathcal{P}^{\frac{1}{2}}$
- For a turbulent boundary layer: $r \simeq \mathcal{P}^{\frac{1}{3}}$

$\mathcal{P}$ being dependent on the point of the boundary layer considered, the correlation has to define the reference value.

Equation (4.37) is therefore written

$$\dot{q} = -\frac{\lambda}{\delta C_p}(h_a - h_p) \tag{4.39}$$

This equation justifies the qualification "adiabatic wall": the flux is zero for $h_p = h_a$.

If we make $\mathcal{P} = 1$ as we did earlier in the simplified boundary layer equations, then $h_a = h_T$.

Moreover, the trivial solution of the equation of mass conservation in the Couette problem gives mass flux normal to the wall

$$J_i = \frac{\rho \mathcal{D}}{\delta}(c_{ip} - c_{ie}) \tag{4.40}$$

If we define the mass transfer coefficient C_M by

$$J_i = -\rho_e u_e C_M (c_{ie} - c_{ip}) \tag{4.41}$$

by comparing the solutions of the Couette problem, it follows immediately

$$\frac{C_M}{C_H} = \frac{\rho \mathcal{D} C_p}{\lambda} = \mathcal{L} \tag{4.42}$$

In the approximation consistent with the previous analysis ($\mathcal{L} = 1$), the Stanton numbers of mass transfer and heat are equal. Of course, this approximation is sketchy; various authors have sought to improve on the following form, sometimes called the Chilton–Colburn relation:

$$\frac{C_M}{C_H} = \mathcal{L}^{\alpha} \tag{4.43}$$

with $\alpha \simeq \frac{2}{3}$.

This coefficient gives very correct values in the absence of injection, but becomes quite poor for significant injection where $\alpha \simeq 1$ [3].

4.4.2 The Entropy Swallowing Phenomenon

In the previous section we wrote the adiabatic enthalpy h_a at the risk of confusion with the inviscid value at stagnation point. Now, the limit of the boundary layer is not a flow line (Fig. 4.3). From upstream to downstream these lines correspond to a crossing of the shock that is more and more oblique, so they correspond to a lower entropy. If one follows the boundary layer to its outer limit, this corresponds to a decrease of entropy, hence the name *entropy swallowing*.

In the methods where the boundary layer is calculated independently of Eulerian flow, this phenomenon is difficult to handle locally and must be

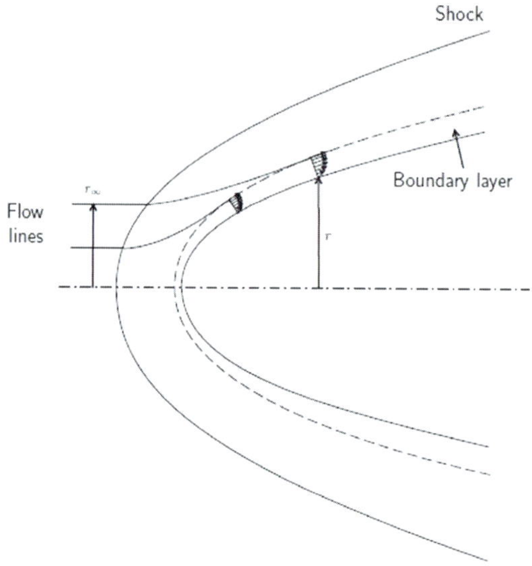

Fig. 4.3 Entropy swallowing.

taken into account by a mass balance. In axisymmetric flow we write (Fig. 4.3)

$$\rho_\infty V_\infty \pi r_\infty^2 = 2\pi r \int_0^\delta \rho u \mathrm{d}y \tag{4.44}$$

where r_∞ is the distance from the axis where the flow line arriving at a point of the body defined by the distance r runs through the shock.

Using modern methods that solve Navier–Stokes equations directly, this problem has vanished.

4.4.3 Effect of Blowing on Exchanges

4.4.3.1 Flat Plate

In what follows, we retain the Couette flow approximation. The species conservation equation [Eq. (4.11)] now is written

$$\frac{\mathrm{d}}{\mathrm{d}y}\left(\rho v c_i - \rho \mathcal{D}\frac{\mathrm{d}c_i}{\mathrm{d}y}\right) = \dot{\omega}_i \tag{4.45}$$

By summation over i it becomes

$$\frac{\mathrm{d}(\rho v)}{\mathrm{d}y} = 0 \tag{4.46}$$

The mass heat flux is constant: $\dot{m} = \rho v = C^{st}$. (The upper wall is supposed to absorb the mass flux.)

Equation (4.45) integrates immediately

$$\rho v c_i - \rho \mathcal{D} \frac{dc_i}{dy} = S_i \tag{4.47}$$

The integration constant representing the mass heat flux of species i can be evaluated at the lower wall

$$S_i = \dot{m} c_{i_p} - \rho \mathcal{D} \left. \frac{dc_i}{dy} \right|_p \tag{4.48}$$

We can rewrite Eq. (4.47) as follows:

$$\frac{dc_i}{\dot{m} c_i - S_i} = \frac{dy}{\rho \mathcal{D}} \tag{4.49}$$

A new integration gives

$$\frac{c_i - \dfrac{S_i}{\dot{m}}}{c_{i_p} - \dfrac{S_i}{\dot{m}}} = \exp\left(\frac{\dot{m} y}{\rho \mathcal{D}}\right) \tag{4.50}$$

Expressing this relation in $y = \delta$, and using the relation in Eq. (4.41) defining C_M

$$\frac{\dot{m} + \rho_e u_e C_M}{\rho_e u_e C_M} = \exp\left(\frac{\dot{m}\delta}{\rho \mathcal{D}}\right) \tag{4.51}$$

It follows

$$\dot{m} = \frac{\rho \mathcal{D}}{\delta} \ell n(1 + \mathcal{B}') \tag{4.52}$$

A simple analysis of the problem without wall injection shows that $\frac{\rho \mathcal{D}}{\delta}$ is the value of $\rho_e u_e C_M$ for $\dot{m} = 0$, which will be noted $\rho_e u_e C_{M_0}$. This gives the desired expression

$$\frac{C_M}{C_{M_0}} = \frac{\ell n(1 + \mathcal{B}')}{\mathcal{B}'} \tag{4.53}$$

We can also express this law as a function of the parameter

$$\mathcal{B}'_0 = \frac{\dot{m}}{\rho_e u_e C_{M_0}} \tag{4.54}$$

This gives an expression equivalent to Eq. (4.53)

$$\frac{C_M}{C_{M_0}} = \frac{\mathcal{B}'_0}{\exp \mathcal{B}'_0 - 1} \tag{4.55}$$

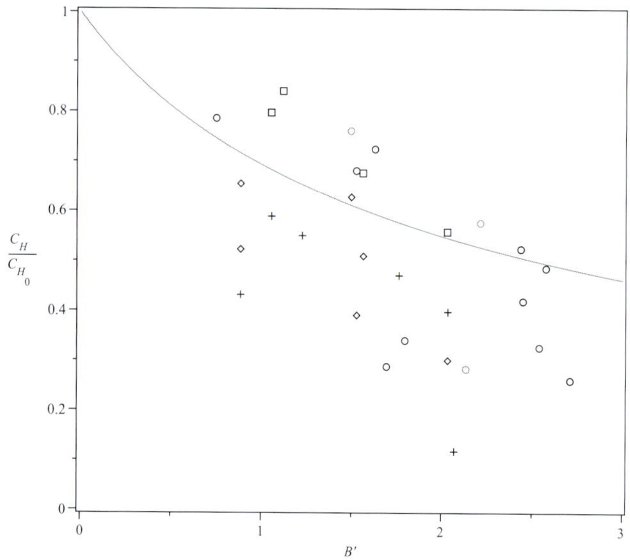

Fig. 4.4 Ablation of PTFE in laminar flow (plasma jet experiments on various devices in flat plate configuration) (3).

Many attempts have been made to validate the expression in Eq. (4.53). The experimental hot wind tunnel validations (Fig. 4.4) have generally been disappointing due to experimental uncertainties. The cold tests on a permeable wall are more accurate, but still cannot show a difference, if any, between laminar and turbulent flows [3]. In fact, this expression is fairly accurate in the case of a flat plate [2].

Note that the physical phenomena have a change of nature with "massive" injections ($\mathcal{B}' > 2.7$ for flat plate, laminar flow). This type of phenomenon, akin to detachment, is not described in the literature.

4.4.3.2 Stagnation Point

The stagnation point on a sphere has been the subject of many studies to validate a particular blowing correlation. Among these, we retain a very complete parameterization of the problem made in [4]. In a wide pressure range (10^3 to 1.52×10^7 Pa) and stagnation enthalpies (11.6 to 30 $\mathrm{MJ \cdot kg^{-1}}$), for various materials (graphite, carbon-phenolic resin, nylon-phenolic resin), for laminar or turbulent flows, and in many conditions (no ablation, stationary ablation, unsteady ablation), various correlations are reviewed, among them:

- Correlation $\mathcal{C}_M/\mathcal{C}_H\mathcal{L}^{\frac{2}{3}} = f(\mathcal{B}'_0)$ is unconvincing. The Chilton–Colburn correlation is indeed put in default by the injection.
- Correlation $\mathcal{C}_M/\mathcal{C}_{H_0} = f(\mathcal{B}'_0)$ gives excellent results (3% error or less). This correlation takes into account the variation of average molecular

weight in the boundary layer

$$\frac{\mathcal{C}_M}{\mathcal{C}_{H_0}} = \frac{\eta \mathcal{B}'_0}{\exp(\eta \mathcal{B}'_0) - 1} \tag{4.56}$$

with

$$\begin{cases} \eta = 2(1.012 + 0.018\mathcal{B}'_0 + 0.0814{\mathcal{B}'_0}^2)(1 - F_1) \\ F_1 = (0.238 + 0.038\mathcal{B}'_0)\left(\dfrac{F_2 - 0.95}{0.6}\right) \\ F_2 = \begin{cases} 0.95 & \text{if } \dfrac{\overline{\mathcal{M}}_p}{\overline{\mathcal{M}}_e} < 0.95 \\ \dfrac{\overline{\mathcal{M}}_p}{\overline{\mathcal{M}}_e} & \text{if } 0.95 < \dfrac{\overline{\mathcal{M}}_p}{\overline{\mathcal{M}}_e} < 1.55 \\ 1.55 & \text{if } \dfrac{\overline{\mathcal{M}}_p}{\overline{\mathcal{M}}_e} > 1.55 \end{cases} \end{cases} \tag{4.57}$$

This correlation was also tested at the stagnation point on an spheroconic geometry [4]. There is a good agreement up to $s/R \simeq 1$ and far downstream of the expansion region ($s/R > 50$), where we find conditions analogous to flat plate conditions. The comparison with flat plate correlation is done in Fig. 4.5.

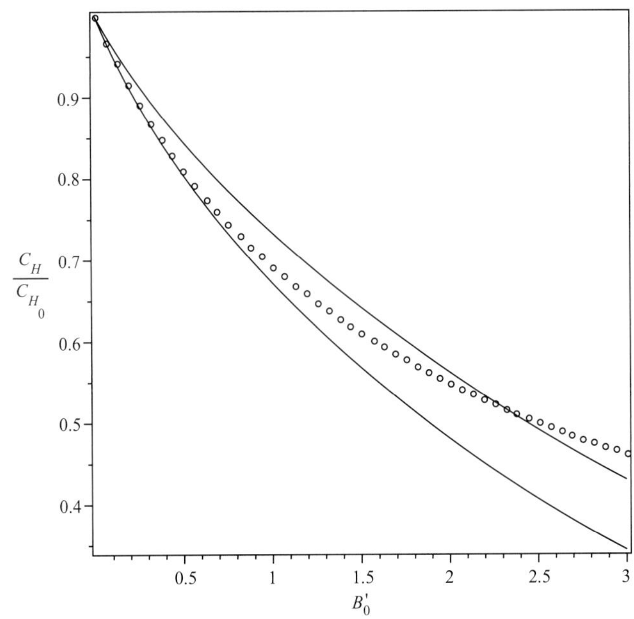

Fig. 4.5 Comparison of flat plate correlation (Eq. (4.55)) with extrema of stagnation point correlation (Eq. (4.57)).

As in the case of flat plate, a massive injection results in a drastic alteration of the flow and thus limits the range of validity of the previous expression.

4.5 Approximate Calculation of Stagnation Point Heat Flux

Many studies have been devoted to calculating the quantities at the stagnation point of an axisymmetric body to gain a better understanding of heat flux. We will deal with the problem without ablation.

4.5.1 Quantities Outside the Boundary Layer

The first problem is to characterize the external flow. For this, we assume that the flow outside the boundary layer corresponds to an isentropic expansion from the stagnation point. With this assumption, the classical isentropic expansion relationship permits us to conclude that, to first order in velocity:

- Velocity is linear in s (curvilinear abscissa): $u_e = \beta s$ where the velocity gradient at stagnation point. The β is given in Eq. (4.58).
- The density ρ and temperature T are quadratic in velocity, to order 1, then $\rho = \rho_a$ and $T = T_a$ where the index a denotes the values at the stagnation point.
- Hence the bulk composition is constant (at first order) and thus the viscosity $\mu = \mu_a$ is as well.

The velocity gradient can be calculated assuming a Newtonian variation of the pressure in the vicinity of the stagnation point. (p is proportional to the cosine of the local angle.) The calculation then gives

$$\beta = \left.\frac{\mathrm{d}u_e}{\mathrm{d}s}\right|_a \simeq \frac{1}{R}\left(\frac{2p_a}{\rho_a}\right)^{\frac{1}{2}} \tag{4.58}$$

where R is the local radius at the stagnation point.

Quantities p_a, ρ_a, and T_a are characterized by the Rankine–Hugoniot relations, followed by isentropic compression to the stagnation point. These relations are analytic for a perfect gas. In the latter case, and for $\mathcal{M}a \gg 1$, it was

$$\left(\frac{2p_a}{\rho_a}\right)^{\frac{1}{2}} \simeq \alpha_1 V_\infty \tag{4.59}$$

Similarly, one can calculate the approximate value of p_a

$$\frac{p_a}{p_\infty} \simeq \alpha_2 V_\infty^2 \tag{4.60}$$

For sphere-torus forms and for very high Mach numbers ($\mathcal{M}a > 3$), there is a correlation of the velocity gradient at the stagnation point issued from the experiment [4]. This type of geometry is found in various test configurations (pressure or flux sensors, ablated shapes in laminar flow).

4.5.2 Flow Similarity

Much work has been devoted to 2-D flows with the property being reduced to a 1-D problem, with the change of variable ad hoc. Among these include the flow near the stagnation point for a perfect gas, on a non-reactive wall. The change of variable is the following (sometimes called the Lees–Doronitsyn transformation)

$$\begin{cases} \xi(s) & = \displaystyle\int_0^s \rho_e u_e \mu_e r^2 ds' \\ \eta(\xi, y) & = \displaystyle\frac{u_e r}{(2\xi)^{\frac{1}{2}}} \int_0^y \rho dy' \end{cases} \tag{4.61}$$

One defines:

- The quantity f by $\partial f / \partial \eta = u/u_e$
- The quantity $c = \rho\mu$
- The quantity $g = h/h_e$

Through the linearization of the velocity near the stagnation point, the boundary layer equations become [1]

$$\begin{cases} (cf'')' + ff'' & = (f')^2 - g \\ \left(\dfrac{c}{\mathcal{P}} g'\right)' + fg' & = 0 \end{cases} \tag{4.62}$$

In these equations, the prime represents the differentiation with respect to η. We note that these equations do not explicitly involve the variable ξ. The solutions in the neighborhood of the stagnation point are similar. The boundary conditions are

- At the wall: $\eta = 0, f = f' = 0, g = g_p$ given
- Outside the boundary layer: $\eta \to \infty, f' = 1, g = 1$

Many authors have numerically solved this system and proposed an analytical representation of their results. Among these solutions we note that of Fay and Riddell [6], obtained for nonequilibrium medium and for $\mathcal{P} = 0.71$, $\mathcal{L} = 1.0, 1.4, 2.0$, $300\ \mathrm{K} < T_p < 3000\ \mathrm{K}$

$$\dot{q}_p = 0.76\mathcal{P}^{-0.6}(\rho_p\mu_p)^{0.1}(\rho_e\mu_e)^{0.4}\left[1 + (\mathcal{L}^{0.52} - 1)\frac{h_D}{h_e}\right](h_a - h_p)\beta^{\frac{1}{2}} \quad (4.63)$$

with

$$h_D = -\sum_i c_{i_e} h^0_{f_i} \quad (4.64)$$

where $h^0_{f_i}$ is the enthalpy of formation of species i at 0 K and $p = p_0$.

This formulation allows us to highlight the dependence of the flux at stagnation point in $R^{-\frac{1}{2}}$ and to the quantity $h_a - h_p$.

The implementation of this equation requires a calculation of unknown quantities in the flow. This is the reason why other, simpler relationships were researched.

4.5.3 Sutton and Graves Method

By plotting Eq. (4.58) in the expression in Eq. (4.63), we find an expression having the following form:

$$\dot{q}_p = \alpha_3\left(\frac{p_a}{R}\right)^{\frac{1}{2}}(h_a - h_p) \quad (4.65)$$

Sutton showed that the dependence of α_3 to various quantities contained in this value could be ignored as a first approximation [7].

This has been tested on many gas mixtures representative of planetary atmospheres (Fig. 4.6). The relative error is generally less than 10%, except for some gases (CH_4, NH_3) or mixtures containing hydrogen. We can further simplify this expression. Indeed, if we look at the cold wall heat flux, obviously $h_a \gg h_p$. Furthermore, conservation of total enthalpy gives $h_a = h_{a_\infty} \simeq \left(V^2_\infty/2\right)$. Using the expression in Eq. (4.60), we find a particularly simple expression

$$\dot{q}_p = \alpha_4\left(\frac{p_\infty}{R}\right)^{\frac{1}{2}} V^3_\infty \quad (4.66)$$

Values of α_4 for different atmospheres were given in Sec. 1.2.

4.6 Mass and Energy Balance at Wall

The energy conservation is obtained by a simple balance at wall (Fig. 4.7). The fact that the wall moves is involved in the expression of certain terms in the equation

$$\dot{q}_{\mathrm{cond}} = \dot{q}_{\mathrm{conv}} - \dot{m}h_g + \dot{m}h_s + \dot{m}_g h_{gp} - \dot{q}_R \quad (4.67)$$

In this expression, h_{gp} is the enthalpy of pyrolysis gas.

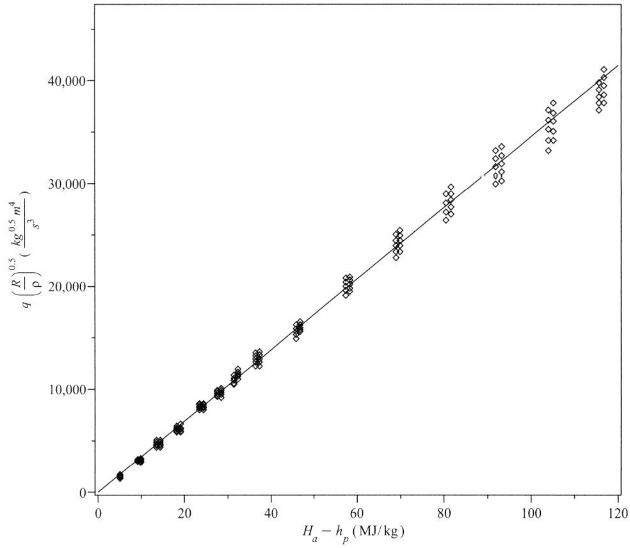

Fig. 4.6 The Sutton and Graves method in the case of air (7).

Considering the relations in Eqs. (4.22) and (4.18), it follows

$$\dot{q}_{\text{cond}} = \rho_e u_e C_H (h_a - h_p) + \dot{m}(h_s - h_p) + \dot{m}_g(h_{gp} - h_p) - \dot{q}_R \quad (4.68)$$

Introduced in the previous expression is the heat flux in the absence of ablation $\rho_e u_e C_H (h_a - h_p^*)$ (h_p^* is the enthalpy of gas without ablation)

$$\dot{q}_{\text{cond}} = \rho_e u_e C_H (h_a - h_p^*) + \rho_e u_e C_H (h_p^* - h_p) + \dot{m}(h_s - h_p) + \dot{m}_g h_{gp} - \dot{q}_R \quad (4.69)$$

This defines the apparent ablation enthalpy Δh by

$$-\dot{m}\Delta h = \rho_e u_e C_H (h_g^* - h_g) + \dot{m}(h_s - h_g) + \dot{m}_g(h_{gp} - h_p) \quad (4.70)$$

This quantity is a kind of black box that gives the overall effect of ablation. Depending on the authors, the last term either is or is not included in this amount.

We can write this equation differently; by dividing Eq. (4.68) by $\rho_e u_e C_H$, we get

$$-\mathcal{B}'\Delta h = h_g^* - (1 + \mathcal{B}' + \mathcal{B}'_g)h_p + \mathcal{B}' h_s + \mathcal{B}'_g h_{gp} \quad (4.71)$$

Equation (4.69) is then written

$$\dot{q}_{\text{cond}} = \rho_e u_e C_H (h_a - h_p^*) - \dot{q}_R - \dot{m}\Delta h \quad (4.72)$$

This equation is what would be obtained without ablation (however, C_H contains the blowing correction) with an additional term.

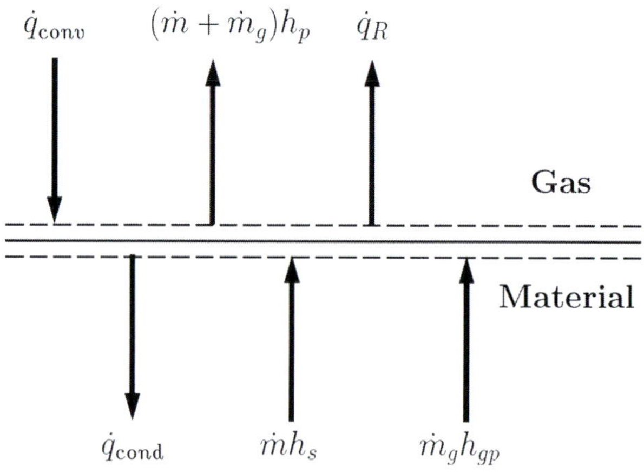

Fig. 4.7 Wall energy balance.

Also found in the literature are much more restrictive definitions of the apparent enthalpy of ablation [8], based on a linearization of the blowing term

$$\rho_e u_e C_H = \rho_e u_e C_{H_0} - \eta \dot{m} \tag{4.73}$$

With this simplification and the expression of $\dot{q}_{\text{cond}} \simeq \dot{m}[h_s(T_p) - h_s(T_0)]$ (see next section), Eq. (4.72) is written

$$Q^* = \frac{\rho_e u_e C_{H_0}(h_a - h_p^*) - \dot{q}_R}{\dot{m}} = \eta(h_a - h_p^*) + \Delta h - h_s(T_p) + h_s(T_0) \tag{4.74}$$

The first term is an apparent enthalpy of ablation [different from the value defined in Eq. (4.70)] that is proportional to $h_a - h_p^*$ if other terms are constant—which is true, for example, for a silica–resin material (see Sec. 9.2.4.3) or PTFE (Sec. 8.1.2). The regression of experimental points in Fig. 4.8 allows access to the quantity Δh after correction of the enthalpy of the solid. The quantities obtained by this method are similar to those obtained by theory (see Chapter 8).

In general, this type of method is used to interpret the tests. Other approximations can be used for a calculation, as described in Sec. 8.7.1.

4.7 Steady State Ablation

When the ablation rate is important and away from singularities in geometry, the system becomes quasi-stationary and 1-D. This gives a simple expression by an energy balance (Fig. 4.9) in a lamina between the back

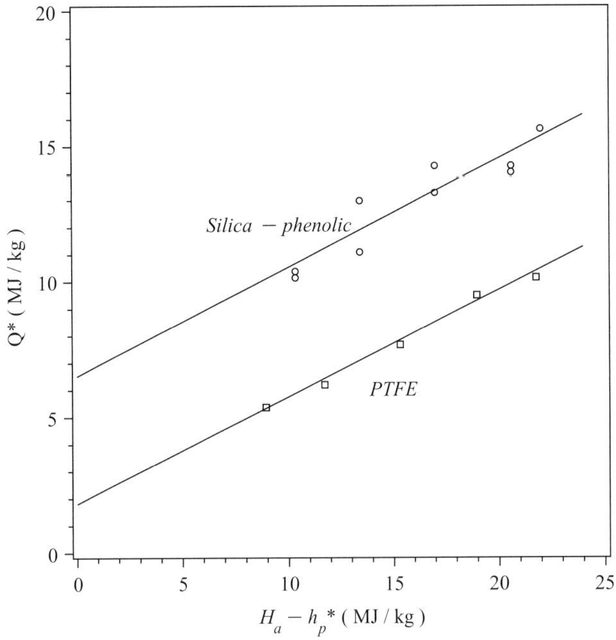

Fig. 4.8 Apparent ablation enthalpy Q^*; plasma jet measurements (8).

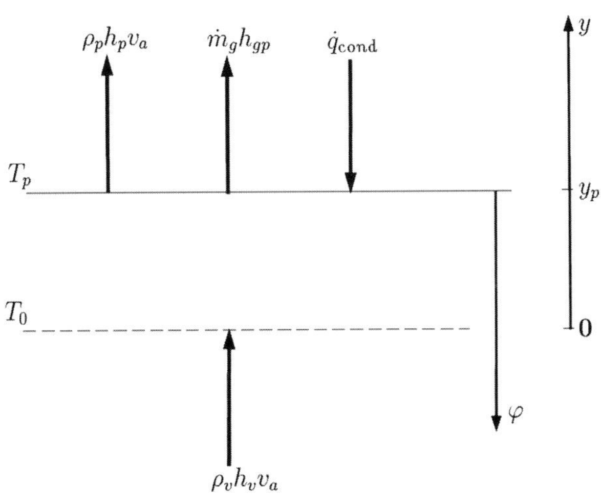

Fig. 4.9 Energy balance for steady-state ablation. In the reference frame bound to the solid, the back face moves to the top, hence the corresponding energy, low as compared to others.

cold region (temperature T_0) and surface (temperature T_p)

$$\dot{q}_{\text{cond}} = \rho_s h_s(T_p)v_a - \rho_v h_v(T_0)v_a - \dot{m}_g h_g(T_p) \tag{4.75}$$

The index v is for the virgin material.

In particular, for a nonpyrolyzable material

$$\dot{q}_{\text{cond}} = \rho_s h_s(T_p)v_a - \rho_s h_s(T_0)v_a \tag{4.76}$$

$v_a = -\frac{\mathrm{d}y}{\mathrm{d}t}$ is positive. Perform the change of variable $\varphi = y_p - y$.

$$\begin{cases} \dfrac{\partial\rho(y,t)}{\partial y} = -\dfrac{\partial\rho(\varphi,t)}{\partial y} \\ \dfrac{\partial\rho(y,t)}{\partial t} = \dfrac{\partial\rho(\varphi,t)}{\partial t} - v_a\dfrac{\partial\rho(\varphi,t)}{\partial y} \end{cases} \tag{4.77}$$

so

$$\dot{m}_g = v_a \int_{\varphi}^{\infty} \frac{\partial\rho}{\partial x}\,dx = v_a(\rho_v - \rho) \tag{4.78}$$

in particular

$$\dot{m}_g = v_a(\rho_v - \rho_p) \tag{4.79}$$

This gives a very good approximation of total ablation (Fig. 4.10) for severe reentries because:

- It is fairly well verified when the ablation velocity is high.

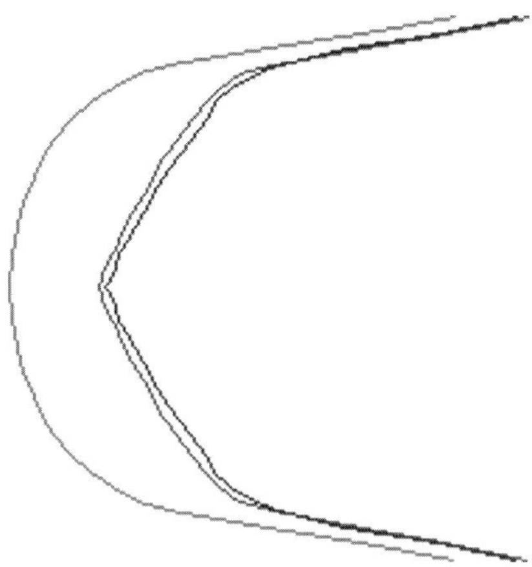

Fig. 4.10 Steady state ablation: comparison with unstationary case for a severe reentry.

- The portions of trajectory (beginning and end) during which the error on velocity ablation is very large do not induce any effect on the total ablation, the corresponding ablation rate being low.

From Eq. (4.79) we obtain a relationship between $\mathcal{B}'$ and $\mathcal{B}'_g$, quantities defined in Sec. 4.3.2

$$\mathcal{B}'_g = \left(\frac{\rho_v}{\rho_p} - 1\right)\mathcal{B}' \tag{4.80}$$

In this approximation, the energy balance [Eq. (4.67)] becomes

$$\rho_e u_e C_H (h_a - h_p) - \dot{q}_R - \frac{\rho_v}{\rho_p}\dot{m}[h_p - h_v(T_0)] = 0 \tag{4.81}$$

Note that the pyrolysis does not appear explicitly; it is only an intermediate in the system.

References

[1] Anderson, J. D. *Hypersonic and High Temperature Gas Dynamics*, McGraw-Hill, New York, 1989.

[2] Thomson, R. A., and Gnoffo, P. A., "Implementation of a Blowing Boundary Condition in the LAURA Code," AIAA Paper 2008-1243, *46th AIAA Aerospace Sciences Meeting and Exhibit*, Reno, Nevada, Jan. 2008.

[3] Lees, L., "Convective Heat Transfer with Mass Addition and Chemical Reactions," *Third AGARD Symposium on Combustion and Propulsion*, Pergamon Press, New York, 1959, pp. 451–498.

[4] Putz, K. E., and Bartlett, E. P. "Heat-Transfer and Ablation-Rate Correlations for Re-Entry Heat-Shield and Nosetip Applications," *Journal of Spacecraft and Rockets*, Vol. **10**, No. 1, 1973, pp. 15–22.

[5] Zoby, E. V., and Sullivan, E. M., "Effects of Corner Radius on Stagnation-Point Velocity Gradients on Blunt Axisymmetric Bodies," NASA Report TM X-1067, March 1965.

[6] Fay, J. A., and Riddell, F. R. "Theory of Stagnation Point Heat Transfer in Dissociated Air," *Journal of the Aerospace Sciences*, Vol. **25**, No. 2, 1958, pp. 83–86.

[7] Sutton, K., and Graves, R. A., Jr., "A General Stagnation-Point Convective-Heating Equation for Arbitrary Gas Mixtures," NASA Technical Report TR R-376, Nov. 1971.

[8] Laub, B., and Curry, D., "Tutorial on Ablative TPS," NASA Ames Conference Center, Second International Planetary Probe Workshop, Moffett Field, CA, August 2004.

Chapter 5 Ablation of Carbon

- List mechanisms of a carbon heat shield
- Describe oxidation
- Understand sublimation
- Model exchanges in the boundary layer
- Calculate partition of incident energy

In this chapter, we list the physico-chemical mechanisms involved in the case of a carbon heat shield. This includes C–C or C–R materials, and for the latter when the surface temperature is sufficient for the pyrolysis of resin to be complete. Chemical reactions will form carbon oxides (CO, CO_2), carbon nitrides (CN, C_2N, C_2N_2, C_4N_2), and linear chains of carbon (C_1, C_2, C_3, C_4, C_5) by sublimation and gas phase reactions [1]. There are many more molecules in smaller quantities that are ignored in the analysis. Similarly, the cyclic molecules of carbon that can exist in small quantities are not taken into account. This approximation would not be feasible in some phenomena such as the chemical vapor infiltration (CVI) process that play a significant role because they affect the physical nature of the carbon deposit formed, as well as its thermomechanical qualities and chemical reactivity.

5.1 Oxidation

The literature contains a compilation of data for oxidation reactions [2, 3]. You will see that one does not necessarily need to know in detail all the chemicals reactions of this type. Indeed, for the type of application that interests us and that is present at high temperatures, oxygen diffusion in the boundary layer is a factor limiting reactions [1]. The system of reactions described in Table 5.1 is sufficient (C_s is a solid C atom). The reactions of ablation (carrying away one or more carbon atoms) are assumed reversible. Such species can be created in a gaseous way, such that their partial pressure becomes greater than the saturation vapor pressure, hence leading to a chemical vapor deposition. This phenomenon is also used for manufacturing some carbon–carbon materials (using CVI). As we said earlier, the

Table 5.1 Oxidation Reactions

Reaction R_1	$2Site + O_2 \rightleftharpoons 2Site - O$
Reaction R_2	$Site - O \rightleftharpoons Site + O$
Reaction R_3	$Site - O + O \rightleftharpoons Site + O_2$
Reaction R_4	$Site - O + CO \rightleftharpoons Site + CO_2$
Reaction R_5	$Site - O\ (+C_s) \rightleftharpoons Site + CO$
Reaction R_6	$Site - O + O\ (+C_s) \rightleftharpoons Site + CO_2$

correct description of such a phenomenon requires a more detailed description of the phenomenon than the one conducted here.

All reactions are described from the concept of an active site on the wall surface. For graphite, these are the edges of the graphene plane (Fig. 5.1), a usually incompletely graphitized carbon ("turbostratic," Fig. 5.1). Note that the various faces have different reactivity. However, it does take into account only one type of site because of the lack of data. The notation *Site*–O in Table 5.1 represents an oxygen "stuck" on a carbon wall site without any notion of the kind of potential involved.

For example, for the first reaction in the table, the experiments conducted [4, 5] suggest the dissociative adsorption of O_2 on two adjacent sites where the probability they will be free is proportional to θ^2. Thermal desorption will be through a mechanism of Langmuir–Hinshelwood between two nearby occupied sites (probability proportional to θ_O^2). The rate of surface sites is about 2.1×10^{19} m^{-2}, the maximum possible for a graphitic configuration (0.142 nm × 0.335 nm) in the plane perpendicular to the graphite planes. This compares with the experimental value cited in [3], 3.5×10^{19} m^{-2}, which is adopted thereafter. Note that this value may result from the surface of the material, at the atomic scale.

The conservation of sites number (Sec. 3.2.2) for all of these reactions can be written

$$\frac{\mathrm{d}}{\mathrm{d}t}(N_s\theta) = -\frac{\mathrm{d}}{\mathrm{d}t}(N_s\theta_O) = \sum_i \tau_i \tag{5.1}$$

where τ_i is the surface terms creation

$$\begin{aligned}
\mathcal{N}\tau_1 &= k_{f_1}\theta^2 p_{O_2} &&- k_{b_1}\theta_O^2\\
\mathcal{N}\tau_2 &= k_{f_2}\theta_O &&- k_{b_2}\theta p_O\\
\mathcal{N}\tau_3 &= k_{f_3}\theta_O p_O &&- k_{b_3}\theta p_{O_2}\\
\mathcal{N}\tau_4 &= k_{f_4}p_{CO}\theta_O &&- k_{b_4}p_{CO_2}\theta\\
\mathcal{N}\tau_5 &= k_{f_5}\theta_O &&- k_{b_5}\theta p_{CO}\\
\mathcal{N}\tau_6 &= k_{f_6}\theta_O p_O &&- k_{b_6}\theta p_{CO_2}
\end{aligned} \tag{5.2}$$

At steady state, the rates of formations of sites, free or occupied, cancel. It is not necessarily the same for the mass rates of the gas species $\dot{\omega}_i$. Indeed, here one works on an open system for the ablation reactions (no conservation in gaseous volume of element C). The flux of ablation for these reactions is

$$\dot{m}_C = \mathcal{M}_C\left(\frac{\tau_5}{\mathcal{M}_{CO}} + \frac{\tau_6}{\mathcal{M}_{CO_2}}\right) \tag{5.3}$$

Some authors [6–8] propose laws of the $\dot{m}_c = f(t)\, p_{02}^{\alpha}$ type with α ranging between 0.5 and 1. It is a point of view resulting from an experiment that reflects a global but imperfect vision of the phenomena because it includes various elementary mechanisms. It is not possible to define an order of reaction (a dependence in p_{O_2}) except in some particular cases [9]. Here is an example: let us suppose that we are at low temperature ($T < 1500$ K) where O is nonexistent and where one can neglect the formation of CO_2 from heterogeneous reaction. Thus, one retains only the reactions R_1 and R_5 of Table 5.1, the R_5 reaction being supposed irreversible. The stationary solution of the problem is written

$$-k_{f_1}\theta^2 p_{O_2} + k_{b_1}(1-\theta)^2 + k_{f_5}(1-\theta^2) = 0 \tag{5.4}$$

Moreover, one can suppose that the sites are almost entirely occupied (see Fig. 5.5 later in the chapter). Let us write $\theta = 1 - \in$. While deferring

Fig. 5.1 Turbostratic carbon.

in the preceding equation and while neglecting the terms in $\in^2$, it follows

$$\dot{m} = \frac{\mathcal{M}_C}{\mathcal{M}_{CO}} k_{f5} \frac{p_{O_2}}{\frac{2p_{O_2}}{k_{f_5}} + \frac{1}{k_{f_1}}} \tag{5.5}$$

The apparent order of the reaction is thus ≤ 1 and depends on the total pressure.

5.2 Reactions with Nitrogen

The reactions of nitrogen with carbon are not well known, particularly at high temperature [11]. Indeed, their effect is weak compared with that related to the reactions of oxidation. A heterogeneous reaction produced CN and later gaseous reactions species such as C_2N, C_2N_2, C_4N_2, and so on. Taking into account the temperature, the existence even of the species N in the vicinity of the wall can occur only when the external flow is far from local thermodynamic equilibrium or for extremely severe conditions. Such conditions are met only in the case of planetary reentries of probes at very high speeds [12].

The system for the heterogeneous reactions is shown in Table 5.2 [2].

The sites being nonspecific for O or N, one will have

$$\begin{aligned} \mathcal{N}\tau_7 &= k_{f_7}\theta p_N - k_{b_7}\theta_N \\ \mathcal{N}\tau_8 &= k_{f_8}\theta_N p_N - k_{b_8}\theta p_{N_2} \\ \mathcal{N}\tau_9 &= k_{f_9}\theta - k_{b_9}\theta p_{CN} \end{aligned} \tag{5.6}$$

We note that

- The data on the reaction of nitrogen with carbon is scarce; in particular, dissociative sticking similar to that of O_2 is unknown.
- The assumption that the sites for oxygen and nitrogen are the same is based on the concept of site itself, in our case a free connection on a graphitic cycle. That supposes the relation of conservation

$$\theta + \theta_O + \theta_N = 1 \tag{5.7}$$

Oxidation and nitridation are not independent phenomena in this model.

Can one really confuse the sticking sites of nitrogen with those fixing oxygen? In the case of carbon, this is probable. Indeed, the sites in the case

Table 5.2 Reactions of Carbon Nitridation

Reaction R_7	*Site* + N $\rightleftharpoons$ *Site* − N
Reaction R_8	*Site* − N + N $\rightleftharpoons$ *Site* + N_2
Reaction R_9	*Site* + N (+C_S) $\rightleftharpoons$ *Site* + CN

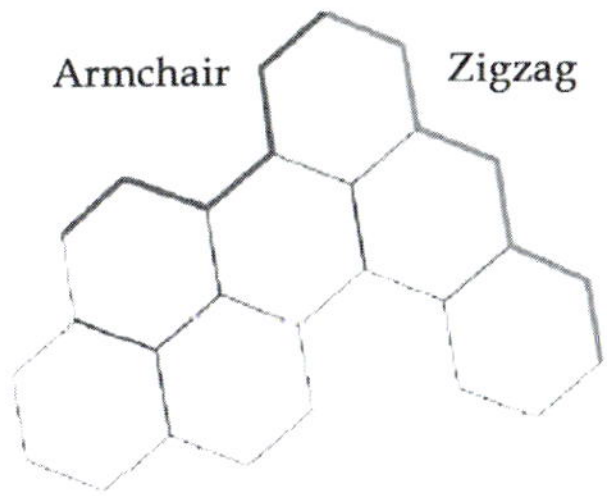

Fig. 5.2 Reactive sites on graphene.

of this material are free ends of the graphene cycles (edges, comprising sites "armchair" or "zigzag" in Fig. 5.2). The number of sites per unit of area will thus, in a disordered material (which is our case), be of the same order as the corresponding face of graphite.

For this group of reactions, the flux of ablation is

$$\dot{m}_C = \mathcal{M}_C \frac{\tau_9}{\mathcal{M}_{CN}} \tag{5.8}$$

This model is unaware of the heterogeneous formation NO from the following reaction:

$$Site - O + N \rightleftharpoons Site + NO$$

The kinetics of this reaction of the Eley–Rideal type is known for metal surfaces, but not as much in the case of carbon [12].

5.3 Sublimation

Carbon is subliming, releasing many different molecules. The most numerous are atomic low-weight linear molecules [1]. For the ranges of temperature and pressure that interest us, one retains the species C_1 to C_3 (Table 5.3), plus C_4 and C_5 produced in gas phase (Table 5.8 later in this chapter). Change of phase can be described by a Knudsen–Langmuir law [13]

$$\dot{m}_i = \alpha_i \left(\frac{\mathcal{M}_{C_i}}{2\pi R T} \right)^{\frac{1}{2}} (p_{C_{i_{eq}}} - p_{C_i}) \tag{5.9}$$

Table 5.3 Sublimation Reactions of Carbon

Reaction R_{10}	$Site\ (+C_S) \rightleftharpoons Site + C$
Reaction R_{11}	$2Site\ (+2C_S) \rightleftharpoons 2Site + C_2$
Reaction R_{12}	$3Site\ (+3C_S) \rightleftharpoons 3Site + C_3$

Table 5.4 Carbon Sublimation: Sticking Coefficients (14)

Species	α_i(2700 K) (base plane)	α_i(2450 K)	α_i(2500 K)
C	0.24	0.37	0.14–0.23
C_2	0.50	0.34	0.26–0.38
C_3	0.023	0.08	0.03–0.04

The α_i is given by various experimental sources, independent of the temperature, but different for each species. Only one author gives values for various faces of a monocrystalline graphite (Table 5.4, last column, in the order base plane, edge). Note that measurements were made at relatively low temperatures compared to the application's need, which requires temperatures up to 4500 K. Indirect measurements allow us to confirm these values at the temperatures of use (Sec. 6.5.3).

The following are the production rates for the equivalent chemical reactions (see Sec. 5.7):

$$\begin{aligned}\mathcal{N}\tau_{10} &= k_{f_{10}}\theta - k_{b_{10}}\theta p_C \\ \mathcal{N}\tau_{11} &= k_{f_{11}}\theta^2 - k_{b_{11}}\theta^2 p_{C_2} \\ \mathcal{N}\tau_{12} &= k_{f_{12}}\theta^3 - k_{b_{12}}\theta^3 p_{C_3}\end{aligned} \tag{5.10}$$

5.4 Relations of Dependence

There are relationships between homogeneous and heterogeneous reactions through constants of equilibrium. This was highlighted in Sec. 3.3.1. We apply the described method below. One arranges the species in the order O_2, O, CO_2, CO, C_s, C, C_2, C_3, N_2, N, followed by sites: *Site* − O, *Site*, *Site* − N. The stoichiometric matrix is written

$$\Omega = \begin{vmatrix} -1 & 0 & 0 & 0 & 0 & 0 & 0 & 0 & 0 & 0 & 2 & -2 & 0 \\ 0 & 1 & 0 & 0 & 0 & 0 & 0 & 0 & 0 & 0 & -1 & 1 & 0 \\ 1 & -1 & 0 & 0 & 0 & 0 & 0 & 0 & 0 & 0 & -1 & 1 & 0 \\ 0 & 0 & 1 & -1 & 0 & 0 & 0 & 0 & 0 & 0 & -1 & 1 & 0 \\ 0 & 0 & 0 & 1 & -1 & 0 & 0 & 0 & 0 & 0 & -1 & 1 & 0 \\ 0 & -1 & 1 & 0 & -1 & 0 & 0 & 0 & 0 & 0 & -1 & 1 & 0 \\ 0 & 0 & 0 & 0 & 0 & 0 & 0 & 0 & 0 & -1 & 0 & -1 & 1 \\ 0 & 0 & 0 & 0 & 0 & 0 & 0 & 0 & 1 & -1 & 0 & 1 & -1 \\ 0 & 0 & 0 & 0 & -1 & 0 & 0 & 0 & -1 & 1 & 0 & 0 & 0 \\ 0 & 0 & 0 & 0 & -1 & 1 & 0 & 0 & 0 & 0 & 0 & 0 & 0 \\ 0 & 0 & 0 & 0 & -2 & 0 & 1 & 0 & 0 & 0 & 0 & 0 & 0 \\ 0 & 0 & 0 & 0 & -3 & 0 & 0 & 1 & 0 & 0 & 0 & 0 & 0 \end{vmatrix} \tag{5.11}$$

The kernel is generated by the n_l eigenvectors

$$l_k = \left\{ \begin{vmatrix} 0 & 0 & 0 & 0 & 0 & 0 & 0 & 1 & 0 & 0 & 0 & 0 \\ 0 & 0 & 0 & 0 & 0 & 0 & 0 & 0 & 1 & 0 & 0 & 0 \\ 0 & 0 & 0 & 0 & 0 & 0 & 0 & 0 & 0 & 1 & 1 & 0 \\ 0 & 0 & 0 & 0 & 0 & 0 & 0 & 0 & 0 & 0 & 0 & 1 \\ 1 & 2 & 0 & 0 & 0 & 0 & 0 & 0 & 0 & 0 & 0 & 0 \\ 0 & -1 & 1 & 0 & 0 & 0 & 0 & 0 & 0 & 0 & 0 & 0 \\ 0 & -1 & 0 & 1 & 0 & 0 & 0 & 0 & 0 & 0 & 0 & 0 \\ 0 & -1 & 0 & 0 & 1 & 0 & 0 & 0 & 0 & 0 & 0 & 0 \\ 0 & -1 & 0 & 0 & 0 & 1 & 0 & 0 & 0 & 0 & 0 & 0 \\ 0 & 0 & 0 & 0 & 0 & 0 & 1 & 0 & 0 & 0 & 0 & 0 \end{vmatrix} \right. \tag{5.12}$$

then the matrix Ω_E

$$\Omega_E = \begin{vmatrix} 0 & 0 & 0 & 0 & 2 & 0 & -1 & 0 & 0 & 0 & 0 \\ 0 & 0 & 0 & 0 & 3 & 0 & 0 & -1 & 0 & 0 & 0 \\ 0 & 0 & 0 & 0 & 0 & 0 & 0 & 0 & -1 & 2 & 0 \\ 0 & 0 & 0 & 0 & 1 & 0 & 0 & 0 & 0 & 1 & -1 \\ 1 & -2 & 0 & 0 & 0 & 0 & 0 & 0 & 0 & 0 & 0 \\ -1 & 2 & 0 & 0 & 0 & 0 & 0 & 0 & 0 & 0 & 0 \\ 0 & 1 & -1 & 1 & 0 & 0 & 0 & 0 & 0 & 0 & 0 \\ 0 & 1 & 0 & -1 & 1 & 0 & 0 & 0 & 0 & 0 & 0 \\ 0 & 2 & -1 & 0 & 1 & 0 & 0 & 0 & 0 & 0 & 0 \\ 0 & 0 & 0 & 0 & 1 & -1 & 0 & 0 & 0 & 0 & 0 \end{vmatrix} \tag{5.13}$$

which represents the chemical pseudo-reactions of Table 5.5 (with the usual conventions for the writing of stoichiometric matrices).

The system of reactions written is not minimal

- Reactions F_5 and F_6 are identical.
- Reactions F_7, F_8, and F_9 are linearly dependent.

Table 5.5 Relations of Compatibility of the System

Reaction F_1	$2C_s \rightleftharpoons C_2$
Reaction F_2	$3C_s \rightleftharpoons C_3$
Reaction F_3	$2N \rightleftharpoons N_2$
Reaction F_4	$C_s + N \rightleftharpoons CN$
Reaction F_5	$O_2 \rightleftharpoons 2O$
Reaction F_6	$2O \rightleftharpoons O_2$
Reaction F_7	$CO + O \rightleftharpoons CO_2$
Reaction F_8	$C_s + O \rightleftharpoons CO$
Reaction F_9	$C_s + 2O \rightleftharpoons CO_2$
Reaction F_{10}	$C_s \rightleftharpoons C$

Table 5.6 Closure Relationship of Equilibrium Constants; by Definition $K_{P_{O_2}} = K_{P_{N_2}} = K_{P_{C_s}} = 1$

Reaction	Closure Relation
F_1	$K_{p_{11}} = K_{p_{C_2}} K_{p_{C_s}}^{-2}$
F_2	$K_{p_{12}} = K_{p_{C_3}} K_{p_{C_s}}^{-3}$
F_3	$K_{p_7} K_{p_8} = K_{p_{N_2}} K_{p_N}^{-2}$
F_4	$K_{p_9} = K_{p_{CN}} K_{p_{C_s}}^{-1} K_{p_N}^{-1}$
F_5	$K_{p_1} K_{p_2}^2 = K_{p_O}^2 K_{p_{O_2}}^{-1}$
F_6	$K_{p_2}^{-1} K_{p_3} = K_{p_{O_2}} K_{p_O}^{-2}$
F_7	$K_{p_2}^{-1} K_{p_4} = K_{p_{CO_2}} K_{p_{CO}}^{-1} K_{p_O}^{-1}$
F_8	$K_{p_2}^{-1} K_{p_5} = K_{p_{CO}} K_{p_{C_s}}^{-1} K_{p_O}^{-1}$
F_9	$K_{p_2}^{-1} K_{p_6} = K_{p_{CO_2}} K_{p_{C_s}}^{-1} K_{p_O}^{-2}$
F_{10}	$K_{p_{10}} = K_{p_C} K_{p_{C_s}}^{-1}$

This justifies the rank of the matrix $rank(\Omega_E) = 8$.

Calculations of linear algebra such as that which has just been done require algorithms using integer numbers. For a Fortran code, one can use the library Netlib [15]. The application for our system is given in Table 5.6.

5.5 Reaction Kinetics

The kinetics of reaction (or pseudo-reaction in the case of sublimation) are given in Table 5.7. The reactions with oxygen and nitrogen result from [2, 3] and the relations of compatibility in Table 5.6 by using the constants of balance of [16], approximated by the method described in Appendix B.

5.6 Homogeneous Reactions

The gas phase will grow rich by carbonaceous species, which can provide many other reactions. In the case of Earth's atmosphere, the reactions of Table 3.1 will have to be supplemented by those utilizing carbonaceous species, such as those of Table 5.8. These reactions reveal new species: CN, C_2N, C_2N_2, C_4N_2, C_4, C_5. The constants of these reactions are given in [17–19].

5.7 Example: Homogeneous Medium

Exchanges can be roughly modeled in the boundary layer as follows:

$$\frac{\partial(\rho c_i)}{\partial t} + \frac{\partial J_i}{\partial y} = \dot{\omega} \tag{5.14}$$

Table 5.7 Reactional Constants

Reaction	k_f	k_b
R_1	$0.2545 \times 10^1\ T^{-1.3488} \exp\left(-\frac{800}{T}\right)$	$0.5834 \times 10^{27}\ T \exp\left(-\frac{30,800}{T}\right)$
R_2	$0.7294 \times 10^{30}\ T \exp\left(-\frac{45,000}{T}\right)$	$0.6587 \times 10^{24}\ T^{-0.5}$
R_3	$0.1601 \times 10^{21}\ T^{0.3488} \exp\left(\frac{800}{T}\right)$	$0.4658 \times 10^{24}\ T^{-0.5} \exp\left(-\frac{14,200}{T}\right)$
R_4	$0.2420 \times 10^{17}\ T^{1.0949} \exp\left(\frac{19,330}{T}\right)$	$0.3575 \times 10^{24}\ T^{-0.5}$
R_5	$0.7294 \times 10^{29}\ T \exp\left(-\frac{40,000}{T}\right)$	$0.8535 \times 10^{18}\ T^{0.5735} \exp\left(-\frac{37,850}{T}\right)$
R_6	$0.5270 \times 10^{24}\ T^{-0.5} \exp\left(-\frac{2000}{T}\right)$	$0.1008 \times 10^{26}\ T^{-1.0214} \exp\left(-\frac{64,180}{T}\right)$
R_7	$0.7042 \times 10^{24}\ T^{-0.5}$	$0.7294 \times 10^{30}\ T \exp\left(-\frac{36,600}{T}\right)$
R_8	$0.1245 \times 10^{22}\ T^{0.1078} \exp\left(\frac{400}{T}\right)$	$0.4979 \times 10^{24}\ T^{-0.5} \exp\left(-\frac{76,600}{T}\right)$
R_9	$0.7117 \times 10^{29}\ T^{-0.5}$	$0.7277 \times 10^{30}\ T^{-1} \exp\left(\frac{36,600}{T}\right)$
R_{10}	$0.3230 \times 10^{27}\ T^{-0.6846} \exp\left(-\frac{37,700}{T}\right)$	$0.1825 \times 10^{24}\ T^{-0.5}$
R_{11}	$0.3395 \times 10^{35}\ T^{-0.9835} \exp\left(-\frac{44,300}{T}\right)$	$0.2689 \times 10^{24}\ T^{-0.5}$
R_{12}	$0.6150 \times 10^{36}\ T^{-1.6563} \exp\left(-\frac{43,610}{T}\right)$	$0.1010 \times 10^{23}\ T^{-0.5}$

where J_i is the vertical mass flux. One defines a volume of homogeneous control (isothermal, isobar) whose wall constitutes the lower boundary. This volume moves with the speed of ablation. With the higher boundary (height δ fixed), one applies a relation giving the gas flux by means of a given coefficient of exchange $\rho_e u_e C_M$ (Sec. 4.4.1). δ can be seen as the size of a mesh close to the wall in a numerical calculation. In the limit case of high speed reactions and small δ, the result tends towards thermodynamic

Table 5.8 Major Gaseous Reactions of Carbonaceous Species

$2CO_2 \rightleftharpoons 2CO + O_2$
$2CO_2 + C_2 \rightleftharpoons 4CO$
$C_2N_2 \rightleftharpoons 2CN$
$C_2N_2 + C_2 \rightleftharpoons C_4N_2$
$2C_2N \rightleftharpoons 2C_2 + N_2$
$3C_4 \rightleftharpoons 4C_3$
$5C_3 \rightleftharpoons 3C_5$
$CO + C \rightleftharpoons C_2 + O$
$CO_2 \rightleftharpoons CO + O$
$C_3 \rightleftharpoons C_2 + C$
$C_2 \rightleftharpoons 2C$
$CN \rightleftharpoons C + N$

equilibrium, and one obtains an equivalent to the approximate methods commonly used in the field, developed in Sec. 8.7.1.

By discretization of Eq. (5.14) and with the simplifying assumptions $\mathcal{L} = \mathcal{P} = 1$, it follows

$$\frac{\mathrm{d}(\rho c_i)}{\mathrm{d}t} + \frac{J_{i_\delta} - J_{i_p}}{\delta} = \dot{\omega}_i \tag{5.15}$$

with

$$J_{i_p} = \dot{m}_i \tag{5.16}$$

$$J_{i_\delta} = -\rho_e u_e C_M (c_{i_e} - c_i) + \dot{m} c_i \tag{5.17}$$

The second term in this last equation is flux related to the gas injection $\rho v c_i = \dot{m} c_i$.

The quantities c_{i_e} are the values outside the boundary layer and are data, such as $\rho_e u_e C_M$, which represents a measure of the flux of mass, corrected for gas injection [Eq. (4.53)].

The basic system led to the results of Figs. 5.3 and 5.4. One finds the traditional ingredients there [1], namely:

- A first augmentation corresponding to oxidation with formation of CO_2.
- A plateau corresponding to the limitation due to the diffusion, which can be calculated simply by assuming that the medium close to the wall consists only of N_2 and CO_2, oxygen being completely consumed. The system in Eq. (4.12) can be written $\tilde{c}_{O_p}/\tilde{c}_{C_p} = 2\mathcal{M}_O/\mathcal{M}_C$. In addition,

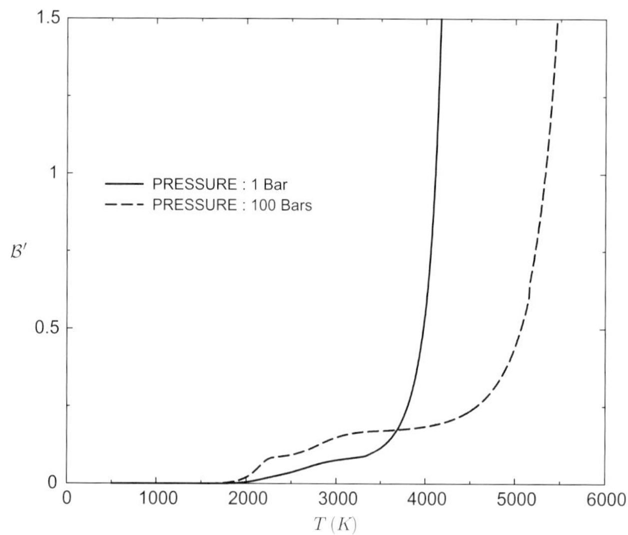

Fig. 5.3 $\mathcal{B}' = f(T, p)$ for carbon.

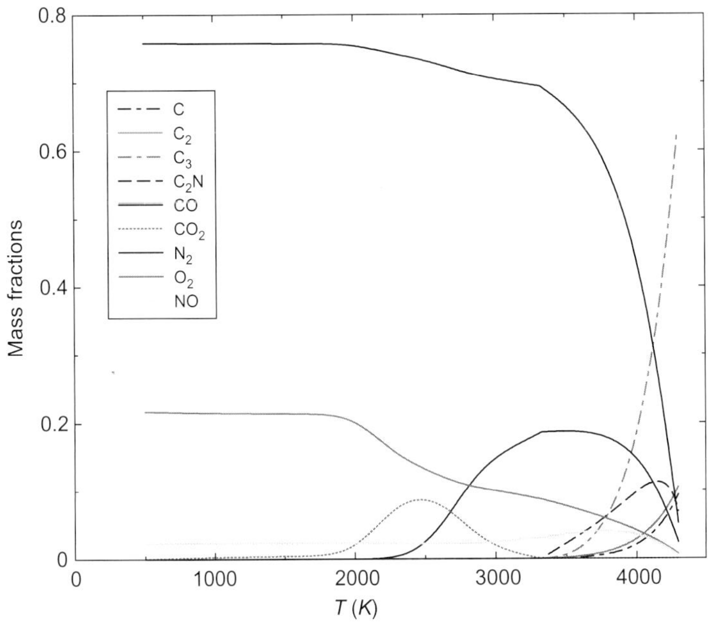

Fig. 5.4 Mass fractions; reduction in N_2 with T is due to numerous other molecules in the medium.

Eq. (4.27) gives $\tilde{c}_{O_p}/\tilde{c}_{C_p} = \tilde{c}_{O_e}/\mathcal{B}'$, then

$$\mathcal{B}' = \frac{\mathcal{M}_C}{2\mathcal{M}_O}\,\tilde{c}_{O_e} \simeq 0.071 \tag{5.18}$$

This numerical value corresponds to Earth's atmosphere.

- A new augmentation due to the formation of CO.
- Again a plateau due to the diffusion, twice as high as the first: the same quantity of oxygen carries twice as much carbon.
- Finally, an exponential augmentation with nitridation and sublimation. Major species are C at low pressure and C_3 at high pressure (>0.1 bar).

Note that, at low temperature, the sites are saturated by oxygen. They are depopulated starting from 1300–1600 K, according to the pressure (Fig. 5.5). This phenomenon is related to thermal agitation. The order of magnitude of the temperature to which sites are free is in conformity with the semi-quantum theory [19].

The nitrogen is not fixed. The molecule is practically not dissociated with temperatures that interest us, so one can deduce from it that the reactions R_7 and R_8 are negligible and that catalytic formation of N_2 is not verified.

The chemical species present in the medium are presented in Fig. 5.4. The variation of sticking coefficients does not change the appearance of the part of the curves corresponding to sublimation, but moves them from 100 to 200 K. This variation, in spite of appearances, is weak. It represents less than 5%, which brings a weak variation of the velocity of ablation (<5%) under given conditions of aerothermal data, through balance of energy. From the experimental point of view, such a variation is difficult to measure, especially because it is not possible to gauge a measuring apparatus for such temperatures. This point is discussed in Sec. 5.10.

The introduction of the heterogeneous reactions with nitrogen does not modify the results notably: the occupation of the sites is negligible. One can conclude from this that the heterogeneous nitrogen–carbon reactions are negligible compared to oxygen–carbon.

A variation of N_S barely modifies the curves. This is understandable:

- Reactions with oxygen are almost always dominated by diffusion.
- As we have just seen, the reactions with nitrogen are negligible.
- For sublimation, the notion of site is an artifact of modeling with no effect on the result.

Note that in all cases, the problem of sublimation of an occupied site does not arise. Indeed, the thermal agitation releases the occupied sites well before the start of sublimation.

Also, note that these calculations were not performed in the hypothesis of a thermodynamic equilibrium in the gas. The results are therefore dependent on δ and $\rho_e u_e C_M$. The influence of this hypothesis is analyzed in [20]. This

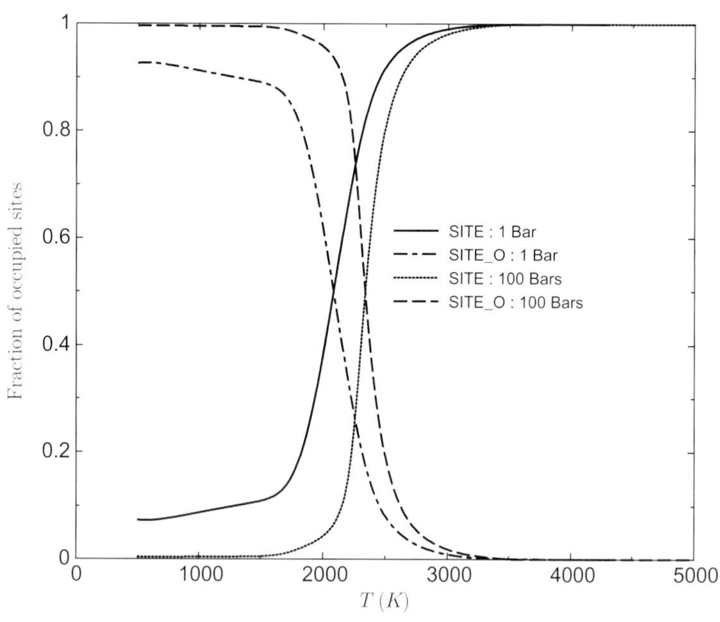

Fig. 5.5 Occupancy rate of sites.

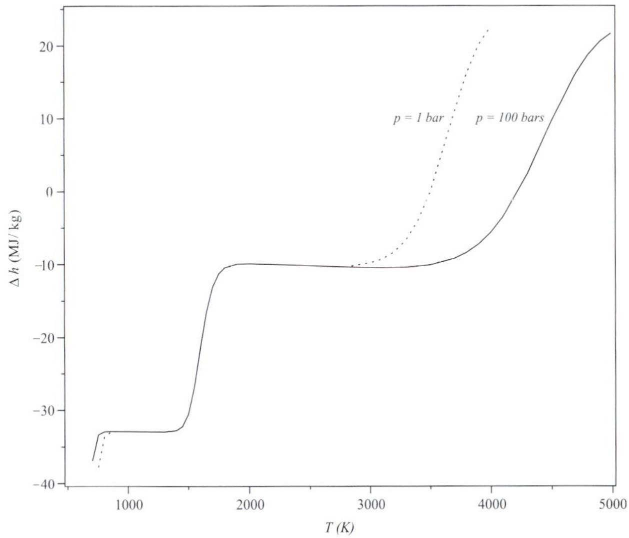

Fig. 5.6 Apparent ablation enthalpy of carbon.

reference shows that this assumption leads to an overvalue of $\mathcal{B}'$ at low temperature, the characteristic times of formation of CO and CO_2 being large compared to those of convection and diffusion.

The exponential curve of mass flux at high temperature can explain the fact that for a given material, the ablation velocity is roughly proportional to the incident flux. Indeed, in the total energy balance, only the quantity $\dot{m}$ varies exponentially, and thus more rapidly than the other quantities, which vary as T^n, approximately. A comparison between calculation and measurements is presented in Sec. 12.2.5.

The energy balance of the ablation process can be understood through the apparent enthalpy of ablation, as defined in Sec. 4.6, Eq. (4.70). This value takes into account the modification of exchanges in the boundary layer due to modifications of chemical composition near the wall. Figure 5.6 shows that:

- At low temperature, the balance is largely exothermic.
- With the regime of sublimation, which in fact involves reactions with N_2 and O_2 (diffusion regime), the balance is very favorable.

5.8 Partition of Energy

It is possible to perform a calculation of orders of magnitude to see what the partition is of the incident energy among conduction, radiation, and chemical reactions, and the influence of gas injection. Take, for example, a wall carbon heated by convection. We study a 1-D model, which assumes the steady state is reached. The energy balance of this system in a reference frame connected to the wall is written [see Eq. (4.72)]

$$\dot{q}_{\text{cond}} = \rho_e u_e C_H (h_a - h_g^*) - \dot{q}_R + \dot{m}\Delta h \tag{5.19}$$

The incident flux is divided among conduction, radiation, and solid–gas phase change performed with a mass flow $\dot{m}$ for the carbon removed. The h_a is the adiabatic enthalpy, Δh the apparent ablation enthalpy defined by Eq. (4.70), and h_g^* the enthalpy of gas at the wall temperature without injection [see Eq. (4.6)]. The Δh not only is the variation of enthalpy due to physico-chemical reactions, but also contains the influence on the convective term by changing the species composition at wall.

Note that this problem has a stationary solution, which is not the case in the absence of ablation. For a semi-infinite monodimensional medium, the flux of conduction can be evaluated simply by noting that it serves to bring the material from T0 = 300 K at "infinity" to T to the wall (the temperature profile is constant in a moving frame bound to the wall)

$$\dot{q}_{\text{cond}} = \dot{m}[h_s(T) - h_s(T_0)] \tag{5.20}$$

The assumption of a thin layer (Sec. 4.4.3) connects the exchange coefficient $\rho_e u_e C_H$ to its value in the absence of injection $\rho_e u_e C_{H_0}$, dependent on external conditions and geometry of the body.

Take, for example, a stagnation point and, for a given atmosphere (we chose to vary the conditions upstream so that the stagnation pressure is constant and equal to p_0), vary V_{inf}. Resolution of system $f(T) = 0$ defined by Eq. (5.15) allows us to obtain a set of values for stationary ablation. For each of them we calculate the fraction of each component of heat flux compared to convective flux calculated in the absence of injection and for the same wall temperature $\dot{q}_0 = \rho_e u_e C_{H_0}(h_a - h_g^*)$. Figure 5.7 shows that when an increase in V_{inf} leads to an increase in T:

- Injection reduces the convective part by about 10%.
- Radiation increases more slowly than other quantities, and its part in energy drained from the system decreases, but it remains important.
- The term of ablation is initially negative (exothermic) and then increases significantly.
- The conduction term is small and substantially constant. Note that the ablation, through chemical reactions, injection, and radiation, has reduced an order of magnitude the flux of conduction compared to the value of the convective flux on an inert wall at current temperature.

Note that these results are independent of the density of carbon.

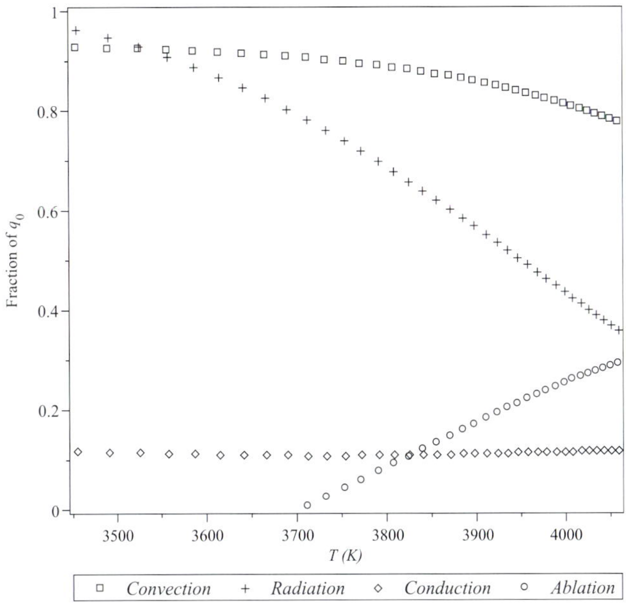

Fig. 5.7 Partition of energy on carbon ablative wall.

5.9 Relation Between Incident Flux and Ablation

Using the previous example of a stationary 1-D ablation, and assuming that the ablation rate is low enough to use a linearization of Eq. (4.53) describing the effect of gas injection on the heat flow (we assume $\mathcal{L} = 1$),

$$\frac{\mathcal{C}_H}{\mathcal{C}_{H_0}} = 1 - \eta\mathcal{B}' \tag{5.21}$$

The balance equation of the preceding paragraph and Eq. (5.21) can express the mass flow versus the flux in the absence of injection $\dot{q}_0 = \rho_e u_e \mathcal{C}_{H_0}(h_a - h_g^*)$

$$\dot{m}[\Delta h - h_s(T) + h_s(T_0) - \rho_e u_e \eta(h_a - h_g^*)] = \dot{q}_0 \tag{5.22}$$

In the oversimplified case of a phase change at constant temperature, and therefore ablation apparent enthalpy also being constant, Eq. (5.22) shows that the rate of ablation is directly proportional to the flux in the absence of ablation.

In the general case (Fig. 5.8), the linear relationship is a good approximation. In the figure, the lag on curves is about 200 K.

5.10 Precision of the Ablation Model

The ablation tests were widely used for the comparison of materials, but there are very few experiences specific and sufficient for use in model updating [21]. We take an experiment analyzed in [22]. These plasma jet tests were made at low pressure on ATJ graphite specimens. The pressure values (from 0.9×10^5 to 1.2×10^5 pa) ensure laminar flow, thus excluding all the difficulties of turbulence and roughness. Figure 5.9 compares the results of these tests to a calculation made with the model described at the beginning of this chapter.

Some conclusions can be drawn from the comparison:

- At low temperature ($T < 3400$ K), ablation is governed by the diffusion of species in the boundary layer. The small differences observed are not attributable to the ablation model.
- At higher temperatures, the difference in temperature for a given $\dot{m}$ is in the order of 200 K. This figure represents 5% in relative terms; it is possible that this is an experimental error. When the comparative measurement with a black body is possible (see, for example, [23] for the dispersion on the measurement of emissivity of a carbon resin or [24] for differences between various carbons), the dispersion is around 2%. Ignorance of the measured value can be higher than that value in our case. It is therefore very difficult to conclude on differences.

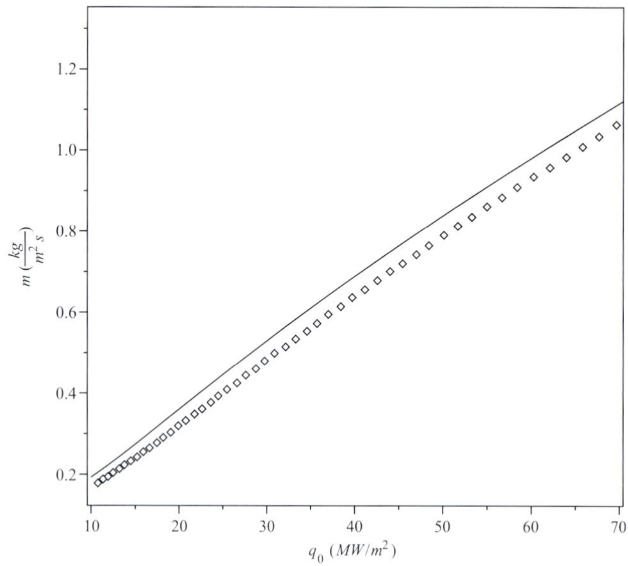

Fig. 5.8 Relation between heat flux and ablation mass rate calculated with and without the hypothesis of Sec. 5.8.

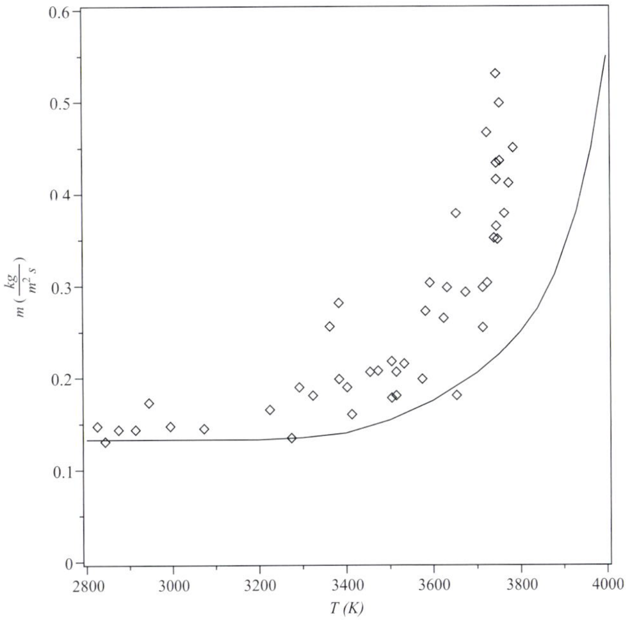

Fig. 5.9 Comparison calculation experiments on graphite.

- From a theoretical perspective, the largest error is certainly on the enthalpies of formation of species C_n, which are known with an uncertainty of about 20%. In our case, this results in temperature error of about 100 K [25].

Let us see what this difference of 200 K induces in ablation. For this, we use the type of calculation described in Sec. 5.8 by shifting the curves $\dot{m} = f(T)$ and $\Delta h = g(T)$ of 200 K to high temperatures. Comparing the result with the nominal case (Fig. 5.8) shows a difference in the mass ablation rate under 5%. This value is lower than the experimental precision, and it is therefore impossible to draw any conclusion regarding the quality of the model. In fact, it can be deduced from this that the error in the data induces a small error in the results.

5.11 Example of Calculation: A Test with Constant Upstream Conditions

The macroscopic simulation is the step by which the engineer captures the key issues of its design. Take the example of the stagnation point of a reentry vehicle. The initial radius of the sphere is 1 in. (1 in. = 25.4 mm), and the angle of the cone is 8 deg. The material is a carbon of density $\rho = 2000\ \mathrm{kg \cdot m^{-3}}$ and high conductivity $\lambda = 30\ \mathrm{W \cdot m^{-1} \cdot K^{-1}}$. This

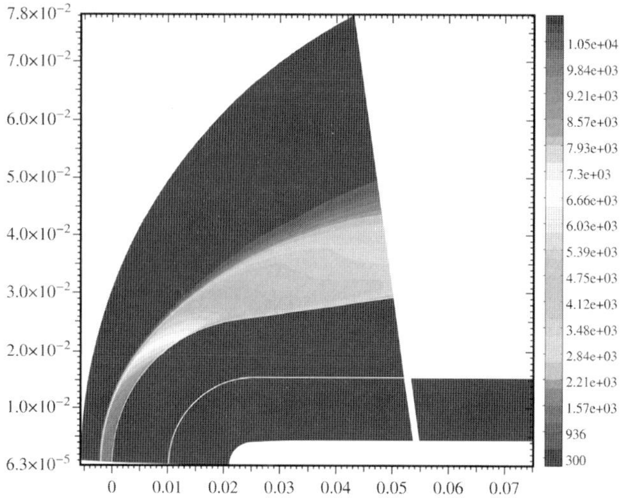

Fig. 5.10 Initial state before ablation (flow at $t = 2$ ms).

specimen is subjected to a pseudo–hot tunnel test in the following conditions: $M_\infty = 15$, $\rho_\infty = 0.186 \text{ kg} \cdot \text{m}^{-3}$, $p_\infty = 16{,}100$ pa. This corresponds, in fact, to reentry conditions, no plasma jet being capable of such performance. The calculation is done on the problem using a four-blocks approach, one of them including the fluid (Fig. 5.10). The code used details the external flow, the heat equation in the solid, and chemical reactions at

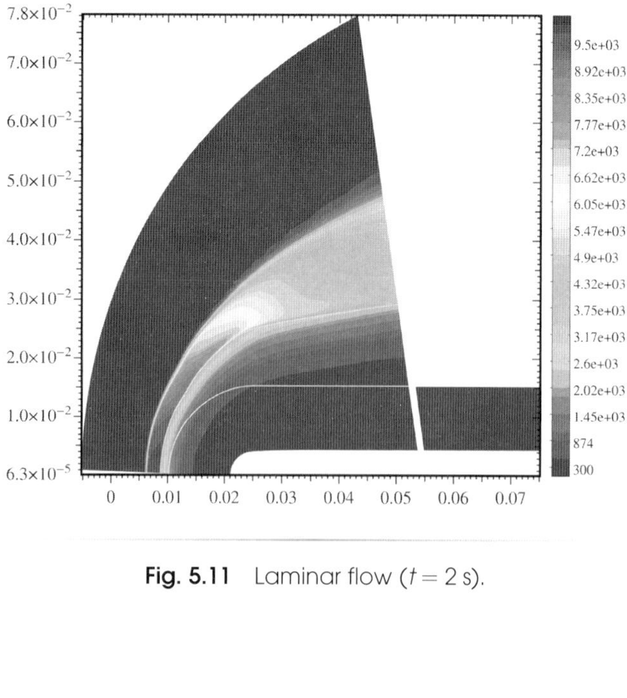

Fig. 5.11 Laminar flow ($t = 2$ s).

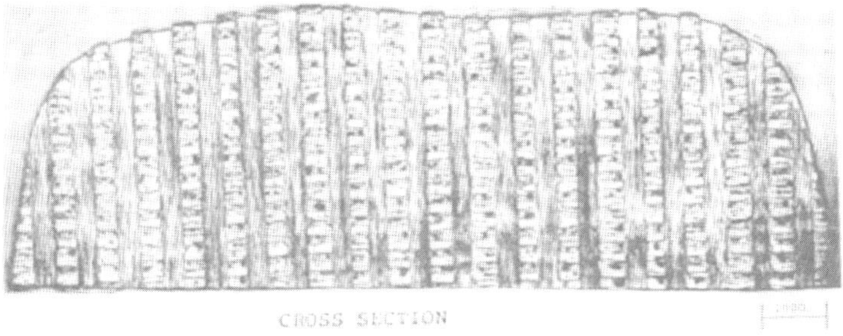

Fig. 5.12 C–C sample after laminar ablation (29).

the wall. It is used for solving these phenomena with numerical methods using a strong coupling [26, 27].

The result obtained in laminar flow (Fig. 5.11) shows the profile obtained after 2 s. This is characteristic of this flow, with its blunt shape associated with a flux distribution decreasing since the stagnation point. This form is characteristic of the resulting shape when arriving on turbulent flow in a ballistic reentry. The stationary profile, flat with a rounded edge (Fig. 5.12), is not yet obtained at 2 s. It should be noted that the cold wall heat flux decreases during the test. Indeed, this value is about three times lower in the stationary form than for the initial profile [28].

The turbulent flow (Fig. 5.13) yields a characteristic conical profile (angle of about 50 deg relative to the axis of symmetry), with a very low blunting at

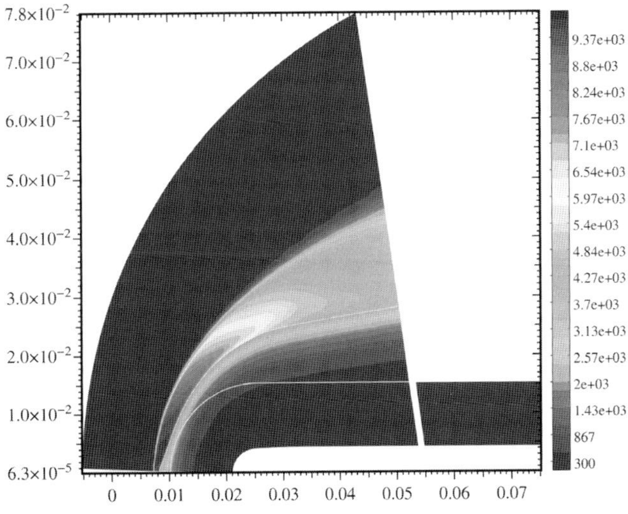

Fig. 5.13 Turbulent flow ($t = 2$ s).

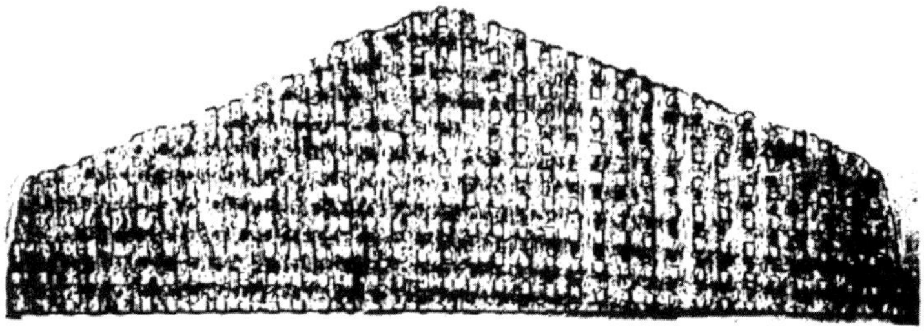

Fig. 5.14 C–C sample after turbulent ablation. (130)

the stagnation point. This profile is stationary, comparable to those obtained by plasma jet testing (Fig. 5.14).

These two characteristic profiles have been used in reference experiments (wind tunnel tests PANT, Sec. 7.1).

Section 12.2.5 compares this model with ablation experiments.

References

[1] Duffa, G., "Ablation," CESTA Monograph, Commissariat a l'Energie Atomique, Le Barp, France, Nov. 1996.

[2] Havstad, M. A., and Ferencz, R. M. "Comparison of Surface Chemical Kinetic Models for Ablative Reentry of Graphite," *Journal of Thermophysics and Heat Transfer*, Vol. 16, No. 4, 2002, pp. 508–515.

[3] Zhluktov, S. V., and Abe, T., "Viscous Shock-Layer Simulation of Airflow Past Ablating Blunt Body with Carbon Surface," *Journal of Thermophysics and Heat Transfer*, Vol. 13, No. 1, 1999, pp. 50–59.

[4] Walker, P. L., Jr., Vastola, F. J., and Hart, P. J., "Oxygen-18 Tracer Studies on the Carbon-Oxygen Reaction," *Fundamentals of Gas-Surface Interactions*, edited by Saltsburg, H., Smith, J. N., and Rogers, M., Academic Press, New York, 1967, pp. 307–317.

[5] Nagle, J., and Strickland-Constable, R. F., "Oxydation of Carbon Between 1000–2000°C," *Proceedings of the Fifth Conference on Carbon*, Pergamon Press, Oxford, UK, 1962, pp. 154–164.

[6] Scala, S. M., and Gilbert, L. M., "Sublimation of Graphite at Hypersonic Speeds," *AIAA Journal*, Vol. 3, No. 9, 1965, pp. 1635–1644.

[7] Park, C., "Effects of Atomic Oxygen on Graphite Ablation," *AIAA Journal*, Vol. 14, No. 11, 1976, pp. 1640–1642.

[8] Metzger, J. W., Engels, M. J., and Diaconis, N. J., "Oxydation and Sublimation of Graphite in Simulated Re-Entry Environments," *AIAA Journal*, Vol. 5, No. 3, 1967, pp. 451–460.

[9] Effron, E., and Hoelscher, H. E., "Graphite Oxydation at Low Temperature," *American Institute of Chemical Engineers Journal*, Vol. 10, No. 3, 1964, pp. 388–392.

[10] Suzuki, T., Fujita, K., and Sakai, T., "Graphite Nitridation in Lower Surface Temperature Regime," *Journal of Thermophysics and Heat Transfer*, Vol. 24, No. 1, 2010, pp. 212–215.

[11] Chen, Y.-K., and Milos, F. S., "Finite-Rate Ablation Boundary Conditions for a Carbon-Phenolic Heatshield," AIAA Paper 2004–2270, *37th AIAA Thermophysics Conference*, Portland, Oregon, June 2004.

[12] Kurotaki, T., "Construction of Catalytic Model on SiO_2-Based Surface and Application to Real Trajectory," AIAA Paper 2000–2366, *34th AIAA Thermophysics Conference*, Denver, Colorado, June 2000.
[13] Schrage, R. W., *A Theoretical Study of Interphase Mass Transfer*, Columbia University Press, New York, 1953.
[14] Palmer, H. B., and Shelef, M., "Vaporization of Carbon," *Chemistry and Physics of Carbon*, Vol. 4, Marcel Dekker, New York, 1968, pp. 85–135.
[15] Springer, J., "Exact Solution of General Systems of Linear Equations," *ACM Transactions on Mathematical Software*, Vol. 12, No. 2, 1986, p. 149.
[16] Chase, M. W., Jr., Davies, C. A., Downey, J. R., Jr., Frurip, D. J., MacDonald, R. A., and Syverud, A. N., "JANAF Thermochemical Tables, Third Edition," *Journal of Physical and Chemical Reference Data*, Vol. 14, Suppl. 1 & 2, 1985.
[17] Kee, R. J., Dixon-Lewis, G., Warnatz, J., Coltrin, M. E., Miller, J. A., and Moffat, H. K., "CHEMKIN III: A Fortran Computer Code Package for the Evaluation of Gas-Phase, Multicomponent Transport Properties," Sandia National Laboratories Report SAND86–8246B, March 1998.
[18] Park, C., "Review of Chemical Kinetics Problems of Future NASA Missions. I: Earth Entries," *Journal of Thermophysics and Heat Transfer*, Vol. 7, No. 3, 1993, pp. 385–398.
[19] Park, C., *Nonequilibrium Hypersonic Aerothermodynamics*, John Wiley & Sons, New York, 1990.
[20] Milos, F. S., and Rasky, D. J., "Review of Numerical Procedures for Computational Surface Thermochemistry," *Journal of Thermophysics and Heat Transfer*, Vol. 8, No. 1, 1994, pp. 24–34.
[21] Chen, Y.-K., Milos, F. S., Reda, D. C., and Stewart, D. A., "Graphite Ablation and Thermal Response Simulation Under Arc-Jet Flow Conditions," AIAA Paper 2003–4042, *36th AIAA Thermophysics Conference*, Orlando, Florida, June 2003.
[22] Lundell, J. H., and Dickey, R. R. "Ablation of ATJ Graphite at High Temperatures," *AIAA Journal*, Vol. 11, No. 2, 1973, pp. 216–222.
[23] Engelke, W. T., Pyron, C. M., Jr., and Pears, C. D., "Thermal and Mechanical Properties of a Nondegraded and Thermally Degraded Phenolic Carbon Composite," NASA Report CR-896, Oct. 1967.
[24] Wilson, R. G., "Hemispherical Spectral Emittance of Ablation Chars, Carbon, and Zirconia to 3700 K." NASA Report TN D-2704, March 1965.
[25] Hurwicz, H., Kratsch, K. M., and Rogan, J. E. "Ablation," AGARDograph No. 161, edited by Wilson, R. E., March 1972.
[26] Dubroca, B., Duffa, G., and Leroy, B., "High Temperature Mass and Heat Transfer Fluid-Solid Coupling," AIAA Paper 2002–5180, *11th Meeting of AIAA/AAAF, "Space Planes and Hypersonic Systems and Technologies," Orléans*, France, Sept. 2002.
[27] Nguyen-Bui, N. T.-H., Dubroca, B., Duffa, G., and Leroy, B., "New Methods for Simulation of Ablative Thermal Protections," *Fifth European Workshop on Thermal Protection Systems and Hot Structures, Noordwijk*, ESA, The Netherlands, May 2006.
[28] Zoby, E. V., and Sullivan, E. M., "Effects of Corner Radius on Stagnation-Point Velocity Gradients on Blunt Axisymmetric Bodies," NASA Report TM X-1067, March 1965.
[29] Eitman, D. A., and DeMichaels, J. E., "Material Characterization of Ground Test Models," Science Applications Incorporated Technical Report TR-80-40, Performance Technology Program (PTP-S II), Vol. X, Jan. 1980.

Chapter 6 Roughness Formation

- Describe responsiveness of composite material
- Know how roughness was created
- Calculate reaction–diffusion

6.1 General Considerations

In some cases roughness has substantial effects on the exchange of mass and heat transfer in the boundary layer. This is discussed in Sec. 7.1. It is therefore important to understand the mechanisms of roughness set-up.

In the case of composite materials, the phenomenon is clearly associated with density contrasts and chemical responsiveness between components. The geometries encountered are of ogival type (Fig. 6.1), the radii of curvature being of the same order of magnitude whatever the size of the unit cell describing the material. The importance of this observation will be seen later in this chapter.

In the case of polycrystalline materials (e.g., carbon), after testing, the scale of grain found in laminar flow was a few micrometers (Fig. 6.2). In contrast, in turbulent flow, the surface geometries are gouge-shaped (scallops). This phenomenon is found on some meteorites; the shapes are also called regmaglypts or thumbprints. We note in passing that some meteors also contain 2-D shapes perpendicular to flow (flutes), a phenomenon reproduced on the ablation materials at low temperatures [2]. The creation of surface roughness is, indeed, a very general phenomenon on surfaces' ablation in many areas:

- Scallops in subglacial caves [3].
- Scallops in snow fields [4]; in this area there are also needle shapes (penitents), similar to the geometry encountered in composite materials.
- Flutes or scallops at solid–liquid interfaces: the lower portions of frozen rivers, pipelines in the chemical industry, or in the cooling circuits of nuclear power plants. This type of phenomenon has been studied in the laboratory by dissolving $CaSO_4$ [5].

The scale of a hundred micrometers (in our case) cannot be connected to a scale of the material or of flow (except for a scale of turbulence, which has no significance, the turbulence spectrum being continuous).

There is no complete work on this problem. We will examine various contributions, obtained most often with very restrictive conditions. All of this work and experimental evidence, however, constitute a pretty clear picture of the phenomena and allow us to reach conclusions with practical utility.

6.2 Scales of the Problem

We find two scales on the surface, denoted l_R, characteristic of the constitution of material (Sec. 1.3.1):

- The scale of fiber, typically a tenth of a micrometer, of the included matrix or fiber–matrix interphase, typically in micrometers; these scales will be classified as micrometric.
- The scale of yarn or fabric, typically just below 1 mm; this will be classified as mesoscopic.

So there is more than one order of magnitude between these scales. Moreover, the characteristic times associated with these two scales are in the ratio of the scales themselves [7]. It is therefore possible to treat each one separately and to introduce the homogenized quantities of the microscopic model in the mesoscopic model (each of these models to be defined) to define a macroscopic model. For a fluid, the situation is much more complex. The examination of scales, absolute or relative, can clarify the situation:

- The Kolmogorov scale measures the size of the smallest vortex that may exist. A smaller roughness than this scale cannot be a source of turbulence. This characteristic length is comparable to the experimental characteristic

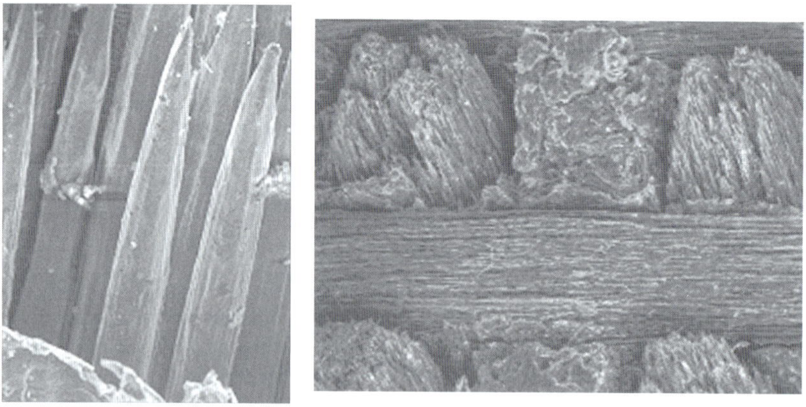

Fig. 6.1 Ablated fibers of composite material at microscopic scale and mesoscopic (millimetric) scale (1).

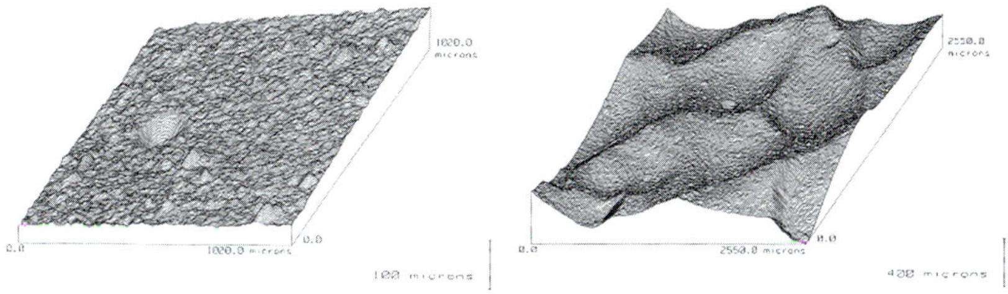

Fig. 6.2 Polycrystalline graphite, laminar and turbulent flow. Note the different scales (6).

dimension for which roughness shows no effect. In this respect we can consider the results of Nikuradze for incompressible flows [8, 9], and those of the PANT program for compressible flows [10] (Sec. 7.1.1).

- The Reynolds number $\mathcal{R}e = \rho u_R l_R / \mu$ with $u_R = u|_{y=l_R}$ (the speed at the height of the roughness for a boundary layer calculated without roughness).
- The Péclet number $\mathcal{P}e = u_R l_R / \mathcal{D}$ measures the relation between convective and diffusive transports.
- The Damköhler number $\mathcal{D}a = v_a l_R / \mathcal{D}$ compares the rates of reaction and diffusion.

Another dimensionless quantity is involved in the problem: the Mach number $\mathcal{M}_a$, the ratio of the speed of sound to the overall speed of the fluid. For $\mathcal{M}_a < 0.4$, we consider the fluid to be quasi-incompressible [8]. This situation still allows us to use scaling techniques [11]. Beyond this value, the pressure waves created by the roughness invade the whole flow and scaling is very difficult [12].

The ratio of recession velocities and diffusion is small enough to ignore any possible coupling in this area, at least in terms of analysis of phenomena. The numerical aspect of this coupling is another problem.

6.3 Reactivity of a Composite Material

Before addressing the problem of the creation of roughness, we consider the problem of responsiveness of a flat composite material of known reactivity of the constituents. Indeed, the different parts of the composite (fiber, matrix, fiber–matrix interface) can be made of different materials. Although they are the same chemical species, such as carbon, its crystal structure will result in different densities, different chemical reactivities, or different phase change kinetics. The first problem is how to give the effects associated with this heterogeneity on a plane surface. Let k_f and k_R be the properties of the two parts of a material in contact with the flow in surface fractions ϕ and $1 - \phi$, respectively. For simple geometries (1-D, for example), these latter quantities represent the volume fractions in the material. In this domain the results were obtained by homogenization [7] under the following restrictive assumptions:

- Steady flow of incompressible laminar boundary layer, whose velocity profile is independent of heterogeneous reactions, given a priori. This type of method is discussed in Sec. 7.1. Reactive chemical species are a supposed minority.
- All diffusion coefficients are equal.
- The heterogeneous chemical system is reduced to a single irreversible reaction of order 1 on the species i (or a phase change).
- The calculation is performed in an elementary cell with periodic boundary conditions (length λ) following the direction of the flow s. The conditions at the upper boundary are given and constant. This problem assumes a large boundary layer height as compared to l_R.

By homogenization, we obtain a system described in Appendix D. Its numerical solution lets us get equivalent reactivity k_{eq} possessing the following properties:

- In the reactive limit ($\mathcal{D}a \gg 1$), equivalent reactivity k_{eq} follows a "parallel" law (algebraic mean)

$$k_{eq} = k_{//} = \phi k_f + (1 - \phi) k_R \tag{6.1}$$

  In this case the amount of material arriving at or leaving from the surface is sufficient for both parties of material not to interfere. In the case of a high-contrast response, the equivalent value is similar in magnitude to the highest reactivity.
- In the diffusive limit ($\mathcal{D}a \ll 1$), equivalent reactivity k_{eq} follows a series law (harmonic mean).

$$k_{eq} = k_{\perp} = \left(\frac{\phi}{k_f} + \frac{1 - \phi}{k_R} \right)^{-1} \tag{6.2}$$

  Equivalent reactivity is of similar magnitude as the lowest reactivity for high contrast of these values. Here, because of the diffusive limitation, both sides of the material use the same reservoir, either to use the required reactant or to evacuate the product(s) formed.
- In the general case, the equivalent reactivity $k_{eq} = f(k_f/k_R)$ is independent of flow for $\mathcal{R}e < 1000$ and increases slowly beyond this value of Reynolds number. We always have $k_{\perp} \leq k_{eq} \leq k_{//}$.

6.4 Roughness Formation

The study discussed in the previous section shows that in a fairly wide area, wall reactions are not dependent on convection of the fluid. This justifies the fact that the following study of the formation of roughness will be done in the limited context of a reaction–diffusion problem.

6.4.1 System Dominated by the Reaction

In this section we assume that $\mathcal{D}a \ll 1$. Diffusion is effective and the problem is local: the species flux lines are vertical. This problem is very simple, the solution is analytic, and as we shall see later, it is quite representative. Consider a fiber or a yarn in the middle of a binder that is much more reactive than it. Ablation will produce the following sequence (Fig. 6.3):

- The binder begins to ablate faster than the fiber or yarn, partially revealing it to the side.
- The top and side of the fiber or the wick ablates. The radius at given altitude is reduced linearly with time, thus giving a conical shape with a flat top.
- This process continues until the geometry becomes conical. We then reach a stationary profile.

If v_{a_f} and v_{a_R} are the ablation rate in a stationary local reference frame for the fiber and matrix, stationarity implies that the half-apex angle of the cone (or the dihedral in a 2-D problem) is

$$\sin\theta = \frac{v_{a_R}}{v_{a_f}} = \frac{\rho_f}{\rho_R}\frac{k_f}{k_R} \tag{6.3}$$

The height of the roughness is $h = r_f/tg\theta$ and the thickness of the ablated matrix at the time it reaches steady state is $h_a = 2r_f/\sin 2\theta$. These

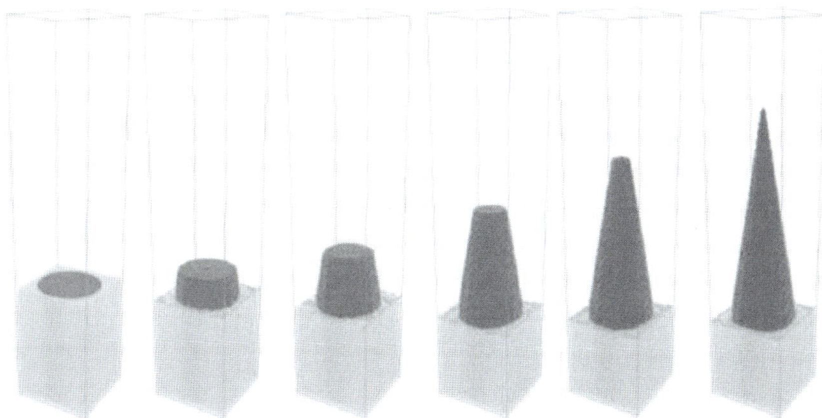

Fig. 6.3 Formation of the roughness of a fiber or a yarn (relative reactivities $\tilde{k} = \rho_R k_R/\rho_f k_f = 6$) (1).

considerations are of considerable practical interest. We can deduce from these results that

- In a reactive regime, a material has a roughness independent of the flow. Note that this assumption is generally used without justification (and often without stating it) in many studies of the effect of roughness.
- This roughness is reached almost instantaneously, the height of ablation required to achieve steady state being just above the roughness height.

We note moreover that the rate of ablation is that of the matrix. But do not mistake the conclusion that can be learned. We have seen that the rate of ablation resulted from a stationary energy balance overall (Sec. 4.7). Reducing the reactivity of the matrix does not significantly alter the ablation velocity, but will cause a slight increase in wall temperature, other things being equal. On the other hand, the roughness of the material will be diminished, which may significantly change the induced effects (transition laminar–turbulent transition or turbulent exchanges).

6.4.2 Reaction–Diffusion System

We take as a simple example a reaction–diffusion system like the ablation of carbon in a sublimation regime. The species are carried vertically by diffusion to a "reservoir" at the reference altitude $y = 0$ (Fig. 6.4). This system is not diffusion-limited. We note J_C the mass flux of carbon. The numerical simulation shows that this hypothesis is verified for the stationary state [7]. The loss of material follows the law of Knudsen–Langmuir [Eq. (3.64)]

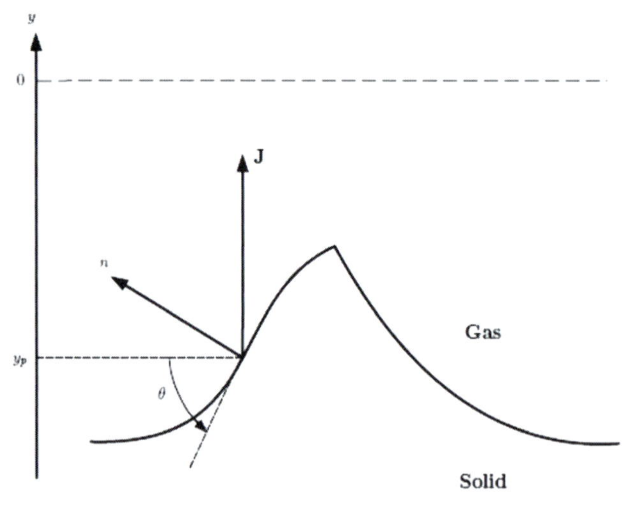

Fig. 6.4 Computational domain and notations.

$$\dot{m}_C = \beta(\xi_{eq} - x_{C_p}) \tag{6.4}$$

with

$$\begin{cases} \beta &= \alpha\left(\dfrac{\mathcal{M}}{2\pi\mathcal{R}T}\right)^{\frac{1}{2}} p \\ \xi_{eq} &= \dfrac{p_{eq}}{p} \end{cases} \tag{6.5}$$

If θ is the local slope of the surface, we have

$$J_C \cos\theta = \dot{m}_C \tag{6.6}$$

The diffusion flux is given by Darcy's law ($\rho\mathcal{D}$ assumed constant)

$$J_C = -\rho\mathcal{D}\frac{dx_C}{dy} \tag{6.7}$$

A steady state J_C is constant, hence giving a linear volume fraction of C

$$x_C = x_{C_p} - \frac{J_C}{\rho\mathcal{D}}(y - y_p) \tag{6.8}$$

The volume fraction of C is assumed to be zero at an arbitrary altitude $y = 0$

$$x_{C_p} = -\frac{J_C}{\rho\mathcal{D}} y_p \tag{6.9}$$

By postponing in Eq. (6.7) and taking into account Eq. (6.6), we get

$$\cos\theta = \frac{\beta\xi_{eq}}{J_C} + \frac{y_p}{l'_R} \tag{6.10}$$

with

$$l'_R = \frac{\rho\mathcal{D}}{\beta} \tag{6.11}$$

l'_R is a characteristic dimension of the problem of reaction–diffusion. Equation (6.10) can be applied to a real or hypothetic point where $\theta = 0$

$$1 = \frac{\beta\xi_{eq}}{J_C} + \frac{y_{p_O}}{l'_R} \tag{6.12}$$

By subtracting Eq. (6.10), it follows

$$1 - \cos\theta = \frac{y_{p_O} - y_p}{l'_R} \tag{6.13}$$

Some basic considerations on trigonometry permit us to see that the geometric solution of this equation is:

- Either $y_p = y_{po} - l'_R + [l'^2_R - (s_p - s_{po})^2]^{1/2}$, describing a circle of radius l'_R tangent to the line $y_p = y_{po}$ at abscissa $s_p = s_{po}$
- Or the line $y_p = y_{po}$ ($\cos\theta = 1$), corresponding to the trivial solution of a plane ablation profile

It is obvious that one obtains the same result in the case of an irreversible chemical reaction of order 1, in which case Eq. (6.7) is written

$$\dot{m}_C = \beta x_C \tag{6.14}$$

with $\beta = \mathcal{M}_C kp/\mathcal{N}$ (see Sec. 3.2.2).

The possible geometries will therefore consist of circular arcs (ogive or sphere in three dimensions) and plane parts. The cases encountered are:

- An ogival profile obtained from two arcs whose concavity is turned downwards. This is obtained for a fiber or a yarn (Figs. 6.1 and 6.5).
- An ogival profile connecting to some higher plane when the radius or the length of side of the object is greater than l'_R. Obtained for a yarn, the size of a fiber seemingly is too low for such a configuration (Fig. 6.5).
- A flat profile obtained for a matrix between fibers or yarns.

The gouge is discussed later, in Sec. 6.5.1.

Note that, as before, it is the "weakest" part of the material that imposes the rate of ablation.

We show that all of these geometries are possible in the sense that they are unconditionally stable [13]. In particular, in the case of a homogeneous material, the final geometry depends only on the initial geometry. This point is developed in the next section.

6.4.3 Direct Numerical Calculation

The numerical calculation of a system of reaction–diffusion offers difficulty in the representation of the geometry of the surface on which may

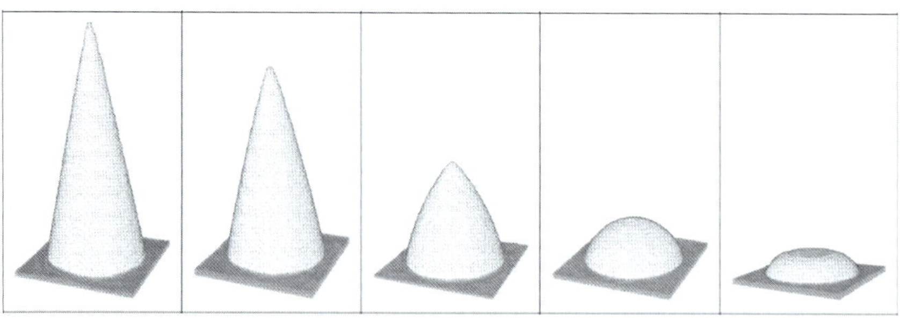

Fig. 6.5 The influence of reaction–diffusion length on the shape [1].

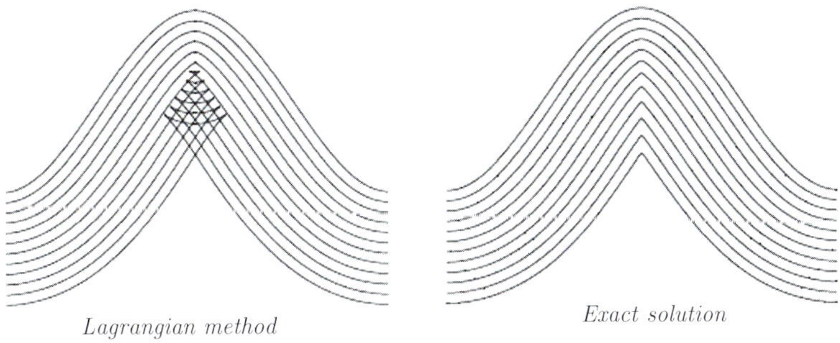

Fig. 6.6 Singularity formation (15).

appear singularities. A Lagrangian description of it is excluded because it leads to artifacts, such as the folding of the surface on itself (Fig. 6.6). This well-known problem led to the description of the surface as a level line of level set function S defined throughout the space [14]. The surface having a local normal speed v_a, its spatio-temporal evolution obeys a Hamilton–Jacobi equation. Here it is written in explicit form, valid at all points where S is differentiable:

$$\frac{\partial S}{\partial t} + \boldsymbol{v_a} \cdot \nabla S = 0 \tag{6.15}$$

The surface, given by $S = 0$ or any other line level, may be irregular. The normal $\boldsymbol{n}$ pointing toward the fluid is defined in regular parts by

$$\boldsymbol{n} = \frac{\nabla S}{\|\nabla S\|} \tag{6.16}$$

Equation (6.15) is a hyperbolic equation whose characteristic equation defines characteristic lines. The focus of these lines led to the creation of a singularity of the geometry. An example is given in Fig. 6.6. A contrario, an initial singularity can completely disappear. The study of various geometric possible configurations is conducted in 2-D in [13].

This type of problem can also be treated by a method called *volume of fluids* [7].

6.5 Applications

Note that the explanations given in this chapter are reduced to a relatively limited scope for theoretical possibilities. It is indeed very difficult to study these phenomena in the case where there is convection. Experience shows that these phenomena do not appear to change in nature in these cases.

In particular, if the surface is dependent on external conditions (except in the case dominated by the reactions), this dependence is not very important. The effects of the surface state are important only in the phase of rapid

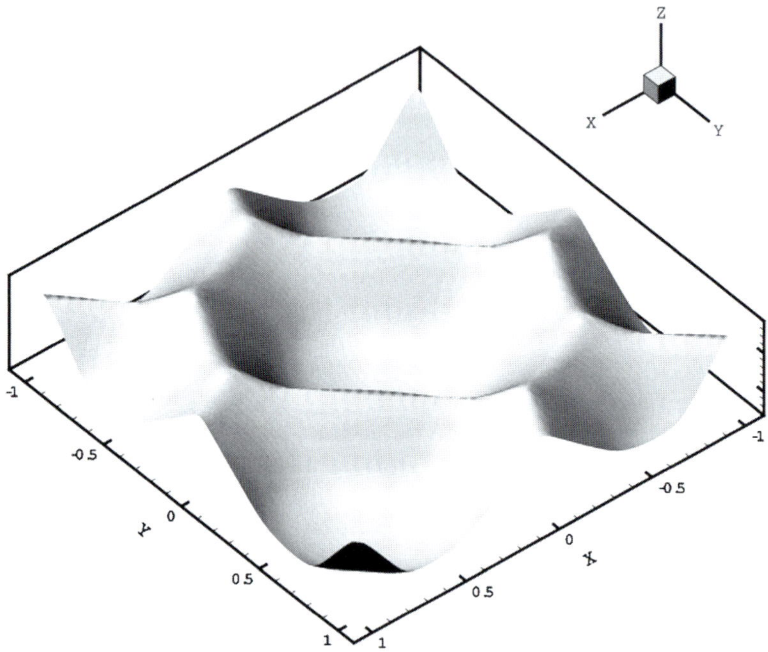

Fig. 6.7 Calculation of a gouge (16). The size of the pattern depends on initial conditions.

ablation, for which one can observe a small variation of external conditions and temperature. It may so speak to a first approximation of the roughness of a material independently of the conditions of use.

6.5.1 Gouges

Gouges appear in turbulent flow on homogeneous materials such as polycrystalline graphite (Fig. 6.2). To try to reproduce them, we must find the "right" initial geometry. For this, one starts with a striated surface by a mechanism of cross-hatching. The geometry so obtained, close to (but not at) steady state (Fig. 6.7), roughly reproduces the test. The difference may be explained by a turbulent flow (the gouges appear only in this case) through an unknown mechanism.

This refers to the explanation of the mechanism in question manifested by cross-striations forming, with the local direction of flow at an angle close to that of the local characteristics Arcsin$(1/\mathcal{M}a_e)$, where $\mathcal{M}a_e$ is the local Mach number outside the boundary layer. Various conditions must be performed for its production:

- A supersonic flow: this is clearly demonstrated in some experiments of cones with an important half-apex large angle (50 deg) so that the windward side is subsonic and the leeward supersonic. Under these conditions the striations appear on the leeward surface only.

- A turbulent flow.
- A boundary layer thin enough as compared to the roughness to show the existence of compressional waves in the flow.

Some experiments show the transition from striated surfaces to gouged ones.

The synthesis of Swigart [17] explains that there is no consensus on this kind of phenomenon.

6.5.2 Fiber

Figure 6.8 shows the influence of various parameters on the geometry of a fiber [18]:

- Size relative to its matrix environment defined by $\tilde{r}_f = r_f/\lambda$, λ being the distance between fibers.
- Relative densities and reactivities of fiber and matrix, defined by $\tilde{k} = k_R/k_f$.
- Damköhler number; note that we find the result obtained in the preceding section when $\mathcal{D}a \ll 1$, and so $l'_R \gg l_R$.

Note that this type of ogival geometry of fiber is found whatever its inclination relative to the surface (Fig. 6.9). The calculation is possible only

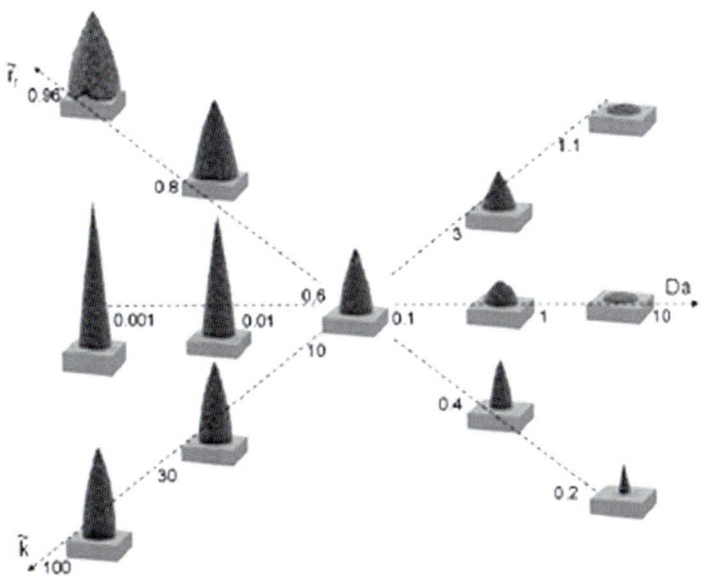

Fig. 6.8 Influence of various parameters on the geometry of the fiber.

numerically in this case, and the difference from a perfect ogival profile is difficult to estimate, certainly low.

6.5.3 Experiments

It should be noted that one can perform representative testing of the surface state in laboratory facilities. It is not necessary to use a heavy test bed as with the plasma jet: a standard oven, an arc image furnace, or a CO_2 laser in a controlled atmosphere is enough to obtain representative results on a material.

On the other hand, the quantity l'_R is experimentally accessible from a microtomography or any measurement of roughness at microscopic level. It is therefore possible to infer reactivity. For example, from plasma jet experiments on a tested carbon material one can estimate a coefficient of vaporization for the major species at the obtained temperature, which may be significantly higher than standard measurements in the laboratory (where values of about 2500 K are generally obtained; see Sec. 5.3).

6.5.4 Materials with Very Low Density

Some materials such as phenolic-impregnated carbon ablator (PICA) have a very low density obtained through a high porosity. Ablation then leads to the disappearance of the matrix before the fibers, leading to the appearance of a fibrous skeleton near the surface (Fig. 6.10). The surface of

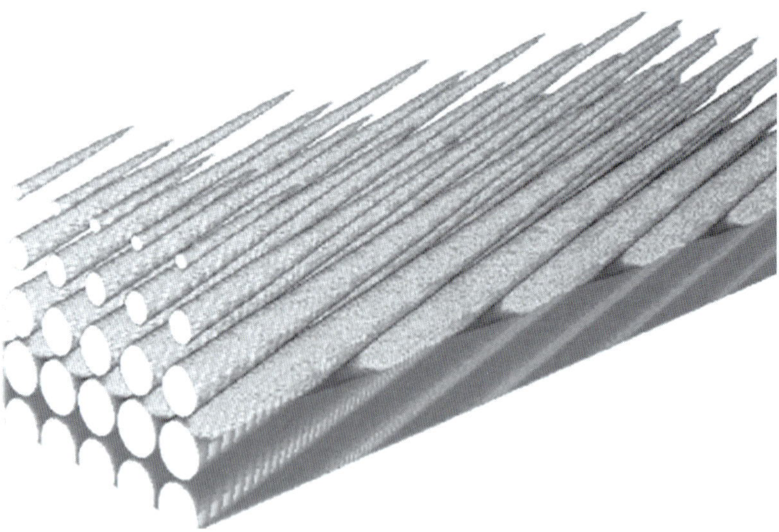

Fig. 6.9 Calculation of ablated fibers tilted 15 deg from the surface of the material (1).

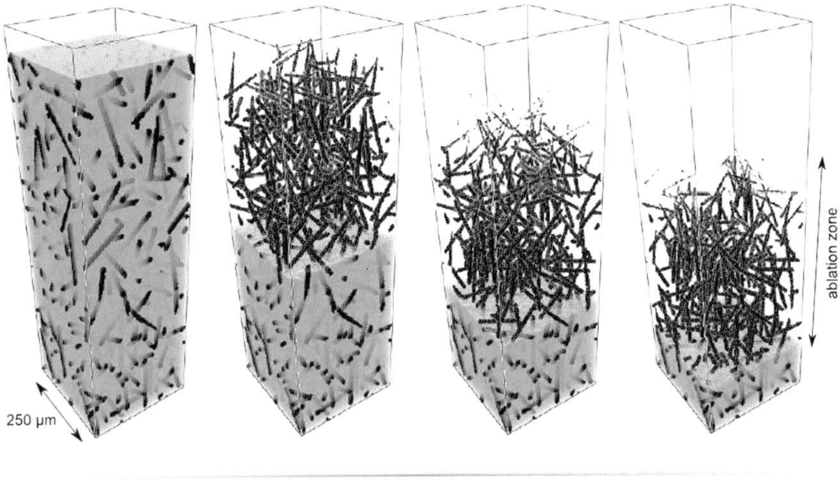

Fig. 6.10 PICA material (19).

the material is not clearly defined. In addition, the flow in the material can be such that the mean free path in gas is the same order of magnitude as the characteristic distance corresponding to the fibers. These issues are addressed in Chapter 8.

References

[1] Lachaud, J., "Modélisation physico-chimique de l'ablation de matériaux composites en carbone," Ph.D. thesis, l'Université Bordeaux, France, Nov. 2006.

[2] Larson, H. K., and Mateer, G. G. "Cross-Hatching. A Coupling of Gas Dynamics with the Ablation Process," AIAA Paper 68-670, 1968.

[3] Kiver, E. P., and Mumma, M. D., "Summit Firn Caves, Mount Rainier, Washington," *Science*, Vol. 173, 1971, pp. 320–322.

[4] Betterton, M. D., "Theory of Structure Formation in Snowfields Motivated by Penitents, Suncups, and Dirt Cones," *Physical Review E*, Vol. 63, No. 056129, 2001.

[5] Villien, B., Zheng, Y., and Lister, D., "Surface Dissolution and the Development of Scallops," *Chemical Engineering Communications*, Vol. 192, 2005, pp. 125–136.

[6] Duffa, G., "Ablation," CESTA monograph, Nov. 1996.

[7] Aspa, Y., "Modélisation de l'ablation des composites C/C dans les tuyères," Ph.D. thesis, l'Institut National Polytechnique de Toulouse, France, No. 2431, Dec. 2006.

[8] Chassaing, P., *Turbulence en Mécanique des Fluides*, Editions Cépaduès, Paris, France, 2000.

[9] Schlichting, H., *Boundary Layer Theory*, McGraw-Hill, New York, 1979.

[10] Wool, M. R., "Summary of Experimental and Analytical Results," Interim Report: Passive Nosetip Technology (PANT) Program, Vol. X, SAMSO-TR-74-86, Jan. 1975.

[11] Achdou, Y., Pironneau, O., and Valentin, F., "Effective Boundary Conditions for Laminar Flows over Periodic Rough Boundaries," *Journal of Computational Physics*, Vol. 147, 1998, pp. 187–218.

[12] Carrau, A., Gallice, G., and Le Tallec, P., "Taking into Account Surface Roughness in Computing Hypersonic Re-entry Flows," *Computing Methods in Applied Sciences*

and Engineering, edited by Glowinsky, R., Nova Science, New York, 1991, pp. 331–343.

[13] Duffa, G., Vignoles, G., Goyhénéche, J.-M., and Aspa, Y., "Ablation of Carbon-Based Materials: Investigation of Roughness Set-up from Heterogeneous Reactions," *International Journal of Heat and Mass Transfer*, Vol. 48, 2005, pp. 3387–3401.

[14] Osher, S., and Fedkiw, R., "Level Set Methods and Dynamic Implicit Surfaces," Springer-Verlag, New York, 2002.

[15] Sethian, J. A., "A Fast Marching Level Set Method for Monotically Advancing Fronts," *Proceedings of National Academy of Sciences of the United States of America (PNAS), Applied Mathematics*, Vol. 93, 1996, pp. 1591–1595.

[16] Velghe, A., Nguyen-Bui, N. T.-H., and Duffa, G., "Modeling the Surface State of Ablatable Material by Reaction-Diffusion Phenomenon," AIAA Paper 2005-3264, *AIAA/CIRA 13th International Space Planes and Hypersonics Systems and Technologies,"* Capua, Italy, May 2005.

[17] Swigart, R. J., "Cross-Hatching Studies. A Critical Review," *AIAA Journal*, Vol. 12, No. 10, 1974, pp. 1301–1318.

[18] Vignoles, G. L., Lachaud, J., and Aspa, Y., "Roughness Evolution in Ablation of Carbon-Based Materials: Multi-scale Modelling and Material Analysis," *Fifth European Workshop on Thermal Protection Systems and Hot Structures, ESA/ESTEC*-Noordwijk, The Netherlands, May 2006.

[19] Kurotaki, T. "Construction of Catalytic Model on SiO_2-Based Surface and Application to Real Trajectory," AIAA Paper 2000-2366, *34th AIAA Thermophysics Conference*Denver, Colorado, June 2000.

Chapter 7 Turbulence and Laminar–Turbulent Transition

- Calculate flow over a rough wall
- Describe the nonlocal effects of turbulence
- Understand transition phenomenon and its modeling

7.1 Coupling Between Turbulence and Surface State

When the characteristic size of the surface roughness becomes of the same order of magnitude as that of smaller eddies present in a turbulent flow (Kolmogorov scale), two local effects will have an impact on the macroscopic quantities: heat flux or apparent shear. (This includes the pressure effects on roughness resulting on a force parallel to the macroscopic smoothed surface; see Secs. 7.1.2.3 and 7.1.2.4).

Recall the classical results obtained by Nikuradse for incompressible flow on a sanded wall, and their interpretation [1]. We define a local characteristic length of flow

$$l = \frac{\mu_p}{\rho_p \tau_p} \tag{7.1}$$

where

$$\tau_p = 2\mu_p \left.\frac{\partial u}{\partial y}\right|_p$$

is the wall shear, calculated in the absence of roughness. The effects of the dimensionless roughness $l_R^* = l_R/l$ are the following:

- For $l_R^* < 3.5$, no effect is observed.
- For $68 > l_R^* > 3.5$, there is an increase of shear with l_R^*, the effects varying with the Reynolds number.

- For $l_R^* > 68$, the phenomenon has a saturation limit.

If one is not interested in the immediate vicinity of the surface, it is assumed there is a mean equivalent smooth flow (on a flat plane). This flow follows the classical logarithmic boundary layer

$$u^+ = \frac{u}{u_\tau} = \frac{1}{\kappa} \ell n\, y^+ + B \tag{7.2}$$

where $u_\tau = (\tau_p/\rho_p)^{1/2}$ is the friction velocity, $y^+ = \rho_p u_\tau y/\mu_p$, $\kappa \simeq 0.41$ is the von Kàrmàn constant, and $B \simeq 5.3$ for a smooth surface [9].

This law is shifted downward by an amount Δu^+ dependent on l_R^* (Fig. 7.1). The shift is expressed by

$$\begin{cases} \Delta u^+ = \dfrac{1}{\kappa} \ell n\, l_R^+ + D \\ l_R^+ = \dfrac{l_R \rho_p u_\tau}{\mu_p} \end{cases} \tag{7.3}$$

D is a coefficient depending on the roughness, and has been the subject of numerous evaluations (see Sec. 7.1.2.2).

Note that the following models assume various general assumptions, which are not always met:

- The part of 3-D flow is confined near the surface. For this, the roughness must be sufficiently small as compared to the boundary layer thickness δ, $l_R/\delta < 0.025$ and $\mathcal{R}e_{l_R} > 4000$ [4], where $\mathcal{R}e_{l_R}$ is the Reynolds number

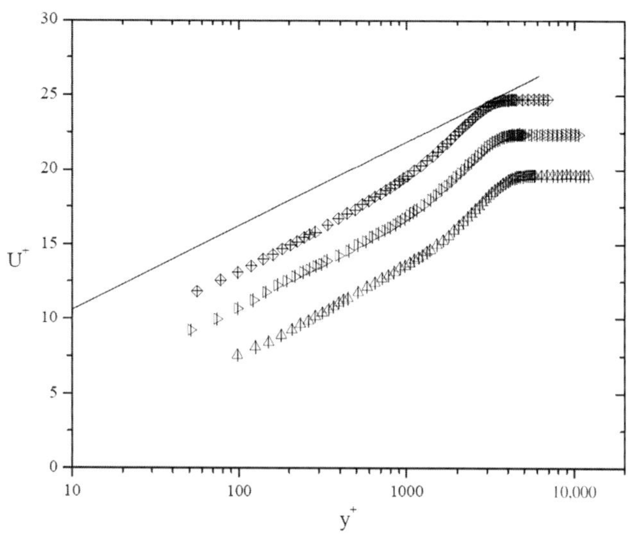

Fig. 7.1 Displacement of wall law on a rough surface (perforated plate, incompressible flow). The line represents the law on a smooth wall (3).

formed with the roughness height and friction velocity

$$\mathcal{R}e_{l_R} = \frac{\rho_p u_\tau l_R}{\mu_p} \tag{7.4}$$

This part is about 5 times the height of roughness [5]. This experimental value is confirmed by numerical simulations [6].

- Outside this region, the flow can be described by standard models of turbulence if the relation $l_R/\delta < 0.012$ is verified [4]. This hypothesis, called Townsend's hypothesis, is confirmed by the results of direct numerical simulation (DNS) [7]. This does not say how to treat the neighborhood of the wall or what boundary condition to impose. This is the subject of Sec. 7.1.3.

Most of the experimental and numerical (DNS) results are obtained in configurations with a small Reynolds number and incompressible flow. We are concerned primarily with highly compressible flows, even if the area of interest is in the vicinity of the wall. It is therefore necessary to go through the experience made in supersonic with measurement of the influence of roughness on heat flux. (The incompressible studies are interested mainly in shear effects.) In this area, Passive Nosetip Technology (PANT) testing constitutes an indispensable reference.

7.1.1 Wind Tunnel Tests

7.1.1.1 PANT Experiments on Rough Walls

The augmentation of heat flux in turbulent flow over rough walls has been the subject of numerous wind tunnel measurements in the PANT program. In the first series of stagnation tests, the models (sphere–cone, ellipsoid–cone, bi-cone) are metallic, and roughness is obtained by sandblasting. The surface roughness is described by (significant) peak to (significant) valley l_R. This notion is rather vague, but may be related, for the roughness described, to quantity R_a (arithmetic mean), which is the first moment of the statistical distribution describing the surface

$$R_a = \frac{1}{L}\int_{s_0}^{s_0+L} |y - y_m| \mathrm{d}s \tag{7.5}$$

Of course, this assumes that the surface is random enough so the result is independent of the length L of the sample and of the definition of the mean surface y_m. For the models used, $l_R \simeq 3.6 R_a$ ($\pm$30%).

The results (Fig. 7.2) show an increase in heat flux up to a factor of 3.

7.1.1.2 Other Tests

Tests performed on long sphere–cones [8] made of sandgrained metal models show that the law defined in the previous section remains valid

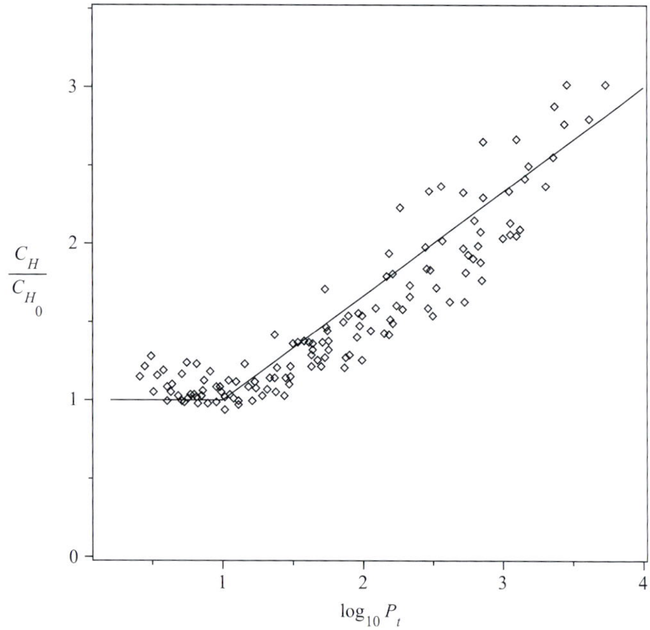

Fig. 7.2 Augmentation of fluxes on a rough wall in turbulent flow.

everywhere. In contrast, another series of tests of the PANT program give conflicting results in the case of scalloped walls (see Fig. 6.2) [9]. In these tests, made on ablated camphor, typical scalloped morphologies appear, with dimensions decreasing from stagnation point to downstream. On these geometries, the heat flux measurements were obtained in two overlapping ways:

- Directly on metal replicas obtained from castings
- Indirectly from measurements of ablation of camphor

The trends are very different than those made on sandblasted metal walls. In particular, the insensitivity to the Reynolds number leads the authors to infer nonviscous effects. In these tests, the visualization highlights many disturbances outside the boundary layer. Note that the roughness does not comply with the criteria given in this chapter's introduction consider that the direct effects are limited in the area adjacent to the surface. No model can reproduce this type of test (see Sec. 6.5.1).

7.1.2 Calculation of Flow Over a Rough Wall

7.1.2.1 Powars's Empirical Law

Powars [9] has proposed a correlation of PANT test results from the parameter l_R/δ_+ where δ_+ is the thickness of the laminar boundary sublayer calculated in the absence of roughness. This parameter is not easily accessible to traditional codes of ablation; it is therefore replaced by an estimate using a "wall law" [1]:

$$\frac{l_R}{\delta_+} \simeq \frac{\mathcal{R}e_{l_R}}{10} \tag{7.6}$$

where $\mathcal{R}e_{l_R}$ is defined by Eq. (7.4). (All of these values are derived from calculation on a smooth wall.) It then uses the Reynolds analogy (with known uncertain validity on a rough wall) with $\mathcal{P} = 1$ to make appear the Stanton number

$$C_{H_0} = \frac{\tau_p}{\rho_e u_e^2} \tag{7.7}$$

With these approximations, we get

$$\frac{l_R}{\delta_+} \simeq \frac{1}{10} \frac{\rho_e u_e l_R}{\mu_e} C_{H_0}^{\frac{1}{2}} \left(\frac{\rho_p}{\rho_e}\right)^{\frac{1}{2}} \frac{\mu_e}{\mu_p} \tag{7.8}$$

In addition, when using an approximate law for viscosity of type $\mu_e/\mu_p = (T_e/T_p)^{0.8}$, it follows

$$\frac{l_R}{\delta_+} = \frac{P_t}{10} \tag{7.9}$$

with $P_t = \mathcal{R}e_{l_R} C_{H_0}^{1/2} (T_e/T_p)^{1.3}$.

This parameter was used by Powars to correlate the experimental results (Fig. 7.2)

$$\begin{cases} P_t < 10 & \dfrac{C_H}{C_{H_0}} = 1 \\ 10 < P_t < 10^4 & \dfrac{C_H}{C_{H_0}} = 1 + \dfrac{2}{3}(\log_{10} P_t - 1) \\ P_t > 10^4 & \dfrac{C_H}{C_{H_0}} = 3 \end{cases} \tag{7.10}$$

7.1.2.2 Equivalent Sand Grain Method

The oldest methods use the concept of "equivalent" sand grain. This method relies on work on walls with random roughness of sand grain type and plan to reduce description of the real geometry to a single parameter, which becomes an adjustment parameter when used in the interpretation of plasma jet experiments. The example of the law given in Sec. 7.1.2.1 has been used extensively, and such settings are found in some material databases.*

Many attempts were made to give an analytical expression of the coefficient D involved in Eq. (7.3). The best known method was developed

* National Aeronautics Space Administration, TPSX Material Properties Database, available online at http://tpsx.arc.nasa.gov [retrieved 30 April 2009].

by Dvorak [10]

$$\begin{cases} D = 12.25\,\ell n\,\Lambda - 17.35, & 1 < \Lambda < 4.68 \\ D = -2.85\,\ell n\,\Lambda + 5.95, & \Lambda > 4.68 \end{cases} \tag{7.11}$$

The parameter Λ describes the roughness. Dirling [11] proposed the following expression for discrete elements on a flat surface:

$$\Lambda = \frac{\lambda}{l_R}\left(\frac{A_1}{A_2}\right)^{\frac{4}{3}} \tag{7.12}$$

A_2 and A_1 are projected areas onto the horizontal surface and a vertical plane perpendicular to the macroscopic flow, respectively, and λ is the periodicity. This was used by Grabow and White [12] in the compressible case to address the effects on composite material. It is obvious that this approach is severely limited by the fact that these ablated surfaces are not random (see Chapter 6). Moreover, on a rough surface, it is difficult to represent simultaneously the effects on friction and heat flux.

7.1.2.3 Method of Discrete Elements

The method of discrete elements was the first attempt to give the problem a realistic physics support [13]. This is an empirical method of volume averaging of the flow with additional terms intended to take into account the effects of roughness on shear and heat flux.

The word *average* concerns the average control volume occupied by the flow at altitude y, counted from the lower part (Fig. 7.3). For a roughness of

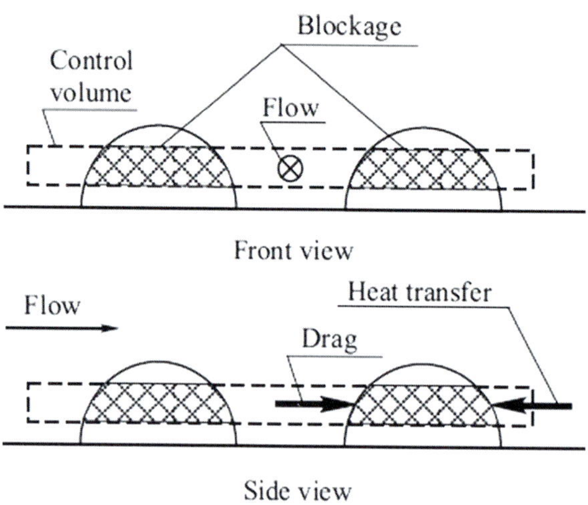

Fig. 7.3 Method of discrete elements.

period λ and area $A(y)$ we define the quantity

$$g = 1 - \frac{A(y)}{\lambda^2} \tag{7.13}$$

The equations are those of the boundary layer in 2-D [Eq. (4.9)]

$$\begin{cases} \dfrac{\partial(g\rho u)}{\partial s} + \dfrac{\partial(g\rho v)}{\partial y} = 0 \\ \rho u \dfrac{\partial(gu)}{\partial s} + \rho v \dfrac{\partial(gu)}{\partial y} = -\dfrac{\partial(gp)}{\partial s} + \dfrac{\partial}{\partial y}\left[(\mu + \mu_t)\dfrac{\partial(gu)}{\partial y}\right] - \mathcal{T}' \\ \rho u \dfrac{\partial(gh_T)}{\partial s} + \rho v \dfrac{\partial(gh_T)}{\partial y} = \dfrac{\partial}{\partial y}\left[g\dfrac{\lambda + \lambda_t}{C_p}\dfrac{\partial h_T}{\partial y} + g(\mu + \mu_t)\left(1 - \dfrac{1}{\mathcal{P}}\right)\dfrac{\partial}{\partial y}\left(\dfrac{u^2}{2}\right)\right] \\ \qquad + \mathcal{T}'u + \mathcal{W} \end{cases} \tag{7.14}$$

$\mathcal{T}'$ is the drag term due to roughness. This term is related to the pressure exerted on the windward side and depression on the leeward side. It may cause an increase of a few percent in the drag of the body and is introduced as an ad hoc term in the equation of momentum. The following expression is that of theoretical drag of a body in free space or a measured value in an experiment, such as the measurement of flows in networks of parallel tubes [14]:

$$\mathcal{T}' = \frac{1}{2}\rho C_A \mathrm{d}(y) \frac{u^2}{\lambda^2} \tag{7.15}$$

$\mathrm{d}(y)$ is a reference length associated with the drag coefficient of the obstacle C_A. This is sometimes expressed in terms of a local Reynolds number [15]. Most often the length of the segment is defined by the part of the flow intercepted by the surface roughness (e.g., the diameter at the height y for hemispherical roughness).

$\mathcal{W}$ is the projection along y of the heat flux term on the roughness. This quantity also is introduced here arbitrarily. Its expression varies among authors. It may, for example, be written as follows:

$$\mathcal{W} = \frac{1}{\lambda^2}\mathcal{N}u_d(h_p - h) \tag{7.16}$$

where h_p is the enthalpy of gas at the wall temperature and $\mathcal{N}u_d$ the Nusselt number formed with d. This value is expressed as a function of the local Reynolds number and Prandtl number in the case generally used (as documented): a network of tubes [14].

This type of approach has been widely and successfully used in the field of incompressible flow (see, e.g., [16]). There are several versions of the model that differ in their method of writing the shear and heat flux. The references are scarce for supersonic flows [15].

Note that this method, by its arbitrary construction, does not see all the physical aspects of the problem, and in particular turbulence models applicable in this case. This is covered in Sec. 7.1.3.

7.1.2.4 Homogenization

The method of homogenization consists of replacing the physical problem at the mesoscopic scale with a jump-type equation on the variable, that is to say a relationship between its value and its gradient at the altitude, where the problem is divided into two regions. This equation results from solving approximately the local problem in the region near the wall. This problem can be 1-D, as in the case of the Knudsen layer (rarefied flow near the wall), or 3-D here. The altitude at which the coupling is performed is generally low enough to be neglected thereafter.

The calculation can be completed in its entirety [17] in the case of an incompressible flow where viscous effects are dominant (Appendix D). At order 0, the microscopic problem is local and periodic in the direction of the flow $f(s+\lambda, y) = f(s, y)$, λ being the period of the phenomenon. The calculation of a higher order shows this is a very good approximation. A balance over a period permits us to note the following result: the flux of input and output in a direction parallel to the macroscopic surface being equal, it is the same for vertical fluxes in a steady-state case. These fluxes are then used as boundary conditions for the local flow (microscopic scale) [6], that is, both components of shear $\mathcal{T}\,\mathbf{n}$ (for a 2-D external flow) and $[\dot{\mathbf{q}} + (\mathcal{T} - p\mathbf{Id})\mathbf{V}] \cdot \mathbf{n}$. To this one must add parietal conditions: temperature and vertical component of mass flux in the case with ablation. Finally, one must tabulate a law giving numerical values of quantities at the upper boundary of the local problem $\langle u \rangle$, $\langle T \rangle$, $\langle p \rangle$, $\langle \dot{m} \rangle$ depending on the quantities defined previously. This can be done by a method of least squares on the basis of orthogonal polynomials, such as Legendre polynomials [6].

The previous remark on the conservation of vertical flux is also important for coupling the material with the external flow at macroscopic scale (see Sec. 2.2.2).

The calculations performed in a laminar flow [6] show that the roughness does not induce any effect on the heat flux. There is a pressure component parallel to the macroscopic smooth wall and a shear component perpendicular to it. This results from the effects of roughness at microscopic scale. There is also a simple translation in altitude of the external flow, not quantifiable in general in the absence of a single definition of altitude 0. This result allows us to question the merits of discrete element models that were never tested on a laminar flow. They tend to prove that modification of the turbulence created by the roughness is the sole cause of macroscopic effects.

This kind of method is, in fact, well known by the aerodynamics domain as "wall law," and is used to remove theoretical or numerical difficulties in calculating the region of space close to the wall [18]. The method has been used for this problem in a form slightly different from the one used previously, consisting of carrying out a series of parametric calculations on the various relevant quantities ($\mathcal{M}a$, $\mathcal{R}e_\infty$, T_∞, T_p, etc.) for a fixed geometry of roughness in a given configuration, such as a flat plate [18].

The turbulence models applicable to internal and external problems are described in the following section.

7.1.3 Turbulence Models for Rough Walls

Turbulence modeling clearly depends on the method used for calculating the flow. In the Powars method, for example, turbulence is not explicit.

7.1.3.1 Models Using Equivalent Roughness

In models of equivalent roughness, the calculation of the boundary layer is obtained as for a smooth wall. Then the turbulence model must represent all the effects of roughness, including the source term created by roughness, and not just those of the turbulence evolution itself. Models of this class give respectable results in view of their low physical content [16]. They are based on the eddy viscosity introduced by Van Driest and Rotta for the inner region of the boundary layer. In the following equations, the lengths y^+ and l_R^+ are the dimensionless values of y and l_R using the characteristic length l given by Eq. (7.1).

$$\mu_t = \rho l^2 \frac{\partial u}{\partial y} \tag{7.17}$$

Mixing length l is given by

$$l = F\kappa(y + \Delta y) \tag{7.18}$$

with

$$\begin{cases} F = 1 - \exp\left(\dfrac{y^+ + \Delta y^+}{A^+}\right) \\ \Delta y = 0.9 \dfrac{\mu_p}{\rho_p u_\tau}\left[l_R^{+\frac{1}{2}} - l_R^+ \exp\left(\dfrac{-l_R^+}{6}\right)\right] \end{cases} \tag{7.19}$$

F is the damping function, Δy is a displacement of the surface representing the effect of geometric roughness height l_R, and $A^+ \simeq 26$ [2].

7.1.3.2 Discrete Elements Turbulence Model

The description of homogenization methods in Sec. 7.1.2.4 allows us to place the model of discrete elements in the class of homogenized problems. A model "away from the wall"—for example, the classic $k - \epsilon$—will be sufficient if you know how to set the right boundary condition represented by the pseudo-wall. This assertion is confirmed by experiment (Fig. 7.4). It is not

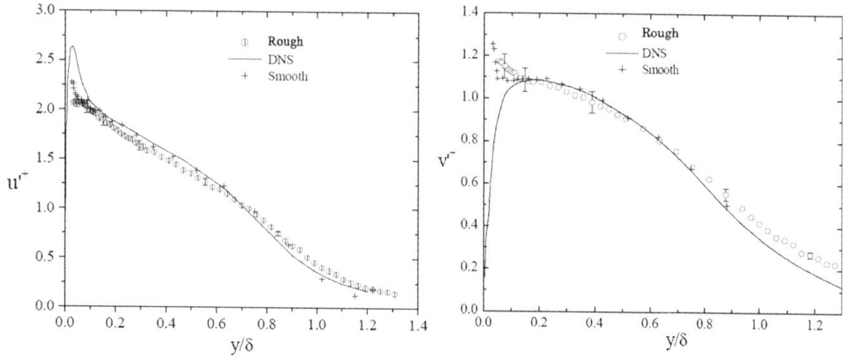

Fig. 7.4 Nondimensional turbulent fluctuation $u'^{+} = (u'/u_\tau)$ and $v'^{+} = (v'/u_\tau)$ (incompressible) [3].

necessary to give a detailed description of flow near the wall; a 1-D model taking into account the effects of roughness is sufficient. This remark authorizes the use of turbulence models such as Durbin's two layers [19]. In this model, the dissipation ϵ and the turbulent viscosity μ_t are given by algebraic relations

$$\begin{cases} \epsilon &= \dfrac{k^{\frac{3}{2}}}{l_\epsilon} \\ \mu_t &= \rho c_\mu k^{\frac{1}{2}} l_\nu \end{cases} \tag{7.20}$$

where l_ϵ and l_ν are characteristic lengths of dissipation, expressed by relations of Van Driest type

$$\begin{cases} l_\epsilon &= 2.5(y+\Delta y)\left[1-\exp\left(\dfrac{\mathcal{R}e_{k_t}}{A_\epsilon}\right)\right] \\ l_\nu &= 2.5(y+\Delta y)\left[1-\exp\left(\dfrac{\mathcal{R}e_{k_t}}{A_\nu}\right)\right] \\ A_\nu &= \max\left[1,\ 62.5\left(1-\dfrac{l_R^+}{90}\right)\right] \end{cases} \tag{7.21}$$

$\mathcal{R}e_{k_t} = (\rho u k_t^{(1/2)}/\mu)$ is the Reynolds number formed with the turbulent kinetic energy. The law $\Delta y^+ = f(y^+)$ is deduced from experience. The turbulent kinetic energy in $y = 0$ is nonzero; it is $(C_\mu = 0.09)$

$$k_t(0) = u_\tau C_\mu^{-\frac{1}{2}} \min\left[1,\ \left(\frac{l_R^+}{90}\right)^2\right] \tag{7.22}$$

7.1.3.3 Homogenized Model

For the actual homogenization, the problem is the need for a realistic model of near-wall to calculatc the internal zone as with Durbin's model $\overline{v^2} - f$. Some references have reported interesting results [18].

7.2 Nonlocal Effects of Turbulence

The nonlocal expression covers a neighborhood of turbulent flows with no developed turbulence, but where the turbulence present upstream or downstream causes a local notable increase in exchanges of momentum and energy. We provide two examples in the present problems that concern us

- The effect of turbulence in the upstream flow, for a stagnation point. One can find this type of problem in particle impacts (Chapter 11) or high pressure plasma jets (Chapter 12).
- The effect of roughness in a transitional flow near a stagnation point.

7.2.1 Effects of Turbulence in Upstream or Downstream Flow

The impact of a turbulent jet has been the subject of numerous experimental studies in incompressible flows that have demonstrated the increase of exchanges at stagnation point. These are often correlated with a turbulent Reynolds number [20]

$$\mathcal{R}e_{Tu_\infty} = \frac{\rho_\infty V_\infty}{\mu_\infty} Tu_\infty \tag{7.23}$$

where $Tu_\infty = [(\overline{V_\infty'^2})^{1/2}/V_\infty]$ is the intensity of upstream turbulence. These studies in the field on incompressible flows neglect the effect of turbulent scales, yet they are important in this problem. More recent theoretical studies [21] show that increasing the incompressible heat flux is in the form

$$\frac{\dot{q} - \dot{q}_0}{\dot{q}_0} \simeq \alpha(\mathcal{P}, \lambda) \frac{Tu_\infty^2 \mathcal{R}e_\infty^{\frac{2}{3}}}{\left(\frac{l_{Tu}}{\mathrm{d}}\right)^{\frac{2}{3}}} \tag{7.24}$$

where d is the diameter of the object, $\mathcal{R}e_\infty = \rho_\infty V_\infty/\mu_\infty$ the upstream unit Reynolds number, l_{Tu} the upstream turbulent scale (integral scale for a homogeneous and isotropic turbulence), and α a value depending on Prandtl number $\mathcal{P}$ and of a parameter λ measuring the relative scales of disturbance and the boundary layer. Important increases of heat flux were observed [20], a factor of up to 2 or more for $\mathcal{R}e_{Tu_\infty} > 10^4$. Large eddy simulation (LES) shows that these effects are related to the creation of vortices of a size close to the boundary layer thickness [21], moving with the stream lines, leading to an important increase for heat transfer $(h_p/h_e = 0.9)$, less for momentum ($h_p/h_e = 0.1$). The topmost curve in Fig. 7.5 is an approximation of the type $C_H/C_{H_0} = \alpha Q_e^{0.27}$. The parameter Q_e is defined by Eq. (7.25). The top curves in the figure are related to heat flux, the bottom to shear.

It should be noted that the standard turbulence models must be adapted to take into account the strong inhomogeneity (vortex stretching) [21, 22]. In this respect, Durbin's model $\overline{v'^2} - f$, which to some extent can take this into account, gives good results [23].

Prior to the advent of this type of model, this effect was modeled numerically by Wassel and Denny [24] in compressible flows for a sphere, with a one-equation turbulence model [2], assuming a gas in local thermodynamic equilibrium and turbulence unaltered by the passage of the shock. This assumption is fairly well verified [21]. The results were grouped by taking as variable the turbulent kinetic energy at the stagnation point $k_{t_a} = (3/2)\overline{u_e'^2}$, dimensionless

$$Q_e = \frac{\frac{1}{2}\rho_e k_{t_a}}{\mu_e \beta} \qquad \beta = \left.\frac{du_e}{ds}\right|_a \tag{7.25}$$

is the velocity gradient at stagnation point, given by Eq. (4.58).

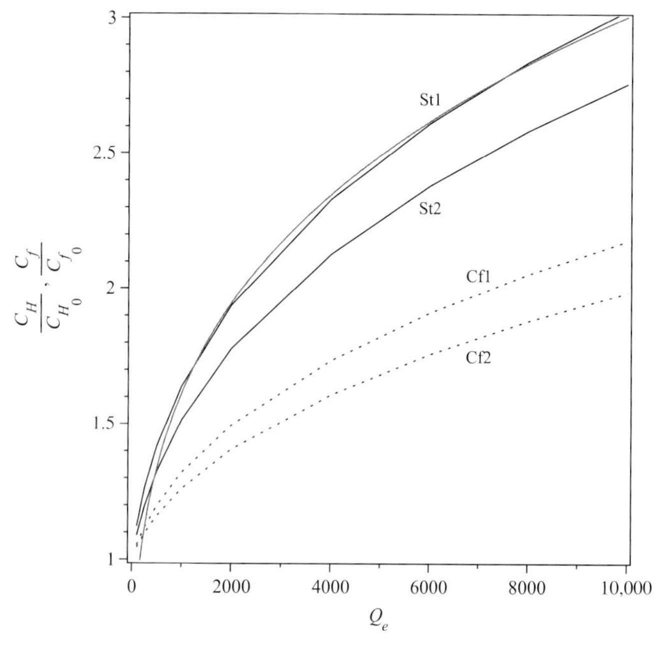

Fig. 7.5 Increases in quantities at stagnation point due to turbulence (20).

The heat flow can be increased in large proportions, up to a factor 3 (Fig. 7.5). Under the same conditions, there is an increase of shear by roughly two times lower.

7.2.2 Roughness and the Transitional Boundary Layer

Increases in heat fluxes were observed in transitional flow when part of a turbulent supersonic flow reached back to the subsonic region near the stagnation point. This phenomenon is observed only on a rough wall [9]. Presumably the explanation of the phenomenon is the same as in the case of the impact of a turbulent jet.

A series of experiments on hemispherical specimens in the PANT program helped to establish an empirical law based on arguments involving an "effective area" subject to the flow, that is to say the surface of gas–solid contact. This correlation uses the following parameter [9]:

$$P_l = \mathcal{R}e_R^{0.2} \frac{l_R}{\theta} \tag{7.26}$$

where the Reynolds number $\mathcal{R}e_R$ is formed with the quantities behind the shock and the radius of the specimen. The l_R is the "peak-to-valley" roughness (Sec. 7.1.1.1) and θ is the momentum thickness calculated on a smooth wall

$$\theta = \int_0^\infty \frac{\rho u}{\rho_e u_e}\left(1 - \frac{\rho u}{\rho_e u_e}\right) \mathrm{d}y \tag{7.27}$$

The empirical law reproduces the experimental results of Fig. 7.6 very well, although its justification seems doubtful given the results of Chapter 6.

$$\frac{C_H}{C_{H_0}} = \begin{cases} 1 & \text{if } P_l < 40 \\ 1.307 \, \ell n \, P_l + 20.17 \, P_l^{-0.606} - 5.9784 & \text{if } P_l > 40 \end{cases} \tag{7.28}$$

7.3 Coupling Between Turbulence and Chemical Reactions

In this section we will focus on a point well known in combustion but little discussed in the hypersonic domain. The problem is the calculations of the multispecies production rate $\dot{\omega}_i = f(c_i, T)$ for each species i [see Eq. (2.2)]. The law f, in general an Arrhenius law, is not linear, so the average $\overline{\dot{\omega}_i}$ is different from $f(\tilde{c}_i, \tilde{T})$, where $\tilde{c}_i$ and $\tilde{T}$ are Reynolds averages of c_i and T, respectively. To solve the problem also requires knowing the statistical fluctuations of these variables. The general problem is extremely complex

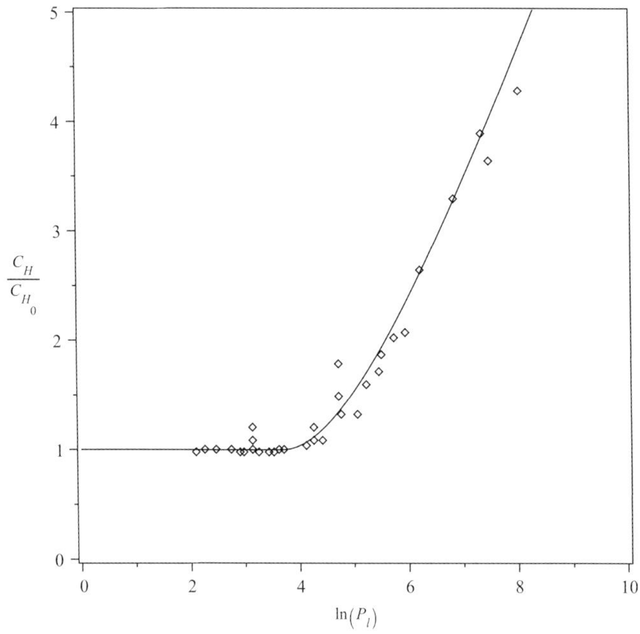

Fig. 7.6 Laminar augmentation heat flux (stagnation point) (9).

and has been addressed in the restrictive case where we assume statistical independence between temperature and mass fractions [25]. Moreover, it is desirable to define probability densities a priori for T and c_i to limit the amount of calculation. In fact, a judicious choice of distributions (Gaussian for T and β-multivariate for c_i) [26] reduces the problem of transport equations on enthalpy and the sum of the variances of the mass fractions of [27]

$$\begin{cases} f(T) = \dfrac{1}{(2\sigma_T)^{\frac{1}{2}}} \exp\left[-\dfrac{(T-\tilde{T})^2}{2\sigma_T}\right] \\ g(c_i) = \dfrac{\Gamma\left(\sum_{i=1}^{n_e} \beta_i\right)}{\prod_{i=1}^{n_e} \Gamma(\beta_i)} \prod_{i=1}^{n_e} c_i^{\beta_i - 1} \end{cases} \tag{7.29}$$

with

$$\beta_i = \tilde{c}_i \left[\frac{\sum_{j=1}^{n_e} \tilde{c}_j (1 - \tilde{c}_j)}{\sigma_c} - 1 \right] \tag{7.30}$$

$\sigma_c = \sum_{j=1}^{n_e} \widetilde{c''_j c''_j}$ is the total variance of mass fractions and $\sigma_T = \widetilde{T''T''}$ that of temperature.

One obtains in this manner two transport equations for the variances of the enthalpy and mass fractions

$$\frac{\partial}{\partial t}(\bar{\rho}\widetilde{h''^2}) + \nabla\cdot(\bar{\rho}\tilde{\mathbf{V}}\widetilde{h''^2}) = \nabla\cdot\left[\left(\frac{\lambda}{C_p}+\frac{\mu_t}{Pr_t}\right)\nabla\widetilde{h''^2}\right] + 2\frac{\mu_t}{\mathcal{P}_t}\|\nabla\tilde{h}\|^2 - C_h\frac{\bar{\rho}\widetilde{h''h''}}{\tau}$$

$$\frac{\partial}{\partial t}(\bar{\rho}\sigma_c) + \nabla\cdot(\bar{\rho}\tilde{\mathbf{V}}\sigma_c) = \nabla\cdot[\bar{\rho}(\mathcal{D}+\mathcal{D}_t)\nabla\sigma_c] + 2\bar{\rho}\mathcal{D}_t\sum_{i=1}^{ne}\|\nabla\tilde{c}_i\|$$
$$-C_{\sigma_c}\frac{\bar{\rho}\sigma_c}{\tau} + 2\sum_{i=1}^{ne}\widetilde{\dot{\omega}_i c''_i} \tag{7.31}$$

where τ is the integral time scale of turbulence, and $C_h = 2$ and $C_{\sigma_c} = 4$ are modeling constants. The last term of the equation on σ_c is calculated from the following identity:

$$\sum_{i=1}^{ne}\widetilde{\dot{\omega}_i c''_i} = \sum_{i=1}^{ne}\overline{\dot{\omega}_i c_i} - \sum_{i=1}^{ne}\bar{\dot{\omega}}\tilde{c}_i = K \tag{7.32}$$

We introduce Eq. (3.3) giving the production rate, getting

$$K = \sum_{i=1}^{n_e}\mathcal{M}_i\sum_{j=1}^{n_r}(\nu''_{ij}-\nu'_{ij})$$
$$\times\left[\overline{k_{f_j}\prod_{l=1}^{n_e}\left(\frac{\bar{\rho}}{\mathcal{M}_l}\right)^{\nu'_{jl}}}\left\langle c_i\prod_{l=1}^{n_e}c_l^{\nu'_{jl}}\right\rangle - \overline{k_{b_j}\prod_{l=1}^{n_e}\left(\frac{\bar{\rho}}{\mathcal{M}_l}\right)^{\nu''_{jl}}}\left\langle c_i\prod_{l=1}^{n_e}c_l^{\nu''_{jl}}\right\rangle\right] \tag{7.33}$$

The variance of temperature is obtained by neglecting the fluctuations of specific heat

$$\widetilde{h''^2} = C_p(\tilde{T},\ \tilde{c}_i)\ \widetilde{T''T''} \tag{7.34}$$

This method was applied to a test case by the European Space Agency, Centre National d'Etudes Spatiales (ESA-CNES), the geometry of which is shown in Fig. 7.7. The calculation conditions correspond to Mars atmosphere at an altitude of 31 km ($p_\infty = 19.6$ Pa and $T_\infty = 141$ K). The velocity is 5321 m $\cdot$ s^{-1} ($\mathcal{M}a = 27.6$). The wall is not catalytic, and the temperatures are 1500 K on windward and 500 K at the rear.

The turbulence model used is the $k-\omega$ of Wilcox, modified to be insensitive to turbulence in upstream flow. In the calculation, the front part of the object is transitional (for the simulation of this type of problem, see Sec. 7.4.5), turbulence being concentrated in the boundary layer at the front and the shear zone at the rear (Fig. 7.8). This last remark explains a small difference between laminar and turbulent fluxes in this area, and the small difference given by the model used, as compared to a standard model, still in this area. On frontshield (Fig. 7.9) there are differences of up to 30% locally. This difference can significantly affect the thickness of the heat

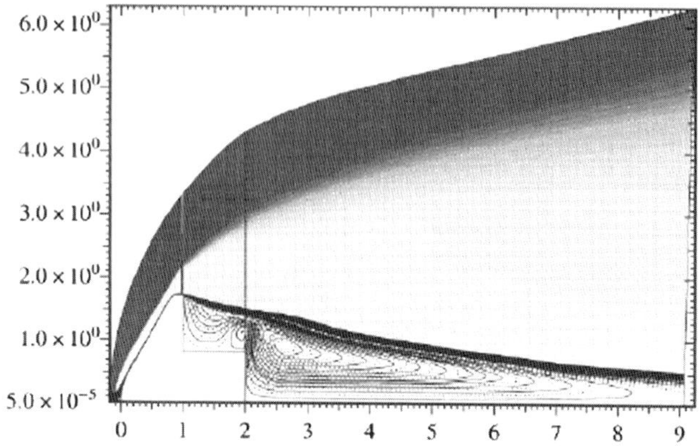

Fig. 7.7 Flow on the ESA study object.

shield or safety margins in design. These differences are present in areas where large fluctuations of mass fractions occur and near the region of laminar–turbulent transition.

Calculations on simple shapes (cones) were performed in DNS that do not include the previously discussed restrictive assumption [28, 29]. These studies show that the hypothesis of decorrelation between fluctuations in temperature and concentrations is not well verified. This study sheds light on the coupling mechanisms and shows that chemical reactions can increase the turbulence, coupling being effected by the change in volume. The heat flux variations that do not exceed 7% in the cases treated are mild for this type of phenomenon. This aspect of the coupling phenomenon should probably be studied more deeply.

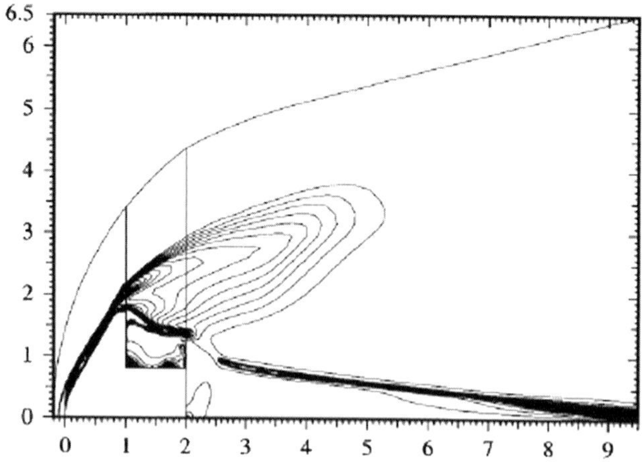

Fig. 7.8 Turbulence $k_t^{\frac{1}{2}} V^{-1}$ of flow on the ESA study object.

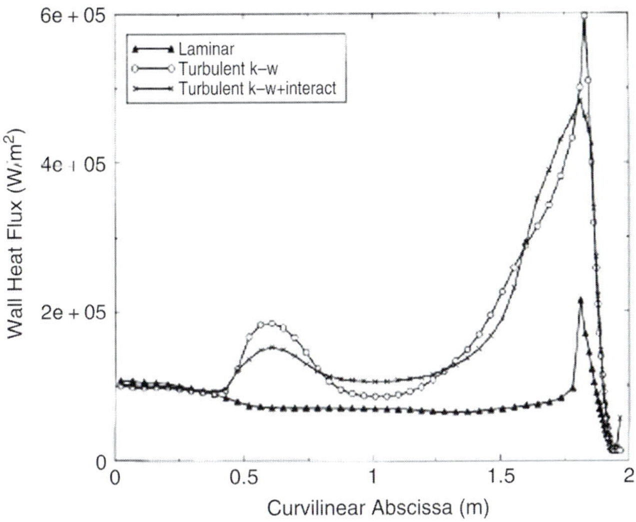

Fig. 7.9 Comparison of fluxes on the ESA object.

7.4 Laminar-Turbulent Transition

In general, the transition in the boundary layer first appears in the wake [30], then at the back of the body, and then reaches forward gradually. The oscillation of incidence, when it exists, causes rapid oscillations on the winward and leeward sides. The transition then suddenly returns to the vicinity of the stagnation point. The apparent rapidity of this phenomenon may be due to a relaminarization of the boundary layer in the violent expansion existing in the region of sphere–cone connection (Fig. 7.10). This relaminarization disappears when the Reynolds number becomes sufficiently large. This phenomenon can cause errors of diagnosis in hot wind tunnels, when one cannot make measurements in this region for topological reasons. Moreover, the phenomenon may be different in flight versus ground tests due to the low Mach number used in the latter case.

An understanding of the transition phenomenon and its modeling is poorly developed despite the amount of work devoted to this subject. This is due to the large number of influential phenomena. Among these are

- The upstream Reynolds number.
- The upstream Mach number.
- Upstream turbulence, which we saw the importance of in Sec. 7.2.1 and which is a crucial point in wind tunnel tests. This type of transition is often referred to as a bypass transition, as opposed to the "natural" transition obtained by instability of the boundary layer. Recall the role of the scale of turbulence, intensity alone being insufficient to characterize the effects.

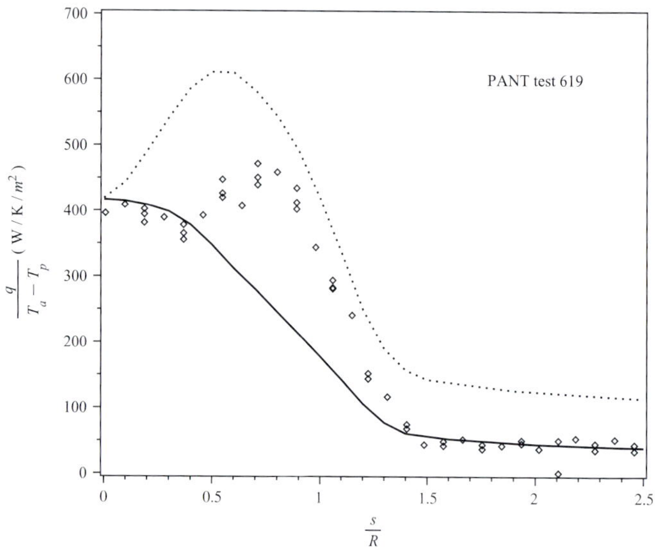

Fig. 7.10 Transitional flow (test $\mathcal{M}a \simeq 5$) (9).

- The surface state of the wall.
- The wall temperature.
- The effect of a velocity gradient.
- High temperature effects in the gas, in particular the chemical reactions.
- The wall blowing.
- Still other effects such as a Görtler-type instability appearing on concave shapes, the tridimensional effects, and so on.

Studies of boundary layer instabilities lead to knowledge of the mechanisms underlying turbulence. This type of study shows the formation of low-frequency vorticity (Tollmien-Schlichting waves or instability of the first mode) or, only in supersonic flow, acoustic waves (instability of the second order). They do not take into account the upstream turbulence or roughness, and cannot match the developed turbulence which is at the origin of thermal effects.

It should be noted that there is no general criterion giving the beginning of transition, and that each of the methods discussed in the following sections is valid in a relatively small area. For a compilation of the problem in hypersonic, see, for example, Morkovin [31], Stetson [32], or Schneider [33–35].

7.4.1 Global Criterion

The large number of parameters governing the transition has led to the proposition of overall criteria. This necessarily leads to a high dispersion of experimental results with respect to the proposed law. This is illustrated in

Fig. 7.11 [36], which compares experiments said to be the most representative flight testing of an empirical law proposed by Vaglio-Laurin

$$\mathcal{R}e_\theta = 200\ e^{0.197\ \mathcal{M}a_e} \tag{7.35}$$

In the figure, points give the beginning of transition on objects with a graphite (G) or carbon–carbon (3DCC) nose, the heat shield being silica–resin 2-D or 3-D, or carbon–resin (TWCP). The figure shows the results obtained for various pairs of materials (front/rear) as a correlation $Re_\theta = f(\mathcal{M}a_e)$, where Re_θ is the Reynolds number formed with the momentum thickness given by Eq. (7.27). These values correspond to the onset of transition. The dispersion of experimental points is considerable.

There are equivalent laws using the Reynolds number based on the curvilinear abscissa ($\mathcal{R}e_\theta$ is proportional to $\mathcal{R}e_s^{1/2}$ in Blasius flow) on the ordinate and the reduced curvilinear abscissa s/R on the x-axis (which is not totally independent of the local Mach number $\mathcal{M}a_e$) [33]. This type of test using a Reynolds number of any kind has been widely used [32, 33]. An example is given in Fig. 7.12, which traces the evolution of the transition over time in reentry F flight [37] and a summary of wind tunnel tests expressed by the following correlation:

$$\log \mathcal{R}e_s = 0.03\mathcal{M}a_e^2 - 0.28\mathcal{M}a_e + 15.37 \tag{7.36}$$

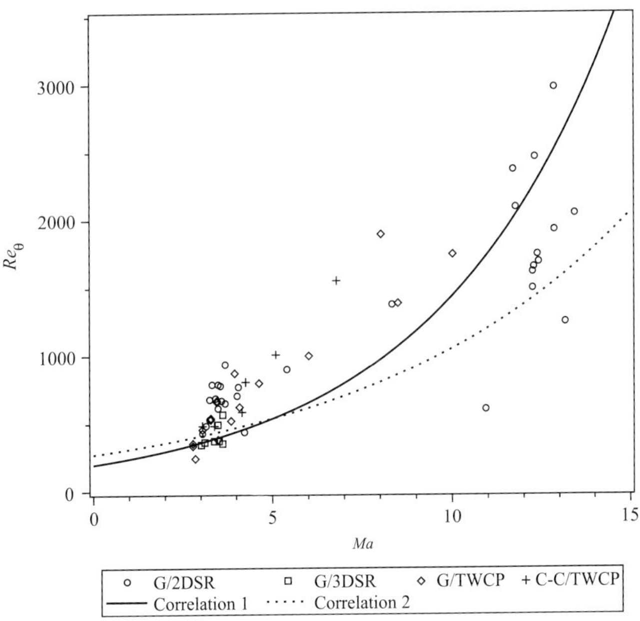

Fig. 7.11 Transition criteria issued from flight tests (36).

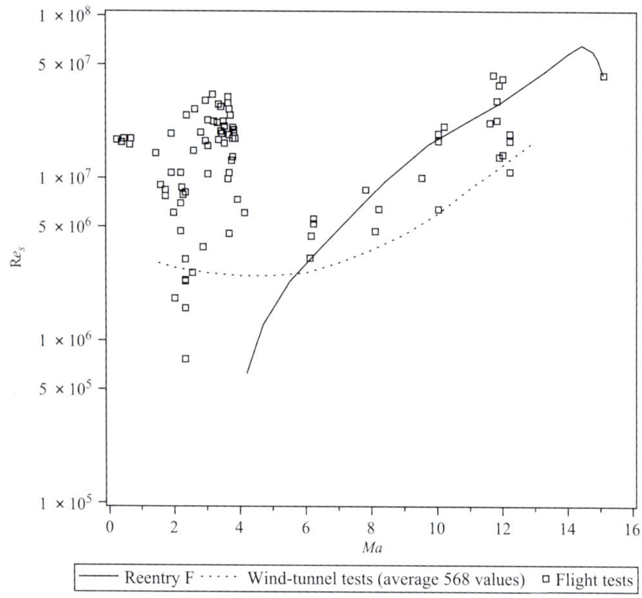

Fig. 7.12 Transition criterion $\mathcal{R}e_s = f(\mathcal{M}a_e)$ (32, 35).

This correlation is valid in the interval [1.5, 13]. This curve shows a minimum of the stability from $\mathcal{M}a = 4$ to 5, a phenomenon predicted by studies of stability of the boundary layer. A high dispersion of points is seen, particularly at low Mach number corresponding to blunt shapes. This curve and the previous one do not take into account effects such as surface roughness, wall temperature, or gas injection.

7.4.2 Method Based on the Study of Instability

Early methods aiming to predict the transition are based on calculating the development of instability. These methods were initially limited to a linear perturbation leading to a differential system on the amplitude of the perturbation (Orr–Sommerfeld). In this case, the problem is local and describes the so-called primary Tollmien–Schlichting waves. Subsequently this method was extended to a nonlocal solution (parabolized stability equations) to more fully describe the phenomenon [38]. In all cases it is necessary to start from a very precise calculation of undisturbed flow to avoid numerical artifacts.

This type of calculation, which predicts the development of instability, does not reach the first turbulent spot, so it is necessary to introduce a heuristic criterion, giving its name to the method known as e^N. The beginning of transition is predicted when the amplitude of the disturbance reaches e^N

times the initial amplitude, N being near 10 for a medium without upstream turbulence (quiet wind tunnel, free flight).

This method gives respectable results on a smooth wall but is rarely used in practice because of the cumbersome implementation. It is primarily a method to measure the influence of various factors such as nonchemical equilibrium. Its use on a rough surface is difficult [38, 39].

7.4.3 PANT Criterion and Derived Methods

The correlations described in Sec. 7.4.1 do not work near stagnation point. The thin boundary layer makes the flow sensitive to the surface roughness. (This does not necessarily mean that this effect is negligible in the experiments of Figs. 7.11 and 7.12 [35].) To make a prevision of the transition that explicitly involves roughness, the most common model uses the results of wind tunnel experiments of the PANT program [9], which performs cold wind tunnel tests on sandblasted metal models. The models' shape is variable and is supposed to represent various configurations of interests: sphere–cone, ellipsoid–cone corresponding to the shape ablated in laminar flow (Fig. 5.12), and bi-conic corresponding to the shape ablated in turbulent regime (Fig. 5.14). These models are equipped with thermocouples on a longitudinal line, which does not allow one to study the intrinsic azimuthal dispersion of the phenomenon. This point is discussed in Sec. 7.4.7. The compilation of these tests led Anderson to propose the following criterion, whose form is based on theoretical considerations [39]:

$$X = \mathcal{R}e_\theta \left(\frac{l_R}{\theta} \frac{T_e}{T_p} \right)^{0.7} = \begin{cases} \text{if 255 at sonic point} \\ \text{occurs at 215} \end{cases} \tag{7.37}$$

l_R is the roughness defined by the average heights of successive peaks and valleys. This quantity is proportional to the usual definition of the roughness average, R_a [40]. This criterion must be understood as follows: the transition will appear at the instant when the criterion is 255 at sonic point, to where the value is 215. The condition at the sonic point is intended to take into account the experimental observation that the transition does not go regularly to the stagnation point, contrary to what happens on a smooth wall. The dispersion results are remarkably low: 20% at two standard deviations (Fig. 7.13). (The physical [intrinsic] part of this dispersion is discussed in Sec. 7.4.7.) The validity of this expression is included in the interval [0.7, 15]:

- Below $X = 0.7$, transition is sensitive to external conditions (turbulence, noise).
- Beyond $X = 15$, roughness is the same order of magnitude as the boundary layer thickness calculated on a smooth surface.

This relationship allows one to show the stabilizing influence of the wall temperature for instability of the second order.

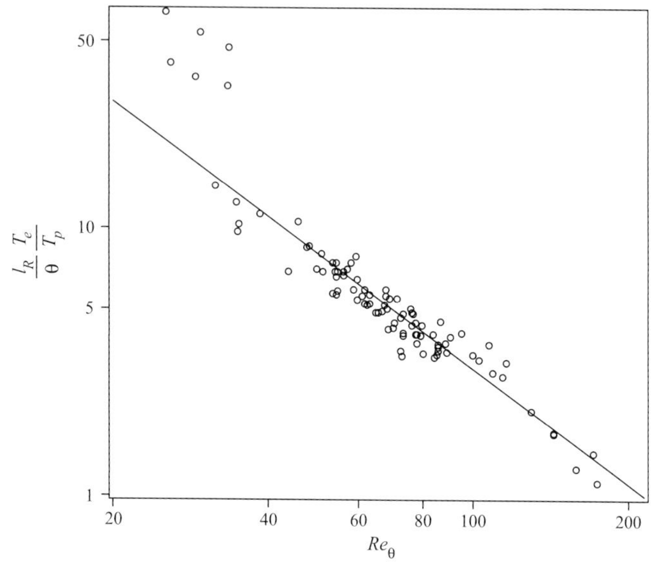

Fig. 7.13 Transition criterion in the vicinity of the stagnation point in a wind tunnel (9).

One could use (ρ_P/ρ_e) instead of (T_e/T_p) in Eq. (7.37) [9]. Similarly, there is another correlation of the same origin giving similar results. It uses the s abscissa and displacement thickness

$$\mathcal{R}e_{l_R}\left(\frac{s}{\delta^*}\right)^{1.3} = \begin{cases} 2300 \text{ at sonic point} \\ \text{occurs at } 2000 \end{cases} \tag{7.38}$$

with

$$\delta^* = \int_0^\infty \left(1 - \frac{\rho u}{\rho_e u_e}\right) \mathrm{d}y \tag{7.39}$$

The dispersion is about 20% and the domain of validity is $0.6 < \frac{s}{\delta^*} < 10$.

These results are imperfectly confirmed by free flight tunnel (Fig. 7.14). In free flight test, detection of the transition models on sphero-conical polycrystalline graphite or C–C is made by infrared imaging. (Polycrystalline graphite models are presented in [41].)

Various authors have tried to find a synthesis of these tests. A reduction of differences was obtained [42], taking into account the stabilizing effect of expansion through a law established by Van Driest and giving a rough equivalent of

$$l'_R = \frac{l_R}{1 + 350\dfrac{l_R}{R}}$$

The value of the roughness l_R used in this approach is redefined from the distribution function of this quantity as the value at 30% of probability to be greater than a given value. The transitional law becomes

$$\mathcal{R}e_\theta = \begin{cases} 500X^{-1.5} & \text{if } X > 1 \\ 500 & \text{if } X < 1 \end{cases} \tag{7.40}$$

with

$$X = \frac{l'_R}{\theta}\frac{T_e}{T_p}$$

The dispersion of experimental points is 30% on X.

Another approach [43] from theoretical considerations gives an expression similar to that of Anderson

$$\mathcal{R}e_\theta \frac{l_R}{\theta}\left(\frac{T_e}{T_p}\right)^{1.27} = 434 \tag{7.41}$$

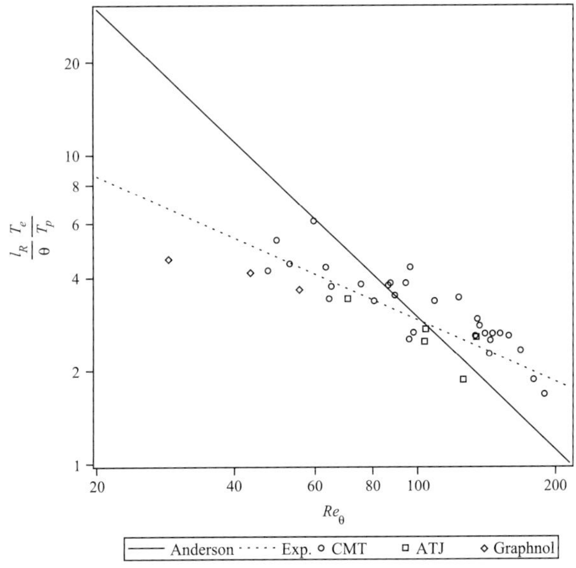

Fig. 7.14 Comparison of transition criteria issued from wind tunnel and free flight tunnel (AEDC Track G).

This expression is used to correct the excessive dependence of (l_R/θ) in Fig. 7.14 while remaining in reasonable agreement with the PANT tests.

It should be remembered that in spite of the success of such criteria, the transition will certainly depend on the geometry of roughness. An approach based primarily on sanded metallic surfaces or graphite, reasonably described by a statistical approach, cannot be applied in the strict sense to a composite for which the surface is almost periodic (Chapter 6).

7.4.4 Effect of Blowing

The injection of gas moves the transition upstream. This effect can be taken into account through physical arguments. Indeed, the choice of perturbation parameter

$$\frac{l_R}{\theta}\frac{T_e}{T_p}$$

in the transition criterion of Eq. (7.37) is based on the correlation of this quantity with reduced kinetic energy of the flow at the height l_R, that is,

$$\frac{(\rho u^2)\big|_{y=l_R}}{\rho_e u_e^2}$$

calculated without blowing. The relationship between these two quantities, trivial in the vicinity of the wall (it is demonstrated easily by linearizing the velocity) is demonstrated in the area of interest by a series of calculations of the boundary layer [9]. The dispersion is very low. The effect of the parietal injection of gas will be taken into account in the same way by replacing the perturbation parameter by its modified value

$$\mathcal{R}e_\theta\left[\frac{l_R}{\theta}\frac{1}{0.1\mathcal{B}' + \dfrac{\rho_e}{\rho_p}(1+0.25\mathcal{B}')}\right]^{0.7} = \begin{cases}255 \text{ at sonic point}\\ \text{occurs at } 215\end{cases} \qquad (7.42)$$

The dispersion in this case is a little stronger, but still very reasonable [9]. However, we note that this relation is obtained in a relatively arbitrary way.

Although many experiments have explored the area, synthesis results were not achieved [35].

7.4.5 Transition Length

The area of laminar–turbulent transition is composed of turbulent regions of space ("spots") developing independently until they merge (Fig. 7.15). The description of this phenomenon involves the function of intermittency γ,

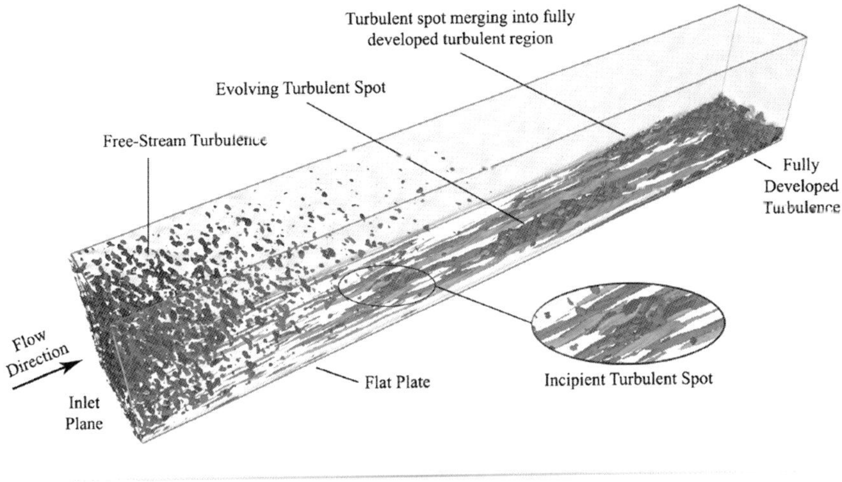

Fig. 7.15 Region of laminar–turbulent transition; measurements made on a flat plate (45).

the time average of the turbulent state at a given point. This amount ranges from 0 in early transition to 1 at the end of it. From geometrical considerations, Emmons has shown [44] that γ follows an exponential in 2-D flows

$$\gamma = 1 - \exp\left[\int_{s_0}^{s} \lambda(s - s')^2 g(s)\, \mathrm{d}s'\right] \tag{7.43}$$

In this equation, s is the curvilinear abscissa, s_0 is the abscissa of the start of transition, g is the rate of formation of turbulent spots, and λ describes the spot shape and geometrical development during its propagation.

In the case where λ and g are supposed constants, this expression simplifies to

$$\gamma = 1 - \exp\left[-\alpha(s - s_0)^3\right] \tag{7.44}$$

with $\alpha = (\lambda g/3)$.

This law correctly describes the experiment (Fig. 7.16). Various hypotheses on g have been made: Dirac distribution in s_0, proportional to s [44]. The results are indistinguishable through a translation in s. However, the origin of the transition s_0 is partially arbitrary, so all authors do not have the same definition and the results are relatively scarce.

This type of law is commonly generalized to hypersonic on a very reduced experimental basis.

Rather than giving the value of the parameter α, the characterization can be done through a length of transitional region Δs. An often-used expression is attributed to Potter and Whitfield. It is valid for rough surfaces and covers the domain $0 < \mathcal{M}a < 8$. Note that the following expression is different from that usually given. It was recalculated from the results given by [49]

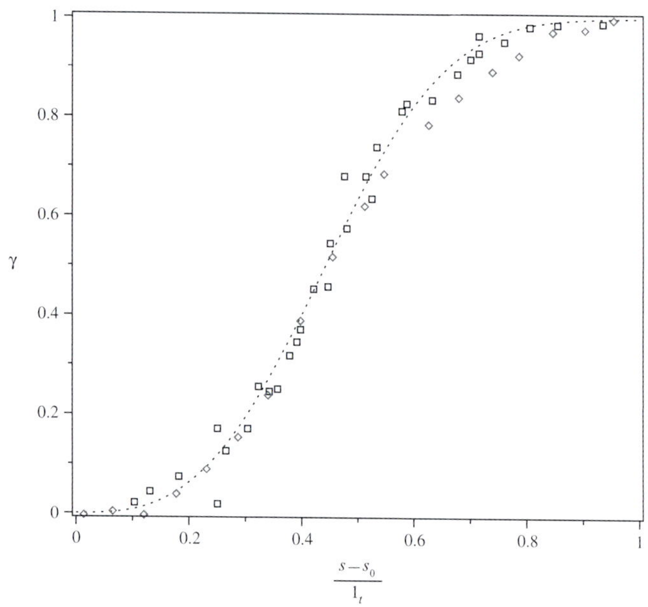

Fig. 7.16 Intermittency function (incompressible and $\mathcal{M}a = 8$); l_t is transition length (46–48).

by choosing a dependence $\mathcal{R}e^{3/4}$ compatible with the dimensional analysis for the Blasius flow [50]

$$\mathcal{R}e_{\Delta s} = \mathcal{R}e_{s_0+\Delta s}^{\frac{3}{4}}(12 + 1.1\mathcal{M}a_e^{1.6}) \tag{7.45}$$

In this expression, $\mathcal{R}e_{\Delta s}$ uses the abscissa of the end of transition $s_0 + \Delta s$, the latter being considered more easily defined experimentally with precision. This choice does not facilitate its practical use with a standard transition criterion. The error of the correlation is less than 15% (Fig. 7.17). The parameter in abscissa for the figure is $p = (\mathcal{R}e_s \times 10^{-6})^{2/3}$ $(42 + 2.2\mathcal{R}a_e^{1.92})$. Note that this expression, recalculated from the tests, does not correspond to that given in [49].

7.4.6 Modeling the Transition Zone and Relaminarization

The transcription of models describing the transition zone can be achieved by an ad hoc method of weighting the turbulent quantities by γ. However, this has no justification other than the fact that the error is in any case limited. We can, however, observe some cases exceeding the turbulent flow when γ is close to unity ("overshoot").

The criteria for the beginning of transition outlined in Secs. 7.4.3 and 7.4.4 are not well adapted to modern codes because they involve integral quantities (momentum thickness, for example) or values at the front of the boundary layer, which are not always easy to evaluate from Navier–Stokes computation in the case of a flow with velocity gradient, for example. Moreover, standard turbulence models are not capable of predicting the onset of transition. They are rarely capable of simulating the transitional zone from an arbitrarily fixed beginning from relations given in Secs. 7.4.3 and 7.4.4 [51]. Their description of a relaminarization (Fig. 7.10) is generally not better [52].

The medium can be described by considering the function of spatio-temporal intermittency $\mathcal{I}$ whose time average is γ, traditional intermittency factor. We define conditional means [53]

$$\begin{cases} \bar{\mathrm{f}}_t = \dfrac{\overline{\mathrm{f}\mathcal{I}}}{\gamma} \\ \bar{\mathrm{f}}_l = \dfrac{\overline{\mathrm{f}(1-\mathcal{I})}}{1-\gamma} \end{cases} \tag{7.46}$$

We can then derive the Navier–Stokes conditional equations, and the modifications of turbulence models. Some writers have pushed the concern of modeling to write a transport equation for γ [54]. Note that the problem is still not fully resolved because this model does not predict the onset of transition. Other authors acknowledge this by writing an additional

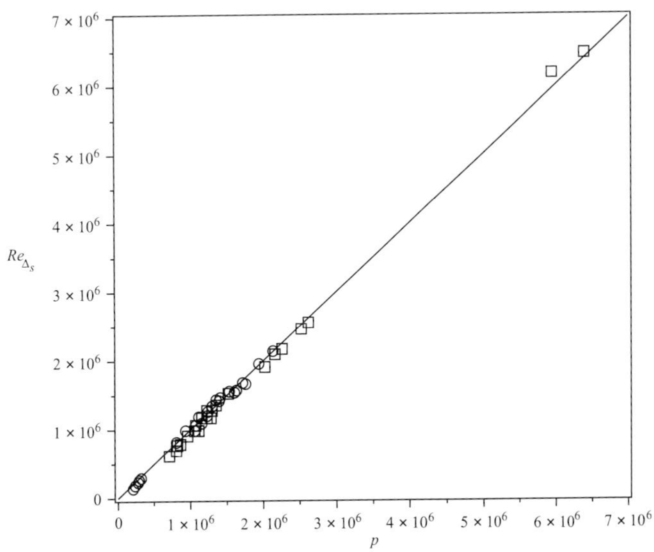

Fig. 7.17 Length of transitional region. The circles correspond to incompressible flows.

transport equation for a local quantity that can be linked to a known criterion. This can be applied to incompressible flow by using the vorticity Reynolds number

$$\mathcal{R}e_V = \frac{\rho y^2}{\mu}\frac{\partial u}{\partial y} \tag{7.47}$$

For a Blasius flow, this quantity, in the vicinity of the wall, has a maximum proportional to $\mathcal{R}e_\theta$, a value often used for the criteria of transition in the subsonic range under the form $\mathcal{R}e_\theta = f(\mathrm{Tu}_\infty)$ [55]. In our case, this type of approach is not directly usable. It seems likely, however, that it is possible to rewrite a new criterion similar to Eq. (7.37) using only local quantities. This criterion would, for example, take the following form

$$\max(\mathcal{R}e_V)\left[\frac{(\rho u^2))\big|_{y=l_R}}{\tau_p}\right]^\alpha = \beta \tag{7.48}$$

7.4.7 Asymmetry of Transition

The wind tunnel tests mentioned in Sec. 7.4.3 [56, 57] allowed us to measure the spatial dispersion of transition on different meridians for the same specimen (see Fig. 7.18). Figure 7.19 shows this dispersion. From these results it is possible to calculate the dispersion of transition σ_s/R (Fig. 7.19). One may wonder whether this dispersion plays an important role in the points that helped establish the correlation in Eq. (7.37), for example. For this, note that near the stagnation point and up to $(s/R) \simeq 0.5$ [9], T_e and θ are substantially constant and $\mathcal{R}e_\theta$ and u_e vary linearly with (s/R) (Sec. 4.5). The quantity

$$\mathcal{R}e_\theta\left(\frac{l_R}{\theta}\frac{T_e}{T_p}\right)^{0.7}$$

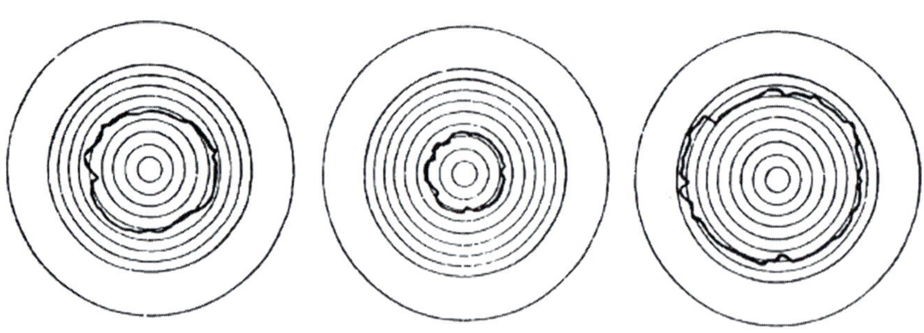

Fig. 7.18 Asymmetry of laminar–turbulent transition (56).

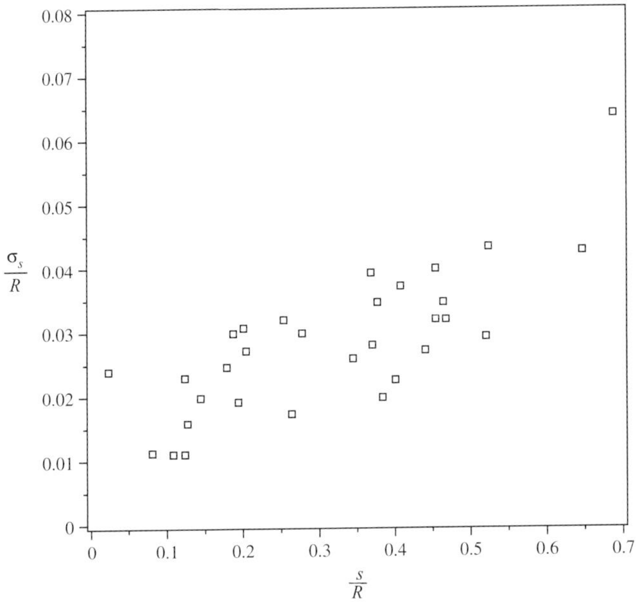

Fig. 7.19 Laminar-turbulent transition dispersion (56).

is roughly proportional to s/R. If you look at the results in Fig. 7.19, you can see that the dispersion at 2σ is close 20 to 30% depending on the region, which is the same order of magnitude as the differences observed in Fig. 7.13. Even if the comparison is skewed by the different nature of the materials used in the corresponding tests, this relativizes the differences

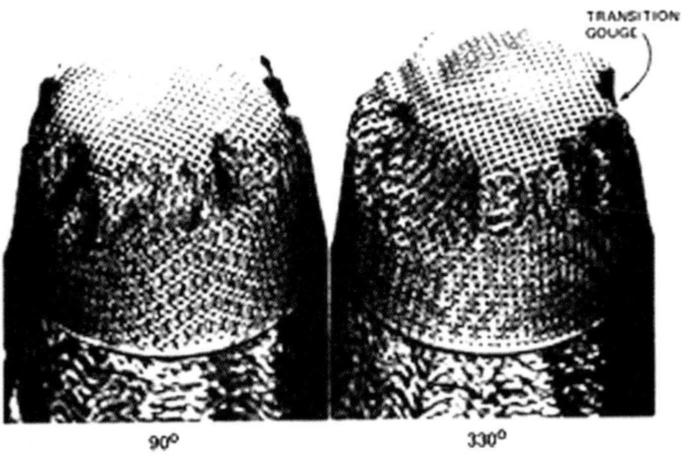

Fig. 7.20 Asymmetric ablation due to laminar-turbulent transition (58).

observed in wind tunnel PANT tests and between these tests and experiments in free flight tunnel.

This asymmetry of transition is perhaps the origin of asymmetries of ablation observed on flight tests or in plasma jet experiments (Figs. 1.19 and 7.20). There is no way of predicting this phenomenon.

References

[1] Schlichting, H., *Boundary Layer Theory*, McGraw-Hill, New York, 1979.

[2] Pope, S. B., *Turbulent Flows*, Cambridge University Press, Cambridge, UK, 2000.

[3] Akinlade, O. G., "Effects of Surface Roughness on the Flow Characteristics in a Turbulent Boundary Layer," Ph.D. Thesis, University of Saskatchewan, Saskatoon, Canada, Dec. 2005.

[4] Jimenez, J., "Turbulent Flows over Rough Wall," *Annual Review of Fluid Mechanics*, Vol. 36, 2004, pp. 173–196.

[5] Raupach, M. R., Antonia, R. A., and Rajagopalan, S., "Rough-Wall Turbulent Boundary Layers," *Applied Mechanics Review*, Vol. 44, No. 1, 1991, pp. 1–25.

[6] Carrau, A., Gallice, G., and Le Tallec, P., "Taking into Account Surface Roughness in Computing Hypersonic Re-entry Flows," *Computing Methods in Applied Sciences and Engineering*, edited by Glowinsky, R., Nova Science, New York, 1991, pp. 331–343.

[7] Ashrafian, A., Andersson, H. I., and Manhart, M., "DNS of Turbulent Flow in a Rod-Roughened Channel," *International Journal of Heat and Fluid Flow*, Vol. 25, 2004, pp. 373–383.

[8] Holden, M. S., "Studies of Boundary Layer Transition and Surface Roughness Effects in Hypersonic Flows," Report AFOSR-TR-34-0251, Dec. 1974.

[9] Wool, M. R., "Summary of Experimental and Analytical Results," Interim Report: Passive Nosetip Technology (PANT) Program, Vol. X, SAMSO-TR-74-86, Jan. 1975.

[10] Dvorak, F. A., "Calculation of Turbulent Boundary Layers on Rough Surfaces in Pressure Gradient," *AIAA Journal*, Vol. 7, No. 9, 1969, pp. 1752–1759.

[11] Dirling, R. B., Jr., "A Method for Computing Roughwall Heat Transfer Rates on Reentry Nosetips," AIAA Paper 73-763, Eighth Thermophysics Conference, California, Palm Springs, July 1973.

[12] Grabow, R. M., and White, C. O., "Surface Roughness Effects on Nosetip Ablation Characteristics," *AIAA Journal*, Vol. 13, 1975, pp. 605–609.

[13] Coleman, H. W., "Generalized Roughness Effects on Turbulent Boundary Layer Heat Transfer. A Discrete Element Predictive Approach for Turbulent Flow over Rough Surfaces," Technical Report AFTL-TR-83-90, Mississippi State University, Nov. 1983.

[14] Žukauskas, A., "Heat Transfer from Tubes in Crossflow," *Advances in Heat Transfer*, J. P. Hartnett and T. F. Irvine, Jr., Eds., Vol. 18, Academic Press, New York, 1972, pp. 87–159.

[15] Glikson, F., "Couche limite sur paroi rugueuse," Ph.D. Thesis, Paris Graduate School of Economics, Statistics and Finance, Dec. 1996.

[16] Havugimana, P.-C., Lutz, C., Saeed, F., Paraschivoiu, I., Kerevanian, G.-K., Sidorenko, A., Benard, E., Cooper, R. K., and Raghunathan, R. S., "A Comparison of Skin Friction and Heat Transfer Prediction by Various Roughness Models," AIAA Paper 2002-3052, *20th AIAA Applied Aerodynamics Conference*, St. Louis, MI, June 2002.

[17] Achdou, Y., Pironneau, O., and Valentin, F., "Effective Boundary Conditions for Laminar Flows over Periodic Rough Boundaries," *Journal of Computational Physics*, Vol. 147, 1998, pp. 187–218.

[18] Mohammadi, B., and Puigt, G., "Generalized Wall Functions for Rough Walls Based on Data Assimilation," *International Journal of Computational Fluid Dynamics*, Vol. 17, No. 6, 2003, pp. 453–465.
[19] Durbin, P. A., Medic, G., Seo, J. M., Eaton, J. K., and Song, S., "Rough Wall Modification of Two-Layer k–ϵ," *Journal of Fluids Engineering*, Vol. 123, No. 1, 2001, pp. 16–21.
[20] Wassel, A. T., and Courtney, J. F., "Environmental Characteristics and Experimental Limitations of Arc-Heated Reentry Test Facilities," Spectron Development Laboratories Report 76–6048, June 1976.
[21] Xiong, Z., and Lele, S. K., "Stagnation Point Flow and Heat Transfer under Free-Stream Turbulence," Stanford University Report AFRL-SR-AR-TR-04, April 2004.
[22] Durbin, P., "On the k–ϵ Stagnation Point Anomaly," *International Journal of Heat and Fluid Flow*, Vol. 17, No. 1, 1996, pp. 89–90.
[23] Behnia, M., Parneix, S., and Durbin, P., "Prediction of Heat Transfer in an Axisymmetric Turbulent Jet Impinging on a Flat Plate," *International Journal of Heat and Mass Transfer*, Vol. 41, No. 12, 1998, pp. 1845–1855.
[24] Wassel, A. T., and Denny, V. E., "Heat Transfer to Axisymmetric Bodies in Super- and Hypersonic Turbulent Streams," *Journal of Spacecraft and Rockets*, Vol. 14, No. 4, 1977, pp. 212–218.
[25] Narayan, J. R., and Girimaji, S. S., "Turbulent Reacting Flow Computations Including Turbulent Chemistry Interactions," AIAA Paper 92-0342, *30th Aerospace Sciences Meeting and Exhibit*, Reno, Nevada, Jan. 1992.
[26] Borghi, R., and Champion, M., "Theory and Modeling of Flames," Editions Technip, Feb. 2000 (in French).
[27] Nguyen-Bui, N. T.-H., and Duffa, G., "Modeling and Simulation of Turbulent Reacting Flow Around a Hypersonic Space Probe," *Journal of Spacecraft and Rockets*, Vol. 43, No. 4, 2006, pp. 919–923.
[28] Duan, L., and Martin, P., "Effect of Turbulence Fluctuations on Surface Heating Rate in Hypersonic Turbulent Boundary Layers," AIAA Paper 2009-4040, *39th AIAA Fluid Dynamics Conference*, San Antonio, Texas, June 2009.
[29] Duan, L., and Martin, P., "Effect of Finite-Rate Chemical Reactions on Turbulence in Hypersonic Turbulent Boundary Layers," AIAA Paper 2009-0588, *47th AIAA Aerospace Sciences Meeting*, Orlando, Florida, Jan. 2009.
[30] Pallone, A. J., Erdos, J. I., and Eckerman, J., "Hypersonic Laminar Wakes and Transition Studies," AVCO Corporation Technical Report RAD-TM-63-33, Rev. 1, Nov. 1963.
[31] Morkovin, M. V., "Critical Evaluation of Transition from Laminar to Turbulent Shear Layers with Emphasis on Hypersonically Travelling Bodies," AFFDL Report TR-68-149, March 1969.
[32] Stetson, K. F., "Hypersonic Boundary Layer Transition," Second Joint Europe/U.S. Short Course in Hypersonics, Colorado Springs, CO, *Advances in Hypersonics*, J. J. Bertin, J. Periaux, and J. Ballman, Eds., Vol. 1, Birkhaüser, Jan. 1989.
[33] Schneider, S. P., "Hypersonic Boundary-Layer Transition on Blunt Bodies with Roughness," AIAA Paper No. 2008-501, *46th Aerospace Sciences Meeting and Exhibit*, Reno, Nevada, Jan. 2008.
[34] Schneider, S. P., "Flight Data for Boundary-Layer Transition at Hypersonic and Supersonic Speeds," *Journal of Spacecraft and Rockets*, Vol. 36, No. 1, 1999.
[35] Schneider, S. P., "Hypersonic Boundary-Layer Transition with Ablation and Blowing," AIAA Paper 2008-3730, *38th Fluid Dynamics Conference and Exhibit*, Seattle, Washington, June 2008.
[36] Berkowitz, A. M., Kyriss, C. L., and Martelluci, A., "Boundary Layer Transition. Flight Test Observations," AIAA Paper 77-125, Jan. 1977.
[37] Wright, M. J., and Zoby, E. V. "Flight Measurements of Boundary-Layer Transition on a 5° Half-Angle Cone at Freestream Mach Number of 20 (Reentry F)," NASA Report TM X-2253, May 1971.

[38] Arnal, D., and Délery, J., "Laminar-Turbulent Transition and Shock Wave/Boundary Layer Interaction," RTO AVT Report RTO-EN-AVT-116, 2004.
[39] Reshotko, E., "Roughness-Induced Transition Transient Growth in 3-D Supersonic Flows," *Advances in Laminar-Turbulent Transition Modelling*, NATO EN-AVT-151-16, Sept. 2009.
[40] Duffa, G., Vignoles, G., Goyhénéche, J.-M., and Aspa, Y., "Ablation of Carbon-Based Materials: Investigation of Roughness Set-up from Heterogeneous Reactions," *International Journal of Heat and Mass Transfer*, Vol. 48, 2005, pp. 3387–3401.
[41] Reda, D. C., "Correlation of Nosetip Boundary-Layer Transition Data Measured in Ballistics-Range Experiments," Sandia Report SAND 79-0649, Nov. 1979.
[42] Batt, R. G., and Legner, H. H., "A Review of Roughness-Induced Nosetip Transition," *AIAA Journal*, Vol. 21, No. 1, 1983, pp. 7–22.
[43] Kontinos, D. A., and Stackpoole, M., "Post-Flight Analysis of the Stardust Sample Return Capsule Earth Entry," AIAA Paper 2008-1197, *46th AIAA Aerospace Sciences Meeting and Exhibit*, Reno, Nevada, Jan. 2008.
[44] Schook, R., "Bypass Transition Experiments in Subsonic Boundary Layers," Ph.D. thesis, Technische Universiteit Eindhoven, Eindhoven, The Netherlands, 2000.
[45] Johnson, G. P., Calo, V. M., and Gaither, K. P., "Interactive Visualization and Analysis of Transitional Flow," *IEEE Transactions on Visualization and Computer Graphics*, Vol. 14, No. 6, 2008, pp. 1420–1427.
[46] Schubauer, G. B., and Klebanoff, P. S., "Contributions on the Mechanics of Boundary-Layer Transition," NASA Technical Note 1289, 1956.
[47] Smits, A. J., Martin, P., and Girimaji, S., "Current Status of Basic Research in Hypersonic Turbulence," AIAA Paper 2009-151, *39th AIAA Fluid Dynamics Conference and Exhibit*, Orlando, Florida, Jan. 2009.
[48] Volino, R. J., and Simon, T. W., "Bypass Transition in Boundary Layers Including Curvature and Favorable Pressure Gradient Effects," NASA Report CR-187187, Aug. 1991.
[49] Potter, J. L., and Whitfield, J. D., "Effects of Slight Nose Bluntness and Roughness on Boundary-Layer Transition in Supersonic Flow," *Journal of Fluid Mechanics*, Vol. 12, No. 4, 1962, pp. 501–535.
[50] Narasimha, R., "The Laminar-Turbulent Transition Zone in the Boundary Layer," *Progress in Aerospace Science*, Vol. 22, 1985, pp. 29–80.
[51] Launder, B., and Sandham, N. (eds.), *Closure Strategies for Turbulent and Transitional Flows*, Cambridge University Press, Cambridge, UK, 2002.
[52] Rumsey, C. L., and Spalart, P. R., "Turbulence Model Behavior in Low Reynolds Number Regions of Aerodynamics Flowfields," *AIAA Journal*, Vol. 47, No. 4, 2009, pp. 982–993.
[53] Steelant, J., and Dick, E., "Modeling of Bypass Transition with Conditioned Navier-Stokes Equations Coupled to an Intermittency Transport Equation," *International Journal for Numerical Methods in Fluids*, Vol. 23, No. 3, 1996, pp. 193–220.
[54] Cho, J. R., and Chung, M. K., "A k–e–y Equation Turbulence Model," *Journal of Fluid Mechanics*, Vol. 237, 1992, pp. 301–322.
[55] Langtry, R. B., and Menter, F. R., "Transition Modeling for General CFD Applications in Aeronautics," AIAA Paper 2005-522, *43rd AIAA Aerospace Sciences Meeting and Exhibit*, Reno, Nevada, Jan. 2005.
[56] Raper, R. M., "Boundary Layer Transition on Large-Scale CMT Graphite Nosetips at Reentry Conditions," AEDC Report TR-79-45, Jan. 1981.
[57] Reda, D. C., "Comparative Transition Performance of Several Nosetips Materials as Defined by Ballistic-Range Testing," *25th International Instrumentation Symposium*, Anaheim, California, May 1979.
[58] Nestler, D. E., "Increased Re-Entry Thermal Protection Requirements Due to Surface Roughness," SAE Technical Paper Series 851381, July 1985.

Chapter 8 Pyrolysis and Pyrolyzable Materials

- Understand how resins undergo pyrolysis
- Show mass loss
- Calculate temperature profiles
- Deduce enthalpy of formation of a compound
- Estimate porosity
- Measure permeability

Many materials are used for reentry, including polymers. The most often used is a resin binder to a fiber network [2-D or 3-D resin–silica materials, or carbon–resin tape-wrapped carbon phenolic (TWCP) or phenolic-impregnated carbon ablator (PICA); see Sec. 1.3.1]. It can also be a material used for tests at moderate temperatures such as polytetrafluoroethylene (PTFE). Resins undergo a thermal degradation phenomenon called pyrolysis, which acts as a thermal barrier.

- Its degradation is endothermic.
- The gases produced migrate to the surface, producing an energy exchange with the solid.
- The porosity of the material increases, causing a decrease in heat conduction.
- Products formed are likely to react in the boundary layer by endo-energetic reactions (but sometimes exo-energetic, such as in the case of hydrogen).
- Being injected into the boundary layer, these gases are opposed to exchanges of mass and energy. This is the phenomenon of "blowing" (see Sec. 4.4.3).

However, these gases do not produce only beneficial effects; they may be responsible for undesired effects like turbulence or delamination phenomena. The optimal weight loss is a compromise: the material must be sufficiently porous after pyrolysis to obtain the various effects listed and dense enough to withstand the thermomechanical stresses. This is particularly true for 2-D composites, which are sensitive to the separation between layers of tissue (ply lift), which can cause significant loss of material on singular points (wedgeout). Moreover, the decrease in density affects, of course, the ablation velocity.

Among all possible resins, one has been widely used and described in the literature. It is phenolic resin (phenol–formaldehyde polymer), which, in addition to the qualities one expects in operation, has the advantage of relatively easy implementation.

Before turning to phenolic composites, we will use PTFE as an example in the following section.

8.1 A Simple Example: PTFE

8.1.1 Pyrolysis of PTFE

Polytetrafluoroethylene (PTFE or Teflon) is used as a low heatflux ablative material. It is a polymer of global formula $(-CF_2 = CF_2-)n$. From the crystalline state at room temperature, it passes into an amorphous state at 600 K. Its density is around $2200\ \mathrm{kg \cdot m^{-3}}$. The change phase enthalpy is $58\ \mathrm{kJ \cdot kg^{-1}}$. Its thermal decomposition [1] leads mainly to the monomer $F_2C = CF_2$ over 95% in mass and to smaller quantities of fragments of chain, at C_3F_6 and other fluorocarbons to be ignored thereafter. Its thermal decomposition, leading to the complete disappearance of the material between 750 K and 780 K, is described by an Arrhenius law [2]

$$\frac{\partial \bar{\rho}}{\partial t} = -A\ \bar{\rho} \exp\left(\frac{T_A}{T}\right) \tag{8.1}$$

where $\bar{\rho}$ is the apparent density including porosity.

The activation temperature is $T_A = 40{,}970\ \mathrm{K}$ and the measured frequency factor is $A = 8.58 \times 10^{20}\ \mathrm{s^{-1}}$.

The low thermal conductivity, less than $0.5\ \mathrm{W \cdot m^{-1} \cdot K^{-1}}$ [3], and the total disappearance of material lead to very large differences in the density near the wall during the ablative phase.

8.1.2 Model of Equivalent Ablation

We are seeking a model that, asymptotically, will replace volume response with surface ablation producing the same mass flux rate. For this we will make the following simplifying assumptions, valid in a thin slice of thickness ϵ in the vicinity of the wall (Fig. 8.1). Modeling involves replacing the domain

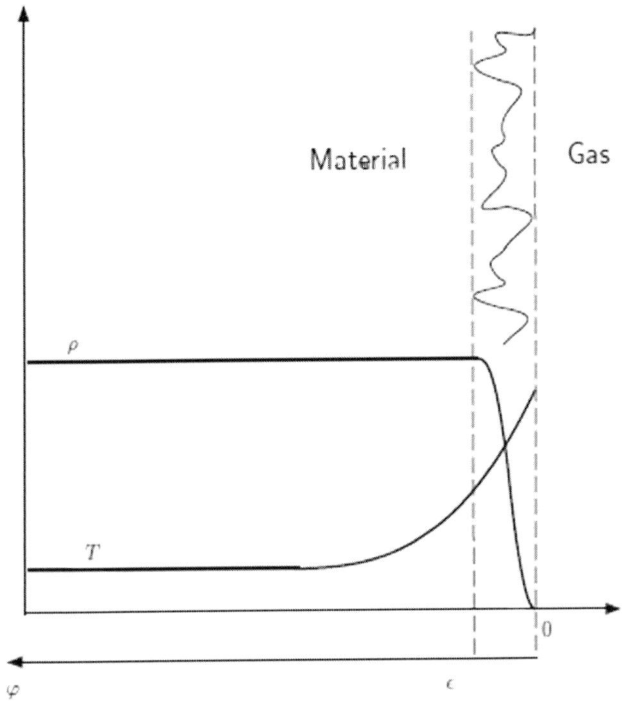

Fig. 8.1 Schematic of asymptotic model.

with a boundary condition that gives the same rate of ablation $v_a(T_p)$. It is assumed that

- The temperature varies linearly

$$T(\varphi) = T_p + \varphi \frac{\mathrm{d}T}{\mathrm{d}\varphi} = T_p - \varphi \frac{\dot{q}_{\text{cond}}}{\lambda(T_p)} \tag{8.2}$$

where

$$\dot{q}_{\text{cond}} = \left.\frac{\mathrm{d}T}{\mathrm{d}\varphi}\right|_p$$

is the conduction wall flux.

- Temperature variation is sufficiently low so that we can linearize $(1/T)$

$$\frac{1}{T} \simeq \frac{1}{T}\left[1 + \frac{\varphi\, \dot{q}_{\text{cond}}}{T_p\, \lambda(T_p)}\right] \tag{8.3}$$

- The problem is locally stationary, and Eq. (4.78) is applicable

$$\frac{\partial \bar{\rho}}{\partial t} = -v_a \frac{\partial \bar{\rho}}{\partial \varphi} \tag{8.4}$$

In this equation, replacing the time variation of $\bar{\rho}$ with its value given by Eq. (8.1), we get

$$v_a \frac{\partial \bar{\rho}}{\partial \varphi} = A \bar{\rho} \exp\left(-\frac{T_A}{T}\right) \tag{8.5}$$

This equation shows that at the material surface, defined by $\bar{\rho} = 0$ (see Fig. 8.1), we have

$$\left.\frac{\partial \bar{\rho}}{\partial \varphi}\right|_p = 0$$

and

$$\int_0^{\varphi} \frac{d\bar{\rho}}{\bar{\rho}} = C^{ste} \times \varphi \quad \text{when } \varphi \to 0 \tag{8.6}$$

By replacing $(1/T)$ in Eq. (8.5) with its expression resulting from Eq. (8.3), we get

$$v_a \frac{d\bar{\rho}}{\bar{\rho}} = A \exp\left(-\frac{T_A}{T}\right) \exp\left[\frac{\varphi T_A \dot{q}_{\text{cond}}}{T_p^2 \lambda(T_p)}\right] \tag{8.7}$$

So by integrating from 0 to ϵ, and given Eq. (8.6)

$$v_a \ln \bar{\rho}_v = A \exp\left(-\frac{T_A}{T}\right) \frac{T_p^2 \lambda(T_p)}{T_A \dot{q}_{\text{cond}}} \exp\left[\frac{\epsilon T_A \dot{q}_{\text{cond}}}{T_p^2 \lambda(T_p)}\right] \tag{8.8}$$

and, by passing to the limit $\epsilon \to 0$

$$v_a = \frac{A T_p^2 \lambda(T_p)}{T_A \dot{q}_{\text{cond}} \ln \bar{\rho}_v} \exp\left(-\frac{T_A}{T}\right) \tag{8.9}$$

where $\bar{\rho}_v$ is the initial density of the material.

Moreover, the conduction flux is expressed simply by Eq. (4.80) in the event of a stationary energy exchange (repeated here)

$$\dot{q}_{\text{cond}} = \bar{\rho}_v v_a [h_g(T_p) - h_s(T_p)] \tag{8.10}$$

So finally

$$v_a = \left[\frac{A T_p^2 \lambda(T_p) \exp\left(-\frac{T_A}{T_p}\right)}{T_A \bar{\rho}_v \ln \bar{\rho}_v [h_g(T_p) - h_s(T_p)]}\right]^{\frac{1}{2}} \tag{8.11}$$

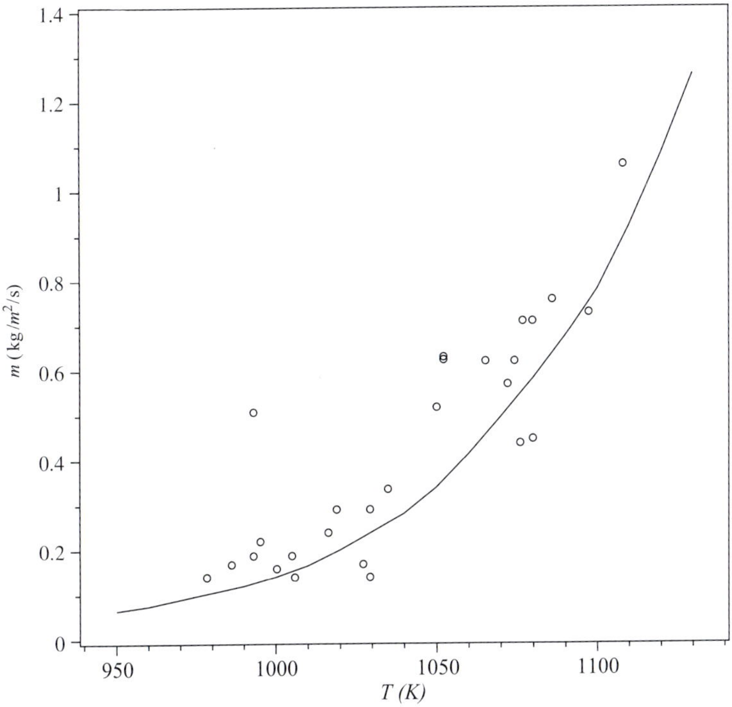

Fig. 8.2 Ablation law of PTFE compared with plasma jet test results (5); assumed emissivity $\epsilon = 0.8$.

This equation connects v_a to T_p and represents all the heterogeneous reactions. It is a "law of ablation" within the meaning given by Munson [4]. It does not depend on boundary layer exchanges and, as such, is not likely to be represented by an expression of type $f(\mathcal{B}', T, p) = 0$ (Sec. 4.3.1). Munson's graph (Fig. 8.2) shows an exponential variation present in the classical ablation problems. The proposed model reproduces the correct results despite the large extrapolation of experimental data on the pyrolysis of this material.

8.1.3 Gaseous Reactions

A thermodynamic analysis demonstrates that, for the temperatures of interest ($T_p < 1100$ K), major species issued from PTFE are C_2F_4, CF_4, COF_2, and CO_2. These species are produced by the following reactions:

$$C_2F_4 + O_2 \rightleftharpoons 2\,COF_2$$
$$2\,COF_2 \rightleftharpoons CF_4 + CO_2$$

This analysis is confirmed by optical measurements made in an ablation test [6].

The resolution of the reaction system by the method described in Sec. 5.7 shows that these reactions are, in fact, limited:

- At low temperatures ($T < 950$ K) by the formation of C_2F_4 by depolymerization
- At high temperatures ($T > 950$ K) by diffusion of oxygen in the boundary layer

Figure 8.3 shows the wall composition versus temperature in a given example ($p = p_0$, $\rho_e u_e C_H = 0.1\ \mathrm{kg \cdot m^{-2} \cdot s^{-1}}$), which illustrates this discussion.

The calculation of the corresponding ablation enthalpy (Fig. 8.4), as defined in Sec. 4.6, Eq. (4.70), shows that:

- At low temperatures ($T < 1000$ K) the phenomenon is dominated by the exothermic reaction with oxygen.
- At high temperatures ($T > 1000$ K) the phenomenon is endothermic, the "sublimation" (actually transformation of the polymer into a monomer)

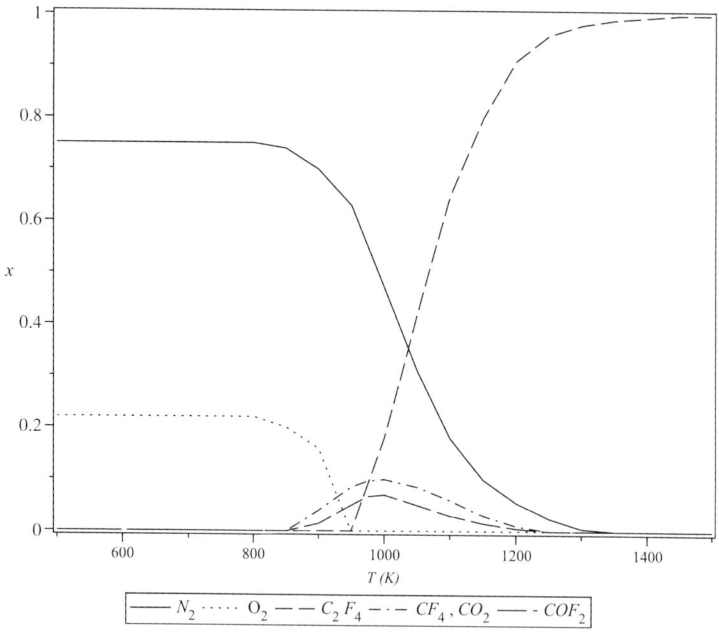

Fig. 8.3 Wall volume composition of the gas over an ablative surface of PTFE.

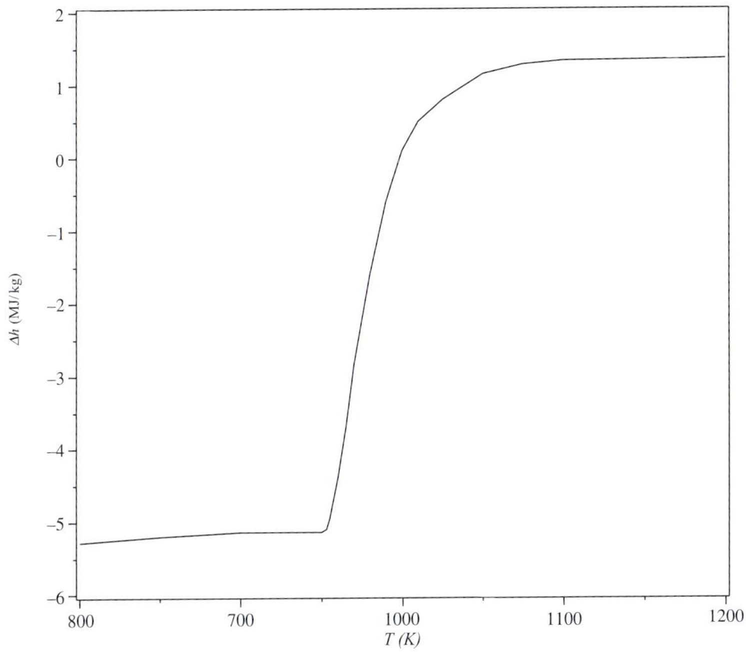

Fig. 8.4 Apparent ablation enthalpy of PTFE.

being the dominant phenomenon. This value is confirmed by experiments (see Fig. 4.8).

8.2 Phenolic Resin

8.2.1 Physicochemical Evolution

The phenolic resin, like the rayon fiber with low thermal conductivity associated with carbon-reinforced plastic materials, is sensitive to water (Appendix F). This is very important for use in nozzles that are known to reside in wetlands (areas of equatorial launch). This is not the case for materials used in reentry, which were degassed in vacuum. However, it should be noted that the samples used in experiments must be treated accordingly. Unfortunately, in the literature these drying times are underestimated, which sometimes leads to problems when interpreting the pyrolysis experiments [7].

8.2.1.1 Phenolic Resin Decomposition

The phenolic resin is typically a substance of density $\rho = 1270\ \mathrm{kg/m^3}$. Elemental mass composition is given in Table 8.1. Some resins may contain inorganic residues.

Table 8.1 Elemental Composition of Phenolic Resin (8)

Element	Mass Fraction (%)
C	71 to 78
O	16 to 18
H	5 to 6
N	0 to 7

The phenolic resin can be described as a first approximation by the following general formula:

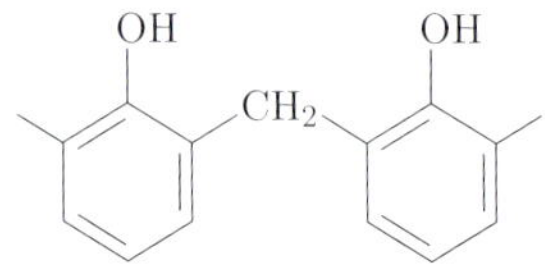

One can verify this by calculating the corresponding composition and comparing it to Table 8.1. The differences can come from a variety of reasons: cross-linking, heteroxy bridges ($-H_2-O-CH_2-$), substitution of an $-H$ by CH_2OH, or the presence of nitrogen. This structure has a polar $-OH$ bond able to stick water molecules.

8.2.1.2 Thermal Decomposition

Note first that the gases formed during pyrolysis are driven to the surface and that, consequently, the decomposition takes place in an atmosphere formed by the pyrolysis gases themselves in the absence of oxygen from the atmosphere. The experiments we will use are conducted in a neutral atmosphere, on samples previously dried. We can distinguish [9] three sets of reactions (temperatures quoted correspond to experiments carried out at a very slow rise in temperature), identified by infrared spectrometry of the gases formed:

- Between 600 K and 800 K, two reactions lead to the depolymerization: creation of the monomers phenol

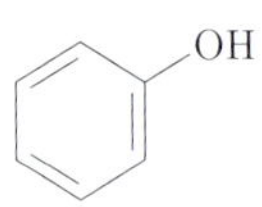

cresol (methylphenol)

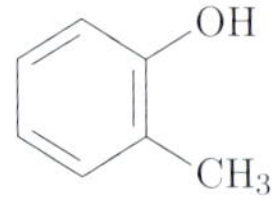

Table 8.2 Pyrolysis Reactions in the Quasi-static Regime; Based on Scheme Proposed by Trick (11)

Temperature (K)	Reaction	Type
$600 < T < 800$	$R_{CH_2} \rightarrow R(CH_3)_n$	Depolymerization
	$2R_{CH_2} \rightarrow R_{CH} + H_2O$	condensation $(-CH_2-) + (-OH)$
$700 < T < 1100$	$2R_{CH_2} \rightarrow R_O + H_2O$	Condensation $(-OH) + (-OH)$
	$R_O \rightarrow R + O$	
	$R_{CH_2} + O \rightarrow R + CO + H_2$	
	$R_{CH_2} + H_2O \rightarrow R + CO + 2H_2$	Reaction $(-CH_2-) + H_2O$
	$R_{CH_2} + H_2 \rightarrow R + CH_4$	Reaction $(-CH_2-) + H_2$
	$R_{CH} \rightarrow R + H_2$	
$T > 900$	$R \rightarrow R_d + H_2$	Hydrogen abstraction

and xylenol (dimethylphenol)

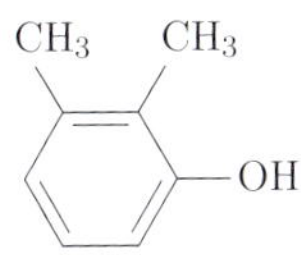

and cross-linking (linking two polymer chains) by reaction of an –OH group with a methylene bridge ($-CH_2-$) and formation of H_2O.

- Between 700 K and 1100 K, condensation occurs by binding –OH with $-CH_2-$, and a series of gas–solid reactions leads to the production of H_2, CO, and CH_4 (Table 8.2). In the table, $R(CH_3)_n$ denotes the species phenol, cresol, and xylenol. R_x is a bond of aromatic rings through X. R is a direct link between aromatic rings and R_d two fused nuclei (graphene). Compounds containing nitrogen ($C_6H_5NH_2$, NH_3) were omitted.
- Between 900 K and 1200 K, there is essentially a dehydrogenation, which produces a more or less graphitized carbon.

However, the chemical evolution cannot fully describe what is happening. Indeed, beyond 1600 K, one can observe [10] a significant decrease in porosity resulting from the rearrangement of carbon at nano-level.

One can determine the production of gas for each group of reactions from the reactions (Table 8.3, where R_2 is an intermediate species mixing R_{CH2} and R_{CH} in proportion given by the mass loss) and the laws of mass

Table 8.3 Pyrolysis Global Reactions; Molar Quantities of Produced Gases Come from (11)

R_{CH2}	$\rightarrow$	$R_2 + 0.5\ H_2O + 0.5\ R(CH_3)_n$
R_2	$\rightarrow$	$R + 0.13\ H_2O + 0.13\ CO + 0.15\ CH_4 + 0.59\ H_2$
R	$\rightarrow$	$R_d + H_2$

loss as defined in Section 8.2.1.3. An example is shown in Fig. 8.5. It uses only approximate values because the reactions contained in a group do not occur exactly at the same time. We note that there are significant differences between authors in the respective amounts of gases produced [7, 11].

Numerical simulation of the phenomena using molecular dynamics remains in its infancy [12], but can help with understanding.

8.2.1.3 Mass Loss Law

The law describing the mass loss is obtained by performing experiments made with fast pyrolysis temperature increases of up to 100 $K \cdot s^{-1}$ or more (Fig. 8.6). One should be wary, however, of the very high heating rates that do not necessarily need the necessary quasi-constant temperature of the sample (see Appendix E.1). Similarly, the high speed of temperature rise is the source of internal pressure in the pores, and its implications for the reaction are unknown and can lead to significant mechanical effects. These experiments highlight the major reactive groups described previously. We observe a strong shift of the peaks with the rate of temperature rise, due to an activation energy (Appendix E.2).

In principle, the description of the reactions is similar to that of a gas mixture (Sec. 3.1.1). The molar mass of solid compounds is unknown

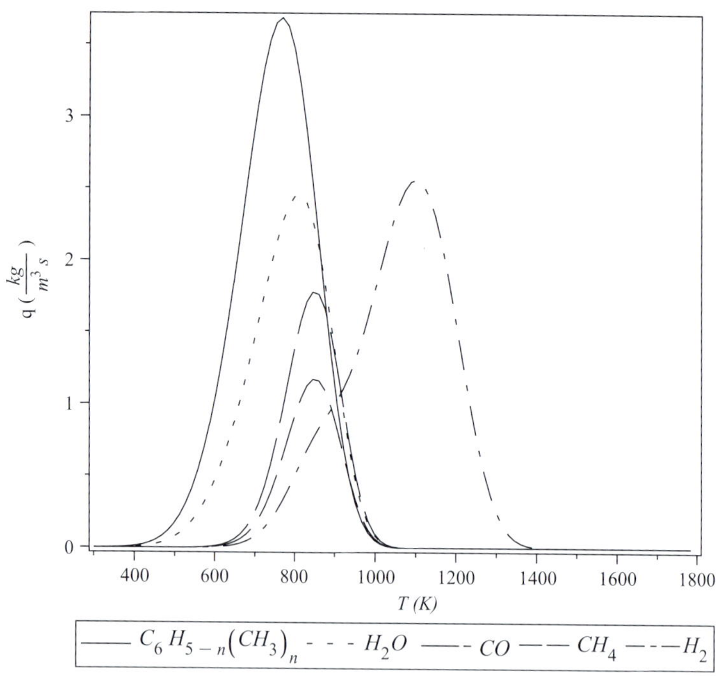

Fig. 8.5 Production of pyrolysis gases for a temperature rise of 1.67 $K \cdot s^{-1}$.

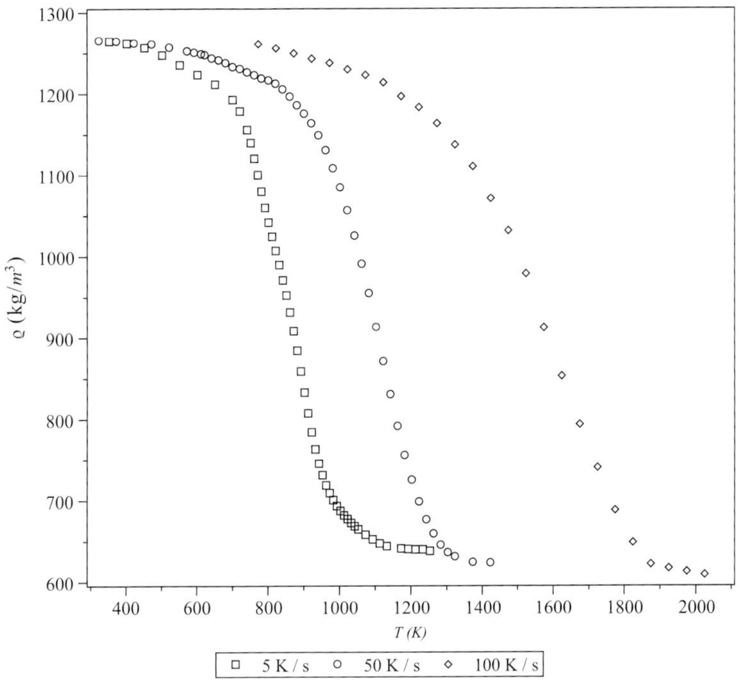

Fig. 8.6 Mass loss for different temperature-rising speeds (5, 50, and 100 K · s^{-1}).

(though large); it is included in the reaction constants. Many of these constants correspond to irreversible reactions and are therefore assumed to be zero. However, this method is not usable due to a lack of data. Modeling is usually done in a limited framework in which we ignore the effect of the gas present and assume irreversible reactions of order n_i. A species is described by its apparent partial density (mass contained in the reference volume containing porosity) $\bar{\rho}_i$, its mass yield, and g_{ij}, defined as the mass fraction of this species transformed in j during its pyrolysis. (We assume that there may be competing reactions.) The true densities of the intermediate products of pyrolysis are difficult to measure; however, they can be expressed taking into account the porosity ϵ by

$$\bar{\rho}_i = (1 - \epsilon)\rho_{s_i} \tag{8.12}$$

The $\bar{\rho}_i$ can be described, in a homogeneous medium, by the following differential system:

$$\frac{\partial \bar{\rho}_i}{\partial t} = \dot{\omega}_{s_i} = -k_i \bar{\rho}_i^{n_i} + \sum_{\substack{j=1 \\ j \neq i}}^{n_e} g_{ji} k_j \bar{\rho}_j^{n_j} \qquad i = 1, n_e \tag{8.13}$$

where n_i is the order of the reaction and g_{ij} the conversion rate of the species i in species j.

The reaction constants are usually expressed by an Arrhenius law

$$k = A \exp\left(-\frac{T_a}{T}\right) \tag{8.14}$$

The final product is a glassy carbon accounting for about 50% of the initial mass of resin. Of course, $k_{n_e} = 0$. We cannot in general give a relation for $(\mathrm{d}\bar{\rho}/\mathrm{d}t)$, except when there are assumed (most of the time implicitly) direct (no competitive reactions) and independent reactions: the reaction $n+1$ begins when the nth is complete. There is then a sum over n_r reactions

$$\frac{\partial \bar{\rho}_R}{\partial t} = \sum_{i=1}^{n_r} k_i \bar{\rho}_i^{n_i} \tag{8.15}$$

Appendix E.3 shows this approximation is valid only when the peak temperature corresponding to the reaction $n+1$ is significantly greater than that of the reaction n. This is not verified in our case.

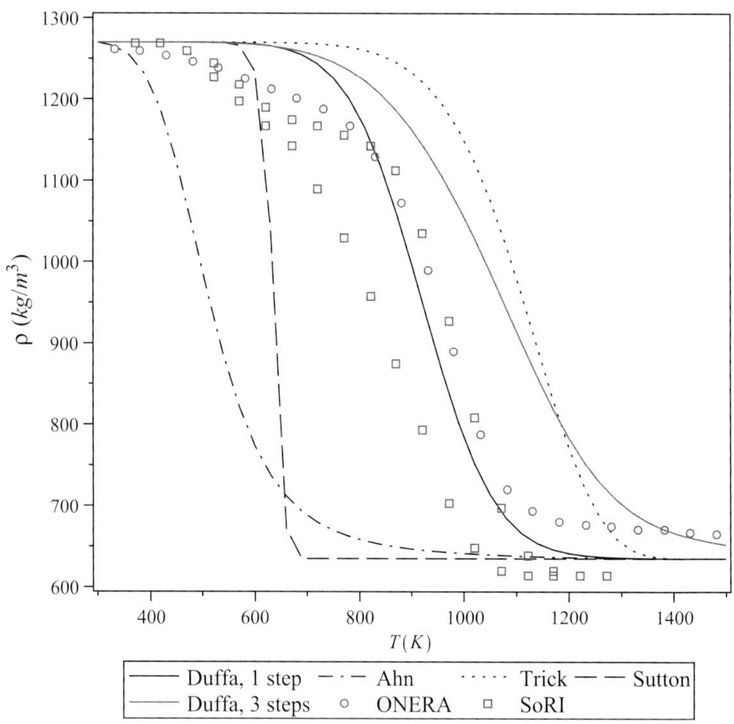

Fig. 8.7 Phenolic resin mass loss (ONERA at 20 $\mathrm{K \cdot s^{-1}}$ and two SORI curves at 17 $\mathrm{K \cdot s^{-1}}$).

Table 8.4 Pyrolysis Mass Loss Law for Phenolic Resin

Author	No. of Steps	g	A	T_A	n
Sutton [17]	1	0.5	4.7×10^{16}	24,536	1
			4.96×10^{14}	25,137	1
Trick et al. [18]	1	0.5	227	8885	1
			1931	11,733	1
			10.55	9270	1
			6.6×10^{6}	23,830	1
Duffa et al. [13]	1	0.5	1390	8070	1.29
Ahn et al. [20]	1	0.5	677	3544	2
			1.64×10^{9}	19,680	2
Duffa	3	0.64	20.1	5500	1
		0.84	115	7000	1
		0.94	122	10,000	1

The experiments are difficult. In Fig. 8.7 we show the results obtained at $20\ K \cdot s^{-1}$ [13] or $17\ K \cdot s^{-1}$ (the domain of uncertainty) [14]. The continuous curves give the results of different models proposed in the literature (see Table 8.4). In these tests, we give the results corresponding to maximum error bounds related to the inhomogeneity of the sample temperature (Appendix E). Note that these curves show an initial loss of mass due to the water present in the material. This is confirmed by experiments on materials previously dried [11, 15, 16] during the long time required for this operation (Appendix F). Interpretations of this type of experience in the form of laws using a single step [Eq. (8.15)] and exponential terms show great heterogeneity (Fig. 8.7). Indeed, many laws have been established on the basis of experiments at low temperature rise and are largely extrapolated. Recall that the operational values of temperature increase can reach several hundred degrees per second. It does not seem possible to obtain a law of mass loss covering the entire area of interest except using kinetic models with no physical basis.

8.2.1.4 Gaseous Reactions

The number of species present in the material is very large. (This does not include species that can be formed within the outer atmosphere.) It has to be limited to those likely to be present in significant quantities, with either of the following:

- Mass fraction is sufficient to influence the energy balance.
- The species plays a significant role even in small amounts. This may be the case for an important reactive intermediate or a species playing a role on the radiation in the external flow.

These include:

- Species directly coming from pyrolysis.
- Those present in local thermodynamic equilibrium such as H, C_2H_4, C_2H_2, C_2H, C_2, or C_3 (Fig. 8.8). Note that it seems there is no kinetic for the reaction leading to C_3H (linear), a species that appears in significant quantities at thermodynamic equilibrium.
- The large number of hydrocarbons from methane and phenol: C_6H_5O, C_5H_5, C_3H_3, C_2H_6, C_2H_4, CH_3, and CH_2 [21–23] (Fig. 8.9).
- Reaction intermediates such OH, O, and C.

This list is not exhaustive. The reaction kinetics can be found in the literature on combustion problems [24] or databases such as NIST.*

8.2.1.5 Thermodynamic Gas State

An order of magnitude permits us to check whether chemical equilibrium is obtained. We can define a reference length from the temperature

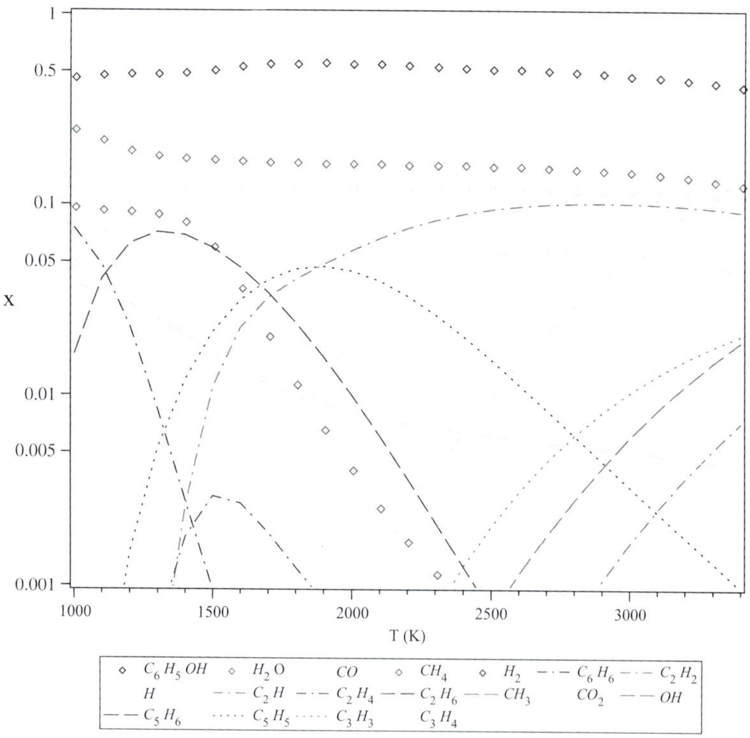

Fig. 8.8 Composition of pyrolysis gases at thermodynamic equilibrium.

* NIST Chemical Kinetics Database, http://kinetics.nist.gov/kinetics/index.jsp [retrieved 1 June 2011].

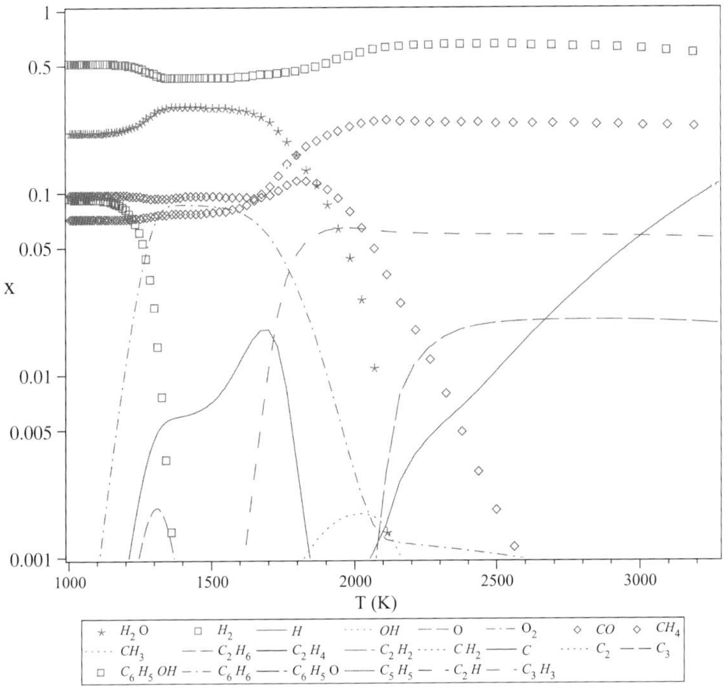

Fig. 8.9 Composition of pyrolysis gases in nonequilibrium.

profile by

$$\left|\frac{\nabla T}{T}\right|^{-1} \in [\sim 10^{-3}\,\text{m}, \infty]$$

The permeation rates are in the interval $[\sim 0, \sim 1\ \text{m}\cdot\text{s}^{-1}]$ (see Fig. 8.21 later in this chapter). The characteristic times will therefore be in the interval $[\sim 1\ \text{ms}, \infty]$. The characteristic time of chemical reaction is very variable in the temperature range that interests us; we will have all the cases, from a frozen state at low temperatures to a regime close to thermodynamic equilibrium at high temperatures. An example is given in Fig. 8.9, corresponding to the case shown in Sec. 8.6, assuming that the macroscopic quantities calculated in local thermodynamic equilibrium is a good approximation for this decoupled calculations, hypothesis justified a posteriori. By comparing the concentrations with those calculated at thermodynamic equilibrium (Fig. 8.8), we see significant differences. These can be measured simply by calculating the enthalpy of the gas. Figure 8.10 clearly shows the transition from a frozen medium to a medium in quasi-equilibrium reached around 2500 K. "Reactive" profiles in the figure correspond to a thermodynamic nonequilibrium calculation made in postprocessing the example treated in Sec. 8.6.

However, many authors make the assumption of equilibrium for the sake of simplicity. Note that this hypothesis alone does not simplify the problem. Indeed, pyrolysis taking place in several stages, the elemental composition is variable in space and time. Here we find a problem already encountered in the boundary layer due to gas injection ablation (Sec. 3.1.3). However, we can simplify the problem, assuming that the pyrolysis is performed in one step. The elemental composition is then known [25]

$$\tilde{c}_i = \frac{\tilde{c}_{V_i} - r\tilde{c}_{C_i}}{1 - r} \tag{8.16}$$

where $\tilde{c}_{V_i}$ and $\tilde{c}_{C_i}$ are the mass fractions of the element i in the virgin resin and coke, respectively. The $r = \rho_c/\rho_v$ is the mass fraction of resin converted into coke ($r \simeq 0.50$ for the phenolic resin).

However, these simplifying assumptions can lead to significant errors on the cold side of material at the end of reentry, as shown in Fig. 8.11, a problem strongly constrained by the boundary conditions [26]. The temperature profiles in the figure were calculated from a pyrolyzed material with a mixture of H_2, CH_4, CO, CO_2, and N_2. The temperatures on both sides were imposed to reconstruct the experience.

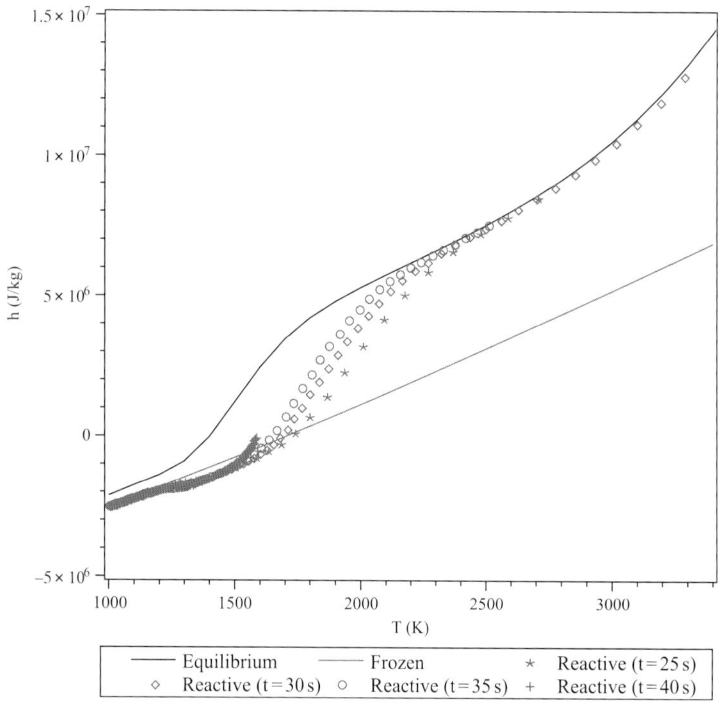

Fig. 8.10 Influence of the thermodynamic state of the pyrolysis gases (enthalpies); an approximate calculation.

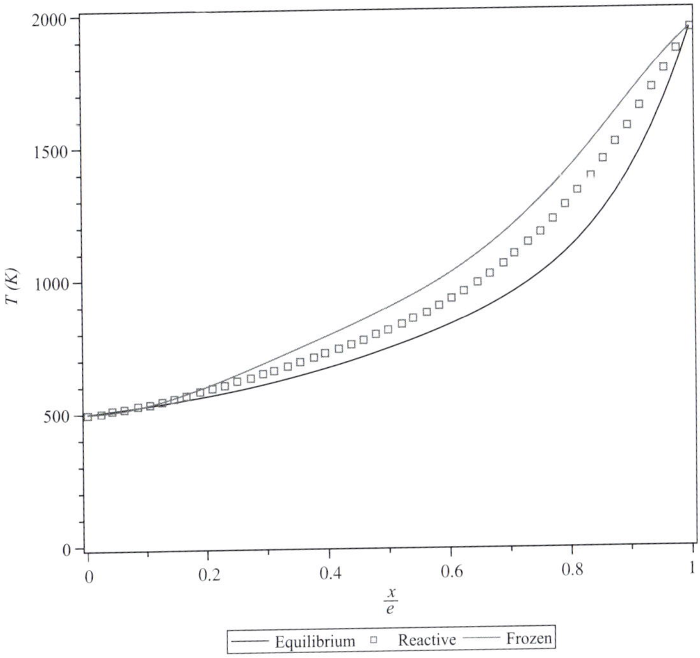

Fig. 8.11 Influence of the thermodynamic state of the pyrolysis gases (temperatures) [27].

8.2.1.6 Carbon Deposition

The observation of a material in the vicinity of the surface shows a chemical deposition of carbon from pyrolysis gases. The amount of carbon created in this way is sufficient for detection by means of physical analysis, but not always high enough to have a measurable effect on the density (with means such as microdensitometry or microtomography; see Fig. 8.12). The deposition mechanism, studied also for the manufacture of materials, is generally a hydrogen abstraction C_2H_2 addition (HACA) mechanism (see Fig. 8.13) [28, 29].This is verified by experiments with a carbon "marked" with ^{14}C [27].

The mechanism will be described using the concept of surface site, the surface density N_s being a parameter of the problem. If θ_1 is the fraction of free sites (available to accommodate a molecule of acetylene) and θ_2 the fraction occupied by $-C_2H\bullet$ ("coverage"), it follows

$$\left\{\begin{aligned} \frac{d\theta_1}{dt} &= k_{f_1}(1-\theta_1-\theta_2)\mathcal{V}_H - k_{b_1}\theta_1\mathcal{V}_{H_2} - k_{f_3}\theta_1\mathcal{V}_{C_2H_2} - k_2\theta_1\mathcal{V}_H \\ &\quad - k_{b_3}\theta_2 + k_4\theta_2\mathcal{V}_{C_2H_2} \\ \frac{d\theta_2}{dt} &= k_{f_3}\theta_1\mathcal{V}_{C_2H_2} - k_{b_3}\theta_2 - k_4\theta_2\mathcal{V}_{C_2H_2} \end{aligned}\right. \tag{8.17}$$

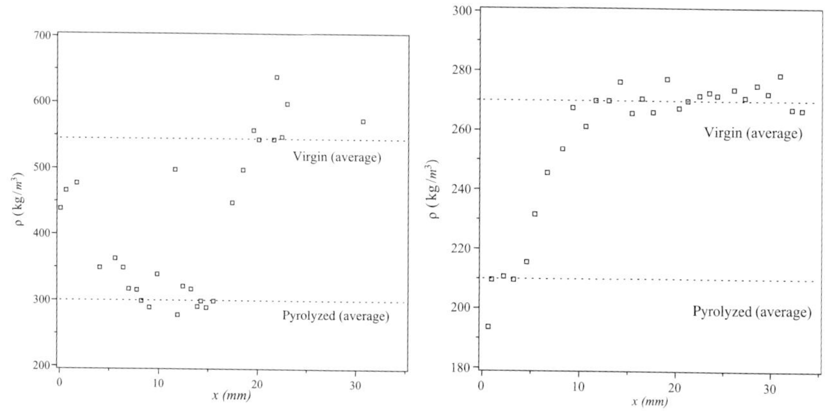

Fig. 8.12 Densities of pyrolyzed materials: Avcoat (left) and PICA (right) (30, 31).

The characteristic time of stationarization is about the *ps*. Then we focus on the stationary solution

$$\begin{cases} \theta_1 = \dfrac{k_{f_1}\mathcal{V}_H}{(k_{f_1}+k_2)\mathcal{V}_H + k_{b_1}\mathcal{V}_{H_2} + k_{f_3}\mathcal{V}_{C_2H_2} + \dfrac{k_{f_1}k_{f_3}\mathcal{V}_H\mathcal{V}_{C_2H_2} - k_{b_3}k_{f_3}\mathcal{V}_{C_2H_2} - k_{f_4}k_{f_3}\mathcal{V}^2_{C_2H_2}}{k_{b_3}+k_{f_4}\mathcal{V}_{C_2H_2}}} \\ \theta_2 = \dfrac{1}{1+\dfrac{k_{b_3}+k_{f_4}\mathcal{V}_{C_2H_2}}{k_{f_1}k_{f_3}\mathcal{V}_H\mathcal{V}_{C_2H_2}}\{(k_{f_1}+k_2)\mathcal{V}_H + k_{b_1}\mathcal{V}_{H_2} + k_{f_3}\mathcal{V}_{C_2H_2}\}} \end{cases} \tag{8.18}$$

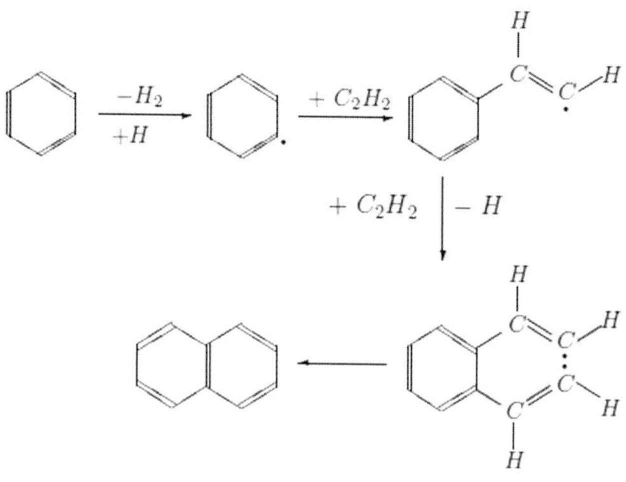

Fig. 8.13 Developed HACA mechanism.

Table 8.5 Kinetics of the Growing Mechanism ($B = 0$ in Any Case); Site• Is a Reaction Site as Defined in Sec. 3.2.1

Reaction	A_f (m^3/mole/s)	E_{A_f} (K)	A_b (m^3/mole/s)	E_{A_b} (K)
$Site + H \Leftrightarrow Site\bullet + H_2$	2.51×10^8	6040	3.98×10^5	3520
$Site\bullet + H \Rightarrow Site$	2.19×10^8	0	—	—
$Site\bullet + C_2H_2 \Leftrightarrow Site\bullet - C_2H_2$	2.00×10^8	2010	5.01×10^7	19,120
$Site\bullet - C_2H_2 + C_2H_2 \Rightarrow Site\bullet$	5.01×10^4	0	—	—

Given the values of the forward and backward constants (Table 8.5), we can simplify the coverage noting that the third reaction is virtually irreversible

$$\begin{cases} \theta_1 = \dfrac{1}{\dfrac{k_{f_3}}{k_{f_4}} + \dfrac{k_{f_2}}{k_{f_1}}} \\ \theta_2 = \dfrac{1}{1 + \dfrac{k_{f_2} k_{f_4}}{k_{f_1} k_{f_3}}} \end{cases} \tag{8.19}$$

At high temperatures of interest, the coverage θ_2 is very close to 1. The mass flow rate approaching carbon deposition is then

$$\dot{m}_C = \frac{2\mathcal{M}_C + \mathcal{M}_H}{\mathcal{M}_C + \mathcal{M}_H} \frac{N_s}{\mathcal{N}} k_{f_4} \rho_{C_2H_2} \tag{8.20}$$

Note that this flow is, in a first approximation, proportional to $\rho_{C_2H_2}$, an assumption made in many models based on experience [32]. In fact, it is as if the radical H was used as a catalyst ($\theta_1 0$).

A numerical calculation at 2000 K with the composition of Fig. 8.8 and kinetics of Table 8.5 gives a growth rate of 3 $\mu m \cdot s^{-1}$. We used a rate of 2.1×10^{19} [28], the maximum possible for a configuration of graphitic type (0.142 nm $\times$ 0.335 nm). The deposition rate decreases with the content of acetylene in the medium and therefore with temperature. For a significant change in the density, we have:

- A temperature not too high, especially because this corresponds to low rates of ablation, which entails the pyrolytic deposit
- A large specific surface area, so small pore sizes for a given porosity
- A low permeation rate

These trends are consistent with the observation of the results presented in Fig. 8.12, but do not constitute a demonstration.

Note that we neglect the possible internal ablation by CO_2 or H_2O [27].

8.2.2 Enthalpies

8.2.2.1 Enthalpy of the Resin

The enthalpy of formation of a compound can be deduced from the calorimetric measurement of the heat of complete combustion ΔH_C

$$C_{xC}H_{xH}N_{xN}O_{xO} + \left(x_C + \tfrac{x_H}{4} - \tfrac{x_O}{2}\right)O_2 \longrightarrow x_C CO_2 + \tfrac{x_H}{2} H_2O + \tfrac{x_N}{2} N_2 + \Delta H_C$$

The value of the heat of formation of phenolic resin can be deduced

$$\Delta H^0_{f298}(\text{resin}) = x_C \Delta H^0_{f298}(CO_2) + \frac{x_H}{2} \Delta H^0_{f298}(H_2O) - \Delta H_C \qquad (8.21)$$

The measured combustion enthalpy is $-28.8\ \text{MJ} \cdot \text{kg}^{-1}$ [8]. Then the heat of formation $\Delta H^0_{f298} = -1.0\ \text{MJ} \cdot \text{kg}^{-1}$. The resulting residue around 1200 K is a carbon not much graphitized. The measurement of its enthalpy of formation by the same method as above gives $\Delta H^0_{f298} \simeq 0.5\ \text{MJ} \cdot \text{kg}^{-1}$. The final product obtained at very high temperatures is more ordered ($\Delta H^0_{f298} \simeq 0$). This implies a phase change, which should be modeled as such. Note that this phase change causes a significant change in conductivity [33].

8.2.2.2 Gas Enthalpy

The problem of knowing the enthalpy of the gas is summarized in Fig. 8.10:

- The medium is generally in chemical nonequilibrium.
- Even in equilibrium, the enthalpy depends on the composition, thus on T, p, and $\tilde{c}_k$.

However, we can simplify the problem by noting that [34]:

- At high temperature and thermodynamic equilibrium, the pyrolysis process is completed, and the composition is only a function of T and p.
- In the same region, the pressure dependence is quite small and can be neglected in first approximation.
- In the low temperature range, the composition can be estimated by assuming that the residence time of gas is small compared with the characteristic time associated with changes in different profiles (temperature, pressure, density, etc.) and the environment is 1-D. Indeed, the flow of the species i at position y can be drawn from a simple quadrature. Then the low temperature properties are those of a gas of known composition

$$\dot{m}_j = -\int_{-\infty}^{y} \sum_i \alpha_{ji} \frac{\partial \rho_i}{\partial t}\, dy' \qquad (8.22)$$

where the values α_{ji} are the mass fractions of solid gasified. We deduce immediately the mass composition of the gas

$$c_j = \frac{\dot{m}_j}{\sum_i \dot{m}_i} \tag{8.23}$$

This allows us to use an enthalpy depending at low temperatures on the temperature and the local composition and at high temperatures on the temperature only. Between these two situations, we see from Fig. 8.10 that the followed trajectories $h(T)$ generally vary little from one situation to another, allowing the definition of a "universal curve" function of T only. This method minimizes the errors mentioned previously in low temperature regions (Fig. 8.11).

We can deduce the heat of pyrolysis of the resin by knowing the composition of the gas formed. This quantity depends on the final state of the gas obtained at a temperature varying with its history. We obtain values in the range [0.8–1.7 $\mathrm{MJ \cdot kg^{-1}}$] [35]. Pyrolysis is a moderately endo-energetic process.

8.3 The General Model

8.3.1 Physical Hypothesis

8.3.1.1 Flow in Porosities

The open porosity of the material increases with the progress of the pyrolysis. It reaches a maximum of the order of 40% in a standard TWCP, before declining significantly to 20% due to the reorganization of the carbon issued from the pyrolysis (excluding any pyrolytic deposition near the surface) [10]. In the case of a less dense material such as PICA, porosity reaches 90%. This porosity is divided into three ranges:

- Porosity at the nanoscale is probably the heart of the inner workings of pyrolysis (it is the essence of the specific surface area); it is ignored in the models. We saw already that the phenomenon was treated in terms of partial density, which implicitly involves homogenization.
- Porosity at the micron scale is involved in the permeation of gases taking place at an average speed of the order of a fraction of $\mathrm{m \cdot s^{-1}}$. The equilibrium or not of local temperature with the solid adjacent is discussed in Sec. 8.3.1.2.
- The porosity is about 100 μm, corresponding to cracking in early degassing of materials like TWCP. This phenomenon is not present in all materials of this type [36] or depends on the orientation of the plies [37].

One difficulty lies in the fact that, at the beginning of pyrolysis, the mean free path in the gas is the same order of magnitude as the size of the porosity or larger, up to 100 μm [38]. The expression of the average velocity of the gas in the pores must be adapted to the physical problem (Sec. 8.4.1).

8.3.1.2 Temperature Homogeneity

Work has been done to describe the inhomogeneity of temperature in the solid phase between fiber and matrix, visible on infrared (IR) thermography. The work [39] led, by homogenization, to a macroscopic model at two temperatures. There are two practical difficulties in their implementation:

- The calculation of the exchange coefficient between solid phases. If this problem is theoretically accessible, it faces the need to know the thermal conductivity of the resin during pyrolysis.
- The surface boundary condition for a material described by two temperatures (see Sec. 8.6).

The differences highlighted in the study of this type are fortunately very low (of the order of K) and hardly justify the modeling efforts to be made to account for this phenomenon.

With regard to the temperature difference between gas and solid, the problem admits no single answer. One can use a study by Florio [40] to predict the magnitude of the problem:

- In the case of a moderately porous material such as TWCP, the temperature difference reaches a few Kelvin [41] and can therefore be legitimately ignored.
- This is not the case for a material such as PICA; the variations may reach several hundred Kelvin [23].

This situation poses the problem of boundary conditions at the wall for a Darcy–Brinkman–type flow with a homogenized nonzero velocity at the wall, parallel to it.

8.3.2 Descriptive Equations

The equations describing the system have been rigorously derived by scaling [42] or by volume averaging [43]. The proof is reproduced in Appendix E.4 and extended to the multispecies case.

1. The equation of conservation of the solid is written as a conservation equation of porosity with the following assumptions:
   - No volume variation
   - No creation of closed porosity
   - True densities of the descendants of the resin equal to that of the resin
   - The same for any carbon deposits; the effects will result in terms of volume formation $S_v \dot{m}_C$, where S_v is the area of solid–gas interface per

unit volume and $\dot{m}_C$ is given by Eq. (8.20), then the porosity of descendants of the resin is:

$$\epsilon_R = \frac{\bar{\rho}_R - \bar{\rho}_{R_0}}{\bar{\rho}_{R_0}} \tag{8.24}$$

hence the total porosity

$$\epsilon = \epsilon_0 + \epsilon_R \tag{8.25}$$

2. Conservation of gaseous products

$$\frac{\partial(\epsilon \rho_{g_i})}{\partial t} + \nabla \cdot (\epsilon \rho_{g_i} \mathbf{V}) = \epsilon\ \dot{\omega}_{g_i} + S_v \dot{m}_{g_i} \tag{8.26}$$

where ρ_{g_i} is the true density (mass of gas by gas volume) and $\dot{m}_{gi}$ is the formation or disappearance rate of gas in a heterogeneous reaction. This term is derived immediately from $\dot{\omega}_{s_i}$ if we know the proportion of gases formed at each stage of the pyrolysis reaction. In the absence of surface reactions, Eq. (8.26) summed over all species is written

$$\frac{\partial(\epsilon \rho_g)}{\partial t} + \nabla \cdot (\epsilon \rho_g \mathbf{V}) = \sum_i \dot{\omega}_{s_i} \tag{8.27}$$

3. Energy conservation

$$\frac{\partial(\rho E)}{\partial t} + \nabla \cdot (\dot{q}_s + \dot{q}_{\mathrm{rad}}) + \nabla \cdot \left[\epsilon(\rho_g h_g \mathbf{V} + \dot{q}_{\mathbf{g}})\right] = 0 \tag{8.28}$$

Note that the terms $\dot{q}_s$ and $\dot{q}_{\mathrm{rad}}$ implicitly contain the porosity.

The volume total energy ρE is given by

$$\rho E = \sum_i \bar{\rho}_{s_i} e_{s_i} + \epsilon \left(\sum_i \rho_{g_i} h_{g_i} + \rho_g \frac{V^2}{2} \right) \tag{8.29}$$

and the heat flow by conduction in the solid is

$$\dot{q}_s = -\nabla \cdot (\lambda_s \nabla T) \tag{8.30}$$

Some remarks on these equations:

- The term corresponding to the kinetic energy in Eq. (8.29) is negligible [44].
- In Eq. (8.28), the conduction term in the gas $\dot{q}_g$ is small before the convective term $\rho_g h_g \mathbf{V}$.
- The equation of momentum and the various terms of transport in the gas are given in Sec. 8.4.1.

- The solid–gas coupling is conducted with the assumption of equal temperatures (Sec. 8.3.1.2). With this condition, one energy equation for the couple solid + gas is sufficient. Otherwise, we write two equations with a coupling term $\rho_g V\, C_H\, S_v\, (h_i - h_s)$, often written as $\alpha(T_g - T_s)$. The local homogenized Stanton number C_H or the transfer volume corresponding to term α belong to a microscopic calculation.
- The conductivity of the solid is, in principle, available if we know the geometry of the medium at all times and the thermal properties of the various constituents. In practice, this raises problems that are difficult to overcome, and more often values deduced from experience are used (Sec. 12.3.1).
- The heat flux $\dot{q}_{\text{rad}}$ corresponding to radiation in the pores (Sec. 10.4) can play an important role for materials with high porosity and temperatures above 1000 K. Again, this term depends on the variation of porosity.
- The advancement rate ζ is often used to describe the evolution of porosity

$$\zeta = \frac{\bar{\rho}_v - \bar{\rho}_R}{\bar{\rho}_v - \bar{\rho}_p} \tag{8.31}$$

where $\bar{\rho}_v \equiv \bar{\rho}_{R_0}$ and $\bar{\rho}_p$ are the partial densities of virgin resin and a completely pyrolyzed one, respectively. The ζ varies from 0 (virgin material) to 1 (fully pyrolyzed material).

8.3.3 Boundary Conditions

The equations are of parabolic type and require boundary conditions of Dirichlet or Neumann type (see Sec. 2.2.2). The permeation rates are low. It results that the pressure in the boundary layer is not affected by the injection of the pyrolysis gases. Hence the condition at the limit for the internal problem $p = p_{\text{ext}}$.

It is important to note that if one assumes the existence of two temperatures in the porous medium, one should not impose any boundary condition on the solid medium; the gas exchange with "external" (a concept without meaning at the microscopic level) takes place through the "internal" gas.

8.4 The Different Levels of Solutions

Different approaches to the problem exist, more or less simplifying it. In general we ignore the terms related to carbon deposition from the pyrolysis gases. There is a significant literature on the subject, from people interested in the problems of manufacturing carbon–carbon materials by chemical vapor infiltration (CVI) [45].

In what follows, we will focus on various aspects of the problem.

8.4.1 Flow Regimes in the Pores

The problem of boundary condition is simple. The equation of conservation of momentum reduces to a constitutive equation: the unsteady term is neglected.

8.4.1.1 Rarefied Regime

In the case of a rarefied medium, the gas–wall collisions play an important role in the transfer. The average speed for a gas in the pores obeys the Darcy–Klinkenberg equation [42]

$$\rho_g \boldsymbol{V} = -\mathcal{D}\nabla\rho_g - \mathcal{D}^T \frac{\nabla T}{T} \tag{8.32}$$

We can rewrite this relationship by showing the gradients of temperature and pressure

$$\rho_g \boldsymbol{V} = -\rho_g \mathcal{D}\left(\frac{\nabla\rho_g}{\rho_g} + \frac{\mathcal{D}^T}{\mathcal{D}\rho_g}\frac{\nabla T}{T}\right) = -\rho_g \mathcal{D}\left[\frac{\nabla p}{p} + \left(\frac{\mathcal{D}^T}{\mathcal{D}\rho_g} - 1\right)\frac{\nabla T}{T}\right] \tag{8.33}$$

Diffusion coefficients $\mathcal{D}(\rho, T)$ and $\mathcal{D}^T(\rho, T)$ can be calculated if one knows the topology of the environment (Fig. 8.14) [42, 45]. Their measurement is also feasible.

At high density (mean free path small compared to the pore size) it is shown that the coefficient of thermal transpiration $\mathcal{D}^T/\mathcal{D}\rho_g$ tends to 0, and

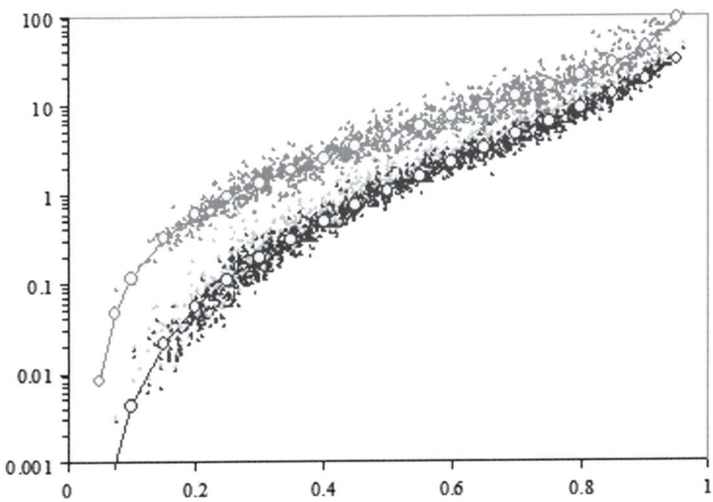

Fig. 8.14 Diffusion coefficient-porosity relationship in a carbon-carbon composite; two axes, microscopic scale, approximate relations (45).

therefore the previous relation becomes

$$\rho_g \mathbf{V} = -\mathcal{D}\frac{\nabla p}{p} \tag{8.34}$$

which is the Darcy relationship

$$\mathbf{K}\nabla p = -\mu_g \mathbf{V} \tag{8.35}$$

The relationships between the coefficients $\mathcal{D}$ and $\mathcal{D}^T$ to the permeability $\mathbf{K}$ can then easily be deduced and are given by

$$\begin{cases} D = \dfrac{Kp}{\mu_g} \\ \mathcal{D}^T = \rho_g \mathcal{D} \end{cases} \tag{8.36}$$

In the general case of a gas mixture the problem is considerably more complex, particularly because of the appearance of a molar mass gradient term. As in the case of a gas in free atmosphere (Sec. 2.3), it is not possible to express explicitly a rate of diffusion in the pores. There are approximate methods ("dusty gas" or "binary friction model") that allow us to write the diffusive flux [46–48]. In this theory of Chapman–Enskog type, the solid is considered a homogeneous assembly of fixed "molecules" (rendered immobile by an external ad hoc force).

8.4.1.2 Continuous Regime

Darcy's law can be found by homogenization in the case of an incompressible gas at low Reynolds number based on porosity, defined with the characteristic length $K^{\frac{1}{2}}$

$$\mathcal{R}e_K = \frac{\rho_g V K^{\frac{1}{2}}}{\mu_g} \tag{8.37}$$

In the case of high rates of permeation, such as 3-D architectured materials, Darcy's equation may not be sufficient due to the inertial terms. We then use a correction term in the equation, which is then called the Darcy–Forchheimer equation. For an arbitrary geometry this result was established in a compressible periodic geometry [44, 49, 50]

$$\mathbf{K}\nabla p = -\mu_g \mathbf{V}(1 + \alpha \mathcal{R}e_K) \tag{8.38}$$

The constant α describing the topology of the medium is sometimes called the Ergun or Ward constant. Its value is about 0.5. The simultaneous measurement of K and α can be made with a graphical method attributable to Cornell and Katz [26], based on a 1-D stationary solution of Eq. (8.38) in a homogeneous medium.

8.4.2 Permeation Time Is Neglected

We suppose that continuous flow is obtained. Introducing the bulk densities $\bar{\rho}_i$, the total energy [Eq. (8.29)] where we hide the gas will be rewritten as follows:

$$\rho E \simeq \bar{\rho} h_s = \sum_i \bar{\rho}_i h_{s_i} \tag{8.39}$$

The first term of Eq. (8.28) is written

$$\frac{\partial}{\partial t}(\bar{\rho} h_s) = \frac{\partial}{\partial T}\left(\sum_i \bar{\rho}_i h_{s_i}\right)\frac{\partial T}{\partial t} = \bar{\rho} C_{p_s}^{\text{tot}} \frac{\partial T}{\partial t} \tag{8.40}$$

where $\bar{\rho}$ is the apparent density of the material and $C_{p_s}^{\text{tot}}$ the specific heat at constant pressure

$$\bar{\rho} C_{p_s}^{\text{tot}} = \sum_i \bar{\rho}_i C_{p_{s_i}} + \sum_i h_{s_i} \frac{\partial \bar{\rho}_i}{\partial T} \tag{8.41}$$

Unlike a traditional solid, $C_{p_s}^{\text{tot}}$ does not depend on T only.

The term corresponding to gas convection is amended as follows:

$$\begin{aligned} \nabla \cdot (\epsilon \rho_g h_g \mathbf{V}) &= h_g \nabla \cdot (\epsilon \rho_g \mathbf{V}) + \epsilon \rho_g \mathbf{V} \cdot \nabla h_g \\ &= h_g \nabla \cdot \left(\dot{m}_g \frac{\mathbf{V}}{V}\right) + \dot{m}_g C_{p_g} \frac{\mathbf{V}}{V} \cdot \nabla T \end{aligned} \tag{8.42}$$

where

$$\dot{m}_g = \epsilon \rho_g V \tag{8.43}$$

is the gas mass flow rate of pyrolysis (for the material as a whole) and $C_{p_g}^{\text{tot}}$ is the total specific heat of the gas, defined by an equation similar to Eq. (8.41)

$$\rho_g C_{p_g}^{\text{tot}} = \sum_i \rho_{g_i} C_{p_{g_i}} + \sum_i h_{g_i} \frac{\partial \rho_{g_i}}{\partial T} \tag{8.44}$$

A common but rather drastic approximation is to assume the gas permeation time is zero, while retaining the gas flow mass rate $\rho_g \mathbf{V}$ finite. We associate another approximation (although it is independent): the problem is 1-D. The gas flow is therefore given by a simple integral

$$\dot{m}_g = -\int_{-\infty}^{y} \frac{\partial \bar{\rho}}{\partial t} \mathrm{d}y' \tag{8.45}$$

In this equation the $-\infty$ represents the depth $y < 0$ corresponding to virgin material. The first term of Eq. (8.42) becomes

$$h_g \nabla \cdot \left(\dot{m}_g \frac{\mathbf{V}}{V} \right) = h_g \frac{\partial \dot{m}_g}{\partial y} = -h_g \frac{\partial \bar{\rho}}{\partial t} \tag{8.46}$$

Then the new energy equation

$$\bar{\rho} C_{p_s}^{\text{tot}} \frac{\partial T}{\partial t} + \dot{m}_g C_{p_g}^{\text{tot}} \frac{\partial T}{\partial y} - h_g \frac{\partial \bar{\rho}}{\partial t} + \nabla \cdot \dot{q} = 0 \tag{8.47}$$

where $\dot{q} = (1 - \epsilon)\dot{q}_S + \epsilon\dot{q}_g + \epsilon\dot{q}_{\text{rad}}$ is the total heat flux in the material, often wrongly called conduction flux, expressed as $\dot{q} = -\lambda_{eq} \nabla T$.

This equation can be written differently by introducing the bulk densities $\bar{\rho}_v$ and $\bar{\rho}_p$ of virgin and completely pyrolyzed material, respectively

$$\bar{\rho} = (1 - \zeta)\bar{\rho}_v + \zeta\bar{\rho}_p \tag{8.48}$$

In the case of a one-step pyrolysis, the second term of Eq. (8.41) becomes

$$\sum_i h_{s_i} \frac{\partial \bar{\rho}_i}{\partial T} = \frac{(1 - \zeta)\bar{\rho}_v h_v - \zeta\bar{\rho}_p h_p}{\bar{\rho}_v - \bar{\rho}_p} \frac{\partial \bar{\rho}}{\partial T} \tag{8.49}$$

where h_v and h_p are the enthalpies of virgin and totally pyrolyzed material, respectively.

By showing the frozen specific heat C_{ps}, the first term of Eq. (8.41) in Eq. (8.47), this equation is written

$$\bar{\rho} C_{p_s} \frac{\partial T}{\partial t} + \dot{m}_g C_{p_g}^{\text{tot}} \frac{\partial T}{\partial y} + \Delta h_p \frac{\partial \bar{\rho}}{\partial T} + \nabla \cdot \dot{q} = 0 \tag{8.50}$$

with

$$\Delta h_p = \frac{(1 - \zeta)\bar{\rho}_v h_v - \zeta\bar{\rho}_p h_p}{\bar{\rho}_v - \bar{\rho}_p} - h_g \tag{8.51}$$

Equation (8.50) is often used in the literature treating different applications.

8.5 Transport Properties

There are three ways to express each of the properties of interest:

- Direct measurement of a series of samples
- A calculation at the microscopic level using a geometry derived from a microtomography [45]
- The use of a geometry model [38], after previous experience on similar problems

In the first two cases, the measurements of various quantities used for pyrolysis are made on samples "frozen" in a given state corresponding to the progress of the pyrolysis of this material, or supposed so.

The quantities are often expressed as a function of porosity or the advancement rate of the reaction (see the following section).

8.5.1 Porosity

It is generally admitted that porosity varies linearly with mass loss. Even assuming a constant volume environment, this is not strictly true. Recall the structural reorganization of carbon at high temperatures, which occurs with significant change in porosity. This is therefore an approximate law, but one that generally describes the phenomenon rather well (Fig. 8.15).

8.5.2 Diffusion Coefficients, Permeability

8.5.2.1 Microtomographies

The use of microtomography is a recent development. We will provide an example to show the effectiveness of this method. Figure 8.15 shows an example of calculation for a material in which the porosity was varied by chemical vapor deposition (CVD). In each voxel (volume of reference) the

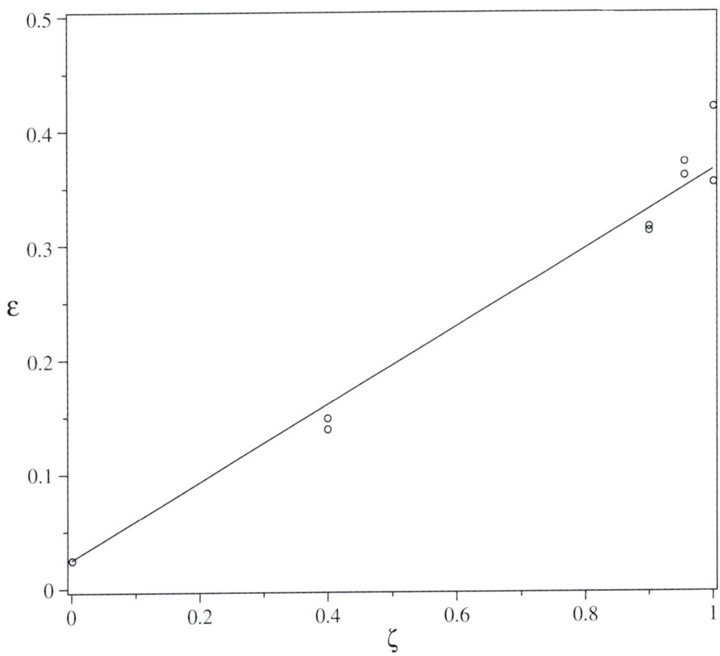

Fig. 8.15 Porosity–mass loss relationship depending on the progress of the pyrolysis (TWCP material) (10).

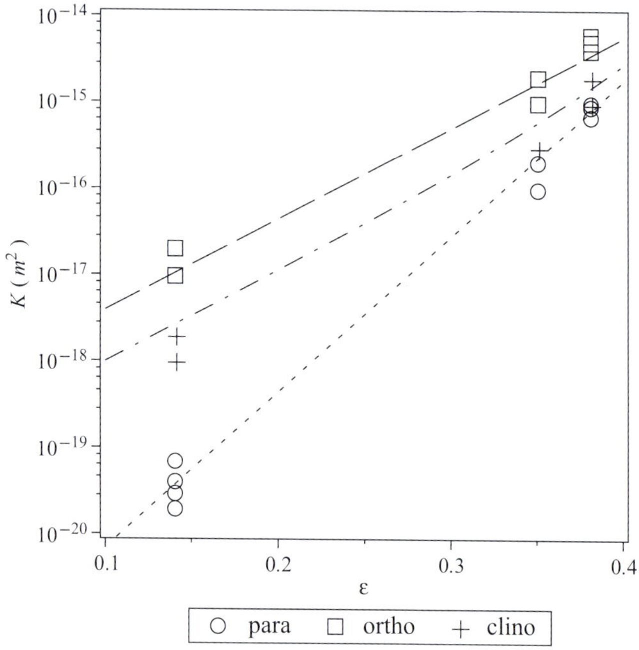

Fig. 8.16 Permeability–porosity relationship.

Knudsen diffusion coefficient was calculated (here on two axes). This evaluation at various points shows that this quantity is randomly distributed. A mean value can be deduced for macroscopic calculations.

The advantage of the method, besides the fact that it provides a remarkable vision of the topology of the environment, is that it is applicable to all transport properties.

8.5.2.2 Direct Measurement of Permeability

The direct measurement of permeability is possible, but faces two challenges:

- The dynamics of the variable are significant because it extends nearly 10 orders of magnitude (Fig. 8.16).
- This quantity is sensitive to the stress state of the material in the case of TWCP.

8.5.2.3 Modeling

At a lower level modeling, we use empirical laws derived from the microscopic calculation [45, 51] or from experience [10, 52, 53]. In the case of permeability, a law linking the permeability to porosity or mass loss is used most often

$$K = K_v^{1-\zeta} K_p^{\zeta} \tag{8.52}$$

where K_v is the permeability of virgin material, and K_p that of the material completely pyrolyzed.

Note that the material TWCP is orthotropic, the layers of fabric having an angle θ with the material surface. If $K_{||}$ and $K_{\perp}$ are the permeabilities in the axes of the material, the permeability K_x and K_y in the material axes are obtained by rotation

$$\begin{vmatrix} K_{xx} & K_{xy} \\ K_{yx} & K_{yy} \end{vmatrix} = R \begin{vmatrix} K_{||} & 0 \\ 0 & K_{\perp} \end{vmatrix} R^T$$
$$= \begin{vmatrix} K_{||}\cos^2\theta + K_{\perp}\sin^2\theta & (K_{||} - K_{\perp})\sin\theta\cos\theta \\ (K_{||} - K_{\perp})\sin\theta\cos\theta & K_{||}\sin^2\theta + K_{\perp}\cos^2\theta \end{vmatrix} \quad (8.53)$$

where R is the Euler rotation matrix

$$R = \begin{vmatrix} \cos\theta & -\sin\theta \\ \sin\theta & \cos\theta \end{vmatrix} \quad (8.54)$$

This relationship is well verified experimentally {Fig. 8.16 and [53]}.

8.5.2.4 Pression Measurements

The onset of pyrolysis occurs in a material with very low porosity. For a material such as TWCP, permeability is very low, $<10^{-20}$ m^2. The pressure generated by the pyrolysis gases (or water in a material not previously dried) often causes cracking of the material. This is probably the cause of various phenomena:

- The measured permeability is sensitive to the stress state of the material [52].
- The pressure variation measured inside the material (Fig. 8.17) is erratic, often poorly reproducible [10, 54]. However, the moment when the rise occurs is clearly correlated with the creation of gas, and the maximum pressure is approximately the same in every test.

8.5.3 Specific Surface

The specific surface area can be obtained through the same methods as the permeability (or diffusion coefficient). Figure 8.18 shows an example; the dotted curve corresponds to a model of nonoverlapping randomly distributed cylinders, $S_v = \alpha(1-\epsilon)$. The results are very sensitive to the measurement method in the following ways:

- The permeation of the fluid probe; for example, mercury porosimetry gives values significantly lower than those obtained using a gas.
- The size of the voxel in the case of a microtomography, which also gives default values.

8.5.4 Conductivity

8.5.4.1 Conduction in Solid

Assessing and measuring the conductivity of the resin during pyrolysis are difficult issues that have never been addressed. We are generally content to introduce an unjustified mixing law

$$\lambda_s = \lambda_v^{\zeta}\, \lambda_p^{1-\zeta} \tag{8.55}$$

There are other laws using a different average conductivity of virgin material λ_v and that of the pyrolyzed material λ_p (Sec. 12.3.4). These averages are applied either on the material or on the resin alone. One can, indeed, calculate the conductivity of the material from the geometry of it and the data of each component [55].

The transformation relations of the permeability tensor in Eq. (8.53) apply equally to the conductivity.

8.5.4.2 Radiation

The high temperatures reached (in excess of 3000 K) and the high total porosity, reaching 40% for a TWCP and much more for a material of PICA type, make playing the radiation in the pores a significant role. If in theory

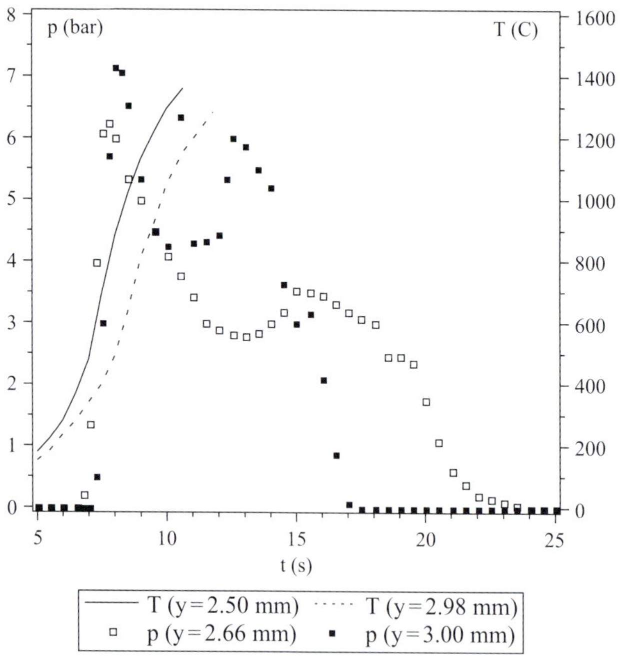

Fig. 8.17 Simultaneous measurement of internal pressure and temperature in a carbon–resin material during pyrolysis (10).

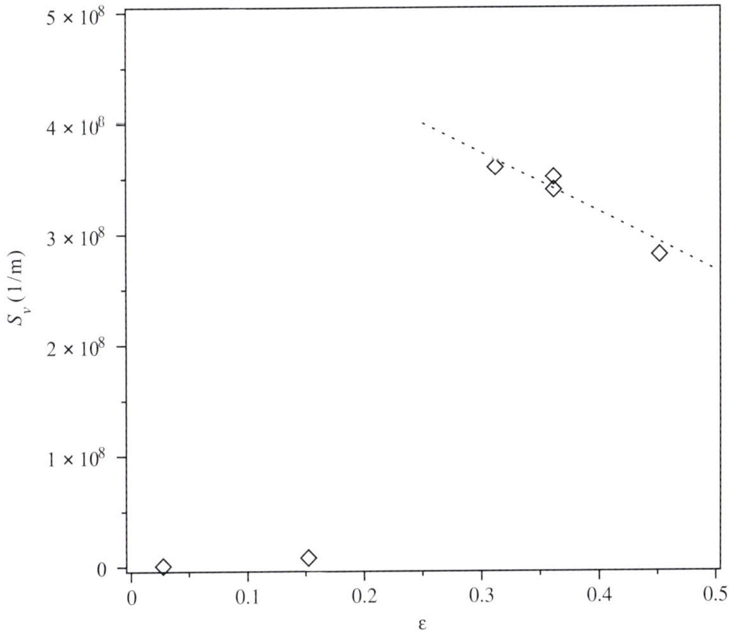

Fig. 8.18 Specific surface for a TWCP, measured by BET method (10, 45).

this mode of energy transport is available when you have a real geometry or model, this type of calculation is rare. We are generally content to introduce a term in σT^3 representing the diffusion limit for radiative transfer. This is discussed in Sec. 10.4.

8.6 Application Example

In this section we show an application example [44, 56]. The material is of TWCP carbon phenolic type on which is applied external conditions representative of a reentry. The problem is assumed 1-D. Pyrolysis is described by a simple mechanism (one-step, no pyrolytic deposition). The gas formed is assumed to be at thermodynamic equilibrium and its permeation described by Darcy's law. In the rear part it is assumed that the Dirichlet conditions apply for temperature and pressure. External conditions (radiative and convective flux, pressure) are prescribed, and a simple model of ablation is used (see Sec. 8.7.1).

The main results are:

- Given the severe external conditions, the temperature reached a maximum level of approximately 3500 K (Fig. 8.19) for an ablation speed of $0.33\ \mathrm{mm \cdot s^{-1}}$. The profiles of temperature and density (Fig. 8.20), shifting to the right due to the ablation, show a quasi-stationary phase. Note that

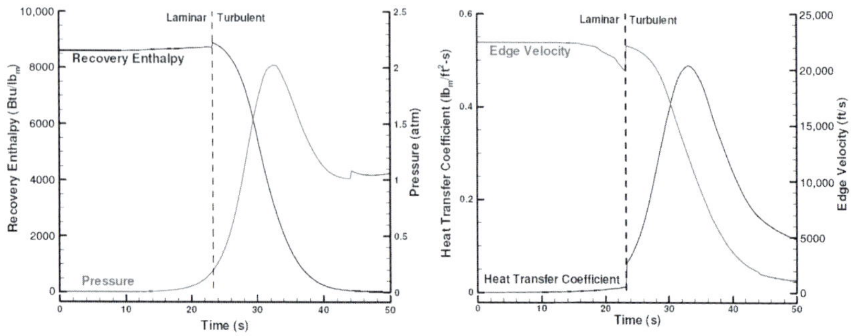

Fig. 8.19 External conditions for calculating ablation and pyrolysis (56).

the temperature reaches 300 K · s^{-1}, a value significantly greater than those used to characterize the mass loss by pyrolysis.

- The pressure curve (Fig. 8.21) shows maximum values of 1.8×10^5 Pa. The back part of the material shows a saturation due to the boundary condition.
- Porosity starts from a value near zero and peaks at 0.35 (Fig. 8.22). The maximum pressure is reached when the porosity is still very low.
- Permeability ranges 6 orders of magnitude or higher during the phenomenon (Fig. 8.22).
- The velocities of the gas (Fig. 8.21) reach 3 m · s^{-1} and then decrease rapidly.

8.7 Ablation of Carbon Phenolics

8.7.1 Approximate Methods

We saw in Sec. 8.2.1.5 that the parietal gaseous medium could be considered at thermodynamic equilibrium for temperatures above

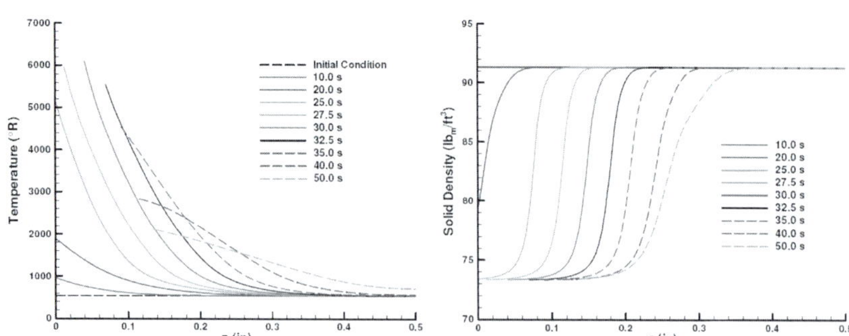

Fig. 8.20 Temperatures and densities during pyrolysis (56).

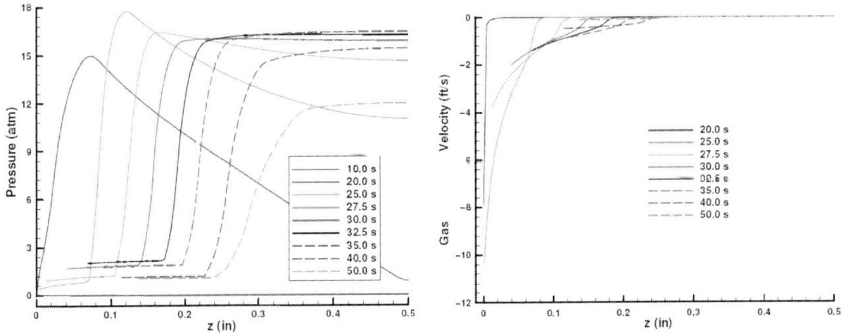

Fig. 8.21 Pressures and velocities of the gas during pyrolysis (56).

2000–2500 K. Below this value, the ablation is very low and has little effect on the final value (Fig. 4.9). For this reason, a method has been developed with the assumption of equilibrium and using the approximation of Sec. 8.4 for changes in the boundary layer.

The first method corresponds to the diagram of Fig. 8.23. In the figure, $c_i = f(T_p, p, \tilde{c}_k)$ is a calculation of thermodynamic equilibrium in the gaseous medium, $\mathcal{B}' \to \rho_e u_e C_M$ is the blockage relationship, and $\mathcal{B}'_g$ is an additional variable that is independent or dependent depending on the assumption of steady-state situation or not. In what follows, we neglect the blockage correction $\rho_e u_e C_M = f(\rho_e u_e C_{M_0}, \mathcal{B}')$ and assume a steady-state phenomenon $\mathcal{B}'_g = g(\mathcal{B}')$. By iteration, it is possible to establish a law a priori $\mathcal{B}' = f(T, p, \rho_e u_e C_M)$, as shown in Fig. 8.24. The calculation uses the database of Burcat and Ruscic [24]. The values of $\rho_e u_e C_M$ are in $kg \cdot m^{-2} \cdot s^{-1}$ and are representative of a severe reentry (see Fig. 8.19).

However, the number of parameters of this law is too big to carry it, especially because in the general case $\mathcal{B}'_g$ is an additional independent variable. The following approximation assumes thermodynamic equilibrium

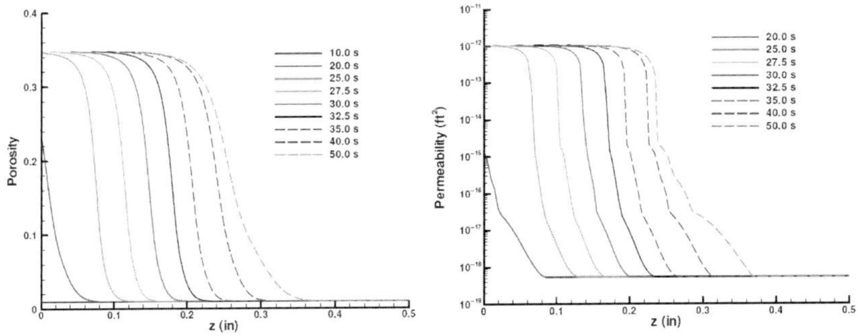

Fig. 8.22 Porosity and permeability during pyrolysis (56).

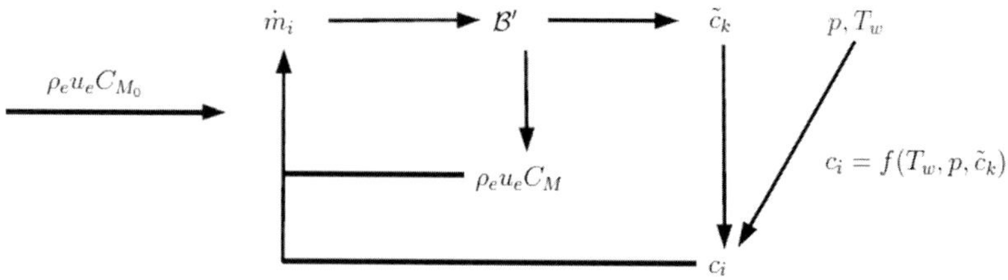

Fig. 8.23 Block diagram for the computation of $\mathcal{B}'$.

between gas and solid. This may seem inconsistent to the extent that this assumption leads to the absence of sublimation. In fact, we must see this method as a passage to the limit $\rho_e u_e C_M \to 0$, the limit value of $\mathcal{B}'$ being nonzero despite the ablation flux being zero. We see in Fig. 8.24 errors induced by this assumption, the consequences of which are discussed in Sec. 5.10. We therefore constructed a law that is relatively easily to use, $\mathcal{B}' = f(T_p, p, \mathcal{B}'_g)$; an example is given in Fig. 8.25. The calculation uses the Gordon and McBride database [58], $\mathcal{B}' = 10$, 5, 2.5, 1, 0.5, 0.1, 0.05,

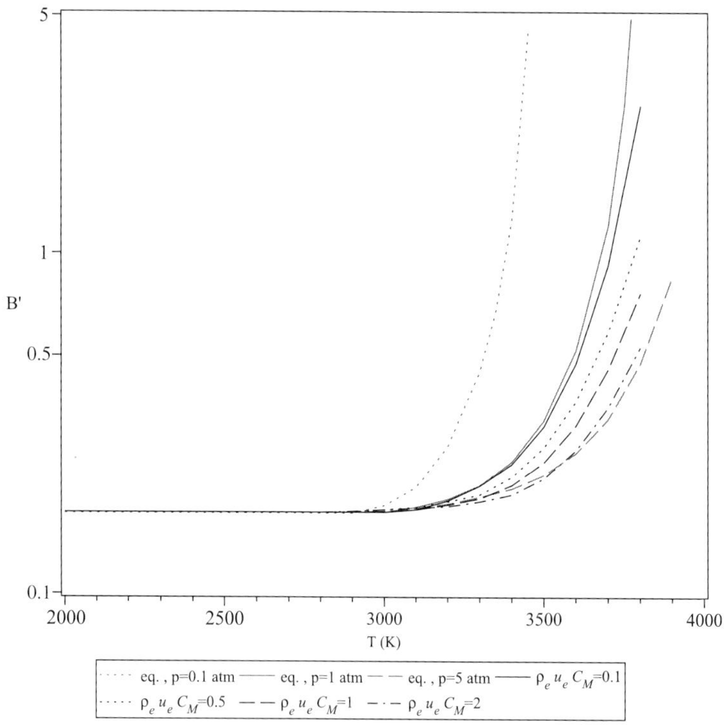

Fig. 8.24 $\mathcal{B}'(T, \rho_e u_e C_M)$ for a TWCP material ($p = p_0$) and the equilibrium curves for various pressures.

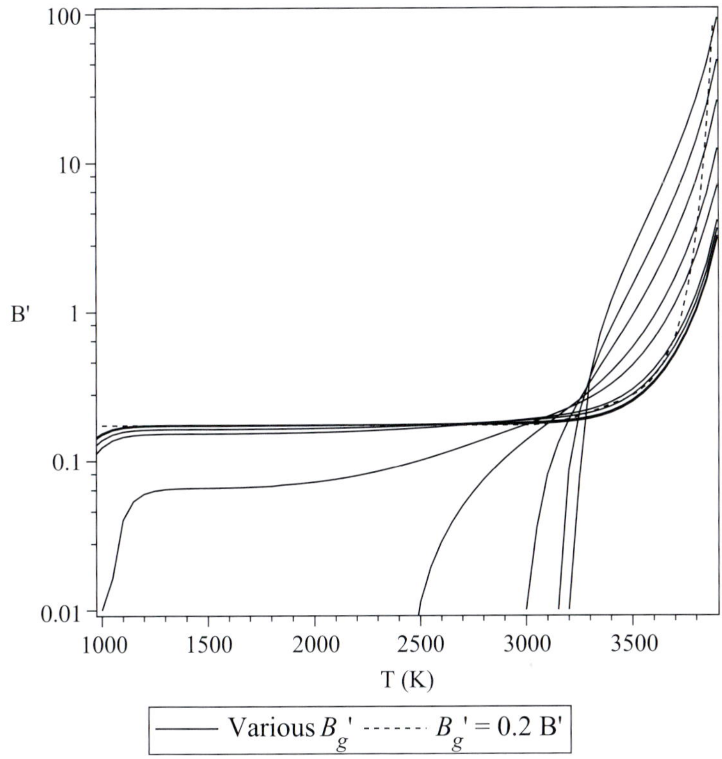

Fig. 8.25 Relation giving $\mathcal{B}'(\mathrm{T}, \mathcal{B}'_g)$ $(p = p_0)$ (57) and $\mathcal{B}'(T)$ for the steady-state case (PICA material).

0.01, 0.001, 0.00001. These curves are not justified below 2000 K, a situation in which thermodynamic equilibrium is not reached.

Assuming that at low temperatures the medium under consideration consists of N_2 and CO only, we can compute, as in Sec. 5.7, the set of curves for Eq. (8.24) or Eq. (8.25)

$$\mathcal{B}' \simeq \frac{\tilde{c}_{O_e}}{\dfrac{\mathcal{M}_O}{\mathcal{M}_C} - \left(\dfrac{\rho_v}{\rho_p} - 1\right)\tilde{c}_{O_g}} \tag{8.56}$$

8.7.2 Influence of Pyrolysis in Ablation

The gaseous medium in the vicinity of the surface will contain a number of major species from the pyrolysis and reactions with the atmosphere. (Figure 8.26 can be compared to Fig. 8.8 for pyrolysis species only.) Note that if the wall temperature does not generally reach the highest values of this curve, these species are present in the warmer part of the boundary layer where species such as OH are also present. To these species we must, of course, add those coming from the atmosphere only. This leads to very heavy calculations, even by reducing the composition to major species [59, 60].

8.7.3 Energy Partition

We can write an energy balance [Eqs. (4.67) and (4.75)] at the wall for a stationary system

$$\rho_e u_e C_H (h_a - h_p) - \frac{\rho_v}{\rho_p} \dot{m} h_p - \dot{q}_R + \frac{\rho_v}{\rho_p} \dot{m} h_v(T_0) = 0 \tag{8.57}$$

From this equation, it is possible to measure the effects of various phenomena as has been done for carbon (Sec. 5.8). Note that here the various quantities calculated are different. We note (Fig. 8.27)

- The nondimensioned convective flux $(\rho_e u_e C_H (h_a - h_p) / \rho_e u_e C_H (h_a - h_p^*))$ is greater than unity. The enthalpy of the gas boundary layer h_p is lower than that of pure air h_p^* except at high temperature.
- The efficiency of the injection is just above that of pure carbon, the increase in the amount of gas being in the ratio (ρ_v / ρ_p).

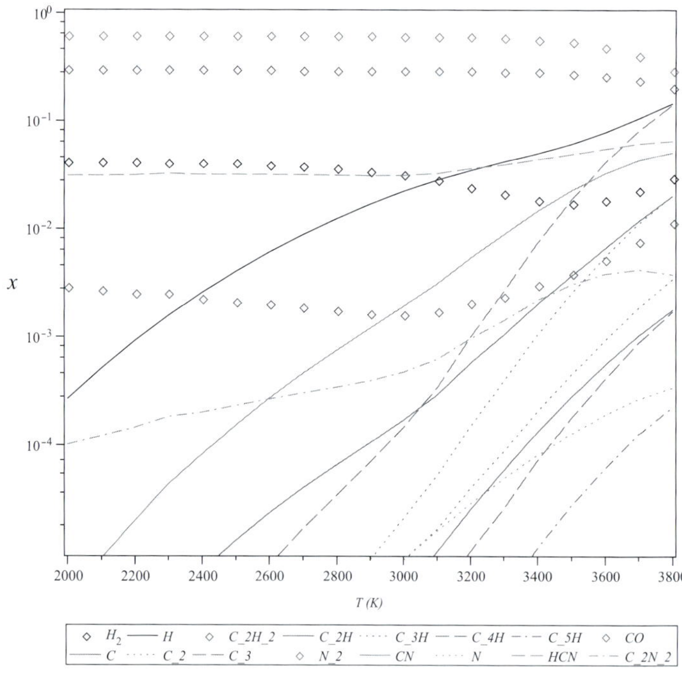

Fig. 8.26 Wall composition for a carbon-resin. Thermodynamic equilibrium calculation ($p = p_0$, $\rho_e u_e C_M = 1 \text{ kg} \cdot \text{m}^{-2} \cdot \text{s}^{-1}$). Argon omitted.

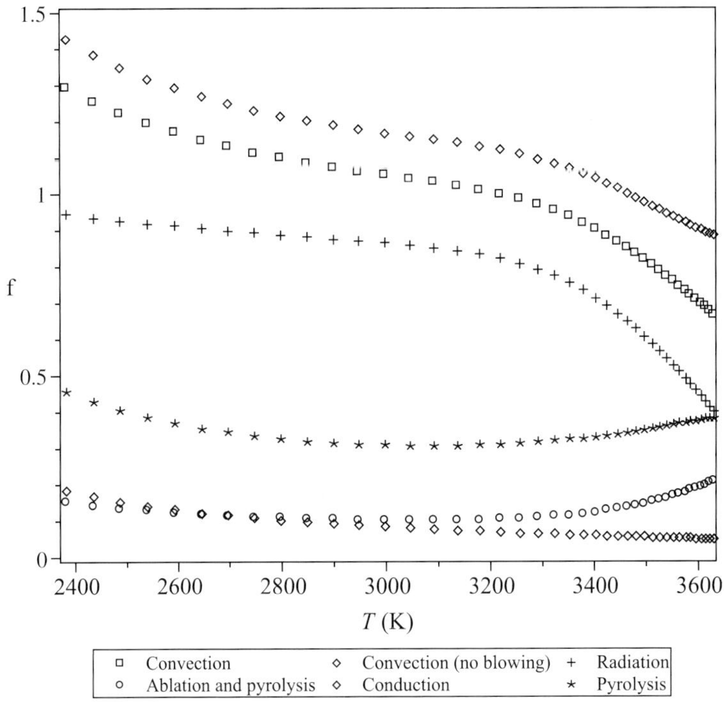

Fig. 8.27 Partition of energy for a TWCP carbon–phenolic material.

- The part of the radiation remains predominant.
- The pyrolysis plays a significant part in the balance (term $\dot{m}_g h_{gp}$).

It results in a part of the conduction of about 10 to 20% versus 10% for carbon. This lower efficiency is, of course, compensated for by the low conductivity of the material.

References

[1] Laux, C., "Physico-Chemical Models for High Enthalpy and Plasma Flows," Von Karman Institute for Fluid Dynamics Lecture Series 2002–07, edited by Fletcher, D., Charbonnier, J.-M., Sarma, G. S. R., and Magin, T., June 2002.

[2] Kline, G. M., "Analytical Chemistry of Polymers, Part 2: Molecular Structure and Chemical Groups," *High Polymers*, Vol. 2, Interscience, 1962.

[3] Schultz, A. W., and Wong, A. K., "Thermal Conductivity of Teflon, KEL-F and Duroid 5600 at Elevated Temperatures," Watertown Arsenal Laboratories Report TR 397/10, March 1958.

[4] Munson, T. R., and Spindler, R. J., "Transient Thermal Behaviour of Decomposing Materials," Institute of Aerospace Science (IAS) Paper 62–30, 1962.

[5] Hiester, N. K., and Clark, C. F., "Feasibility of Standard Evaluation Procedure for Ablating Materials," Stanford Research Institute Report NASR-49(15), SA CR-379, 1966.

[6] Greenberg, R. A., Kemp, N. H., and Wray, K. L., "Structure of the Laminar Ablating Air Teflon Boundary Layer," *AIAA Journal*, Vol. 8, No. 4, 1970, pp. 619–626.
[7] Sykes, G. F., "Decomposition Characteristics of a Char-Forming Phenolic Polymer Used for Ablative Composites," NASA Technical Note TN D-3810, Feb. 1967.
[8] Ladacki, M., Hamilton, J. V., and Cohz, S. N., "Heat of Pyrolysis of Resin in Silica-Phenolic Ablator," *AIAA Journal*, Vol. 4, No. 10, 1966, pp. 1798–1802.
[9] Lincoln, K. A., "Experimental Determination of Vapor Species from Laser-Ablated Carbon Phenolic Composites," *AIAA Journal*, Vol. 21, No. 8, 1983, pp. 1204–1207.
[10] Ducamp, V., "Transferts thermiques dans un matériau composite carbone-résine," Ph.D. Thesis, l'Université Bordeaux 1, No. 2505, April 2002.
[11] Trick, K. A., "Reaction Kinetics of the Pyrolysis of Phenolic/Carbon Composite Material," Ph.D. Thesis, University of Dayton, Florida, 1994.
[12] Desai, T. G., Lawson, J. W., and Keblinsky, P., "Modeling Initial Stage of Ablation Material Pyrolysis: Graphitic Precursor Formation and Interfacial Effects," *Polymer*, Vol. 52, 2011, pp. 577–585.
[13] Duffa, G., "Ablation," CESTA monograph, Commissariat a l'Energie Atomique, Le Barp, France, Nov. 1996.
[14] Wool, M. R., "Summary of Experimental and Analytical Results," Interim Report, Passive Nosetip Technology (PANT) Program, Vol. X, SAMSO-TR-74-86, Jan. 1975.
[15] Epherre, J.-F., and Laborde, L., "Pyrolysis of Carbon Phenolic Composites," *Fourth International Symposium Atmospheric Reentry Vehicles and Systems*, Arcachon, France, March 2005.
[16] Sullivan, R. M., and Stokes, E. H., "A Model for the Effusion of Water in Carbon Phenolic Composites," *Application of Porous Media Methods for Engined Materials*, American Society of Mechanical Engineers, Applied Mechanics Division, Vol. 233, 1999.
[17] Henderson, J. B., Tant, M. R., Wiebelt, J. A., and Moore, G. R., "Determination of Kinetic Parameters for Thermal Decomposition of Phenolic Ablative Materials by Multiple Heating Rate Method," NSWC Report TR-80-024, July 1980.
[18] Trick, K. A., Saliba, T. E., and Sandhu, S. S. "A Kinetic Model of the Pyrolysis of Phenolic Resin in a Carbon/Phenolic Composite," *Carbon*, Vol. 35, No. 3, 1997, pp. 393–401.
[19] Duffa, G., Vignoles, G., Goyhénéche, J.-M.,, and Aspa, Y. "Ablation of Carbon-Based Materials: Investigation of Roughness Set-up from Heterogeneous Reactions," *International Journal of Heat and Mass Transfer*, Vol. 48, 2005, pp. 3387–3401.
[20] Ahn, H.-K., Park, C., and Sawada, K., "Response of Heatshield Material at Stagnation Point of Pioneer-Venus," *Journal of Thermophysics and Heat Transfer*, Vol. 16, No. 3, 2002, pp. 432–439.
[21] Hoshizaki, H., and Lasher, L. E., "Convective and Radiative Heat Transfer to an Ablating Body," *AIAA Journal*, Vol. 6, No. 8, 1968, pp. 1441–1449.
[22] Pike, R. W., Farmer, R. C., and Balhoff, J. F., "Coupled Radiating Shock Layers with Finite Rate Chemistry Effects," Reacting Fluid Laboratory, Department of Chemical Engineering, Louisiana State University, Final Report Part II, June 1975.
[23] Scoggins, J. B., "The Development of a Thermochemical Nonequilibrium Ablation and Pyrolysis Model for Carbon-Phenolic Thermal Protection Systems," Master's thesis, North Carolina State University, Raleigh, NC, March 2011.
[24] Warnatz, J., "Rate Coefficients in the C/H/O System," *Combustion Chemistry*, edited by Gardiner, W. C., Jr., Springer-Verlag, New York, 1984, pp. 197–360.
[25] "Tutorial on Ablative TPS," NASA Ames Conference Center, Moffett Field, California, Aug. 2004.
[26] Pike, R. W., April, G. C., and del Valle, E. G., "Non-Equilibrium Flow and the Kinetics of Chemical Reactions in the Char Zone," NASA Report RFL-7, July 1967.

[27] April, G. C., Pike, R. W., and del Valle, E. G., "On Chemical Reactions in the Char Zone During Ablation," NASA Report CR 1903, Oct. 1971.
[28] Frenklach, M., and Wang, H., "Detailed Mechanism and Modeling of Soot Particle Formation," *Soot Formation in Combustion*, edited by Bockhorn, H., Springer-Verlag, Berlin, 1994, pp. 165–190.
[29] Vignoles, G. L., Langlais, F., Descamps, C., Mouchon, A., Le Poche, H., Reuge, N., and Bertrand, N., "CVD and CVI of Pyrocarbon from Various Precursors," *Surface and Coatings Technology*, Vol. 188–189, 2004, pp. 241–249.
[30] Kontinos, D. A., and Stackpoole, M., "Post-Flight Analysis of the Stardust Sample Return Capsule Earth Entry," AIAA Paper 2008–1197, *46th AIAA Aerospace Sciences Meeting and Exhibit*, Reno, Nevada, Jan. 2008.
[31] Park, C., "Stagnation-Point Radiation for Apollo 4," *Journal of Thermophysics and Heat Transfer*, Vol. 18, No. 34, 2004, pp. 349–357.
[32] Linstedt, P. R., "Simplified Soot Nucleation and Surface Growth Steps for Non-premixed Flames," *Soot Formation in Combustion*, edited by Bockhorn, H., Springer-Verlag, Berlin, 1994, pp. 417–438.
[33] Pradere, C., Batsale, J.-C., Goyhénèche, J.-M., Pailler, R., and Dilhaire, S., "Thermal Properties of Carbon Fibers at Very High Temperature," *Carbon*, Vol. 47, No. 3, 2009, pp. 737–743.
[34] Brewer, W. D., Stroud, C. W., and Clark, R. K., "Effect of the Chemical State of Pyrolysis Gases on Heat-Shield Mass," NASA Report TN D-4975, Dec. 1968.
[35] Lino da Silva, M., Guerra, V., Loureiro, J., and Dudeck, M., "Simulation of Plasma Radiation in Earth and Mars Atmospheric Entries," *Fourth International Symposium on Atmospheric Reentry Vehicles and Systems*, Arcachon, France, March 2005.
[36] Stokes, E. H., "Equilibrated Moisture Component of Several Carbon Phenolic Composites," *AIAA Journal*, Vol. 30, No. 6, 1992, pp. 1597–1601.
[37] Sutton, K., "An Experimental Study of a Carbon-Phenolic Ablation Material," NASA Technical Report D-5930, Sept. 1970.
[38] Preux, C., "Modélisation et calcul du transfert de masse et de chaleur dans un milieu poreux réactif en évolution structurale et applications," Ph.D. Thesis, l'Université Bordeaux 1, No. 3250, Dec. 2006.
[39] Puiroux, N., Prat, M., Quintard, M., and Laturelle, F., "Macro-Scale Non-Equilibrium Heat Transfer in Ablative Composite Layers," AIAA Paper 2002–3336, *Eighth AIAA/ASME Thermophysics and Heat Transfer Conference*, St Louis, Missouri, June 2002.
[40] Fay, J. A., and Riddell, F. R., "Theory of Stagnation Point Heat Transfer in Dissociated Air," *Journal of the Aerospace Sciences*, Vol. 25, No. 2, 1958, pp. 83–86.
[41] Clark, R. K., "A Numerical Analysis of the Transient Response of an Ablation System Including Effects of Thermal Non-equilibrium, Mass Transfer and Chemical Kinetics," Ph.D. thesis, Virginia Polytechnic Institute and State University, Blacksburg, VA, May 1972.
[42] Charrier, P., and Dubroca, B., "Asymptotic Transport Model for Heat and Mass Transfer in Reactive Porous Media," *SIAM Journal on Multiscale Modeling and Simulation*, Vol. 2, No. 1, 2003, pp. 124–157.
[43] Whitaker, S., *The Method of Volume Averaging*, Kluwer Academic, Dordrecht, The Netherlands, 1999.
[44] Martin, A., and Boyd, I. D., "Non-Darcian Behavior of Pyrolysis Gas in a Thermal Protection System," *Journal of Thermophysics and Heat Transfer*, Vol. 24, No. 1, 2010, pp. 60–68.
[45] Coindreau, O., "Etude 3D de préformes fibreuses: interaction entre phénomenès physico- chimiques et géométrie," Ph.D. thesis, l'Université Bordeaux 1, No. 2800, Dec. 2003.
[46] Mason, E. A., Malinauskas, A. P., and Evans, R. B., III, "Flow and Diffusion of Gases in Porous Media," *Journal of Chemical Physics*, Vol. 46, No. 8, 1967, pp. 3199–3216.
[47] Mason, E. A., and Malinauskas, A. P., *Gas Transport in Porous Media: The Dusty-Gas Model*, Elsevier Scientific, Amsterdam, 1983.

[48] Kerkhof, J. P. A. M., "New Light on Some Old Problems: Revisiting the Stefan Tube, Graham's Law, and the Bosanquet Equation," *Industrial and Engineering Chemical Research*, Vol. 36, No. 3, 1997, pp. 915–922.

[49] Kaviany, M., *Principles of Heat Transfer in Porous Media*, Springer-Verlag, New York, 1995.

[50] Nield, D. A., and Bejan, A., *Convection in Porous Media*, 3rd ed., Springer, New York, 2006.

[51] Dullien, J. A. L., *Porous Media. Fluid Transport and Pore Structure*, 2nd ed., Academic Press, San Diego, CA, 1992.

[52] Stokes, E. H., "Permeability of Carbonized Rayon Based Polymer Composites," *Computational Mechanics of Porous Materials and Their Thermal Decomposition*, American Society of Mechanical Engineers, Applied Mechanics Division, Vol. 136, 1992, pp. 145–156.

[53] Suzuki, T., and Sawada, K., "Gas Permeability of Oblique-Layered Carbon-Cloth Ablator," *Journal of Thermophysics and Heat Transfer*, Vol. 18, No. 4, 2004, pp. 548–550.

[54] Harharan, R., Test, F. L., Florio, J., and Henderson, J. B., "Internal Pressure Distribution in Decomposing Polymer Composites," *Ninth Heat Transfer Conference*Jerusalem, Israel, 1990, pp. 335–340.

[55] Bigaud, D., Goyheneche, J.-M., and Hamelin, P., "A Global-Local Non-Linear Modeling of Effective Thermal Conductivity Tensor of Textile-Reinforced Composites," *Composites*, part A, No. 32, Elsevier, 2001, pp. 1443–1453.

[56] Amar, A. J., Blackwell, B. F., and Edwards, J. R., "One-Dimensional Ablation with Pyrolysis Gas Flow Using Finite Control Volume Procedure," AIAA Paper 2007–4535, *39th AIAA Thermophysics Conference*, Miami, Florida, June 2007.

[57] de Muelenaere, J., Lachaud, J., Mansour, N. N., and Magin, T. E., "Stagnation Line Approximation for Ablation Thermochemistry," AIAA Paper 2011–3616, *42nd Thermophysics Conference*, Honolulu, Hawaii, June 2011.

[58] McBride, B. J., Gordon, S., and Reno, M. A., "Coefficients for Calculating Thermodynamics and Transport Properties of Individual Species," NASA Technical Memorandum 4513, 1993.

[59] Martin, A., and Boyd, I. D., "Assessment of Carbon-Phenolic-in-Air Thermochemistry Models for Atmospheric Re-entry," AIAA Paper 2010–4656, *10th AIAA/ASME Joint Thermophysics and Heat Transfer Conference*, Chicago, Illinois, June 2010.

[60] Martin, A., Boyd, I. D., Cozmuta, I., and Wright, M. J., "Chemistry Model for Ablating Carbon-Phenolic Material During Atmospheric Re-entry," AIAA Paper 2010-1175, *48th Aerospace Sciences Meeting Including the New Horizons Forum and Aerospace Exposition*, Orlando, Florida, Jan. 2010.

Chapter 9 Materials Developing a Liquid Layer

- Describe the properties of an ideal surface liquid layer
- Model a liquid layer
- Analyze the influence of change of material composition

Some materials develop a liquid surface layer. This is the case for metals that have been used because of their refractoriness (e.g., tungsten [1, 2]) or materials containing glass fibers. The liquid layer is almost never obtained by fusion when oxygen is present in the medium. Tungsten, for example, form an oxide, whose melting point is lower than that of the metal, which will form the liquid layer. A rare example of direct fusion is pure silica, the most stable of Si oxides. Given this fact, the liquid layer present on the silica–resin materials was attributed to this phenomenon. We will see later that this does not correspond to reality. Before developing the material aspect, we will examine the properties of an ideal surface liquid layer, that is to say covering a flat surface.

One should not forgot that the liquid formed will migrate back under the effect of shear of the boundary layer and may freeze in important thickness in expansion areas [2]. This can have significant effects on the aerodynamics of the body. This phenomenon will not be treated in this book.

9.1 Hydrodynamics of the Liquid Layer

9.1.1 Descriptive Equations

Modeling of such a liquid layer leads to the following issues:

- Mass exchange between solid and liquid on the inner side, and between liquid and gas on the outside and inside of the same fluid
- Energy exchange by radiation in a semi-transparent medium so constituted

The model will be developed under the following restrictions:

- The lamina thickness varies slowly and is low as compared to the radius of curvature of the wall.
- There is no source of mass or energy in the volume, in particular associated with gas bubbles, these having a small effect [3].
- The liquid viscosity is sufficient to make possible the description of a hydrodynamic layer limited to low Reynolds number equations (creep flow).
- We will neglect the longitudinal gradients of velocity and temperature gradients as compared to normal ones.

With the notations of Fig. 9.1, the conservation equation of steady flow is written

$$\frac{\partial(ru)}{\partial s} + \frac{\partial(ru)}{\partial y} = 0 \tag{9.1}$$

with the following boundary conditions:

- At $y = 0$ (solid–liquid interface)

$$\begin{cases} u(s, 0) = 0 \\ v(s, 0) = \dfrac{\dot{m}_l(0)}{\rho_l} \end{cases} \tag{9.2}$$

- At $y = e$ (liquid–gas interface)

$$v(s, e) = -\frac{\dot{m}_l(e)}{\rho_l} + u(s, e)\frac{\mathrm{de}}{\mathrm{ds}} \tag{9.3}$$

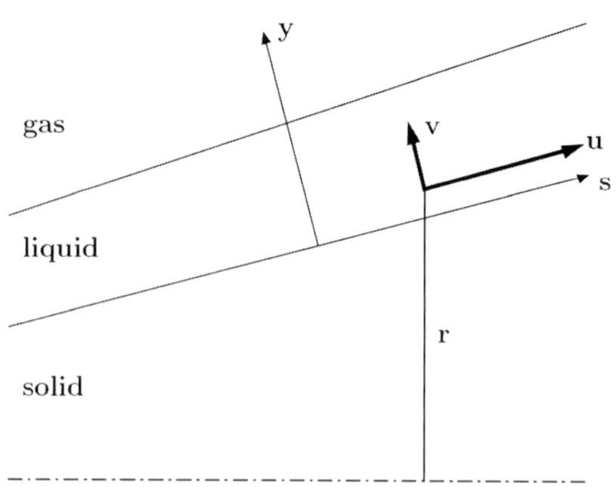

Fig. 9.1 Schematic model of liquid layer.

Moreover, the equation of longitudinal momentum is written

$$\frac{\partial}{\partial y}\left(\mu_l \frac{du}{dy}\right) = \frac{dp}{ds} - \rho_l \Gamma \cos\theta \tag{9.4}$$

where Γ is the accelerometric measure (acceleration minus gravity) of the vehicle supposed to slow down. The second part of this equation, noted S_m, is known.

The condition in $y = e$ is

$$\mu_l \frac{du}{dy} = \tau \tag{9.5}$$

where τ is the shear in the outer flow. Assume it is not modified by the liquid layer, the speed being small compared to the external flow. This hypothesis is consistent with the hypothesis of creep flow.

The second conservation equation on the momentum normal is written

$$\frac{dp}{dy} = 0 \tag{9.6}$$

whose solution is $p = p_e$.

Finally, the energy equation is written, neglecting the dissipative terms and assuming a constant specific heat

$$\rho_l C_{p_l}(v + v_a)\frac{dT}{dy} = \frac{d}{dy}\left(\lambda_l \frac{dT}{dy}\right) \tag{9.7}$$

The thermal conductivity λ_l contains a term relative to the radiative transfer also evaluated (Appendix G).

9.1.2 Approximate Resolution

The solution for v is obtained immediately by noting that with the assumption of absence of volumic source, the solution is $v(s, y) = v(s, 0)$, value noted v_0. For the following we adopt the approach taken in [4].

If diffusivity $\alpha_l = \lambda_l / \rho_l C_{p_l}$ is assumed constant, the energy equation integrates immediately

$$T(s, y) = T(s, e)\exp\left[\frac{(v_a + v_0)(y - e)}{\alpha_l}\right] \tag{9.8}$$

Hence the solution of Eq. (9.4):

$$u(s, y) = s_m \int_0^y \frac{\sigma d\sigma}{\mu_l} + (\tau - S_m e)\int_0^y \frac{d\sigma}{\mu_l} \tag{9.9}$$

where S_m is the second member of Eq. (9.4).

This equation is integrable for a viscosity following a power law of temperature (Sec. 9.1.3)

$$\mu_l = \mu_{le}\left(\frac{T}{T_e}\right)^{-m} \tag{9.10}$$

with $\mu_{l_e} = \mu_l(s, e)$ and $T_e = T(s, e)$.

The solution is written

$$u(s, y) = (\gamma_1 + \gamma_2 y)\exp\left(\frac{y}{\delta}\right) - \gamma_1 \tag{9.11}$$

with

$$\begin{cases} \delta = \dfrac{\alpha_l}{m(v_a + v_0)} \\ \gamma_1 = \dfrac{\delta}{\mu_{le}}[\tau - S_m(e + \delta)]\exp\left(-\dfrac{e}{\delta}\right) \\ \gamma_2 = \dfrac{\delta S_m}{\mu_{le}}\exp\left(-\dfrac{e}{\delta}\right) \end{cases} \tag{9.12}$$

In bringing this term into the equation of mass flux conservation, we get

$$v(s, e) = \alpha_1\beta_1 + \alpha_2\beta_2 + \alpha_3\beta_3 + \beta_1 e + v_0 \tag{9.13}$$

with

$$\begin{cases} \alpha_1 = -\delta\left[\exp\left(\dfrac{e}{\delta}\right) - 1\right] \\ \alpha_2 = -\delta\left[(e - \delta)\exp\left(\dfrac{e}{\delta}\right) + \delta\right] \\ \alpha_3 = -\delta\left[(e^2 - 2e\delta + 2\delta^2)\exp\left(\dfrac{e}{\delta}\right) - 2\delta^2\right] \\ \beta_1 = \dfrac{\mathrm{d}\gamma_1}{\mathrm{d}s} + \gamma_1\dfrac{\sin\theta}{r} \\ \beta_2 = \dfrac{\mathrm{d}\gamma_2}{\mathrm{d}s} + \gamma_2\dfrac{\sin\theta}{r} + \dfrac{\gamma_1 m}{\alpha_l}\dfrac{\mathrm{d}}{\mathrm{d}s}(v_a + v_0) \\ \beta_3 = \dfrac{\gamma_2 m}{\alpha_l}\dfrac{\mathrm{d}}{\mathrm{d}s}(v_a + v_0) \end{cases} \tag{9.14}$$

Introducing this last expression with Eq. (9.11) in Eq. (9.3), and using the variable change $\frac{\mathrm{d}\cdot}{\mathrm{d}s} = \frac{\mathrm{d}\cdot}{\mathrm{d}e}\frac{\mathrm{d}e}{\mathrm{d}s}$, we obtain the differential equation verified by the thickness e

$$\frac{\mathrm{d}e}{\mathrm{d}s} = \frac{\alpha_1\gamma_1\frac{\sin\theta}{r} + \alpha_2\left[\gamma_2\frac{\sin\theta}{r} + \frac{\gamma_1 m}{\alpha_l}\frac{\mathrm{d}}{\mathrm{d}s}(v_a + v_0)\right] + \alpha_3\beta_3 + \beta_1 e + v_0}{(\gamma_1 + \gamma_2 e)\exp\left(\frac{e}{\delta}\right) - \gamma_1 - \alpha_1\frac{\mathrm{d}\gamma_1}{\mathrm{d}e} - \alpha_2\frac{\mathrm{d}\gamma_2}{\mathrm{d}e}} \tag{9.15}$$

9.1.3 Silica

Subsequently, we will focus on silica, which is present in many materials. We can approach various viscosity measurements made on this material by a power law of T (Fig. 9.2), showing a rapid variation with temperature

$$\mu_1 \simeq 2.91 \cdot 10^7 \left(\frac{T}{2000}\right)^{-35} \tag{9.16}$$

Equation (9.13) determines a characteristic length for the establishment of a constant thickness over a material when conditions in the flow over the silica layer are constant. Under the conditions of interest, this length is of the order of centimeters. We can therefore define a local ablation model (which depends only on local conditions) on this type of material when the radius of curvature of the object is much greater than this value. This is consistent with the observation of test results that show only the beginning of the sample always has a higher ablation in constant conditions.

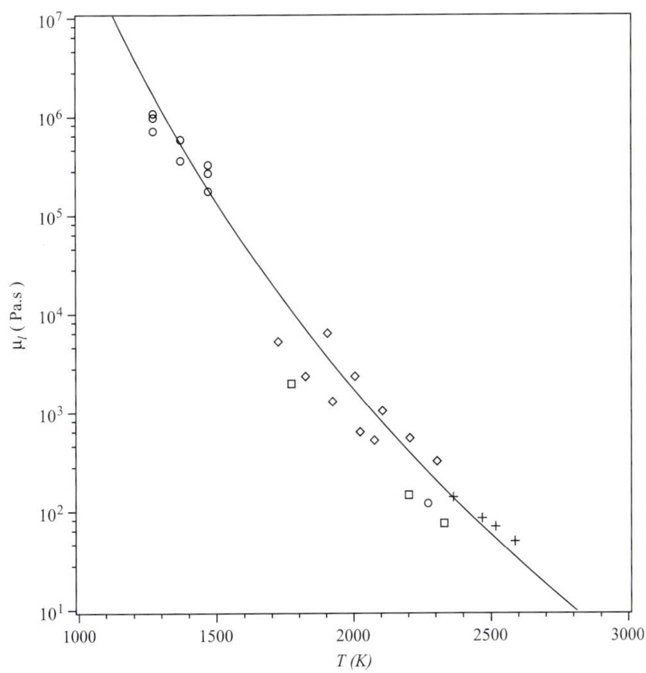

Fig. 9.2 Viscosity of silica: measurements (5–8) and power law of T.

9.2 Silica-Resin Materials

9.2.1 Physicochemical Reactions

The mixture of SiO_2 and carbon produced by pyrolysis has the following overall reaction:

$$SiO_2 + C \rightarrow SiO + CO$$

Many articles also provide a reaction of carburization

$$SiO_2 + 3C \rightarrow SiC + 2CO$$

The analysis of test samples cannot find SiC in significant amounts, at least in the given tests. This may be an intermediate reaction that we will ignore. The first reaction is global in the sense that it is a complex reaction between the gases issued from solids. It begins with the decomposition of silica

$$SiO_2 \rightarrow SiO + \frac{1}{2}O_2$$

The oxygen formed reacts with the carbon. Diffusion may play a role in this type of reaction, which varies according to the state of division of the solid. Various values of activation temperature are given, often close to the value used thereafter, so 57,000 K. However, it should be mentioned that experiments show a change of apparent kinetics at high temperatures [9], perhaps related to the diffusion. The system is described by the law giving the rate of C consumed by the reaction

$$\dot{m}_C = A \exp\left(\frac{-T_A}{T_0}\right) \tag{9.17}$$

where T_0 is the temperature of the solid-liquid interface. The form chosen for this law is justified in Sec. 8.1.2 [Eq. (8.9)]. In fact, A is a numerical value that averages the correct term.

Outside the liquid surface, the SiO will recombine with oxygen (in the case of air) to restore the SiO_2 under gaseous form. This will be present in large quantities, with a partial pressure above the equilibrium vapor pressure, hence its condensate in the form of SiO_2 liquid. Note that this phenomenon is crucial because it explains the presence of a liquid layer while the molar ratio SiO_2/C in the solid material is usually less than 1. This reaction is described by a Knudsen–Langmuir law [Sec. 3.2.3, Eq. (3.64)]

$$\dot{m}_l = \beta\left(\frac{\mathcal{M}_{SiO_2}}{2\pi RT}\right)^{\frac{1}{2}}\left(p_{eq_{SiO_2}} - p_{SiO_2}\right) \tag{9.18}$$

The value of β is unknown. However, a numerical study shows that the results are somewhat independent of its value when $\beta > 0.1$.

It may be noted that at higher temperatures, one can observe the limitation of reactions due to one of the following reasons:

- By the limitation of oxygen diffusion in the boundary layer
- By the limitation of the input of SiO by the material

9.2.2 Ablation of Silica–Resin

The problem thus posed has been solved:

- Assuming a gas flow in local thermodynamic equilibrium
- Exchanges in the boundary layer described by the approximation of a "thin layer" (Sec. 4.4)
- The assumption of stationary ablation
- Assuming the asymptotic thickness of liquid

The results, described in Fig. 9.3, show the dependence of $\rho_p v_a / \rho_e u_e C_H$ with surface temperature T_e. (The portions of curves less than 2700 K for one bar and 2800 K for five bars are not physically realistic.) There are two

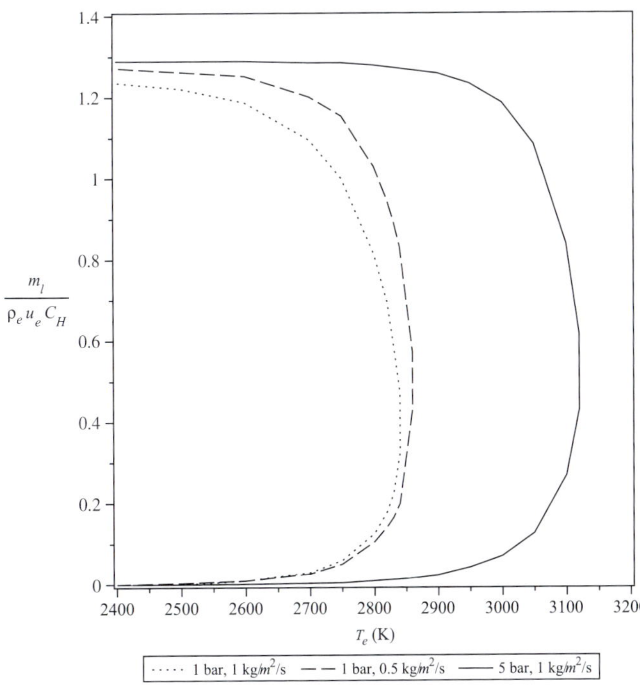

Fig. 9.3 Analytical ablation model of silica–resin materials ($\beta = 1$, $A = 1.25 \times 10^8$ kg · m^{-2} · s^{-1}).

possible solutions for a given value of T_e, the choice between these two solutions given by the energy conservation. The lower branch curves correspond to high values of the thickness of silica. We find a dependence on pressure: the translation of phenomena to higher temperatures for higher pressures. Dependence on exchanges within the boundary layer is small at the T_e given. However, it should not be forgotten that the latter value depends on the exchange through the equation of energy conservation.

9.2.3 Mass Balance and Energy Conservation

The material fluxes are (Fig. 9.4)

- The mass flux of carbon from pyrolysis, consumed by reaction with silica

$$\dot{m}_C = g\rho_p v_a \tag{9.19}$$

where g is the mass fraction of C in the pyrolyzed material.

- The mass flux rate of appearance (or disappearance) of silica

$$\dot{m}_{SiO_2} = -\frac{\mathcal{M}_{SiO_2}}{\mathcal{M}_C}\dot{m}_C + (1-g)\rho_p v_a \tag{9.20}$$

The first term corresponds to the reaction SiO_2/C and the second to the contribution of the liquid layer. Given this equation, we get

$$\dot{m}_{SiO_2} = \rho_l v_0 = (1-6g)\rho_p v_a \tag{9.21}$$

- A gas mass flux of SiO and CO equal to $6g\rho_p v_a$ corresponds to these values, which, added to pyrolysis mass flux $(\rho_p - \rho_v)v_a$, gives a total gas mass flux

$$\dot{m}_g = [\rho_v + (6g-1)\rho_p]v_a \tag{9.22}$$

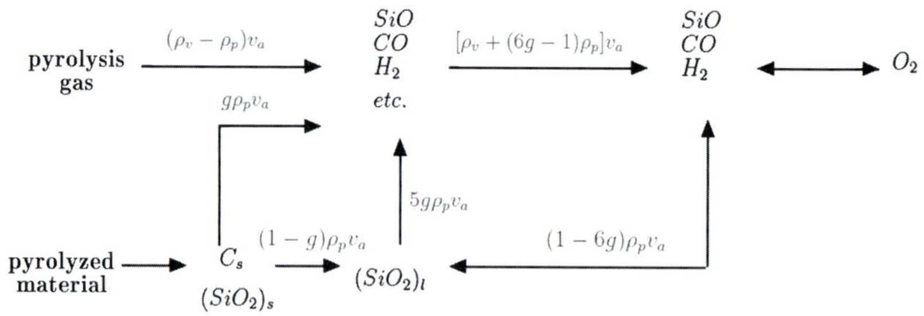

Fig. 9.4 Mass fluxes in a silica-resin composite in stationary regime ("etc." represents products issued from pyrolysis).

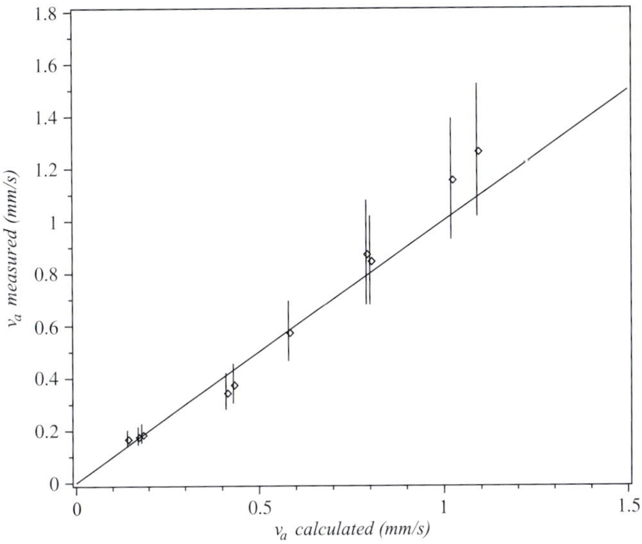

Fig. 9.5 Silica–resin composite ablation tests; comparison of measurements theory with roughness effects (9).

From the calculated mass fluxes we obtain immediately the equation of energy conservation for the stationary state

$$\rho_p h_p(T_0) - \rho_v h_v(T_{\text{int}}) + (\rho_v - \rho_p) h_g(T_e) + 6g\rho_p[h_{g_R}(T_e) - h_{g_R}(T_0)] + (1 - 6g)\rho_p[h_l(T_e) - h_l(T_0)] = \frac{\dot{\mathbf{q}}_{\text{cond}}}{v_a} \tag{9.23}$$

where h_l is the enthalpy of liquid SiO_2, h_g the pyrolysis gas enthalpy, h_{gR} the enthalpy of the mix (SiO + CO), and T_{int} the initial (internal) material temperature.

9.2.4 Validation and Use of the Model

9.2.4.1 Validation

Such a model has been tested on a 3-D silica–resin material. On this type of material, the yarn perpendicular to the surface gives a rough surface that makes the radiative model unreliable. Moreover, this tormented surface leads to a modification of exchanges in the boundary layer. A comparison with plasma jet experiments (Fig. 9.5), however, shows a very reasonable agreement on the ablation velocity, poorer on temperatures, whose values are dependent on the apparent emissivity [9].

9.2.4.2 Influence of the Composition of the Material

The model previously described can be used to analyze the influence of a change of composition of material. Figure 9.6 shows the effect of variations in the volume fraction of silica or porosity.

- The fiber fraction plays only a minor role in the performance—a variation of 10% (technologically very difficult to obtain) gives an equivalent variation on the ablation mass flux. This is confirmed by the tests. Such varying material has a consequence on its conductivity, which increases significantly.
- A variation from 0 to 20% in porosity causes a slight increase in the ablation velocity (4%) but a decrease of 10% of the mass flux rate of ablation. This effect comes from changing the reaction balance, itself due to the change in the balance of SiO_2/C in the pyrolyzed material. Such a change comes over a significant decrease in conductivity.

9.2.4.3 Global Model

As we saw in the case of PTFE, the law $v_a(T_e)$ is sometimes used to describe the ablation. In the case of silica–resin, a law similar to the following was often used [10]:

$$v_a = \alpha T + \beta * \mathrm{T}^{\gamma} \exp\left(-\frac{\delta}{\mathrm{T}}\right) \tag{9.24}$$

One can build this kind of law with test results (Fig. 9.7). (In the figure, the boxes are tests, temperature being obtained with a hypothetical emissivity

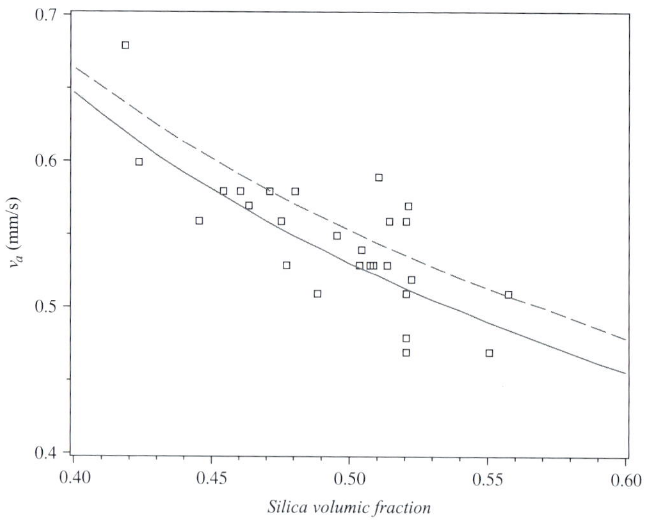

Fig. 9.6 Influence of silica-resin composition on ablation velocity (curves 0 and 20% porosity).

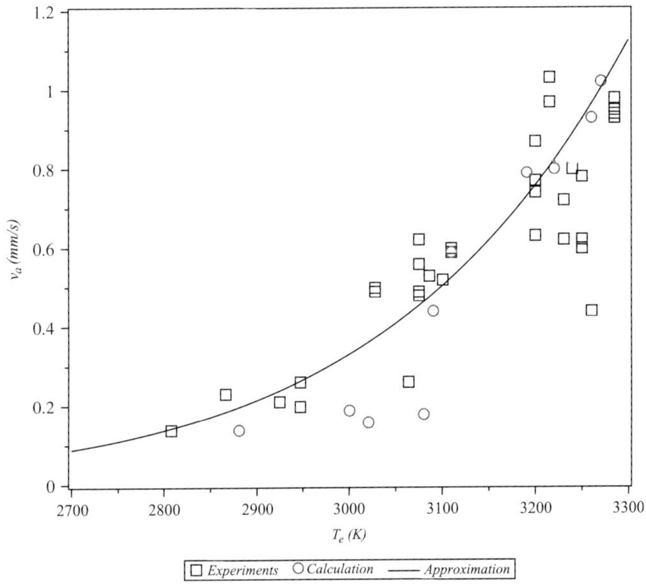

Fig. 9.7 Silica–resin ablation global model.

of 0.3. The circles are calculations with the analytical model.) This type of curve requires an assumption about the emissivity of the material and gives the best approach to results with a dispersion attributed to experience. If we did the same with theoretical results, we would obtain a nearly equivalent dispersion. This highlights the caution with which we must use a global model that hides very complex physical phenomena.

The model must be supplemented by an apparent ablation enthalpy [Sec. 4.6, Eq. (4.70)] on which one can make the same comments as those on the ablation "law." In the case of silica–resin material, an enthalpy of $5.0 \pm 1.4\ \mathrm{MJ \cdot kg^{-1}}$ deduced from the analytical model is confirmed by the tests [10].

References

[1] Auerbach, I., "Ablation Performance of Tungsten, Copper Infiltrated Tungsten, and Other Metal Systems in Arc Heated Jets," AIAA Paper 77-239, *15th AIAA Aerospace Sciences Meeting*, Los Angeles, CA, Jan. 1977.

[2] Moody, H. L., Smith, D. H., Haddock, R. L., and Dunn, S. S., "Tungsten and Molybdenum Ablation Modeling for Re-entry Applications," *Journal of Spacecraft and Rockets*, Vol. 13, No. 12, 1976, pp. 746–753.

[3] Hsieh, C. L., and Seader, J. D., "Surface Ablation of Silica-Reinforced Composites," *AIAA Journal*, Vol. 11, No. 8, 1973, pp. 1181–1187.

[4] Bethe, H. A., and MacAdams, C., "A Theory for Ablation of Glassy Materials," *Journal of the Aerospace Sciences*, Vol. 26, No. 6, 1959, pp. 321–328.

[5] Bowen, D. W., and Taylor, R. W., "Silica Viscosity from 2300 to 2600 K," *Ceramic Bulletin*, Vol. 57, 1978, pp. 818–819.
[6] Pascal, P., "Silicium," *New Presentation of Mineral Chemistry*, Vol. 2, Masson, Paris, 1965.
[7] Schick, H. L., "A Thermodynamic Analysis of the High-Temperature Vaporization Properties of Silica," *Chemical Review*, Vol. 60, No. 8, 1960, pp. 331–362.
[8] Weiss, W., "A Rapid Torsion Method for Measuring the Viscosity of Silica Glasses up to 2200°C," *Journal of the American Ceramic Society*, Vol. 67, No. 3, 1958, pp. 213–222.
[9] Duffa, G., "Ablation," CESTA monograph, Commissariat a l'Energie Atomique, Le Barp, France, Nov. 1996.
[10] "Tutorial on Ablative TPS," NASA Ames Conference Center, Moffett Field, CA, Aug. 2004.

Chapter 10 Radiation

- Describe transitions
- Calculate state populations
- Deduce relative transfer
- Determine effects of coupling
- Solve transfer equation

10.1 Introduction

10.1.1 Convection and Radiation

The relative importance of convection and radiation were discussed in Sec. 1.3.4. The convection is dominant at low speed (typically $V_\infty < 12\ \mathrm{km \cdot s^{-1}}$ in Earth's atmosphere). This must, of course, be weighted by the radius R at the stagnation point, especially if one is interested in testing in a tunnel facility for which the models are very small, which promotes convection but greatly disadvantages the radiation (see Sec. 10.2.2.2). It is also important to note that this depends on the type of atmosphere we're dealing with. For example, in the case of the reentry of Huygens on Titan, the species *CN*, formed from the components of the atmosphere N_2 and CH_4, is created behind the shock in a very hot region. This species is very active in the ultraviolet and leads to a radiative heat flux at stagnation point, which is very important for low velocities (typically $6\ \mathrm{km \cdot s^{-1}}$). In the case of an Earth reentry, this species, also created by ablation of carbon, remains confined to the boundary layer. Its effect is a priori not obvious because it will at once radiate to the wall and screen for radiation from the shock layer.

High temperatures are linked to the presence of shock, which will generate an important thermal nonequilibrium of molecules in the case of low-density environments. Unlike most of the problems addressed in this book, this nonequilibrium plays a role in modeling. A counterexample is the modeling of the electric arc in a plasma jet. In this case, the high pressures (several bars or tens of bars, see Table 12.1) lead to an environment adequately described by local thermodynamic equilibrium, except in the vicinity of the electrodes.

10.1.2 Emission and Absorption in a Gas

In this section we provide some basic information necessary for the use of databases. We discuss the elements necessary for complete description of phenomena and resolution methods in [1–7]. We also summarize some important points for understanding the problem.

10.1.2.1 Generalities

A photon is characterized by its frequency ν, the wavelength $\lambda = c/\nu$, the "wavenumber" (of spectroscopists) $\omega = \lambda^{-1}$, or energy $h\nu$. The domain of spectroscopy uses the wavelength (in micrometers or Angstroms) or wavenumber (in reciprocal centimeters); in atomic physics the energies are in electronvolts ($1\ \text{eV} = 1.60217649 \times 10^{-19}$ J).* They are sometimes expressed as reduced values E/hc, again in reciprocal centimeters. The polarization plays no role in the field of interest. Similarly, the momentum $(2\pi\nu/c)\Omega$ (where Ω is the unit vector direction described by two components) plays no role, the radiative pressure being negligible compared with the kinetic pressure.

The radiation is fully described by the knowledge of the spectral radiance $I_\nu\ (x, t, \Omega)$ (also called spectral intensity by physicists). I_ν is the energy flux in the direction Ω, in the solid angle $d\Omega$ for the spectral range $d\nu$ centered at ν.

Part of the radiation we are interested in lies in the infrared $\lambda \leq 10\ \mu\text{m}$, the visible, and the ultraviolet $\lambda \geq 0.08\ \mu\text{m}$ (15 eV). It is linked to the transitions discussed in the following sections.

Bound–Bound Transitions

These are the electronic, vibrational, or rotational transitions between different bound states of a molecule, an atom, or an ion, resulting from spontaneous emission, absorption (the reverse phenomenon), or stimulated emission by a photon of the same energy. These transitions can be written formally (A_i denotes the species A in state i of lower energy in the transition) by

$$\begin{array}{llll} A_j & \rightarrow & A_i + h\nu_{ij} & \text{spontaneous emission} \\ A_i + h\nu_{ij} & \rightarrow & A_j & \text{absorption} \\ A_j + h\nu_{ij} & \rightarrow & A_i + 2h\nu_{ij} & \text{induced emission} \end{array}$$

Bound–bound transitions are characterized by lines:

- Whose energy is the difference in the energy of bound levels

$$h\nu_{ij} = E_j - E_i \tag{10.1}$$

States can be slightly modified by the environment (collisions), leading to a change of high energy levels conducting to a line broadening, which thereby becomes a distribution of ν.

* National Institute on Standards and Technology, NIST Reference on Constants, Units and Uncertainty, 2006, http://physics.nist.gov/cuu/Constants/ [retrieved 24 April 2008].

- Whose width is related to microscopic phenomena: collisions (Lorentz effect) or local electric field (Stark effect). The probability distribution in frequency (Lorentzian distribution) is fully characterized by half-linewidth λ_L. This is dependent on the collisional partner; dependence can be approximated by the following empirical expression:

$$\gamma_L = \frac{p}{p_0} \sum_i \alpha_i \left(\frac{T_0}{T} \right)^{\beta_i} x_i \tag{10.2}$$

T is the heavy particle translational temperature, and T_0 and p_0 are reference standard values (see nomenclature).

We place ourselves in a fixed frame of reference, so we also will have observational effects associated with microscopic particle motion (Doppler effect), fully characterized by a half-linewidth given, assuming a Maxwellian distribution of velocities, by

$$\gamma_D = \frac{\nu_0}{c} \left(\frac{2\mathcal{R}T}{\mathcal{M}} \ln 2 \right) \tag{10.3}$$

where ν_0 is the central value of the line.

In total, the probability density describing the line is given by the convolution of the previous profiles, called the Voigt profile

$$f(\nu) = \left(\frac{\ln 2}{\pi} \right) \frac{y}{\pi \gamma_D} \int_{-\infty}^{+\infty} \frac{e^{-t^2} \mathrm{d}t}{y^2 + (x-t)^2} \tag{10.4}$$

with

$$\begin{cases} x = (\ln 2)^{\frac{1}{2}} \dfrac{\nu - \nu_0}{\gamma_D} \\ y = (\ln 2)^{\frac{1}{2}} \dfrac{\gamma_L}{\gamma_D} \end{cases} \tag{10.5}$$

It is possible to approximate this profile by an analytical form [8].

- Whose density power in frequency results from the transition probability between these two levels, characterized by the Einstein coefficients A_{ji} for the spontaneous emission, B_{ji} for the induced emission, and B_{ij} for absorption. This latter quantity is related to the strength of line $\mathcal{M}_{ij}$ (a quantity used in some databases, expressed in debye squared) issued from atomic physics and to oscillator strength f_{ij} (multiplication factor of the classical mechanical dipole oscillator) by

$$\begin{cases} B_{ji} = \dfrac{4\pi^2 q_e^2}{chm_e \epsilon_0 \nu_{ij}} \dfrac{g_j}{g_i} f_{ij} \\ f_{ij} = \dfrac{8\pi^2 m_e \nu_{ij}}{3hq_e^2} \dfrac{\mathcal{M}_{ij}}{g_j} \end{cases} \tag{10.6}$$

The Einstein coefficients are connected by the following relations:

$$\begin{cases} g_i B_{ij} & = g_j B_{ji} \\ A_{ji} & = B_{ji} \dfrac{8\pi h \nu_{ij}^3}{c^2} \end{cases} \tag{10.7}$$

Degeneracies g_i are known characteristics. These relationships, demonstrated by statistical physics for thermodynamic equilibrium only (from the microreversibility of populations) are de facto general.

The equation on n_i population levels is written by taking into account only the radiative transition $i \to j$

$$\frac{\partial n_i}{\partial t} = -\frac{\partial n_j}{\partial t} = A_{ji} n_j + \bar{I}_\nu \left(B_{ji} n_j - B_{ij} n_i \right) \tag{10.8}$$

where

$$\bar{I}_\nu = \frac{1}{4\pi} \int_{4\pi} I_\nu \mathrm{d}\Omega \tag{10.9}$$

represents the mean spectral intensity at a given point.

The calculation of the quantities n_i is discussed in Sec. 10.1.2.2.

From the foregoing, one can obtain the luminance variation for the frequencies corresponding to the line along a given direction Ω, in stationary state

$$\Omega \cdot \nabla I_\nu = \left[\frac{A_{ji}}{4\pi} n_j + B_{ji} \left(n_j - \frac{g_j}{g_i} n_i \right) \frac{I_\nu}{c} \right] h \nu_{ij} f(\nu - \nu_{ij}) \tag{10.10}$$

This equation will be written by showing macroscopic quantities

$$\Omega \cdot \nabla I_\nu = \mathcal{J}_\nu - \kappa_\nu I_\nu \tag{10.11}$$

We have introduced

1. The spontaneous emission coefficient

$$\mathcal{J}_\nu = \frac{A_{ji}}{4\pi} n_j h \nu_{ij} f(\nu - \nu_{ij}) \tag{10.12}$$

2. The absorption coefficient

$$\kappa_\nu = B_{ji} \frac{h \nu_{ij}}{c} \left(\frac{g_j}{g_i} n_i - n_j \right) f(\nu - \nu_{ij}) \tag{10.13}$$

Note that this quantity can be negative because of the induced emission; it is then spoken of as the gain of the amplifier medium. This concerns only the laser cavities.

We also define the source function

$$S_\nu = \frac{\mathcal{J}_\nu}{\kappa_\nu} = \frac{2h\nu_{ij}}{c^2}\left[\frac{g_j n_i}{g_i n_j} - 1\right]^{-1} \tag{10.14}$$

In the case of atoms, assuming steady state [$\partial./\partial t = 0$ in Eq. (10.4)] and Boltzmann distribution of energy levels at temperature T_e, then

$$\frac{g_j\, n_i}{g_i\, n_j} = \exp\left(\frac{h\nu_{ij}}{kT_e}\right) \tag{10.15}$$

In this case, $S_\nu = B_\nu$, thus defining the Planck distribution

$$B_\nu(T_e) = \frac{2h\nu^3}{c^2}\left[\exp\left(\frac{h\nu}{kT_e}\right) - 1\right]^{-1} \tag{10.16}$$

The radiative energy is then written

$$E_{R_\nu} = \frac{4\pi B_\nu}{c} \tag{10.17}$$

The number of these lines is extremely high in the media of interest (Fig. 10.1), typically several hundred thousand for an atom, and several thousand or tens of thousands for a molecule or molecular ion [6, 9]. In the latter case, the lines are grouped by vibrational bands (transitions of rotational states between two electronic/vibrational states). One can describe a band so approached by various methods: equidistant spacing of lines of given type, statistical distribution of heights of lines of random position, and so on [1]. For the figure, the typical composition for Fire II is $T_{vib_i} = T_e = 10{,}000$ K, $T = T_R = 15{,}000$ K. It is a Spartan calculation performed with the GPRD database.

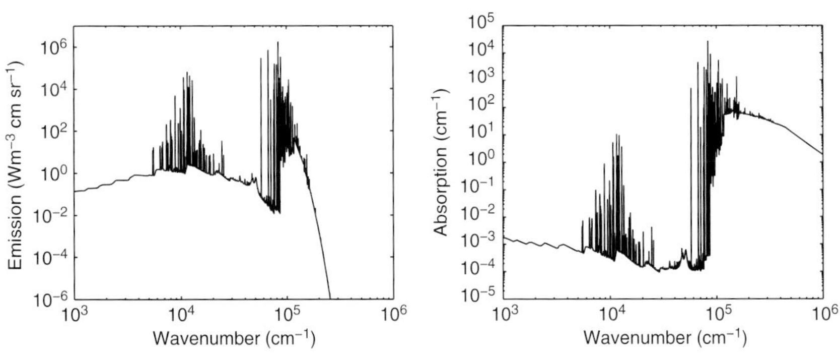

Fig. 10.1 Emission and absorption spectra of air.

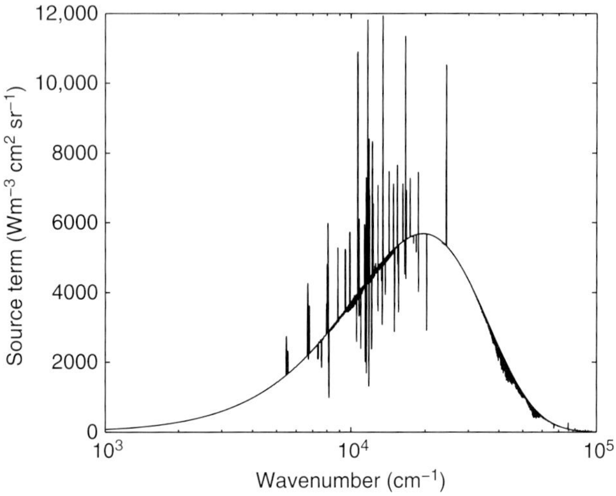

Fig. 10.2 Source function for air corresponding to Fig. 10.1.

Using the source function can be interesting even in nonequilibrium thermodynamics because it is close to a Planckian distribution (Fig. 10.2). The energy gap between the integral of the exact distribution and aT_e^4 is indeed only a few percent (3% in the case of Fig. 10.2).

The number of possible transitions of an atom is theoretically infinite (energy levels associated with n^{-2}, $n \in [1, \infty]$). It is, in fact, limited by collisions or the local electric field causing a decrease in ionization energy E_0. The precise calculation of this complex phenomenon is hardly justified by the low share in the radiation levels of higher energy. So we use a hydrogenic approximation for the potential and a simple criterion to determine the highest level $n^*_{\max}$. It will be like "obstruction" $r_{\max}$ (as given by Bohr's semi-quantum approach) of the atom is equal to a characteristic length: Debye length l_D for electrons (Sec. 2.1.2.2) or average distance between collisions for a neutral particle

$$l = \left(\frac{3}{4\pi n}\right)^{\frac{1}{3}} \tag{10.18}$$

where n is here the particle density of neutrals.

$r_{\max}$ is expressed as a function of the Bohr radius a_0 by

$$r_{\max} = a_0 n^*_{\max} = \min(l, l_D) \tag{10.19}$$

We deduce the correction to the ionization energy

$$\Delta E_0 = -\frac{(Z+1)^2}{n^*_{\max}} E_{0_H} \tag{10.20}$$

where Z is the charge number ($Z = 0$ for a neutral particle) and E_{0_H} the ionization energy of hydrogen given by

$$E_{0_H} = \frac{q_e^2}{8\pi\epsilon_0 a_0} = 13.598 \text{ eV} \tag{10.21}$$

The number of transitions is limited by the molecular dissociation energy and set a priori [10].

Bound-Free Transitions

This corresponds to transitions between bound states and electronic continuum transitions and by inverse transitions corresponding to

$A + h\nu \rightleftharpoons A^+ + e^-$ photoionization and radiative recombination
$A^- + h\nu \rightleftharpoons A + e^-$ photodetachment and photoattachment

These transitions can be written formally (the index indicating the state)

$$A_i + h\nu_{ij} \rightleftharpoons B_j + e^- \tag{10.22}$$

The energy conservation is expressed as follows:

$$h\nu_{ij} = E_0 - E_i + E_j + \frac{1}{2} m_e v_e^2 \tag{10.23}$$

The energies E_i are counted from the ground state; E_0 is the ionization or detachment energy counted from the ground state, corrected for the effect of an electric field [Eq. (10.20)]. In the problems discussed here the electrons are assumed to obey Maxwell statistics at temperature T_e. Because of the relative mass of the electron and ion or neutral particles and the conservation of momentum, we retain only the kinetic energy of the electron. From Eq. (10.23) we deduce that there can be no transition for the pair of species (A, B) in the ground state ($E_i = E_j = 0$) at an energy such that $h\nu < E_0$. The various components being in the ground state or in a state close thereof, this results in discontinuities in the properties (Fig. 10.1). We obtain the coefficients of spontaneous emission and absorption by a similar approach to that adopted for bound–bound transitions

$$\begin{cases} \mathcal{J}_\nu = \dfrac{2h\nu_{ij}^3}{c^2} n_i \mathcal{S}_{ij} \alpha \\ \kappa_\nu = n_i \mathcal{S}_{ij}(1 - \alpha) \end{cases} \tag{10.24}$$

with

$$\alpha = \frac{n_e n_j g_i}{n_i g_e g_j Q_e(T_e)} \exp\left(\frac{E_0 - E_i + E_j - h\nu}{kT_e}\right) \tag{10.25}$$

$Q_e = \left(2\pi m_e k T_e / h^2\right)^{3/2}$ is the partition function of the electron and $\mathcal{S}_{ij}(\nu)$ is the differential cross-section of absorption [see Chapter 2, Eqs. (2.5) and

(2.12)]. n_i and n_j are not known at this point. Their calculation is discussed in Sec. 10.1.2.2.

The special case of thermodynamic equilibrium is described by Saha's law

$$\frac{n_e n_j}{n_i} = \frac{g_e g_j}{g_i} Q_e \exp\left(-\frac{E_0 - E_i + E_j}{kT_e}\right) \tag{10.26}$$

In this case, $\alpha = \exp(-h\nu/kT_e)$ and we then find the Planck distribution.

Free–Free Transitions or Bremsstrahlung Radiation

These transitions correspond to the radiation emitted by the interaction of electrons with essentially ions when they are present in the medium and with neutral particles at lower temperature. Electron–electron interactions are negligible, the pair $e^- - e^-$ having a zero dipole moment. These transitions are characterized by a continuous spectrum.

This spectrum, Planckian when assuming a Maxwell–Boltzmann distribution of electrons at temperature T_e, is characterized by an emission proportional to the electron population and those of the heavy particle n_L. The law is expressed based on total cross-sections $\mathcal{S}_{\text{tot}}$

$$\begin{cases} \kappa_\nu = n_L n_e \mathcal{S}_{\text{tot}}(\nu, T_e)\left[1 - \exp\left(-\frac{h\nu}{kT_e}\right)\right] \\ \mathcal{J}_\nu = \kappa_\nu B_\nu \end{cases} \tag{10.27}$$

For interaction with the ion, the expression of the cross-section is analytic in the context of semi-quantum mechanics when we neglect the ion velocity of the electron (Kramer's law). It must be corrected for quantum effects by a term called Gaunt factor $\mathcal{G}(\nu, T_e)$ close to unity

$$\mathcal{S}_{\text{tot}} = \mathcal{G}\frac{q_e^6}{48\pi^{\frac{5}{2}}\epsilon_0^3 m_e^{\frac{3}{2}} hc\nu^3}\left(\frac{2}{3kT_e}\right)^{\frac{1}{2}} Z^2 \tag{10.28}$$

Other Processes

Photodissociation $AB + h\nu \leftrightharpoons A + B$ and the reverse reaction (chemiluminescence), the result of the radiative decay of the reaction products, negligible at high temperatures, are sometimes taken into account [6].

The relative part of each of these phenomena is shown in Fig. 10.3. Up to 6000 or 7000 K at normal pressure, the greater part of the emission is due to bound–bound transitions of molecules. Beyond this temperature, they are bound–bound transitions of atoms that dominate in the air or other gas mixtures such as CO_2–N_2 corresponding to the atmospheres of Mars or Venus [11]. In the case of bound–bound processes, emission and absorption are proportional to the pressure at a given composition. The latter changes only slowly with pressure; media properties are roughly proportional to p. The temperature variation is considerable.

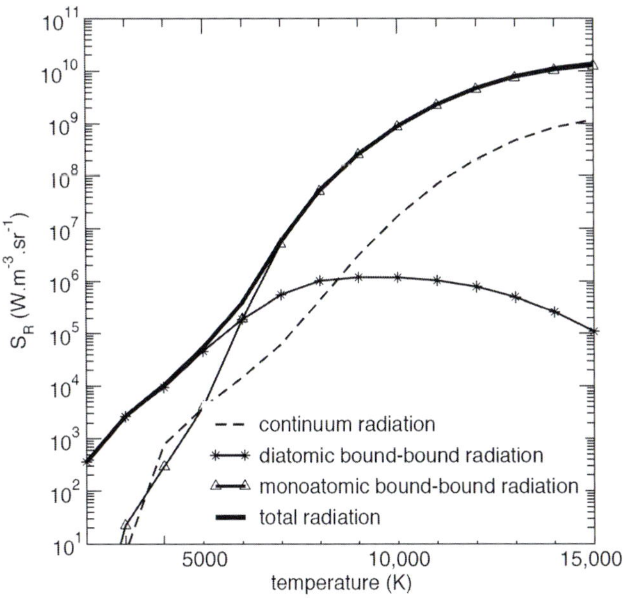

Fig. 10.3 Emission $\int_{V_1}^{V_2} \mathcal{J}_\nu \mathrm{d}\nu$, between 1000 and 50,000 cm^{-1}, in air at thermodynamic equilibrium ($p = p_0$) (6).

10.1.2.2 Calculation of State Populations

General Case

The phenomenon is completely known when we can give anywhere and at any time the populations of each energy state of each molecule, atom, or ion. The most general method is to make no hypothesis a priori on these populations. This method accounts for all possible transitions, collisional or radiative, and is called the collisional–radiative model. Given the number of states involved for various species, we are facing a formidable problem involving more than 10^5 unknown, but also the need to characterize all collisional formation rates and corresponding transport properties. It is also necessary to know the radiation intensity at any point. This problem is usually unattainable, and one is therefore forced to reduce its ambition:

- For the numerical nature of the problem we focus only on 1-D problems, for example, a stagnation line in Eulerian flow (no transport terms).
- To solve the hydrodynamic problem we assume that the macroscopic data (density, pressure, velocity) issued from the resolution of a global approximate problem is a good approximation.
- We use the general assumption that some physical degrees of freedom follow a Boltzmann distribution energy. The most detailed models assume a rotational temperature that differs from T [12]. At the other end of the approximation [13, 14], we are interested only in atoms, which we saw were

playing a leading role in the radiative transfer at high temperature, the other degrees of freedom being described by different temperatures (see Sec. 2.6.1).

- To solve the radiative problem we assume that the solution of a simplified problem provides a good approximation or assume the medium is transparent. In some studies, it is assumed the medium is homogeneous and 1-D. Radiative transfer is then very simple to solve and leads to the evaluation of terms measuring the energy leaving the system ("escape factors") [14].

Local, Stationary Approximation

An important method for the approximation assumes the quasi–steady state approximation (QSS). This hypothesis implies that we face a local problem (for the hydrodynamics and radiation) or a 1-D problem (for the hydrodynamics only) and a steady state compatible at any time of the evolution of the flow. For a local problem we write

$$\frac{\partial n_i}{\partial t} = \dot{\omega}_i = 0 \tag{10.29}$$

For a 1-D problem, we write a Lagrangian equation

$$\frac{\partial n_i}{\partial t} = \dot{\omega}_i - \nabla \cdot (n_i \mathbf{V}) = 0 \tag{10.30}$$

This is a problem where **V** is assumed to be known as computed by a method using a less detailed model. This assumption is justified by the magnitude of the characteristic time to obtain a stationary population of a level, typically 0.1 μs [15]. It should be noted that it is better to solve numerically the unsteady system rather than the stationary, nonlinear system.

For example, a complete simulation behind a shock shows that the electronic population of low levels of atoms is described by a Boltzmann distribution at temperature T_e, and the distribution of weakly bound levels obeys the Saha–Langmuir equation (Sec. 3.1.6.3), both distributions being different because of chemical nonequilibrium. For the problem of a stagnation line with slow variations of all variables, the QSS method gives good results compared to a more complete collisional–radiative mode [14].

Boltzmann Distributions of Internal Degrees of Freedom

At a higher level of modeling (and therefore a poorer description of physics), we assume a multitemperature environment, each temperature describing a Boltzmann distribution: T_i^{vib} for the vibration levels of the molecules, $T = T_{\mathrm{rot}}$ for the rotational levels of molecules or molecular ions and their translation, and T_e for electrons and electronic levels of atoms and molecules (see Sec. 2.6). This type of model can be reduced to two temperatures: $T_i^{\mathrm{vib}} = T_{\mathrm{vib}} = T_e$ and $T = T_{\mathrm{rot}}$.

Thus, in the case of atoms, the populations are given by the Boltzmann statistics (index m is relative to the species)

$$n_{im} = n_m \frac{Q_{im}(T_e)}{Q_m^{\text{tot}}(T_e)} \tag{10.31}$$

where

$$\begin{cases} n_m = \dfrac{px_m}{kT_e} \\ Q_{im} = g_{im} \exp\left(-\dfrac{E_{im}}{kT_e}\right) \\ Q_m^{\text{tot}} = \sum_i Q_{im} \end{cases} \tag{10.32}$$

In the case of diatomic molecules, we obtain a similar expression for the state (e, v, J) (here we use the conventions usually used for the vibrational quantum number and angular momentum)

$$n_{evJm} = n_m \frac{Q_{evJm}(T_e, T_m^{\text{vib}}, T_{\text{rot}})}{Q_m^{\text{tot}}(T_e, T_m^{\text{vib}}, T_{\text{rot}})} \tag{10.33}$$

where Q_{evJm} is the partition function given by

$$Q_{evJm} = g_{evJm} \exp\left(-\frac{E_{em}}{kT_e}\right) \exp\left(-\frac{E_{\text{evm}}^{\text{vib}}}{kT_m^{\text{vib}}}\right) \exp\left(-\frac{E_{evJm}^{\text{rot}}}{kT_{\text{rot}}}\right) \tag{10.34}$$

This expression is obtained by assuming that the vibration–rotation interaction is characterized by T_{rot} [16]. The total partition function is obtained by summation on the levels

$$Q_m^{\text{tot}} = \sum_e \sum_v \sum_J Q_{evJm} \tag{10.35}$$

Medium in Local Thermodynamic Equilibrium

Denser media can be described by a single temperature. Emission obeys Planck's law. Knowledge of populations is not necessary, and the absorption coefficient is a function of T, p (or any other pair of thermodynamic variables), and mass fractions. One then finds the same problems as those related to the equation of state or transport properties (Sec. 3.1.3).

10.1.2.3 Databases

The data is contained in databases such as:

- Gas and Plasma Radiation Database (GPRD).†

† Gas & Plasma Radiation Database, http://cfp.ist.utl.pt/radiation/ [retrieved 9 Oct. 2009].

- High-temperature Aerothermodynamic Radiation Algorithm (HARA) [17].
- High-resolution transmission molecular absorption (HITRAN),‡ high temperature spectroscopic absorption parameters (HITEMP), High Temperature Low Resolution (HITELOR), and Gestion et Etude des Informations Spectroscopiques Atmosphériques (GEISA)§ concern the physics of the atmosphere and combustion and provide comprehensive data in the IR domain [18, 19].
- High Temperature Gas Radiation (HTGR) [6].
- Langley Optimized Radiative Nonequilibrium (LORAN) [20].
- Monitor System for Transfer of Electromagnetic Radiation (MONSTER) [21].
- Nonequilibrium Air Reaction (NEQAIR, origin NASA) [22].
- Plasma Radiation Database (PARADE, origin ESA) [23].
- RADICAL (also known as RAD/EQUIL) [24].
- Structured package for radiation analysis (SPRADIAN) [25].
- SPECAIR (based on NEQAIR) [2].
- TIPTOPbase, containing atomic transitions used here, although the project concerns astrophysics.¶

With the exception of two of them, these databases are made for hypersonic reentry and hot gas generators, and are typically usable up to 25,000 K. Note that some important species of ablation, such C_3, C_nH, and HCN, are not always present. One can find a number of them elsewhere [26–28].

The accuracy of this data type ranges from 10% on average for molecular infrared (IR) spectra [18] to 50% in the ultraviolet (UV) [29]. It is about 30% for atomic spectra, and an imprecision of 50 to 100% must be added to the linewidth due to the Stark effect [29]. The continuous spectrum is known with a much lower accuracy [2, 23, 30]. An attempt at evaluation [29] gives 20 to 50% for photoionization cross-sections, and up to an order of magnitude for the cross-sections for electron impact on atoms and molecules.

This evaluation is insufficient in that it says nothing about the fact that deviations are more or less systematic and therefore cannot quantify the error in a spectral band. This is not anecdotal because quantification of the error on heat flux provides the following for a detailed calculation for Earth reentry [29]:

- 6% error for random errors
- 30% for systematic errors

‡ HITRAN database, www.cfa.harvard.edu/hitran/ [retrieved 5 Nov. 2009].
§ GEISA Spectroscopic Database, http:/ether.ipsl.jussieu.fr/etherTypo/index.php?id=950 [retrieved 19 July 2010].
¶ TIPTOPbase, http://www.nist.gov/pml/data/asd.cfm [retrieved 31 Jan. 2013].

Some databases like NEQAIR or PARADE are in the form of codes providing the coefficients of emission and absorption point by point, but also can solve a monodimensional radiation problem. This possibility can be used in a Monte Carlo code, for example, but is unsuited for a multidimensional numerical calculation. These codes use the QSS approximation for atoms and a hypothesis of a transparent medium. This obviously raises a problem of consistency.

These databases will be used later in this chapter; here we summarize the following forms of emission factors and absorption. The form factors f_{ijm} (m being the index corresponding to species) are retained, although they are not involved in the averages used.

1. The emission coefficient and the source function for the bound–bound atomic transitions are expressed very simply [Eqs. (10.7, 10.12, 10.13, 10.31)]

$$\begin{cases} \mathcal{J}_\nu(T_e) &= \dfrac{p}{kT_e}\sum_m x_m C_m^{(1)} \\ \kappa_\nu(T_e) &= \dfrac{p}{kT_e}\sum_m x_m C_m^{(2)} \end{cases} \tag{10.36}$$

with

$$\begin{cases} C_m^{(1)}(\nu, T_e, E_{0m}) &= \sum_{j=1}^{j_{\max}(E_{0m})} \dfrac{Q_{jm}}{Q_m^{\text{tot}}} \sum_i 2B_{jim} \dfrac{h^2\nu_{ijm}^4}{c^2} f_{ijm} \\ C_m^{(2)}(\nu, T_e, E_{0m}) &= \dfrac{1}{Q_m}\sum_{j=1}^{j_{\max}(E_{0m})} \sum_i B_{jim}\dfrac{h\nu_{ijm}}{c} f_{ijm}\left(\dfrac{g_{jm}}{g_{im}}Q_{im} - Q_{jm}\right) \end{cases} \tag{10.37}$$

We take care that the result is in any event dependent $j_{\max}$, which is related to the ionization energy E_0, so the particle density or the Debye length [Eq. (10.20)].

2. The corresponding expressions for the molecular transitions are formally identical to Eq. (10.37), but this time $j_{\max}$ is known a priori and Q_{jm} and Q_m^{tot} are functions of T_e, T_{vib}, and T_{rot} [Eq. (10.33)].
3. The coefficients of emission and absorption of bound-free transitions, given by Eq. (10.24), will be obtained by summation over all pairs i, j [the indices refer to Eq. (10.22)]

$$\begin{cases} \mathcal{J}_\nu(T_e) = \left(\dfrac{p}{kT_e}\right)^2 x_e \sum_m x_m C_m^{(3)} \\ \kappa_\nu(T_e) = \dfrac{p}{kT_e}\sum_m x_m C_m^{(4)} - \left(\dfrac{p}{kT_e}\right)^2 x_e \sum_m x_m C_m^{(5)} \end{cases} \tag{10.38}$$

with

$$\begin{cases} C_m^{(3)}(\nu, T_e) &= \sum_i \sum_j \dfrac{Q_{jm}}{Q_m} \dfrac{2h\nu_{ijm}^3}{c^2} \beta_{ijm}(\nu, T_e) \\ C_m^{(4)}(\nu, T_e) &= \sum_i \sum_j \dfrac{Q_{ijm}}{Q_m} S_{ijm}(\nu) \\ C_m^{(5)}(\nu, T_e) &= \sum_i \sum_j \dfrac{Q_{jm}}{Q_m} \beta_{ijm}(\nu, T_e) \\ \beta_{ijm}(\nu, T_e) &= \dfrac{g_{im}}{g_e g_{jm} Q_e(T_e)} S_{ijm}(\nu) \exp\left(\dfrac{E_{0m} - E_{im} + E_{jm} - h\nu}{kT_e}\right) \end{cases} \tag{10.39}$$

4. The coefficients of emission and absorption of free–free transitions, given by Eq. (10.27), are written (the summation being performed on the heavy species)

$$\begin{cases} \kappa_\nu(T_e) &= \left(\dfrac{p}{kT_e}\right)^2 x_e \sum_m x_m C_m^{(6)} \\ \mathcal{J}_\nu(T_e) &= \kappa_\nu B_\nu \end{cases} \tag{10.40}$$

with

$$C_m^{(6)}(\nu, T_e) = S_m^{\text{tot}}(\nu, T_e)\left[1 - \exp\left(-\frac{h\nu}{kT_e}\right)\right] \tag{10.41}$$

Also, in the case of local thermodynamic equilibrium, the absorption coefficient κ_ν is a function of T and p only (or any other pair of thermodynamic variables), and elemental fractions. Modeling this coefficient is similar to that of a state equation, which is not easy in an environment with injection at wall where elemental fractions vary spatially and temporally (see Sec. 3.1.3).

10.2 Radiative Transfer Equation

10.2.1 Basic Equation

The temporal aspect of radiation can be physically relevant or not, but we will discuss it due to the benefits of the numerical solution. Moreover, there is no coupling frequency: every problem for a given frequency is independent of others. The problems of reentry flows do not generally involve phenomena of angular dispersion (also called diffusion or scattering), except in highly porous materials.

If one includes I_ν in all directions, we obtain the spectral energy

$$E_{R_\nu} = \frac{1}{c}\int_{4\pi} I_\nu \mathrm{d}\boldsymbol{\Omega} \tag{10.42}$$

By integration of these quantities over the entire spectrum, we obtain the total intensity

$$I_R = \int_0^{\infty} I_\nu \mathrm{d}\nu \tag{10.43}$$

and total energy

$$E_R = \int_0^{\infty} E_{R_\nu} \mathrm{d}\nu \tag{10.44}$$

The spectral intensity obeys the following equation attributed to Schwarzschild [3, 7, 31]:

$$\frac{1}{c}\frac{\partial I_\nu}{\partial t} + \mathbf{\Omega} \cdot \nabla I_\nu = \mathcal{J}_\nu - (\kappa_\nu + \kappa_\nu^S) I_\nu + \frac{\kappa_\nu^S}{4\pi} \int_{4\pi} P_\nu(\mathbf{\Omega}' \to \mathbf{\Omega}) I_\nu \mathrm{d}\mathbf{\Omega}\, \mathrm{d}\nu \tag{10.45}$$

$\mathcal{J}_\nu$ is the emission term. The last terms of the equation describe the scattering through the coefficient absorption of κ_ν^S and the law of redistribution or angular phase function P_ν $(\mathbf{\Omega}' \to \mathbf{\Omega})$. This term is negligible in gases and will therefore be omitted in the following sections. They will be discussed in Sec. 10.4 regarding radiation in porous solids.

We can also write this equation by setting $\mathcal{J}_\nu = \kappa_\nu S_\nu$; the quantity S_ν is more regular than is $\mathcal{J}_\nu$ (Fig. 10.2). Particularly in the case of equilibrium thermodynamics, the source term S_ν is given by Planck's law

$$S_\nu = B_\nu \tag{10.46}$$

In these equations, κ_ν is the spectral coefficient of extinction (or absorption in the absence of diffusion), sometimes referred to as spectral opacity, although some prefer to reserve this term for the quantity $\kappa^* = \kappa_\nu/\rho$. κ is a property of the medium, and κ^* of constitutive species. These terms include not only the absorption, but also the induced emission.

Despite its apparent simplicity, the equation of radiation is very difficult to solve because it is defined in a space of dimension 7, in general ($\mathbf{x}$, Ω, ν, and t). Modern methods of numerical analysis allow a resolution in a reasonable amount of time when it is a unique problem, but remain very expensive when one wishes to solve this problem coupled with the simultaneous solution of the flow. This coupling itself is difficult because of the very different nature of equations (hyperbolic for the flow, kinetic for radiation).

Another difficulty is the possible coexistence in the same problem of regions with very different mean free paths due to the absorptions themselves being very different. For this reason, a nondimensional approach is often done by introducing the optical depth τ_ν, defined by $\mathrm{d}\tau_\nu = \kappa_\nu \mathrm{d}x$. This method is useful in 1-D (or multidimensional problems treated as a series of 1-D problems, Monte Carlo for example), an area in which the use of a variable step method for the numerical solution is common. Its disadvantage is it makes the spectral averaging, which actually becomes a simultaneous operation in frequency and space (band models [1]), very complex. Conventional methods of numerical solution often use this change of variable and sometimes refer to the integro-differential equation obtained by angular integration of Eq. (10.45) {S_n method, also called discrete ordinates [31]}. Modern methods of numerical solution of the problem use finite volume for the spatial discretization [32].

For all these reasons, approximate methods have been developed, which are the subject of the following sections. In particular, we note that in the problems that interest us, the details of the spectrum are unimportant, except for diagnostics, which can be treated independently from the rest of the problem.

The numerical issues usually justify the fact that we retain in Eq. (10.45) the term depending on the time, which physically can be neglected. The temporal term supply indeed to use a natural method of convergence to steady state using very large time steps (CFL numbers $> 10^{10}$) [33]. In this case, the temporal consistency of the solution is not achieved but the steady state is correctly evaluated.

10.2.2 Approximate Radiative Fluxes in the Absence of Feedback on the Fluid

Various approximate methods are used within transport (optically thin medium) or diffusion (optically thick medium) approximations. This diagnostic is made by comparing the mean free path in the medium $1/\kappa$ with its thickness δ. This test gives different results depending on the type of transition [34]:

- Molecular transitions in the visible and IR ranges are optically thin.
- Those corresponding to the UV range are optically thick.
- Atomic transitions are optically thick.
- Bound–free transitions can be either optically thin (photodetachment) or thick (photoionization).
- Free–free transitions are optically thin.

It is clear from this classification that no asymptotic method can provide good results. We discuss them, however, because they provide orders of magnitude of some macroscopic phenomena.

10.2.2.1 Transparent Medium

One commonly used hypothesis consists of neglecting the absorption in Eq. (10.45). The stationary problem is reduced to a geometric assessment of the impact of an isotropic source of volumic power density

$$J_R = \int_0^\infty \mathcal{J}_\nu \mathrm{d}\nu \tag{10.47}$$

If one takes a 1-D plane model, and moreover it is assumed that the temperature is constant near the stagnation point, a simple geometric calculation allows one to know the incident flux at the stagnation point

$$\dot{q}_R = \frac{E_R}{4\pi} \int_0^{\frac{\pi}{2}} \int_0^{2\pi} \int_0^{\frac{\delta}{\cos\beta}} \sin\beta \cos\beta \, dr d\psi d\beta = 2\pi J_R \delta \tag{10.48}$$

where δ is the thickness of the slab (the detachment distance of the shock). This value is proportional to the radius R of the stagnation point, and so $\dot{q}_R = \alpha R$. This is a consequence of the invariance by homothety of the Euler equations. The dependence is linear with radius, in contrast to the convective flow in $R^{-1/2}$ (laminar flow). This dependence implies difficulties for wind tunnel testing or free flight tunnel, which generally require models of small size. For numerical values, we replace δ by an estimate, for example, that of Serbin [35] for a gas in local thermodynamic equilibrium

$$\delta \simeq \frac{2R}{3\left(\dfrac{\rho_2}{\rho_\infty} - 1\right)} \tag{10.49}$$

where the density behind the shock ρ_2 is given by the Rankine–Hugoniot equations.

Moreover, in the case of constant extinction coefficient in frequency κ, the radiative power is given by

$$J_R = \frac{\kappa(T)\sigma T^4}{\pi} \tag{10.50}$$

The approximation of the transparent medium allows a numerical calculation with strong coupling between flow and radiative transfer, and has been used for detailed calculations of complex spectra [36]. However, it should be mentioned that this approximation induces significant errors in the order of 30% or more [36] due to the low mean free path of UV.

We can go further in the approximation of the stagnation line [7], making the following assumptions:

- The medium is a 1-D plane with constant velocity gradient $V = -\alpha y$.
- It is incompressible (density ρ_2) and at thermodynamic equilibrium.
- Its kinetic energy is negligible compared to the internal energy.
- The wall does not participate in the problem.
- There is no influence of radiation on the geometry of the flow.

The conservation of momentum can be written

$$\rho_\infty V_\infty = \rho_2 V_2 \tag{10.51}$$

Hence the expression of velocity

$$V = \frac{\rho_\infty V_\infty}{\rho_2} \frac{y}{\delta} \tag{10.52}$$

The conservation of total energy allows us to write the radiative balance fluid-altitude y

$$-\rho_2 V \frac{dh}{dy} = 4\pi J_R \tag{10.53}$$

so

$$-\rho_\infty V_\infty C_p \frac{dT}{dy} \frac{y}{\delta} = 4\kappa\sigma T^4 \tag{10.54}$$

This equation on temperature is rewritten showing the relationship between incident energy flux of a lamina of thickness δ whose characteristics are those behind the shock, given by Eq. (10.48), and incident energy flux $-\rho_\infty V_\infty h_s$ $[h_s = h_\infty + (V_\infty^2/2)]$. This ratio Γ is called the Goulard number

$$\Gamma = \frac{-4\kappa_s \sigma T_s^4 \delta}{\rho_\infty V_\infty h_s} = -\frac{\dot{q}_{R_s}}{\rho_\infty V_\infty h_s} \tag{10.55}$$

Equation (10.54) is then written

$$\frac{dy}{y} = \Gamma^{-1} \frac{\kappa_s}{\kappa} \left(\frac{T_s}{T}\right)^4 \frac{C_p dT}{h_s} \tag{10.56}$$

In the case $\Gamma \ll 1$ consistent with the hypothesis of unmodified geometry of the flow, and assuming a variation of the absorption coefficient of the form $\kappa/\kappa_s = (T/T_s)^\beta$ and the same kind of variation for the specific heat $C_p/C_{p_s} = (T/T_s)^\alpha$, this system can be integrated. It becomes [37]

$$\begin{cases} \dfrac{T}{T_s} & = 1 + \Gamma \ln\left(\dfrac{y}{\delta}\right) \\ \dfrac{\dot{q}_R}{\dot{q}_{R_s}} & = 1 - (\beta + 3)\Gamma \end{cases} \tag{10.57}$$

10.2.2.2 Absorbing Medium

In the case of a medium in local thermodynamic equilibrium, without coupling between gas and radiation with a nonparticipating wall (absorbent, without injection), Tauber and Sutton [38] constructed a correlation giving the radiative flux at stagnation point of a sphere with an error of less than 20%

$$\dot{q}_{\text{rad}} = \alpha R^a \rho_\infty^b f(V_\infty) \tag{10.58}$$

with (SI units):

- For Earth atmosphere

$$\begin{cases} \alpha = 4.736\times 10^8 \\ a = 1.072\ 10^6\ V_\infty^{-1.38}\rho^{0.325} \\ a = \begin{cases} \leq 0.6\ si\ 1 < R < 2 \\ \leq 0.5\ si\ 2 < R < 3 \end{cases} \\ b = 1.22 \\ f = -496.56 + 0.82013 V_\infty - 0.18649\times 10^{-3} V_\infty^2 \\ \quad + 0.14146\times 10^{-7} V_\infty^3 - 0.31705\times 10^{-12} V_\infty^4 \end{cases} \tag{10.59}$$

  This expression is valid for 72 km > h > 54 km, an area in which radiative fluxes are maximum for speeds of 16 $\text{km}\cdot\text{s}^{-1} > V_\infty > 10\ \text{km}\cdot\text{s}^{-1}$.
- For Mars atmosphere

$$\begin{cases} \alpha = 2.35\times 10^8 \\ a = 0.526 \\ b = 1.19 \\ f = 1850.2 - 1.0080 * V_\infty + 0.20284\times 10^{-3} V_\infty^2 \\ \quad - 0.17937\times 10^{-7} V_\infty^3 + 0.59448\times 10^{-12} V_\infty^4 \end{cases} \tag{10.60}$$

  This expression is valid for 51 km > h > 30 km, an area in which radiative fluxes are maximum for speeds of 9 $\text{km}\cdot\text{s}^{-1} > V_\infty > 6.5\ \text{km}\cdot\text{s}^{-1}$.

One can also find values in graphical form in the case of Earth or Mars reentry, but with coupling fluid-radiation [39, 40].

10.2.3 Moments Method

For the sake of clarity, we will first describe the method in the case where the extinction coefficient κ is independent of frequency. We define the moment of a scalar function or vector $g(\nu, \Omega)$ on the space of frequencies and directions by

$$\langle g\rangle = \frac{1}{c}\int_\nu\int_{4\pi} g(\nu, \mathbf{\Omega})\mathrm{d}\nu\,\mathrm{d}\mathbf{\Omega} \tag{10.61}$$

We define:

- Radiative energy density by $E_R = \langle I_\nu\rangle$
- Radiative flux vector by $\mathbf{F}_R = \langle c\mathbf{\Omega} I_\nu\rangle$
- Dimensionless flux $\mathbf{f} = \mathbf{F}_R/cE_R$ and its norm $f = ||\mathbf{f}||$ representing the anisotropy of the radiation
- Main propagation vector $\mathbf{k} = \mathbf{f}/f$
- Radiative pressure tensor $\mathbf{P}_R = \langle\mathbf{\Omega}\otimes\mathbf{\Omega} I_\nu\rangle$
- Eddington tensor $\mathbf{D}_R$ by $\mathbf{P}_R = \mathbf{D}_R E_R$

In the particular case of a Planck distribution of intensity (isotropic) we have

$$\begin{cases} E_R &= aT^4 \\ \mathbf{F}_R &= 0 \\ \mathbf{P}_R &= \dfrac{aT^4}{3}\mathbf{Id} \end{cases} \tag{10.62}$$

In the case of any radiative intensity, one must define the radiative temperature T_R from energy by $E_R = aT_R^4$. In the case of a Planck distribution, the temperature is, of course, $T_R = T$.

By premultiplying the radiative transfer in Eq. (10.45) by 1 and Ω and by taking the moments, we get

$$\begin{cases} \dfrac{\partial E_R}{\partial t} + \nabla \cdot \mathbf{F}_R &= c(\langle J_\nu \rangle - \langle \kappa_\nu I_\nu \rangle) \\ \dfrac{\partial \boldsymbol{F}_R}{\partial t} + c^2 \nabla \cdot (E_R \mathbf{D}_R) &= c(\langle \mathbf{\Omega} J_\nu \rangle - \langle \kappa_\nu \mathbf{\Omega} I_\nu \rangle) \end{cases} \tag{10.63}$$

In the case of κ constant, it follows

$$\begin{cases} \langle J_\nu \rangle &= \kappa \langle B_\nu \rangle &= \kappa a T^4 \\ \langle \kappa_\nu I_\nu \rangle &= \kappa \langle I_\nu \rangle &= \kappa E_R \\ \langle \kappa_\nu \mathbf{\Omega} I_\nu \rangle &= \kappa \langle \mathbf{\Omega} I_\nu \rangle &= \dfrac{\kappa}{c}\mathbf{F}_R \\ \langle \mathbf{\Omega} J_\nu \rangle &= \kappa \langle \mathbf{\Omega} B_\nu \rangle &= 0 \end{cases} \tag{10.64}$$

This system is obviously not closed: as in any system at moments, each of the equations on a moment includes a term of higher rank. The second equation of the flux makes the radiative pressure tensor appear. It will therefore be necessary to add a hypothesis for the resolution. This problem being solved, we can see the considerable interest of the method: besides the fact that the unknown quantities include only a small number of variables (x and t), the hyperbolic structure allows the use of numerical solvers identical to those used for fluid mechanics and, consequently, to couple strongly the resolution of various problems. Typically, the computational cost of such a model is about one or two orders of magnitude lower than the direct solution of the equation of radiative transfer. The error induced depends, of course, on the closure of the system chosen and the configuration studied.

10.2.3.1 Eddington Approximation

For the record, and as an introduction to the method of moments, we now mention the Eddington approximation (or model P_1) of developing I_R as follows:

$$I_R = \frac{1}{4\pi}(I_{R_0} + 3\mathbf{\Omega} \cdot \boldsymbol{I}_{R_1}) \tag{10.65}$$

By multiplying this equation by 1 and $\mathbf{\Omega}$ angularly and integrating, we identify I_{R_0} and I_{R_1}

$$\begin{cases} I_{R_0} = cE_R \\ \mathbf{I}_{R_1} = \mathbf{F}_R \end{cases} \tag{10.66}$$

This development, carried in the radiative transfer of Eq. (10.45), allows us to find the system of moments in Eq. (10.63), taking $\mathbf{D}_R = (1/3)\mathbf{Id}$. Due to the type of development used to determine I_R, the Eddington approximation properly represents only minor deviations from an isotropic radiation. Generalization {spherical harmonic expansion method or P_n [7]} slowly improves the result with the increasing order of development, which limits the usefulness of the approach.

10.2.3.2 Entropic Closure

Entropy closure, introduced by Dubroca and Feugeas [41], is based not on an assumption about the form of the solution, but on a property that must verify the system: its solution must be the maximum of the total entropy on matter plus radiation. The latter is defined by [42]

$$\mathcal{S}_R(I_\nu) = -\left\langle [n \log\, n - (n+1)\log(n+1)]\frac{2k\nu^2}{c^3} \right\rangle \tag{10.67}$$

where the occupation number n is

$$n = \frac{c^2}{2h\nu^3} I_\nu \tag{10.68}$$

In particular, a medium in local thermodynamic equilibrium satisfies the Bose–Einstein statistics

$$n = \left[\exp\left(\frac{h\nu}{kT}\right) - 1\right]^{-1} \tag{10.69}$$

The model M_1, obtained by the method of Lagrange multipliers (with entropy maximization constraint achieving the first two moments E_R and $\mathbf{F}_R$) is analytic. The intensity is

$$I_\nu(\Omega, E_R, \mathbf{F}_R) = \frac{2h\nu^3}{c^2}\left\{\exp\left[\frac{h\nu}{kT_R}\alpha(1-\beta\, \boldsymbol{k}\cdot\mathbf{\Omega})\right] - 1\right\}^{-1} \tag{10.70}$$

with

$$\begin{cases} \beta = \dfrac{2-\gamma}{f} \\ \alpha = \dfrac{f}{2(\gamma-1)^{\frac{1}{4}}(f^2-2+\gamma)^{\frac{1}{2}}} \\ \gamma = (4-3f^2)^{\frac{1}{2}} \end{cases} \tag{10.71}$$

The corresponding Eddington tensor is

$$\begin{cases} \mathbf{D}_R = \dfrac{1-\chi}{2}\mathbf{Id} + \dfrac{3\chi - 1}{2}\mathrm{k} \otimes \mathrm{k} \\ \chi = \dfrac{3+4f^2}{5+2(4-3f^2)^{\frac{1}{2}}} \end{cases} \tag{10.72}$$

χ is called the Eddington factor.

This solution allows both the isotropy ($f = 0$) and the extreme anisotropy ($f = 1$) that represent a Dirac of I_ν (a ray). Between these two extremes that are transport (transparent medium) and diffusion (very opaque medium), it is a reasonable approximation when the angular distribution is regular (Fig. 10.4).

This method can be generalized by splitting the space into quadrants [43]. This allows us, for example, to treat more precisely the boundary conditions in problems of interest.

Note also that one can hybridize this method with the kinetic equation for simultaneous processing of spectral broadband and spectral thin areas containing a lot of energy (e.g., laser beam) [44].

As an example we will use a gas layer of assumed constant opacity to test the method error. (A realistic case is discussed in Section 10.2.4.3.) The error is 3% on the radiative flux, in the case corresponding to the previous

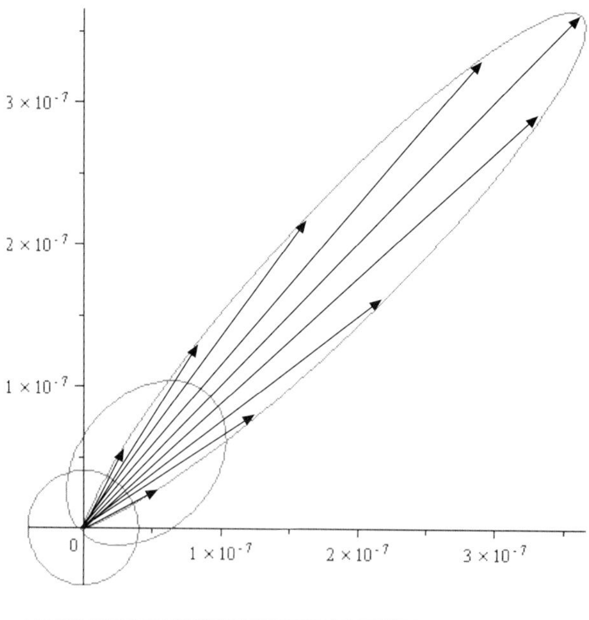

Fig. 10.4 M_1 angular distribution for $f = 0$, 0.7, and 0.99, respectively; propagation vector k points to $\pi/4$.

equations. Note that the case mentioned, simple in absolute terms, is an interesting test because the exact angular distribution has a singularity in the plane parallel to the slab [45]. Note that this error is less than the one induced by the precision of data (Sec. 10.1.2.3).

10.2.3.3 Boundary Conditions

The boundary conditions for the moments equations are also derived from those of the radiation transport equation at any point of the boundary

$$I_\nu(\nu,\ \Omega,\ t) = \Gamma(\nu,\ \mathbf{\Omega}), \quad \mathbf{\Omega}\cdot\mathbf{n} < 0 \tag{10.73}$$

Γ is any distribution defined on the boundary, and $\mathbf{n}$ is the outward normal. However, insofar as the underlying distribution model for the moments model was chosen to approach the solution of transport equation, it can satisfy the boundary condition of Eq. (10.73) only in an integral sense.

Thus, multiplying Eq. (10.73) by $\mathbf{\Omega}\cdot\mathbf{n}$ and integrating over all incoming directions, one obtains

$$\int_{\Omega\cdot n<0} \mathbf{\Omega}\cdot\mathbf{n}\, I_\nu(\nu,\ \Omega)\mathrm{d}\Omega = \int_{\Omega\cdot n<0} \mathbf{\Omega}\cdot\mathbf{n}\,\Gamma(\nu,\ \mathbf{\Omega})\mathrm{d}\mathbf{\Omega} \tag{10.74}$$

For the moments solution, this boundary condition generalizes the Marshak type condition at boundary [7], associated with the P_1 model.

10.2.3.4 General Case

Except for the important special case of the Rayleigh diffusion $\kappa\nu = \alpha\nu$ [46] encountered in combustion problems (soot), there is no analytical solution to the problem when the absorption coefficient depends on the frequency. It is then necessary to divide the spectrum into groups, each centered on a frequency ν_i (multigroup approximation). For every group, the transfer equation will be written similarly to Eq. (10.63)

$$\begin{cases} \dfrac{\partial E_{R_i}}{\partial t} + \nabla\cdot\mathbf{F}_{R_i} = c(\langle J_\nu\rangle_i - \langle\kappa_\nu I_\nu\rangle_i) \\ \dfrac{\partial \mathbf{F}_{R_i}}{\partial t} + c^2\nabla\cdot(\mathbf{D}_{R_i}E_{R_i}) = c(\langle\mathbf{\Omega} J_\nu\rangle_i - \langle\kappa_\nu\mathbf{\Omega} I_\nu\rangle_i) \end{cases} \tag{10.75}$$

We can rewrite this system in another form (J_ν is isotropic)

$$\begin{cases} \dfrac{\partial E_{R_i}}{\partial t} + \nabla\cdot\mathbf{F}_{R_i} = c(\kappa_{P_i} a T_{R_i}^4 - \kappa_{E_i}E_{R_i}) \\ \dfrac{\partial \mathbf{F}_{R_i}}{\partial t} + c^2\nabla\cdot(\mathbf{D}_{R_i}E_{R_i}) = -c\kappa_{F_i}F_{R_i} \end{cases} \tag{10.76}$$

in which were introduced the following various coefficients:

$$\begin{cases} \kappa_{P_i} = \dfrac{\langle J_\nu \rangle_i}{\langle S_\nu \rangle_i} \\ \kappa_{E_i} = \dfrac{\langle \kappa_\nu I_\nu \rangle_i}{\langle I_\nu \rangle_i} \\ \kappa_{F_{ix}} = \dfrac{\langle \kappa_\nu \Omega_x I_\nu \rangle_i}{\langle \Omega_x I_\nu \rangle_i} \end{cases} \tag{10.77}$$

Moments previously defined by Eq. (10.61) were redefined as

$$\langle g \rangle_i = \frac{1}{c} \int_{\Delta\nu_i} \int_{4\pi} g(\nu, \mathbf{\Omega}) \mathrm{d}\nu \, \mathrm{d}\mathbf{\Omega} \tag{10.78}$$

T_{R_i} is a radiative temperature defined by

$$\langle S_\nu \rangle_i = a T_{R_i}^4 \tag{10.79}$$

This system has a solution close to that of a constant absorption coefficient system [45]. In particular, the Eddington factor $\chi_i = g(f_i, E_{R_i})$ is numerically close to that defined by Eq. (10.72), except for groups containing significant energy or for high anisotropy. In any case, this factor is calculable a priori.

Of all opacities, only the Planck one, κ_{P_i}, is calculable a priori because it does not depend on the solution I_ν. This is not true for opacity energy κ_{E_i} and flux opacity (diagonal tensor) κ_{F_i}.

This observation leads us to address more generally the problem of spectral averages before returning to the case-particular model M_1, for which we know the form of the solution.

10.2.4 Spectral Averages for the Kinetic Equation

10.2.4.1 Multigroup Frequency Method

A general method is to cut the frequency space into selected intervals chosen a priori and take averages. We leave aside “narrowband” methods related to an integral formulation of the transfer equation [1].

The general idea is to note that the exact position of a line or any other part of the spectrum does not play an important role, and therefore that the frequencies are close and thus S_ν varies slowly in the interval considered. This is immediately understandable in the case of thermodynamic equilibrium with the slow variation of B_ν with frequency.

We will find the average intensity $\bar{I}_i$ that satisfies in an approximate way the transfer equation for this set of frequencies of thickness $\Delta\nu_i$

$$\frac{1}{c}\frac{\partial \bar{I}_i}{\partial t} + \mathbf{\Omega} \cdot \nabla \bar{I}_i = \frac{1}{\Delta\nu_i} \int_{[\nu_i]} (\mathcal{J}_\nu - \kappa_\nu I_\nu) \mathrm{d}\nu \tag{10.80}$$

Define the average emission

$$\bar{\mathcal{J}}_i = \frac{1}{\Delta\nu_i}\int_{[\nu_i]} \mathcal{J}_\nu \, \mathrm{d}\nu \tag{10.81}$$

Define similarly the average opacity

$$\bar{\kappa}_i = \frac{1}{\Delta\nu_i}\int_{[\nu_i]} \kappa_\nu \, \mathrm{d}\nu \tag{10.82}$$

We will close the system by assuming that one can write

$$\int_{[\nu_i]} \kappa_\nu I_\nu \mathrm{d}\nu \simeq \bar{\kappa}_i \bar{I}_i \tag{10.83}$$

Equation (10.80) is then written in a standard form

$$\frac{1}{c}\frac{\partial \bar{I}_i}{\partial t} + \mathbf{\Omega} \cdot \nabla \bar{I}_i = \bar{\mathcal{J}}_i - \bar{\kappa}_i \bar{I}_i \tag{10.84}$$

The simple method consists of performing a simple average in the interval

$$[\nu_{j-\frac{1}{2}},\ \nu_{j+\frac{1}{2}}]$$

is very inefficient. There must be an order of almost 10^4 bands for a precision of the order of a few percent on the intensity in a typical problem [9]. However, we find in the literature applications with two bands only [47], a method poorly justified by the shape of the spectrum of air at high pressure and high temperature, as we encounter in plasma jets (100 bars, 20,000 K).

For the remainder of the approximation, we simply cite the opacity sampling method, which is very difficult to work with because it assumes a very good knowledge of the spectrum. Moreover, they are not particularly advantageous because it typically takes a thousand points to get a reasonable accuracy (of the order of 1%) on the macroscopic quantities [48]. In Sec. 10.2.4.2 we will further discuss a method to work on the spectrum considered a random distribution of absorption values.

10.2.4.2 Method of Opacities Multigroup

Let $[\nu_i]$ be the set of frequencies corresponding to the absorption coefficient

$$\kappa_i \in [\kappa_{i-\frac{1}{2}},\ \kappa_{i+\frac{1}{2}}]$$

called group (or bin, or picket). This produces an approximation of the probability distribution of opacity throughout the spectrum (Fig. 10.5; the curve corresponds to the data from Fig. 10.1). For this reason, the method is called the opacity distribution function (ODF) method or the cumulative K-distribution (CK) method.

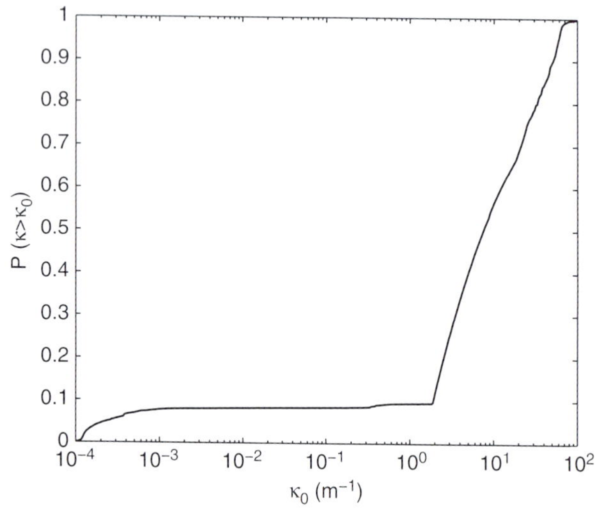

Fig. 10.5 Probability law for the absorption coefficient.

The equations are identical to those of the previous section, replacing mean values over a spectral band with those on a group of lines whose heights are included between two given values.

This method provides reasonable accuracy (a few percent on the intensity) with a dozen groups only [9]. Increasing this number does not significantly improve the result, Eq. (10.84) not being accurate.

For a mixture, the summation of opacities results in the calculation of joint probabilities [49].

Of course, one can combine this approach with the frequency multigroup approach.

10.2.4.3 Approximation for M_1 Method

The case of the M_1 method is different in that we know the form of the solution I_ν given by Eq. (10.70). In the interval

$$[\nu_{i-\frac{1}{2}}, \nu_{i+\frac{1}{2}}]$$

of width $\Delta\nu_i$, integrals defined by Eq. (10.78) are easily computed

$$\begin{cases} \langle \mathcal{J}_\nu \rangle_i &= \dfrac{4\pi}{c\Delta\nu_i} \displaystyle\int_{\Delta\nu_i} \mathcal{J}_\nu \, \mathrm{d}\nu \\ \langle \kappa_\nu I_\nu \rangle_i &= \dfrac{1}{c\Delta\nu_i} \displaystyle\int_{\Delta\nu_i} \kappa_\nu K_\nu^{(0)} \mathrm{d}\nu \\ \langle \mathbf{\Omega} I_\nu \rangle_i &= 0 \\ \langle \mathbf{\Omega} \kappa_\nu I_\nu \rangle_i &= \dfrac{1}{c\Delta\nu_i} \displaystyle\int_{\Delta\nu_i} \kappa_\nu K_\nu^{(1)} \mathrm{d}\nu \end{cases} \tag{10.85}$$

$K_\nu^{(0)}$ and $K_\nu^{(1)}$ are the angular momentum of I_ν defined by

$$\begin{cases} K_\nu^{(0)} = \int_{4\pi} I_\nu(\Omega)\mathrm{d}\Omega \\ K_\nu^{(1)} = \int_{4\pi} I_\nu(\Omega)\Omega\mathrm{d}\Omega \end{cases} \tag{10.86}$$

These quantities are easily calculable and can be approximated numerically as a function of two parameters only.

The precision of this method was measured on the case of a slab. This example is close to the problems we are interested in (Sec. 10.2.2). The error is around 5% for a decomposition into 15 groups in frequency [45], half about the spectral decomposition, the other half for the error due to the method M_1 itself.

10.3 Effects of Coupling Between Flow and Radiation

10.3.1 Modification of Shock Layer

The radiation of the flow behind the shock, between it and the wall, reduces its enthalpy, which decreases from the shock to the wall because the fluid energy is lost (Fig. 10.6; this calculation assumes a perfectly absorbing and nonemitting wall in $y = 0$). In contrast, the enthalpy is essentially constant

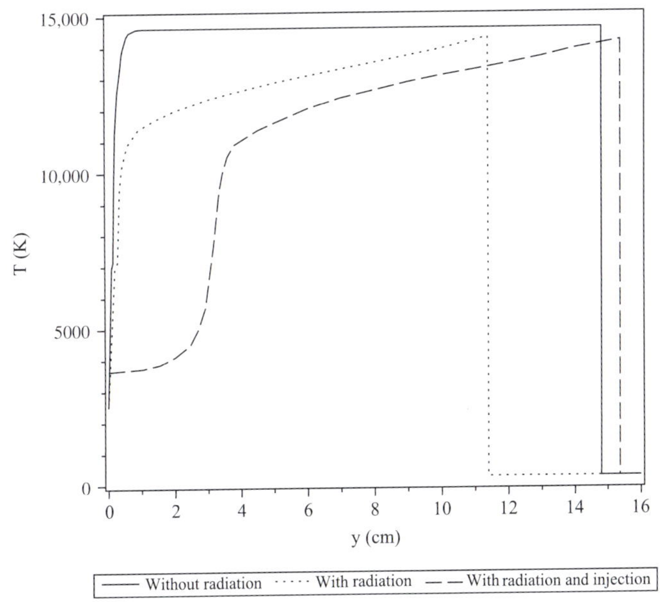

Fig. 10.6 Fluid-radiation coupling with and without injection; Earth atmosphere (54).

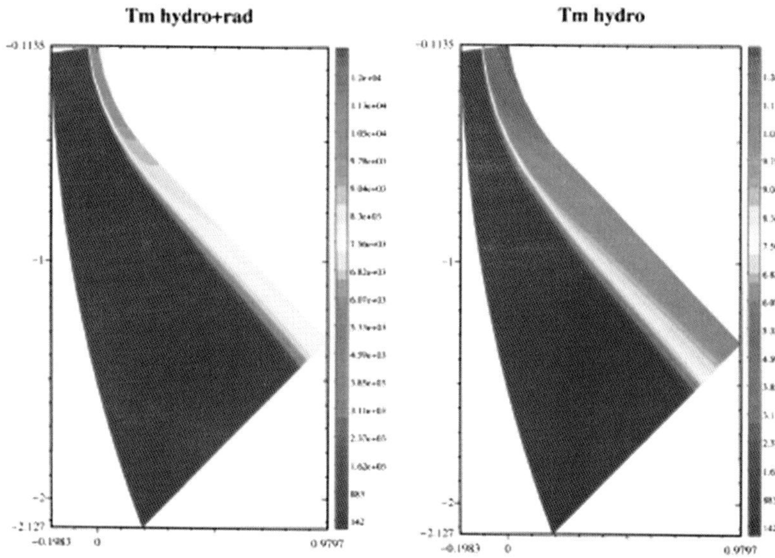

Fig. 10.7 Temperatures with and without radiative coupling; atmosphere of Venus at 80 km altitude, reentry velocity is 11 km · s^{-1}, method M_1, 12 spectral bands (45).

in the nonemissive layer of shock, the kinetic energy being small compared to the internal energy. The pressure is little affected [50]. The density is increased and the detachment distance reduced accordingly due to the conservation of mass flux (Fig. 10.7) [51]. A few examples are given in [52, 53].

This has the effect of substantially reducing the radiative and convective fluxes, compared to a calculation without coupling, due to the decrease of the enthalpy near the wall. The decrease of both can reach a few tens of percent [50, 52, 55]. For the radiation, we can estimate the influence of coupling at the stagnation point from a quantity similar to the Goulard number, defined by

$$\Gamma = \frac{2\dot{q}_{\text{rad}}}{\frac{1}{2}\rho_\infty V_\infty^3} \tag{10.87}$$

$\dot{q}_{\text{rad}}$ is the heat flux in the absence of coupling. The quantity in the denominator is approximately the convective flux [Eq. (1.1)]. The approximation of Tauber and Wakefield [56] permits us to estimate the coupled radiative flux $\dot{q}'_{\text{rad}}$

$$\frac{\dot{q}'_{\text{rad}}}{\dot{q}_{\text{rad}}} = \frac{1}{1 + \alpha\,\Gamma^{0.7}} \tag{10.88}$$

where α is a constant dependent on the atmosphere ($\alpha = 3.45$ for air, 3 for the atmosphere of Jupiter, 2 for that of Titan).

10.3.2 Effect of Wall Injection

The injection of gas through the wall influences the convective flow (see Sec. 4.4.3) and the radiative flux. For the latter, the effect is complex. First, the temperature in the boundary layer decreases due to injection (Fig. 10.3). [Figure 10.3 shows two air layers ($p = 5 \times 10^4$ Pa): a shock layer ($T =$ 10,000 K, $e = 15$ cm) and a boundary layer ($e = 1$ cm, temperature T_2.)] This is superimposed on the potential effects described previously, where they exist. The effect will tend to increase or decrease the flux depending on the radiative mean free path; that is to say, the issuance from hot regions will be more or less absorbed by the cold regions. For air and ablation product species issued from phenolic resin, it is the large energies that absorb radiation and tend to decrease it in the region of 7–10 eV [55] via the species C, C_3, CO, H_2, C_2H, and so on (Sec. 8.2.1.4). In contrast, the flux in IR regions is enhanced. In total, the effects are relatively moderate in a representative case of a lunar return to Earth (Fig. 10.8). Various authors have reported very small changes in the case of Earth reentries [57, 58]; others noted decreases ranging up to 20% [54]. In the extreme case of Galileo reentry on Jupiter, calculations (uncoupled fluid-radiation) show a decrease by a factor of 2 [59].

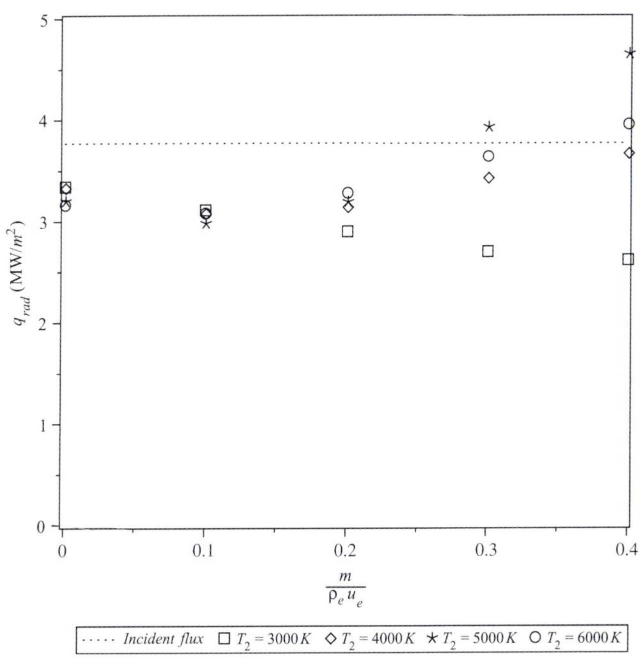

Fig. 10.8 Influence of injection on radiative flux (54).

10.3.3 Turbulence–Radiation Coupling

Section 7.3 examined the effects of coupling between turbulence and chemical reactions. One can expect similar effects for radiation. A simple evaluation can assess this effect. Let us start from Eq. (10.45) in the case of a medium at thermal equilibrium, without scattering. We take a time average

$$\frac{1}{c}\frac{\partial \bar{I}_\nu}{\partial t} = \overline{\sigma_\nu B_\nu} - \overline{\sigma_\nu I_\nu} \tag{10.89}$$

By applying the operator defined by Eq. (10.61), we find an equation of radiant energy similar to Eq. (10.63)

$$\frac{\partial \bar{E}}{\partial t} + \nabla \cdot \bar{\mathbf{F}} = c(\langle \overline{\sigma_\nu \mathrm{B}_\nu} \rangle - \langle \overline{\sigma_\nu I_\nu} \rangle) \tag{10.90}$$

The source term can be calculated in the case of a Gaussian temperature distribution $g(T)$ (see Sec. 7.3) and an absorption coefficient $\sigma_\nu = \alpha\nu$. This is typical of the Rayleigh diffusion. Although belonging to the domain of combustion (soot), this example will be used because it allows an analytical calculation [46]

$$\langle \overline{\sigma_\nu B \nu} \rangle = \frac{192\pi\zeta_5 k^5 A}{h^4 c^3} \int_0^\infty g(T) T^5 \mathrm{d}T' \tag{10.91}$$

where ζ is a Riemann zeta function $\zeta_5 = \zeta(5) \simeq 1.03692$.

The integral of Eq. (10.91) is analytic. The following is a development of the solution:

$$\int_0^\infty g(T) T^5 \mathrm{d}T' = \tilde{T}^5 \left[1 + \frac{1}{2\pi}\frac{\sigma_T}{\tilde{T}} + 10\left(\frac{\sigma_T}{\tilde{T}}\right)^2 + \frac{9}{2\pi}\left(\frac{\sigma_T}{\tilde{T}}\right)^3 + \cdots \right] \tag{10.92}$$

In the case of interest, $\sigma_T/\tilde{T}$ is the order of a few percent. The correction term is close to unity: the contribution of turbulence is very low. This is verified by DNS calculations with radiative emission (transparent medium) [60, 61], which demonstrate moreover that the emission term produces a turbulence damping by energy loss of the hottest regions.

10.4 Radiation in Porous Media

10.4.1 Equivalent Conductivity and Emissivity

Porous media are involved in several types of applications:

- As a highly porous foam-like material for thermal insulation at high temperatures [62], for example, used in the heat shield of Genesis (C–C dual layer).

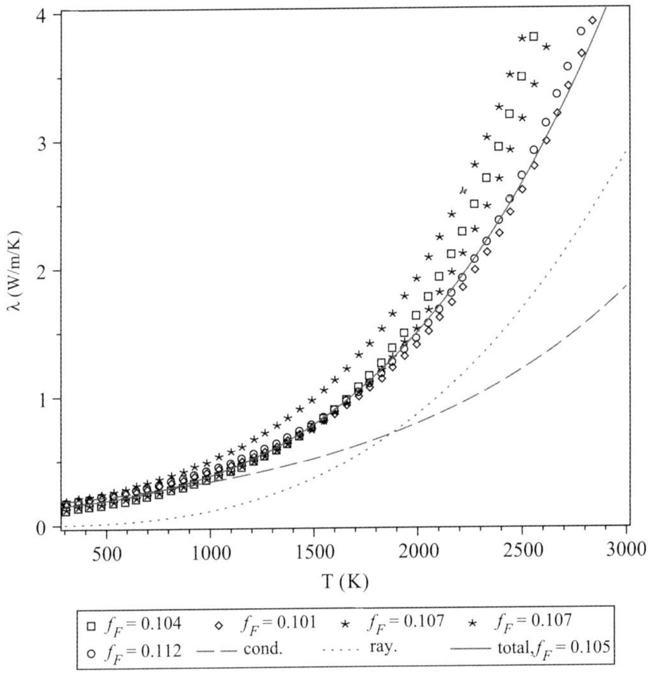

Fig. 10.9 Apparent conductivity of PICA preform; f_F is the fiber volume fraction.

- As a material in which pyrolyzable thermal decomposition induces an important porosity, typically 40% for a TWCP, 90% for a PICA-like material (Chapter 8). In this type of material, radiation plays an important part in energy transfer at high temperature. For example, in the PICA material radiation becomes predominant at 2000 K in the preform or the pyrolyzed material (Fig. 10.9).

It is common to describe the radiative transfer in a porous material with an equivalent conductivity. We can justify this assumption with two hypotheses:

- The equation of radiative transfer is valid, the mean free path being large compared with the wavelength and diffraction effects limited to a very small region of space and therefore representable by an absorption and a phase function.
- The mean free path is small compared to any characteristic dimension of the problem at macroscopic level. This concerns in particular the thermal gradient characterized by the length

$$l_T = \frac{T}{||\nabla T||} \tag{10.93}$$

If, moreover, we neglect the scattering, we can write the transfer Eq. (10.45) in diffusive form [3, 7], here stationary, 1-D, and integrated in frequency

$$\frac{\mu}{\kappa + \kappa^R} \frac{dI_R}{dy} = \frac{\epsilon \sigma T^4}{\pi} - I_R \tag{10.94}$$

y is the direction normal to the surface of the material and $\mu = \cos\theta$, where θ represents the angle to the surface.

It follows, considering than in a first approximation we can replace the derivative of I with that of its approximation of order 0 (i.e., $\epsilon\sigma T^4/\pi$)

$$I_R = \frac{\epsilon \sigma T^4}{\pi} - \frac{4\mu\epsilon\sigma T^3}{\pi \kappa + \kappa^R} \frac{dT}{dy} \tag{10.95}$$

Integrating in 4π gives heat flux

$$\dot{\mathbf{q}}_s = \epsilon \sigma T_s^4 - \frac{8\epsilon\sigma T_s^3}{3\kappa + \kappa^R} \frac{dT}{dy}\bigg|_s \tag{10.96}$$

We can then identify the equivalent conductivity

$$\lambda_{eq} = \frac{16\epsilon\sigma T^3}{3\kappa + \kappa^R} \tag{10.97}$$

This expression, classical in opaque gaseous medium, is sometimes called the Rosseland approximation. It is indeed generalized to any medium by replacing $\kappa + \kappa^R$ by the average Rosseland mean [3, 63].

This result can be obtained more rigorously by asymptotic analysis of a periodic medium [64].

This expression gives pretty good results (Fig. 10.9). (The model in Fig. 13.9 is constituted by a random anisotropic assembly of fibers and neglects scattering [67, 68].) This poses a problem for the conductivity measurement; however, it must be noted that this approximation is at fault in the vicinity of the surface to a depth of a few mean free paths. In particular, the radiation emitted to the outside comes in part from inside the material, which is at a lower temperature than the surface. The radiative flux to the outside of the material is then reduced. We can consider this phenomenon by a method similar to that used to express the equivalent conductivity, noticing the outgoing rays are not affected by the presence of the surface. We can therefore use the expression of I given previously and integrate in a half-space

$$\dot{\mathbf{q}}_s = \epsilon \sigma T^4 - \frac{8\epsilon\sigma T^3}{3\kappa + \kappa^R} \frac{dT}{dy} \tag{10.98}$$

Then we have the equivalent emissivity

$$\epsilon_{\text{eq}} = \epsilon\left(1 - \frac{8}{3}\frac{l_R}{l_T}\right) \tag{10.99}$$

where $l_R = (\kappa + \kappa^R)^{-1}$ is the mean free path. The characteristic dimension corresponding to the temperature gradient l_T [Eq. (10.93)] can reach a value of the order of millimeters, compared to the hundreds of micrometers for the mean free path $\{l_R \simeq 160$ μm in the case of Fig. 13.9, much as in the case of foams [65]$\}$. This implies a significant effect on ϵ_{eq}. This has already been met in the case of materials with a semitransparent liquid layer on the surface (Appendix G). It should be noted that the use of this correction requires a good evaluation of the temperature gradient, a condition not really verified using equivalent conductivity. (The model in Fig. 13.9 is constituted by a random anisotropic assembly of fibers and neglects scattering [66, 67].)

Note that the gas plays no role on the radiation. On the contrary, its conduction is not completely negligible and is difficult to calculate for rarefied media (mean free path of the same order of magnitude as the pore size) [64, 65].

10.4.2 Real Radiative Problem

The best solution is to solve the transfer equation, especially because the problem is mostly independent of the frequency for the carbon (fiber, pyrolysis products, foams).

One approach is to consider the material as a periodic or dispersed medium, partially transparent and diffusive. We can then have access to properties in one of two ways:

- By direct measurements [68]
- By calculation from a geometrical model derived from the observation, in particular the microtomography. The calculation is then very simple (analytic in some cases), or much more complicated (classic optics [64, 65] or electromagnetism, assuming independent scattering of several parts of the geometry [69]).

The phase function describing this type of environment is simple and can be described by a model with a single parameter, such as the Henyey–Greenstein law obtained by analogy with the theory of Mie [1]

$$P_\nu(\mathbf{\Omega}' \to \mathbf{\Omega}) = \sum_{i=1}^{\infty} (2i+1)\alpha^i P_i(\mathbf{\Omega}' \cdot \mathbf{\Omega}) = \frac{1-\alpha^2}{(1+\alpha^2 - 2\alpha\mathbf{\Omega}' \cdot \mathbf{\Omega})^{\frac{3}{2}}} \tag{10.100}$$

where $P_i(\bullet)$ is the Legendre polynomial of order i. P_i is regular (Fig. 10.10) and can represent a diffusion both predominantly forward ($\alpha > 0$) or backward ($\alpha < 0$). The foam in Fig. 10.10 is reconstituted from measurement

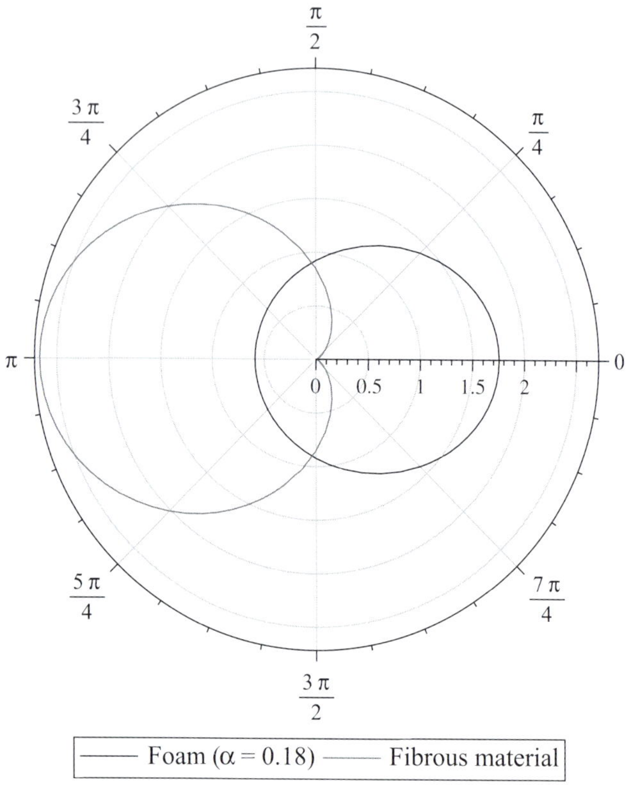

Fig. 10.10 Phase function of a metallic foam (Al, IR domain, $\alpha = 0.18$).

[69] and of a fibrous material with diffuse reflection obtained by calculation (optical model). The radiation is coming from the left.

10.4.3 M1 Model with Scattering

We suppose that

$$p_\nu(\Omega' \to \Omega) = P_\nu(\Omega' \cdot \Omega) \tag{10.101}$$

This relationship expresses the cylindrical symmetry of the interaction of the ray with the solid diffuser. An example of this type of phase function is given in the preceding section.

The model is particularly effective in this framework. The moments system M_1 with scattering is obtained as before. The calculation involves two integrals related to the phase function. The first is

$$\frac{1}{4\pi c}\int_0^\infty \int_{4\pi} I_\nu(\Omega') \int_{4\pi} P_\nu(\Omega' \to \Omega)\, \mathrm{d}\Omega' \mathrm{d}\Omega\, \mathrm{d}\nu \tag{10.102}$$

Taking into account the relationship $\int_{4\pi} P_\nu\ (\Omega' \rightarrow \Omega)\mathrm{d}\Omega' = 4\pi$, the quantity in Eq. (10.102) is none other than the radiative energy E_R.

The second integral involves the angular moment of order 1

$$g_\nu = \int_{4\pi} P_\nu(\Omega' \rightarrow \Omega)\Omega \mathrm{d}\Omega' \tag{10.103}$$

The system with scattering is written

$$\begin{cases} \dfrac{\partial E_{R_\nu}}{\partial t} + \nabla \cdot \mathbf{F}_{R_\nu} = c(\kappa_{P_\nu} a T^4_{R_\nu} - \kappa_{E_\nu} E_{R_\nu}) \\ \dfrac{\partial \mathbf{F}_{R_\nu}}{\partial t} + c^2 \nabla \cdot (\boldsymbol{D}_{R_\nu} E_{R_\nu}) = -c[\kappa_{F\nu} + \kappa^s_{F\nu}(1 - g_\nu)]\mathbf{F}_{R_\nu} \end{cases} \tag{10.104}$$

10.4.3.1 Example of Severe Radiative Flux on PICA-Like Material

An example in given in Fig. 10.11 using a radiative incident heat flux equivalent to convective conditions defined in Sec. 8.6 applied to a pyrolyzed PICA-like material. The boundary condition is an isotropic radiative flux. The anisotropy is important on a great part of the material because:

- The rise under the surface (typically 2 mean free paths) due to penetration depth depending on incidence
- The effect of the important temperature gradient (see the explanation scheme in Fig. 10.11 showing irradiation at a given point)

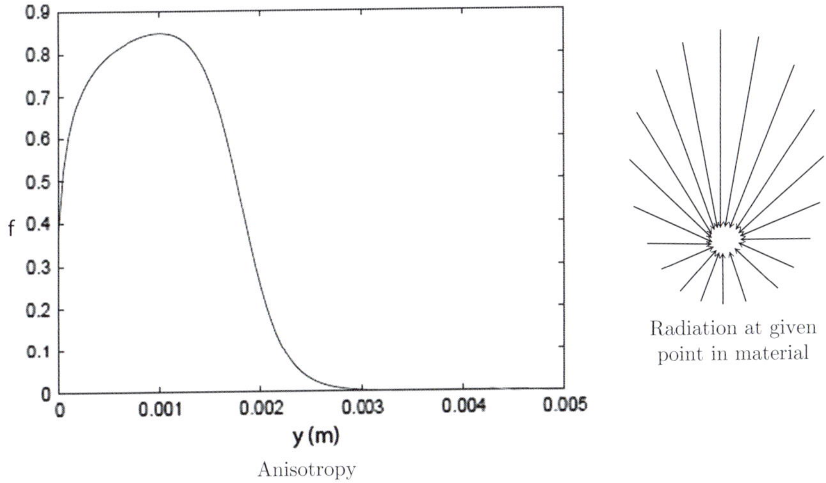

Fig. 10.11 Radiation anisotropy in a fibrous material in a severe case.

In this type of situation, the diffusion approximation is poor:

- Directly because of anisotropy
- Indirectly because the temperature gradient near the surface is poorly evaluated, giving an important error on equivalent emissivity.

An important error regarding temperatures results, in particular in a low-temperature region. In contrast, the error made by the M_1 model is less than 1%.

Note that, for the same reason, the spectrum is not perfectly described by a Planck distribution. This fact seems of secondary importance.

References

[1] Goody, R. M., and Yung, Y. L. *Atmospheric Radiation. Theoretical Basis*, 2nd ed., Oxford University Press, Oxford, UK, 1989.

[2] Laux, C., "Physico-Chemical Models for High Enthalpy and Plasma Flows," *Von Karman Institute for Fluid Dynamics Lecture Series 2002–07*, edited by Fletcher, D., Charbonnier, J.-M., Sarma, G. S. R., and Magin, T., VKI, Rhode Saint Genese, Belgium, June 2002.

[3] Mihalas, B., and Weibel-Mihalas, B., *Foundations of Radiation Hydrodynamics*, Oxford University Press, New York, 1964.

[4] Park, C., *Nonequilibrium Hypersonic Aerothermodynamics*, John Wiley & Sons, New York, 1990.

[5] Penner, S. S., *Quantitative Molecular Spectroscopy and Gas Emissivities*, Addison-Wesley, Boston, MA, 1959.

[6] Perrin, M.-Y., Rivière, P., and Soufiani, A., "Radiation Database for Earth and Mars Entry," NATO Research and Technology Organization Report RTO-EN-AVT-162, Nov. 2008.

[7] Pomraning, G. C., *The Equations of Radiation Hydrodynamics*, International Series of Monographs in Natural Philosophy, Vol. 54, Pergamon Press, Oxford, UK, 1973.

[8] Whiting, E. E., "An Empirical Approximation to the Voigt Profile," *Journal of Quantitative Spectroscopy and Radiative Transfer*, Vol. 8, No. 6, 1968, pp. 1379–1384.

[9] Wray, A. A., Ripoll, J.-F., and Prabhu, D., "Computation of Radiation in the Apollo AS-501 Reentry Using Opacity Distribution Functions," *AIAA Journal*, Vol. 45, No. 9, Sept. 2007, pp. 2359–2363.

[10] Babou, Y. Rivière, P., Perrin, M.-Y., and Soufiani, A., "High-Temperature and Nonequilibrium Partition Function and Thermodynamic Data of Diatomic Molecules," *International Journal of Thermophysics*, Vol. 30, 2009, pp. 416–438.

[11] Babou, Y. "Transferts radiatifs dans les plasmas de mélange CO_2-N_2: base de données spectroscopiques, étude expérimentale et application aux rentrées atmophériques," Ph.D. thesis, l'Université Orsay, Paris XI, No. 8704, July 2007.

[12] Gökçen, T., "Computation of Nonequilibrium Radiating Shock Layers," *Journal of Thermophysics and Heat Transfer*, Vol. 9, No. 1, 1995, pp. 34–40.

[13] Johnston, C. O., "Nonequilibrium Shock-Layer Radiative Heating for Earth and Titan Entry," Ph.D. thesis, Virginia Polytechnic Institute and State University, Blacksburg, Virginia, Nov. 2006.

[14] Panesi, N., Magin, T., Bourdon, A., Bultel, A., Chazot, O., and Babou, Y., "Collisional-Radiative Modeling in Flow Simulations," NATO Research and Technology Organization Report RTO-EN-AVT-162, Sept. 2009.

[15] Lino da Silva, M., Guerra, V., Loureiro, J., and Dudeck, M., "Simulation of Plasma Radiation in Earth and Mars Atmospheric Entries," *Fourth International*

*Symposium on Atmospheric Reentry Vehicles and Systems*Arcachon, France, March 2005.

[16] Jaffe, R. L., "The Calculation of High-Temperature Equilibrium and Nonequilibrium Specific Heat Data for N_2, O_2 and NO," AIAA Paper 1987–1633, *22nd Thermophysics Conference*Honolulu, Hawaii, June 1987.

[17] Johnston, C. O., Hollis, B. R., and Sutton, K., "Spectrum Modeling for Air Shock-Layer Radiation at Lunar-Return Conditions," *Journal of Spacecraft and Rockets*, Vol. 45, No. 5, 2008, pp. 865–878.

[18] Rothmans, L. S., and Schroeder, J. "Millenium HITRAN Compilation," *12th Atmospheric Radiation Measurement Science Team Meeting Proceedings*, St Petersburg, Florida, 2002.

[19] Scutaru, D., Rosenmann, L., and Taine, J., "Approximate Intensities of CO_2 Hot Bands at 2.7, 4.3 and 12 μm for High Temperature and Medium Resolution Applications," *Journal of Quantitative Spectroscopy and Radiative Transfer*, Vol. 2, No. 6, 1994, pp. 765–781.

[20] Chambers, L. H., "Predicting Radiative Heat Transfer in Thermochemical Nonequilibrium Flow Fields. Theory and User's Manual for LORAN Code," NASA Technical Memorandum 4564, Sept. 1994.

[21] Surzhikov, S. T., "Radiation Modeling and Spectral Data," *Von Karman Institute for Fluid Dynamics Lecture Series 2002–07*, edited by Fletcher, D., Charbonnier, J.-M., Sarma, G. S. R., and Magin, T., VKI, Rhode Saint Genese, Belgium, June 2002.

[22] Whiting, E. E., Park, C., Liu, Y., Arnold, J. O., and Paterson, J. A., "NEQAIR96, Nonequilibrium and Equilibrium Radiative Transport and Spectra Program: User's Manual," NASA Reference Publication 1389, Dec. 1996.

[23] Smith, A. J., Gogel, T., and Vandevelde, P., "Plasma Radiation Database PARADE, Final Report," Fluid Gravity Engineering Ltd Report Number TR28/96, 1996.

[24] Nicolet, W. E., "User's Manual for the Generalized Radiation Transfer Code RAD/EQUIL," NASA Report CR-116353, Oct. 1969.

[25] Fujita, K., and Abe, T., "SPRADIAN, Structured Package for Radiation Analysis: Theory and Application," Institute of Space and Astronautical Science Report No. 669, Sept. 1997.

[26] Jones, J. J., "The Optical Absorption of Triatomic Carbon C_3 for the Wavelength Range 260 to 560 nm," NASA Technical Paper 1141, March 1978.

[27] Prakash, S. G., and Park, C., "Shock Tube Spectroscopy of C3 + C2H Mixture in the 140 to 700 nm Range for Jovian Entry Probe Ablation Layer Simulation," AIAA Paper 1979–0094, *17th Aerospace Sciences Meeting*, New Orleans, Louisiana, Jan. 1979.

[28] Shinn, J. L., "Optical Absorption of Carbon and Hydrocarbon Species from Shock Heated Acetylene and Methane in the 135–220 nm Wavelength Range," AIAA Paper 81–1189, *16th Thermophysics Conference*Palo Alto, California, June 1981.

[29] Kleb, B., and Johnston, C. O., "Uncertainty Analysis of Air Radiation for Lunar Return Shock Layers," AIAA Paper 2008–6388, *AIAA Atmospheric Flight Dynamics Conference and Exhibit*, Honolulu, Hawaii, Aug. 2008.

[30] Bose, D., MacCorkle, E., Bogdanoff, D., and Allen, G. A., Jr., "Comparisons of Air Radiation Model with Shock Tube Measurements," AIAA Paper 2009–1030, *47th AIAA Aerospace Sciences Meeting Including the New Horizons Forum and Aerospace Exposition*Orlando, Florida, Jan. 2009.

[31] Modest, M. F., *Radiative Heat Transfer*, 2nd ed., Academic Press, San Diego, CA, 2003.

[32] Chai, J. C., Lee, H. S., and Patankar, S. V., "Finite-Volume Method for Radiation Heat Transfer," *Journal of Thermophysics and Heat Transfer*, Vol. 8, No. 3, pp. 419–425.

[33] Turpault, R., "Modélisation, approximation numérique et applications du transfert radiatif en déséquilibre spectral couple avec l'hydrodynamique," Ph.D. thesis, Université Bordeaux 1, Bordeaux, France, Sept. 2003.

[34] Lamet, J.-M., "Transferts radiatifs dans les écoulements hypersoniques de rentrée atmo- sphérique terrestre," Ph.D. thesis, Ecole Centrale de Paris, France, Sept. 2009.
[35] Serbin, H., "Supersonic Flow Around Blunt Bodies," *Journal of Aeronautical Sciences*, Readers' Forum, Vol. 25, No. 58, 1958.
[36] Wright, M. J., Bose, D., and Olejnicsak, J., "The Impact of Flowfield-Radiation Coupling on Aeroheating for Titan Aerocapture," AIAA Paper 2004–0484, *42nd AIAA Aerospace Sciences Meeting and Exhibit*Reno, Nevada, Jan. 2004.
[37] Goulard, R., "The Coupling of Radiation and Convection in Detached Shock Layers," *Journal of Quantitative Spectroscopy and Radiative Transfer*, Vol. 1, 1961, pp. 249–257.
[38] Tauber, M. E., and Sutton, K., "Stagnation Point Radiative Heating Relations for Earth and Mars," *Journal of Spacecraft and Rockets*, Vol. 28, No. 1, 1991, pp. 40–42.
[39] Sutton, K., and Hartung, L. C., "Equilibrium Radiative Heating Tables for Earth Entry," NASA Report TM-102652, May 1990.
[40] Hartung, L. C., Sutton, K., and Brauns, F., "Equilibrium Radiative Heating Tables for Aero-bracking in the Martian Atmosphere," NASA Technical Report TM-102659, May 1990.
[41] Dubroca, B., and Feugeas, J.-L., "Entropic Moment Closure Hierarchy for the Radiative Transfer Equation," *Comptes-rendus de l'Académie des Sciences, Paris*, Vol. I, No. 329, 1999, pp. 915–920.
[42] Oxenius, J., "Kinetic Theory of Particles and Photons," *Springer Series in Electrophysics*, Vol. 20, Springer-Verlag, Berlin, 1986.
[43] Dubroca, B., and Klar, A., "Half Moment Closure for Radiative Transfer Equations," *Journal of Computational Physics*, Vol. 180, No. 2, 2002, pp. 584–596.
[44] Turpault, T., "Properties and Frequential Hybridisation of the Multigroup Ml Model for Radiative Transfer," *Nonlinear Analysis: Real World Applications*, Vol. 11, No. 4, 2010, pp. 2514–2528.
[45] Turpault, T., "A Consistent Multigroup Model for Radiative Transfer and Its Underlying Mean Opacities," *Journal of Quantitative Spectroscopy and Radiative Transfer*, Vol. 94, 2005, pp. 357–371.
[46] Ripoll, J.-F., Dubroca, B., and Duffa, G., "Modelling Radiative Mean Absorption Coefficients," *Combustion Theory and Modelling*, Vol. 5, 2001, pp. 261–274.
[47] Nicolet, W. E., Shepard, C. E., Clark, K. J., Balakrishnan, A., Kesselring, J. P., Suchsland, K. E., and Reese, J. J., Jr., "Analytical and Design Study for a High-Pressure, High-Enthalpy Constricted Arc Heater," AEDC TR-75-47, July 1975.
[48] Helling, C., "Application of Opacity Sampling (OS) and Opacity Distribution Function (ODF) to Model Calculation of Cool Giants," Diploma thesis, Institute for Astronomy and Astrophysics, Technical University of Berlin, Germany, July 1996.
[49] Saxner, M., and Gustafsson, B., "A Method for Adding Opacity Distribution Functions," *Astronomy and Astrophysics*, Vol. 140, 1984, pp. 334–340.
[50] Chisnell, R. F., Hoshizaki, H., and Lasher, L. E., "Stagnation Point Flow with Radiation," NASA Report CR-70171, Oct. 1965.
[51] Hayes, W. D., and Probstein, R. F., *Hypersonic Flow Theory*, Vol. 1, Academic Press, New York, 1959.
[52] Anderson, J. D., *Hypersonic and High Temperature Gas Dynamics*, McGraw-Hill, New York, 1989.
[53] Gollan, R. J., Jacobs, P. A., Karl, S., and Smith, S. C., "Numerical Modelling of Radiating Superorbital Flows," *ANZIAM Journal*, Vol. 45(E), 2004, pp. C248–C268.
[54] Johnston, C. O., Gnoffo, P. A., and Sutton, K., "The Influence of Ablation on Radiative Heating for Earth Entry," *Journal of Spacecraft and Rockets*, Vol. 46, No. 3, 2009, pp. 481–191.
[55] Mazoué, F., and Marraffa, L., "Flowfield/Radiation Coupling Analysis for Huygens Probe Entry into Titan Atmosphere," AIAA Paper 2005-5392, *38th AIAA Thermophysics Conference*, Toronto, Ontario, June 2005.
[56] Raupach, M. R., Antonia, R. A., and Rajagopalan, S., "Rough-Wall Turbulent Boundary Layers," *Applied Mechanics Review*, Vol. 44, No. 1, pp. 1–25.

[57] Gupta, R. N., "Aerothermodynamics Analysis of Stardust Sample Return Capsule with Coupled Radiation and Ablation," AIAA Paper 99-0227, *37th Aerospace Science Meeting*, Reno, Nevada, Jan. 1999.

[58] Hoshizaki, H., and Lasher, L. E., "Convective and Radiative Heat Transfer to an Ablating Body," *AIAA Journal*, Vol. 6, No. 8, 1968, pp. 1441–1449.

[59] Kumar, A., and Tiwari, S. N., "Laminar and Turbulent Flow Solutions with Radiation and Ablation Injection for Jovian Entry," NASA Report CR-163-215, May 1980.

[60] Duan, L., Sohn, I., Grube, N., Martin, M. P., Levin, D. A., and Modest, M. F., "Study of Emission Turbulence-Radiation Interaction in Hypersonic Boundary Layers," AIAA Paper 2010-354, *48th AIAA Aerospace Sciences Meeting*, Orlando, Florida, Jan. 2010.

[61] Feldick, A. M., Duan, L., Modest, M. F., Martin, M. P., and Levin, D. A., "Influence of Interactions Between Turbulence and Radiation Transmissivities in Hypersonic Turbulent Boundary Layers," AIAA Paper 2010-1185, *48th AIAA Aerospace Sciences Meeting*, Orlando, Florida, Jan. 2010.

[62] Fort, C., Goyhénèche, J.-M., and Duffa, G., "Transferts thermiques dans un milieu poreux," *CHOCS*, Vol. 19, 1998, pp. 25–34.

[63] Pope, S. B., *Turbulent Flows*, Cambridge University Press, Cambridge, UK, 2000.

[64] Charrier, P., and Dubroca, B., "Asymptotic Transport Model for Heat and Mass Transfer in Reactive Porous Media," *SIAM Journal on Multiscale Modeling and Simulation*, Vol. 2, No. 1, 2003, pp. 124–157.

[65] Bourret, F., Fort, C., and Duffa, G., "Conductivity thermique de mousses cellulaires de carbone," *Revue Générale de Thermique*, Vol. 36, No. 7, 1997, pp. 510–519.

[66] Pradere, C., Batsale, J.-C., Goyhénèche, J.-M., Pailler, R., and Dilhaire, S., "Thermal Properties of Carbon Fibers at Very High Temperature," *Carbon*, Vol. 47, No. 3, 2009, pp. 737–743.

[67] Tran, H. K., Johnson, C. E., Rasky, D. J., Hui, F. C. L., Hsu, M.-T., Chen, T., Chen, Y. K., Paragas, D., and Koyabachi, L., "Phenolic Impregnated Carbon Ablators (PICA) as Thermal Protection Systems for Discovery Missions," NASA Technical Memorandum 110440, April 1997.

[68] Zhao, C. Y., Lun, T. J., and Hodson, H. P., "Thermal Radiation in Ultralight Metal Foams with Open Cells," *International Journal of Heat and Mass Transfer*, Vol. 47, No. 2, 2004, pp. 2927–2939.

[69] Doermann, D., and Sacadura, J.-F., "Heat Transfer in Open Cell Foam Insulation," *Journal of Heat Transfer*, Vol. 118, No. 1, 1996, pp. 88–93.

Chapter 11 Erosion by Particle Impact

- Describe the atmosphere on Mars
- Describe the atmosphere on Earth
- Determine particle-flow interactions
- Characterize the hypervelocity impact

11.1 Introduction: Phenomenology

Some atmospheres can be loaded with particles. This is the case for Earth's atmosphere, which has particles of ice and water in the form of clouds and precipitation. In the case of Mars's atmosphere, clouds of dust can be raised to high altitude by winds. A description of these media is given in Sec. 11.2.

These particles reach the detached shock when the vehicle still has a very high speed, typically several kilometers per second. The trajectories and fate of these particles in the shock layer (fractionation, heating, vaporization) are discussed in Sec. 11.3.

The flow is profoundly affected by the passage of the particle, and not only locally: the shock layer can be considerably distorted. Instantaneous or mean wall heat fluxes are significantly increased, up to several hundred percent. This effect results from particle–flow interaction, which is particularly complex. The study of this phenomenon is provided in Sec. 11.4.

Finally, the impact of the particle that occurs when it is not too small (and therefore destroyed or deflected) induces the formation of a crater, with a departure of material corresponding to several tens of times the mass of the incident particle. Part is vaporized, and another part removed and injected into the flow as ejecta of small dimensions, with relatively high speeds, of the order of several hundreds of meters per second. These ejecta contribute to modifying the flow. The crater also can play a role in the flow, in particular the laminar–turbulent transition. All these phenomena are discussed in Sec. 11.5.

11.2 Atmospheres

11.2.1 Mars Atmosphere

11.2.1.1 Nature and Dimensions of Particles

The spectroscopic observations by orbiting instruments or sensors on the ground (Viking, Mariner 9, Phobos, Pathfinder, Mars Global Surveyor, Mars Reconnaissance Orbiter) show the presence of not very dense clouds of solid particles, ice water, and carbon dioxide. The size of these particles is about 1 micron. They are generally confined to high latitudes. In addition, periodically, the atmosphere becomes charged with particles raised by winds to an altitude over 60 km. Such events cover large geographic areas and, even as often as a few times per decade, arrive to cover the entire planet (Fig. 11.1). The size of these particles is several tens of micrometers, so it is therefore these on which we focus.

Knowledge of the particles is mainly based on spectroscopic measurements and measures in situ [1]. They consist of oblate spheroids made of palagonite

$$NaAl_2(Si_3Al)O_{10}(OH)_2$$

of density 2780 kg · m^{-3}, or montmorillonite

$$(Na, Ca)_{0.33}(Al, Mg)_2(Si_4O_{10})(OH)_2, nH_2O$$

of density between 2000 and 2700 kg · m^{-3}, species whose optical spectrum is the best suited for measurements in the visible or infrared (IR).

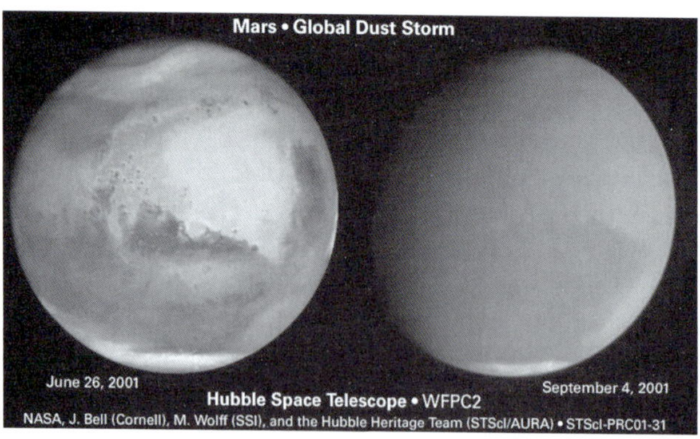

Fig. 11.1 Global dust storm on Mars.*

* NASA, "The Perfect Dust Storm Strikes Mars," http://science.nasa.gov/science-news/science-at-nasa/2001/ast11oct_2/ [retrieved 30 May 2008].

Given the type of measurement, the particle content is often expressed through the optical depth at zenith

$$\tau_\nu = \int_{h=0}^{\infty} \sigma_\nu \mathrm{d}h' \tag{11.1}$$

where σ_ν is the extinction coefficient at frequency ν (see Sec. 10.2.1). This quantity, calculated in the visible spectrum, ranges from 0.1 for an atmosphere not much charged in particles, to 4 for a storm, and up to 17 for the most important events, such as that of Fig. 11.1.

The reverse problem of restitution of particle profiles from measurements is ill conditioned and must be done using the regularization assumptions [2]. The size of these particles is deduced from these optical measurements, assuming:

- Spatial and time invariance for each series of measurements, hence the independence to line of sight.
- A mass fraction of particles independent of the altitude considered; this assumption has been verified by numerical simulations of cloud development.
- A distribution of a priori spherical particles. The choice of a given distribution influences only slightly results in this type of process of creation and coalescence [3]. A log-normal distribution is commonly used [4]

$$f(r) = \frac{1}{(2\pi)^{\frac{1}{2}} r\sigma} \exp\left[-\frac{(\ell n\, r - \mu)^2}{2\sigma^2}\right] \tag{11.2}$$

μ and σ are the mean and standard deviation of $\ell n\, r$, respectively.

These assumptions are probably wrong at high altitude because of the poor efficiency of convection as compared to the problems of gravity. However, this concerns only a small fraction of the total content of the atmosphere particles and is therefore not detectable by optical measurements including the entire line of sight. This obviously does not induce a big effect of total erosion.

Sizes thus determined are typically less than 50 µm, with a median value of some micrometers [4] (see Fig. 11.2, data Mariner 9, 1991). The particle density is approximately 3.5×10^6 m^{-3} in this severe case, so a volume fraction is 4×10^{-10}. The medium is particularly diluted.

11.2.1.2 Altitude Distribution

The values reconstructed with measurements and physical models give an approximate expression for the partial density vs altitude [5]

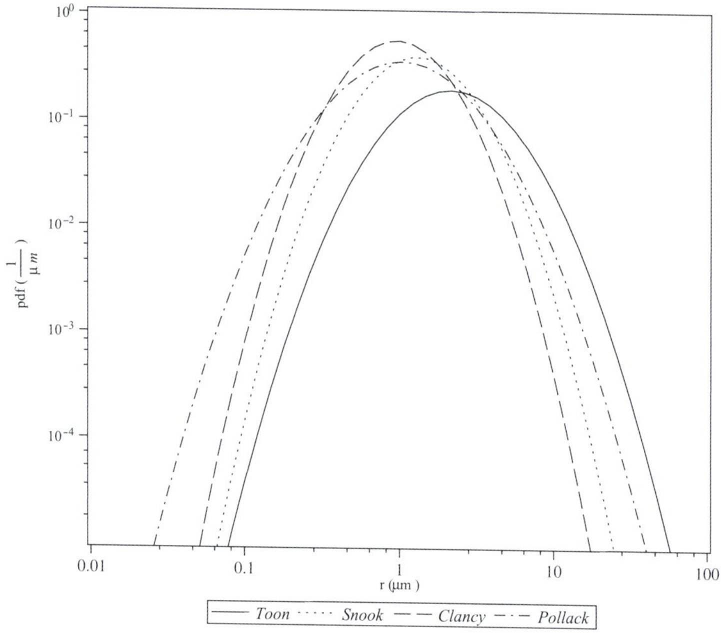

Fig. 11.2 Clay particle size distributions for a heavily charged atmosphere; median radius is 1.4–3.7 µm (4).

$$\begin{cases} \dfrac{\rho}{\rho_0} = \exp\left\{0.007\left(1 - \left(\dfrac{p}{p_0}\right)^{\frac{h_0}{h_{\max}}}\right)\right\} & \text{if } p < p_0 \\ \rho = \rho_0 & \text{if } p > p_0 \end{cases} \tag{11.3}$$

where $p_0 = 700$ Pa is the reference pressure at altitude 0, ρ_0 is the density at reference altitude 0 and a parameter of the problem, $h_0 = 70{,}000$ m, and $h_{\max}$ is the maximum altitude given by the following expression (L is the longitude, $L_0 = 160$, and l is the latitude; values are given in degrees and meters):

$$h_{\max} = 60{,}000 + 18 \sin(L - L_0) - 22 \sin^2 l \tag{11.4}$$

11.2.2 Earth Atmosphere

In Earth's atmosphere, the hydrometeors are present at low altitude, up to about 18 km at low latitudes, 12 kilometers in temperate latitudes. The highest altitude clouds, stratospheric (polar) and mesospheric (noctilucent) clouds, have no effects in our case.

Historically, measurements were first made in situ using airplanes, balloons, or sounding rockets. Measurements ex situ such as weather radar or lidar (ground-based or in orbit) are more recent. This induces a difference in the terms used to measure phenomena. Here we speak of ice water content (IWC, in kilograms per cubic meter) or liquid water content (LWC), which represent the partial densities of ice and liquid water, respectively. Studies of weather statistics can give average or extreme profiles for any given cloud type or for all cloud types combined [6, 7].

The ice particles in high-altitude cirrus (240 K $< T <$ 293 K) are sized between 25 μm and some centimeters, with median values of about 200 μm to 1 mm. The diameter distribution is usually represented by a gamma function

$$f(\mathrm{d}) = \alpha\, \mathrm{d}^{\mu}\, e^{-\lambda \mathrm{d}} \tag{11.5}$$

The shape of the crystals is highly variable (hexagonal flat, needle-shaped, columnar, dendritic), and their density ranges from 300 to 900 $\mathrm{kg \cdot m^{-3}}$ [8]. Larger objects are phenomena of low spatial extent (storm hail), and can be omitted hereafter. In contrast, we cannot omit aggregates formed (the snow). The ICW is between 5×10^{-7} and 1×10^{-3} $\mathrm{kg \cdot m^{-3}}$ [9]. If, as an order of magnitude, we take the example of Fig. 11.3 approached by a population of spherical particles of size $\mathrm{d}_m = 217$ μm and density $\rho_s = 800\ \mathrm{kg \cdot m^{-3}}$, with an ICW of $\rho_{\mathrm{ICW}} = 10^{-4}\ \mathrm{kg \cdot m^{-3}}$ (typical value), we find an average

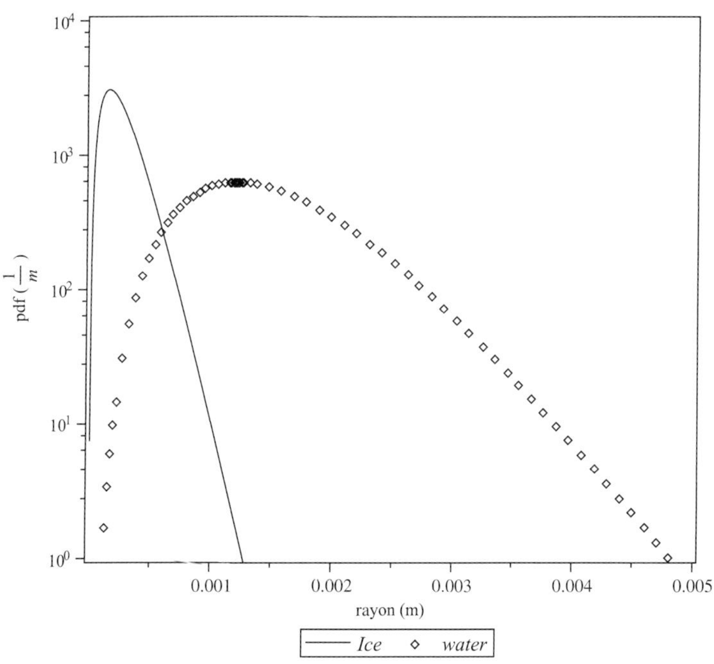

Fig. 11.3 Size distribution of ice and water particles for a cirrus-type altitude cloud (8, 9).

distance between particles $L = [\pi\rho_s(\mathrm{d}_m/2)^3/\rho_{ICW}] \simeq 3$ cm. It is therefore an extremely dilute environment. We can also calculate this value from d_m from the quantity $(L/\mathrm{d}_m) \simeq 150$, which leads to the same conclusion. The volume fraction in the example of Fig. 11.3 is 2.2×10^{-7}. The median value is 217 μm for ice and 1.4 mm for water.

Water drops that form the low clouds and precipitation have a spherical shape, the surface tension dominant efforts being of aerodynamic origin for radii below the millimeter. For larger particles, a slight deformation is present, but it does not exceed a value of thickness-to-diameter ratio of 0.7 for drops of 5 mm diameter [8], which are among the largest encountered (see Fig. 11.3).

Water or ice, the speed of these particles in the air is obviously negligible compared to that of the reentering object.

11.3 Effect of Flow on the Particles

11.3.1 Density of Particles in the Flow

Besides the clay particles, ice, or water described in Sec. 11.2, the flow may also contain ejecta from the impact (Sec. 11.5), whose size is between a few micrometers and some dozens of micrometers for carbon. Only the largest particles coming from upstream, typically 100 μm or more, reach the wall in the case of Earth reentry, virtually without decelerating (Sec. 11.3.4). There is no significant increase in the density of particles in the shock layer due to slowdown. We saw that the medium consisting of these particles was extremely diluted; this is also true in the flow. On a sphere, shock detachment distance is approximately $R/5$ for high Mach numbers. Let $R = 200$ mm, and the distance from detachment will be 40 mm. In the cylinder flow around the stagnation point of radius equal to the diameter of the sphere, on average there will be less than a dozen particles present in the shock layer. This observation allows us to deduce two important characteristics of this phenomenon:

- Intermittence: Unlike ejecta, one cannot describe the incident solid or liquid phase by a continuous model. A Lagrangian treatment is therefore required.
- There is an absence of interaction between the incident particles in the shock layer, given the low Reynolds number (see the next section).

11.3.2 Flow Regimes

In the vicinity of the stagnation point, the gas is at an elevated temperature in excess of 5000 K. This leads to a noteworthy mean free path despite the high pressure. The Knudsen number formed with the diameter of the particle shows that the flow around it covers continuous and transitional regimes.

Furthermore, although the relative velocity of incident particles is high, the Mach number in the shock layer is small (Mach 3 to 4) because of very high sound speed due to temperature. The Reynolds number (formed with local conditions and the diameter of the particle) is also low, less than 1000. This viscous flow will result in limiting the development of the wake of the particle.

Finally, the characteristic length of flow, measured, for example, by $l_\rho = \rho/\|\nabla\rho\|$ is few hundred micrometers. It is not large compared with the size of larger particles, particularly the drops. This poses a problem for calculating the drag of these particles.

11.3.3 Modification of Particles

The particle will be changed by its transit in the shock layer:

- It can be split during the passage of the shock wave by internal reflection of the latter and Richtmyer–Meshkov instability, particularly for a drop of water (Fig. 11.4) [10].
- It can also be vaporized during its trajectory; however, this phenomenon only affects smaller particles. As we will see in Sec. 11.3.4, they are sufficiently deflected to not impact the material.

The mechanism of deformation and fragmentation that relates mainly to water drops involves the antagonistic effects of surface tension forces and aerodynamic forces. This is measured through a Weber number

$$We = \frac{\rho_l V_l^2 \mathrm{d}_l}{\sigma} \qquad (11.6)$$

where σ is the surface tension and d_l the diameter.

The drop is, according to the value of We (Fig. 11.4) [12]:

- Distorted by Richtmyer–Meshkov instability from $We \simeq 30$
- Fragmented from $We \simeq 1000–2000$

In our case, the Weber number is about 10^5 to 10^6, which corresponds to a very rapid process. However, the distance characteristic of the phenomenon is $l_\Delta \simeq \mathrm{d}_l\,(\rho_l/\rho_g)^{1/2}$, about a centimeter. The process is far from total for low values of the vehicle radius, and the phase distortion only will be observed in this case (Fig. 11.4).

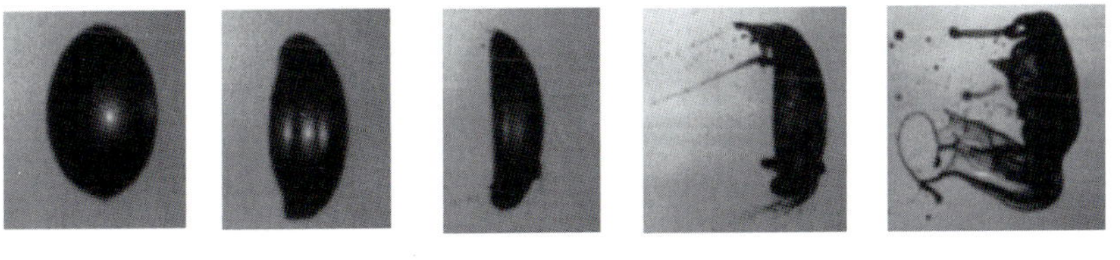

Fig. 11.4 Deformation of a drop (12).

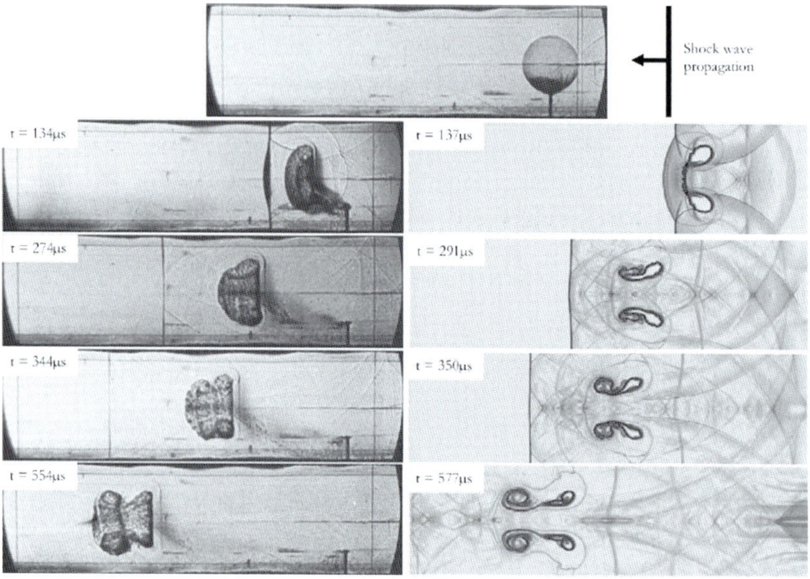

Fig. 11.5 Interaction of a shock with a drop; comparison experiment and calculation [11].

Numerical simulation of the first microsecond of the phenomenon gives very satisfactory results (Fig. 11.5). The phase of splitting is less precise, the size of fragments being of the same order of magnitude as the mesh calculation. This is not very annoying in our case because this phenomenon is mainly involved in particle ovalization, which has an effect on trajectory (see Sec. 11.3.4.2).

Note that the phenomenon of fragmentation can also affect ice crystals with a high porosity [13].

11.3.4 Particle Trajectories

11.3.4.1 Solid Particles

For simplicity, one assumes spherical particles initially, a reasonable approximation for clay and ice particles after crossing the shock and the fragmentation induced by it. We neglect the time variation of the difference in speed between particle and flow (effect of "added mass"), gradients in the flow measured across the particle (Basset force), and the weight and force of Archimedes [14]. The calculation of trajectories requires knowledge of the drag coefficient of the particle. This is generally described by an expression attributed to Henderson, valid throughout the field from rarefied to continuous [15, 16]:

- For $\mathcal{M}a < 1$

$$C_{A_1} = 24\left\{\mathcal{R}e_\mathrm{d} + S\left[4.33 + \frac{3.65 - 1.53\frac{T_p}{T}}{1 + 0.353\frac{T_p}{T}}\exp\left(-0.247\frac{\mathcal{R}e_\mathrm{d}}{S}\right)\right]\right\}^{-1}$$
$$+ \exp\left(-0.5\frac{\mathcal{M}a}{\mathcal{R}e_\mathrm{d}^{\frac{1}{2}}}\right)$$
$$\times\left[\frac{4.5 + 0.38(0.03\mathcal{R}e_\mathrm{d} + 0.48\mathcal{R}e_\mathrm{d}^{\frac{1}{2}})}{1 + 0.03\mathcal{R}e_\mathrm{d} + 0.48\mathcal{R}e_\mathrm{d}^{\frac{1}{2}}} + 0.1\mathcal{M}a^2 + 0.2\mathcal{M}a^8\right]$$
$$+ 0.6S\left[1 - \exp\left(-\frac{\mathcal{M}a}{\mathcal{R}e_\mathrm{d}}\right)\right] \tag{11.7}$$

with $\mathcal{R}e_\mathrm{d} = \rho ud/\mu$ and $S = \mathcal{M}a(C_p/2C_V)^{1/2}$.

- For $1.75 < \mathcal{M}a < 6$

$$C_{A_2} = \frac{0.9 + \frac{0.34}{\mathcal{M}a^2} + 1.86\left(\frac{\mathcal{M}a}{\mathcal{R}e_\infty}\right)^{\frac{1}{2}}\left(2 + \frac{2}{S^2} + \frac{1.058}{S}\frac{T_p}{T} - \frac{1}{S^4}\right)}{1 + 1.86\left(\frac{\mathcal{M}a}{\mathcal{R}e_\infty}\right)^{\frac{1}{2}}} \tag{11.8}$$

with $\mathcal{R}e_\infty = \rho u/\mu$.

- For $1 < \mathcal{M}a < 1.75$, one uses a linear fitting equation

$$C_A(\mathcal{M}a, \mathcal{R}e_\mathrm{d}, \mathcal{R}e) = C_{A_1}(1, \mathcal{R}e_\mathrm{d}) + \frac{4}{3}(\mathcal{M}a - 1)$$
$$\times\left[C_{A_2}(1.75, \mathcal{R}e_\infty) - C_{A_1}(1, \mathcal{R}e_\mathrm{d})\right] \tag{11.9}$$

This expression is based on various theoretical and experimental results. Its shape has been chosen to find classical behavior, such as the Stokes–Oseen equation for $\mathcal{M}a \to 0$ and sufficiently small $\mathcal{P}e$

$$C_A = 24\mathcal{R}e^{-1} + 4.5 \tag{11.10}$$

11.3.4.2 Water Drops

The case of water drops is different. The velocity in the shock layer is sufficient to deform these particles, increasing its drag. The problem then couples the deformation problem and the unsteady calculation trajectory through the drag.

In the case of dominant inertial effects, the calculation of the deformation is made by solving in an approximate way the tranverse momentum equation (perpendicular to the velocity) [13, 17].

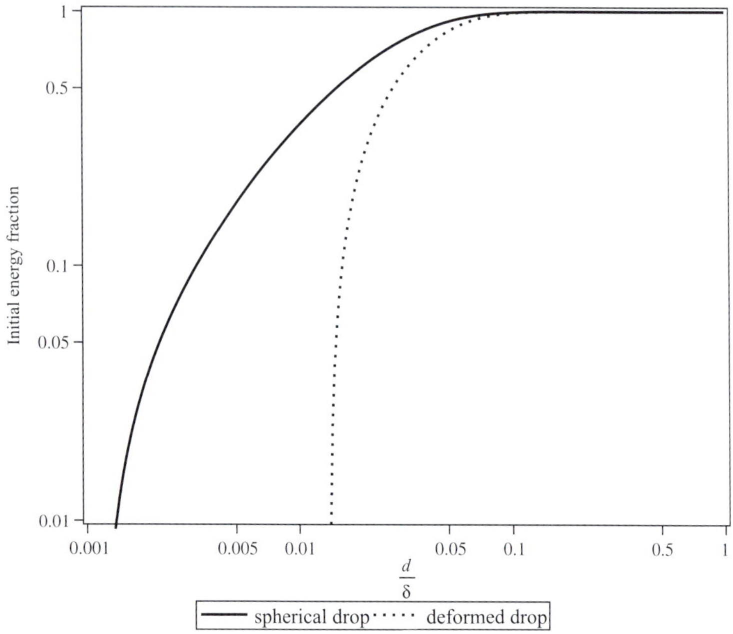

Fig. 11.6 Particle slowdown in the shock layer; effect of drop deformation on the fraction of initial kinetic energy saved at impact (13).

The results [Fig. 11.6, (d: diameter; δ: shock detachment distance)] show that drops of small diameter reach the surface with low kinetic energy [13, 17]. This phenomenon is coupled with the vaporization that is effective only on very small drop diameters [18], which is therefore a secondary effect.

It can therefore be considered as a first approximation that:

- Drops of small diameter have no effect on erosion.
- Drops of larger diameter are not affected by the shock layer—except for the shape, which, as we will see, plays only a secondary role during the impact.

11.4 Effect of Particles on the Flow

11.4.1 Experimental Observations

11.4.1.1 Experiments

It is necessary to describe the wind tunnel tests (hot or cold) that were used to assess the phenomenon, insofar as the interpretation of these can cause confusion.

In reality, for the phenomenon we are interested in, the relative velocity of the particles is about 5000 to 7000 $\mathrm{m \cdot s^{-1}}$. In the tests, the speeds are between 1000 and 2000 $m \cdot s^{-1}$ [19, 20]. Moreover, they do not have the speed of the flow because of the insufficient duration of their acceleration. They reach about 80 to 90% of the gas velocity [19]. Thus, the particle is preceded (in the laboratory system) by its wake, which will reach the shock layer first.

With the impact velocities that result from these test conditions, at least some of the particles bounce back out of the shock layer. This rebound effect is seen in Fig. 12.4 in Chapter 12. It is different from what is observed during a hypervelocity impact (Sec. 11.5), which poses the problem of representativeness of these tests.

11.4.1.2 Deformation of the Shock

Under the effect of any disturbance, the shock detachment may briefly adopt a tapered shape ("spiked"). In the case of wind tunnel tests with seeding, the disturbance may be due to the passage of a particle from upstream to downstream or from downstream to upstream for particles bouncing on the model [20]. This second assumption seems to be confirmed by investigators from high-speed visualizations. This observation led to experiments in which particles were launched from a cannon located within the model placed in a wind tunnel [21]. This experiment replicated the phenomenon, and shows that:

- If the particle velocity is low enough that it stays in the shock layer, we observe a simple modification of the transient flux (1.5 times the value steady for a few milliseconds) and pressure.
- If the particle crosses the shock but does not travel a distance greater than the diameter of the model, a spike emerges, which disappears immediately.
- For a displacement of a particle of larger magnitude, there is a pulsation of the spike with a peak flow (around five times the stationary value).

The same effect of deformation of the shock is obtained by any other form of disturbance in the subsonic region, for example, a laser deposition [2] or heavy air injection from the model in the subsonic region [23, 24]. We can compare this phenomenon to that of a "carbuncle," a stationary numerical solution leading to this type of geometry (Fig. 11.7).

There is no definitive explanation for this type of phenomenon.

11.4.1.3 Wall Heat Fluxes

Many hot and cold wind tunnel experiments have shown a very large increase in the mean wall fluxes in the presence of particles. The observations are as follows:

- On metallic titanium models with solid particle seeding (SiC, MgO) in a wind tunnel, the fraction of the kinetic energy transformed into heat in adiabatic conditions has been measured. The yield is about $0.85 \pm 15\%$ [26] or $0.6 \pm 20\%$ [20] in these experiences. The same quantity, measured

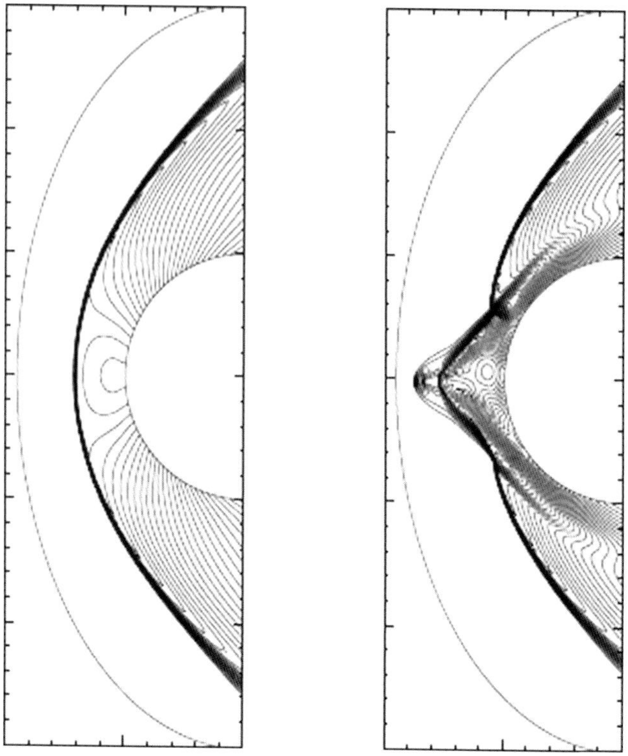

Fig. 11.7 Perturbation of strong shock obtained numerically ("spiked flow") (25).

on single impacts of silica balls on a titanium plate (particles of 100 to 200 μm, result independent of speed of between 900 and 2200 $m \cdot s^{-1}$) [27], is 70 ± 15%.

- Heat flux measured increases with the Reynolds number, all other things being equal [20].
- These values also increase with the mass flow of incident particles [28].
- On the graphitic models used in two-phase flow, subsequent tests in monophasic flow on the same models show that the increased heat flux due to wall roughness is only a small proportion of the total [28] (Fig. 11.8; Boeing Hypersonic Wind Tunnel experiments on graphite).
- Equivalent tests conducted on metal and graphite samples [28] show that the last ones see a much larger increase of heat flux {Fig. 11.9; Arnold Engineering Development Center (AEDC) Dust Erosion Tunnel (DET) experiments on titanium and graphite discs [28]: (1) without seeding, (2) titanium target, and (3) graphite target}. These tests also show a decrease of the flux with the wall temperature, reflecting the convective behavior of the phenomenon.

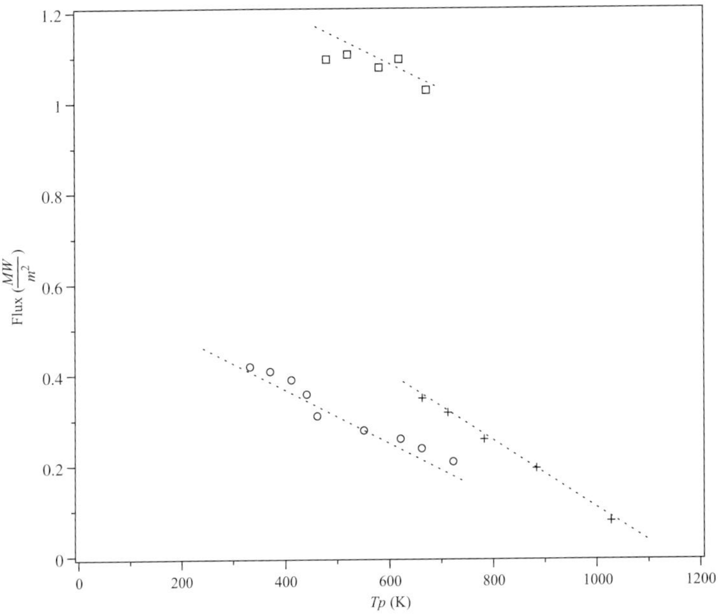

Fig. 11.8 Heat flux augmentation (smooth, rough, two-phase flow) (28).

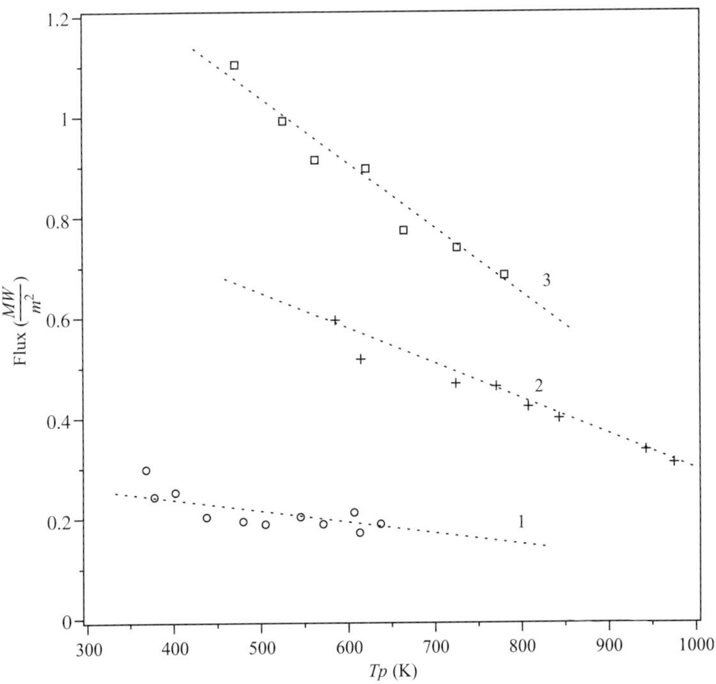

Fig. 11.9 Heat flux augmentation at stagnation point: influence of material.

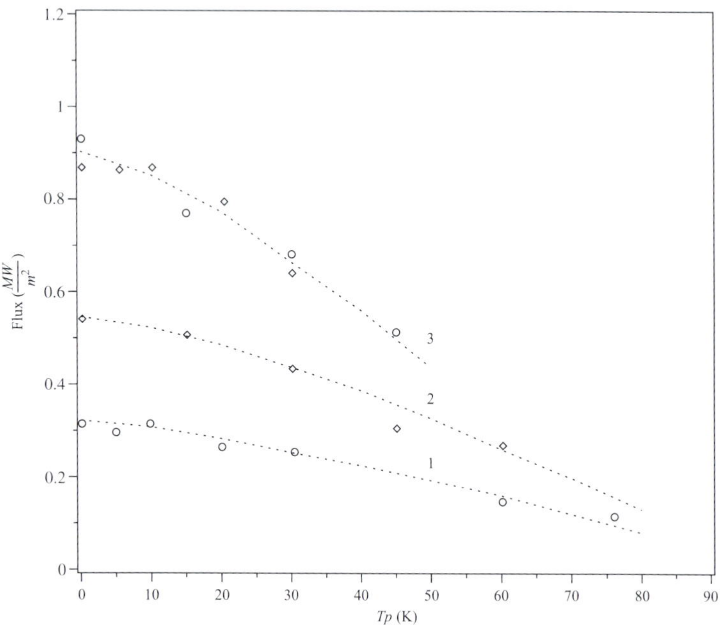

Fig. 11.10 Increase of heat flux at the running point: influence of the radius of specimen.

- Equivalent tests performed on a metal model [28] show that heat flux in the presence of particles does not change when doubling the radius of the nose {Fig. 11.10; AEDC DET experiments on titanium hemispherical specimen [28]: (1) radius R_1 without seeding, (2) radius $R_2 = 2R_1$ without seeding, and (3) R_1 and R_2 with seeding}.

11.4.2 Explanations and Models

Various attempts at explanation and prediction have been proposed, which are described in the following:

1. The first attempts of explanation are based on the kinetic energy transfer to the wall. This explanation would have the advantage of allowing one to understand the same fluxes on specimen of different radii (Fig. 11.10). In fact, the calculation of heat fluxes related to energy transferred upon impact explains only a small part of the difference with the flux in the absence of particles.

   Moreover, all of these tests show that a substantial part of this energy is of convective origin (see the decrease of the flow with T_p in Figs. 11.8 and 11.9). Figure 11.8 permits us to refute an increase in the convective flow based on the creation of roughness of the material: the measurements made on the same specimen in "clear" flow shows only small increases.

2. A first approach of a predictive model is a simple correlation of tests based on observations: independence on geometry, increase with the amount of material injected into the upstream flow or the amount of ejecta, and with the Reynolds number:

$$\begin{aligned} \dot{q} &= C_H \rho_\infty V_\infty (h_\infty - h_p) \\ C_H &= 0.098\ \xi^{0.317} \sin\theta \\ \xi &= \frac{\rho_{s_\infty} V_{s\infty}}{\rho_\infty V_\infty}(1+G) \end{aligned} \tag{11.11}$$

ρ_{s_∞} and $V_{s\infty}$ are the partial density of particles and their upstream velocity (in tests $V_{s\infty} < V_\infty$ due to the incomplete velocity implementation) and G is the ratio of the mass removed as compared to the incident mass (or the fluxes of these quantities) as defined in Sec. 11.5. The θ is the impact angle, relative to the normal of the surface.

This expression, which covers both trials with a dispersion of about $\pm 50\%$ (Fig. 11.11), has no theoretical justification. Note that in this expression the impactor and ejecta play the same role.

3. The deformations of the shock, sometimes mentioned as a possible explanation, have been numerically studied in detail [29, 30], describing mechanisms of quasi-periodic pulsations. These studies are linked to the

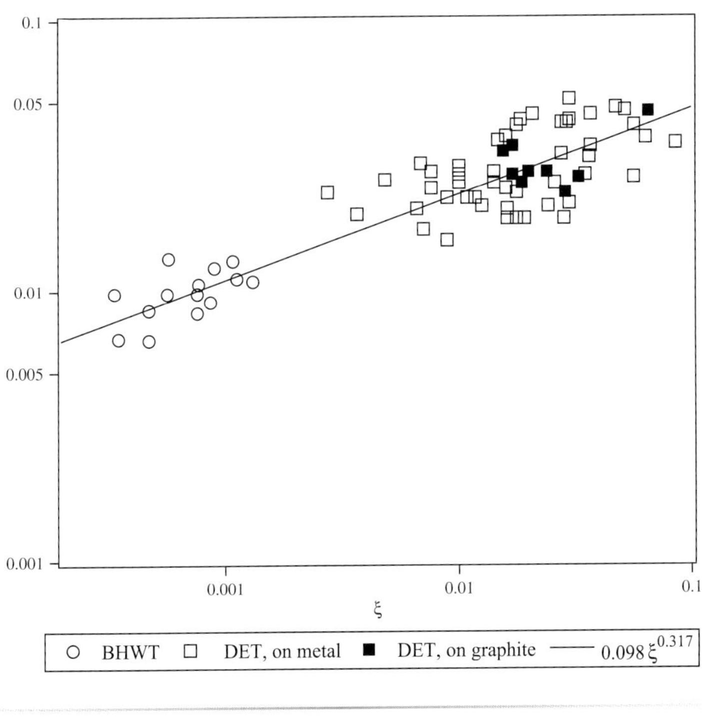

Fig. 11.11 Stanton number for erosion (28).

case of a particle flowing back to a diameter less than the diameter of the object, and only tested where one can observe a pulse of this type. Wind tunnel measurements with seeding certainly show deformation shock but no pulsation [28, 31]. Consequently, and given the important interval between two impacts, possible effects seem at odds with the observed values.

4. The only explanation offered based on a physical explanation is by Shih and Hove [2]. It is based on an analogy with the effect of the impact of a turbulent flow (see Sec. 7.2.1). Analysis of tests described previously leads us to propose an equivalent turbulent intensity (in the entire upstream flow) given by:

$$Tu_{\infty_{eq}} = 0.71\left(\frac{\rho_p}{\rho_\infty}\right)^{\frac{1}{2}}\left(\frac{V_{p\infty}}{V_\infty}\right)^{\frac{3}{2}} \tag{11.12}$$

Note that this model does not reproduce all the phenomena listed, especially the fact that in some experiments the heat flux is the same for different radii, relative scales of turbulence in these two cases being insufficient to explain the phenomenon.

11.5 Particle–Wall Interaction

11.5.1 Phenomenology

The hypervelocity impact is characterized by the following successive phenomena [33–35] (Fig. 11.12):

- Creation in the impactor and the target of shock and expansion waves, the latter obtained by reflection on free surfaces.
- After a few microseconds, the pressure is almost constant and the system behavior is of a hydrodynamic type with partial evaporation of the incident particle and material. Note that in all experiments (representative impact velocities on cold material or low-speed impacts on hot material) the evaporative fraction is low, but these experiments probably are not representative of what happens in real conditions.
- The pressure decreases then and the material returns to a behavior in conformity with its mechanical properties. In this phase, the crater acquires its final size, after a few tens of milliseconds. The ejecta are created by material removal from the bottom of the crater. One also observes the creation of cracks in the material and "lips" on the edges of the crater.

The impact is reflected thus:

- In the material by creating a crater and a local mechanical degradation.

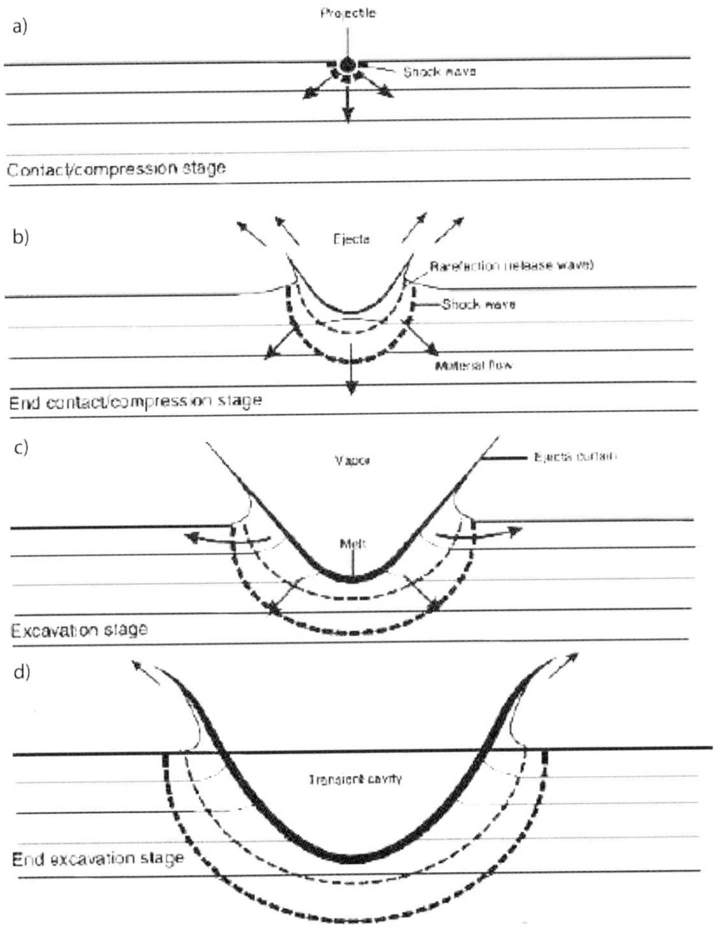

Fig. 11.12 Hypervelocity impact phenomenology (33).

- In the flow by the injection of gas and ejecta in large quantities, during a very short period of time, a few tens of microseconds. These are ejected following a cone at a speed of a few hundred meters per second.

The temperature of the material will play an important role at the end of the cratering. Indeed, the carbon material has a very different behavior depending on temperature:

- At room temperature, it has a brittle performance: impact creates numerous cracks.
- Beyond 3000 K, its behavior becomes ductile [36]. The cracks are minimized. The difficulty of conducting impact tests at high temperature explains the strong misunderstanding of the subject.

Fig. 11.13 Impact crater in a metal plate and impactor identical to that used for the experiment.†

Regarding the energy transferred to the medium and remaining in it as heat, it should be noted that its proportion must decrease with temperature, at least in the case of a carbon material whose surface is subliming and the temperature gradient across the crater low. Indeed, low heating of material near the surface will lead to its immediate vaporization, or vaporization after a short latency.

The crater formed is approximately of spherical shape [33], at least for high speeds, typically over $2\ \mathrm{km} \cdot \mathrm{s}^{-1}$. Its size is somewhat independent of the shape of the impactor (Fig. 11.13), insofar as the ratio of the largest dimension to the smallest is not too important. This form of crater remains unchanged for moderately oblique impacts, particularly if the impact velocity is high.

11.5.2 Mass Loss

Mass losses are characterized by the ratio $G = (m_s/m_p)$ of impactor mass m_s to eroded mass m_p. This implicitly assumes that the surface is eroding regularly on average.

Impact measurements were performed on ATJ graphite [34]. The particles are water ice, polyethylene, or glass, with dimensions between 300 and 1500 μm. The speeds range from 1280 to $4000\ \mathrm{m} \cdot \mathrm{s}^{-1}$. The results show (Fig. 11.14):

- A very high mass loss factor G, increasing approximately as V^{α}, α ranging between 1.5 and 2

† ESA Multimedia, http://esamultimedia.esa.int/images/hvi_impact_sample_xlrg.jpg [retrieved 2 Dec. 2009].

- A very large dispersion ($\pm 40\%$), hardly understandable for a relatively homogeneous material such as graphite and incompatible with the experimental uncertainties
- An inconsistent temperature dependence, due to the dispersion of tests, which is especially high at elevated temperature target.

Various studies of this type of material permit one to write an empirical law of the type [37]

$$G \simeq \beta \left(\frac{\rho_p}{\rho_s} \right)^3 \left(\frac{V_p \cos\theta}{a_s} \right)^\alpha \tag{11.13}$$

where θ is the impact angle counted from the vertical, a_s is the speed of sound in the material, and α is between 1.7 and 2. The constant β is close to 24.

Note that the mass loss is very sensitive to the plasticity of the material, measured through a_s. An increase of the latter leads to an increase of G, and vice versa. In this regard, the impacts of aluminum balls on a plate of similar material give fairly representative results, the values of sound speed in this material and graphite being similar. This is confirmed by the values of G measured in the range $2000 \text{ m} \cdot \text{s}^{-1} < V < 8000 \text{ m} \cdot \text{s}^{-1}$ [38]. There was thus a way of studying the influence of passage in the elastic–plastic domain that is not readily accessible to carbon, for which this phenomenon

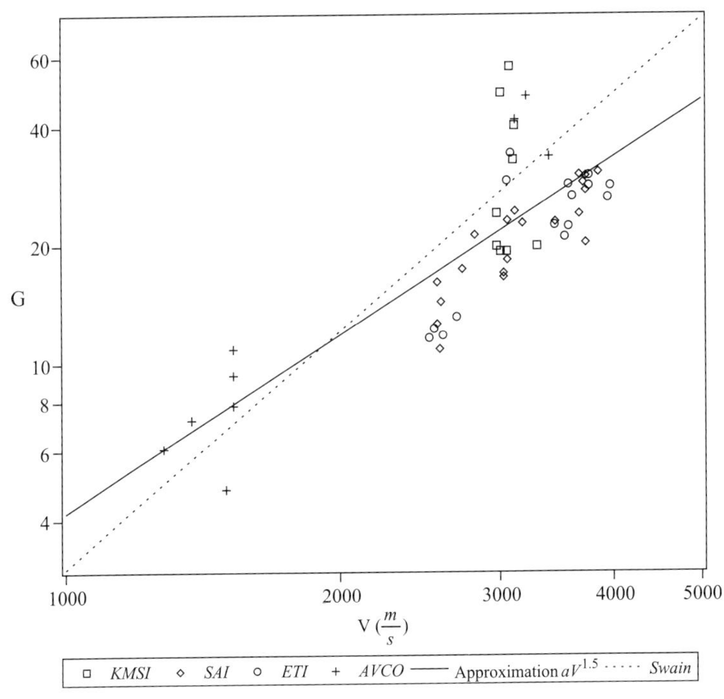

Fig. 11.14 Mass loss by impact on ATJ graphite (34, 39).

appears beyond 3000 K [36]. In the case of aluminum, there is an increase of G by a factor of 2 to 4, depending on the alloys used, when going from room temperature to a temperature near the melting point [37]. Also note that the amount of ejecta decreases significantly with increasing temperature. This analogy between aluminum and graphite is used with caution, however, in the absence of results on carbon itself.

11.5.3 Partition of Energy and Ejecta

The ejecta are created during the final phase of impact, when the crater creation is complete [40]. The first particles come primarily from the impactor and are ejected at a small angle relative to the surface: 10 to 40 deg [40, 41]. The bulk of the mass consists of particles of a few micrometers to a few hundred micrometers, depending on target material (Fig. 11.15; impacts of nylon spheres on pumice, $V = 4400\ \mathrm{m \cdot s^{-1}}$), ejected at an angle of around 40 to 50 deg [40, 41] and mainly from the target material [41]. There is a strong correlation between speed and angle of ejection [42] (Fig. 11.16; impacts of aluminum spheres on basalt). For impact velocities of interest, typically $V > 5\ \mathrm{km \cdot s^{-1}}$, the ejection speeds are the order of several hundreds of meters per second [41, 43] [Fig. 11.17; impacts of aluminum spheres on basalt ($6.1 < V < 6.4\ \mathrm{km \cdot s^{-1}}$), fraction F of the mass having a speed in excess of V]. These velocities imply that a significant portion of the energy of the incident particle is transferred to the ejecta, up 40 to 50% in the case of brittle materials [42].

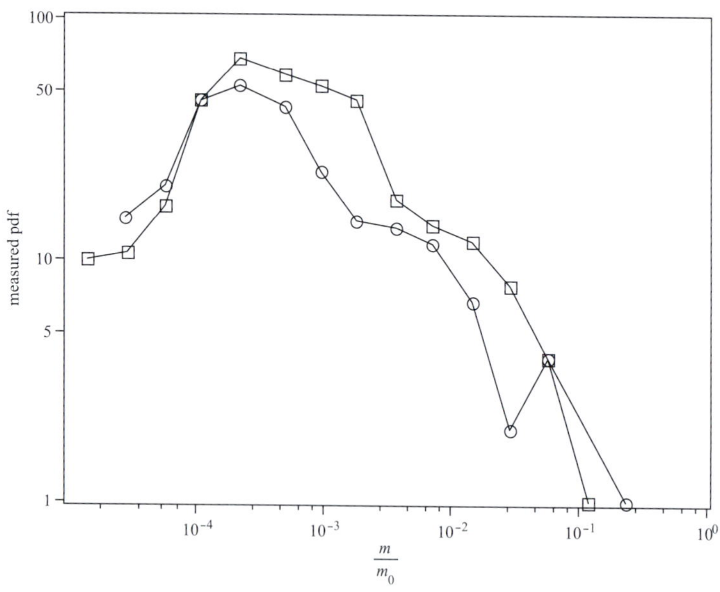

Fig. 11.15 Mass of ejecta relative to those of the impactor (44).

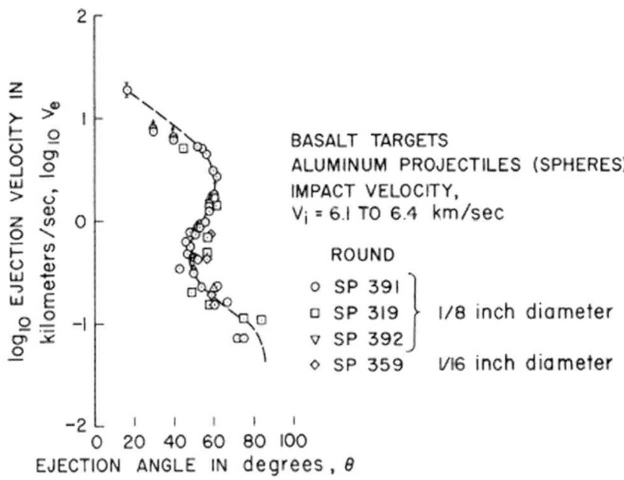

Fig. 11.16 Angle-velocity relationship for the ejecta (42).

The remaining energy is divided among (numerical results):

- Comminution of the material into fragments; this part may exceed 10% of the total.
- Degradation by thermal energy, with values significantly lower than the values listed in Sec. 11.4.1.3 for tests at low speed—20% in such a fragile material [42].

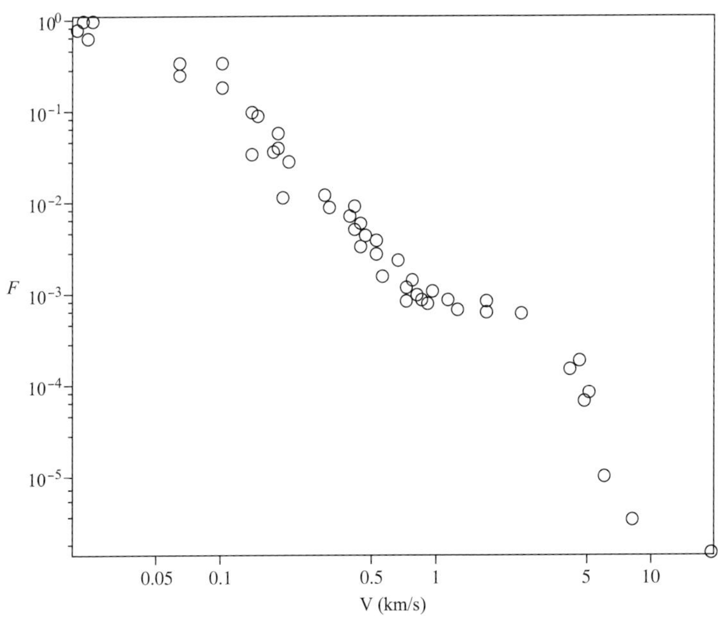

Fig. 11.17 Speed of ejecta (42).

- Fusion, or in the case of carbon, vaporization.
- The cracking of the material under the crater and the residual stresses, some percent at most.
- Radiative energy lost in the flash created because of the high temperatures of the gas phase: weak again [44, 45].

The ejecta of small size do not flow back far, despite their high initial velocity. The largest (100 μm or more), injected with a speed much greater than the flow near the stagnation point, thus have a significant probability of reaching the shock and thus cause the phenomena described in Sec. 11.4.1.

Note that these ejecta are likely to reimpact the surface, but with a much slower speed than primary particles and at larger angles, thus not having a large effect.

11.5.4 Surface State Protection by the Ejecta

Few orders of magnitude show that the surface is dominated by the impacts. Take the example of a reentry to Earth with a cloud crossing. The impact of an ice particle of 100 μm at 4300 $\mathrm{m \cdot s^{-1}}$ (value below the median value) causes a crater in the graphite 5 mm in diameter and 2 mm in depth. A particle density of $3 \times 10^6\ \mathrm{m^{-3}}$ will create a new impact on the previous surface after an average time of 0.6 μs. Moreover, suppose a severe ablation speed of 5 $\mathrm{mm \cdot s^{-1}}$ and a "healing" of the surface when the ablation is approximately two times the depth of the crater (Chapter 6). This happens in our example after about 1 s. This example shows that the surface of the material results from impacts, even for the crossing of low-density clouds.

Another phenomenon is sometimes referred to regarding the protection provided by the ejecta against another incident particle in the vicinity the region of the impact of the first. A calculation of the same order of magnitude as in the previous paragraph does not exclude this phenomenon, at least in the case of flat specimens and high rates of particles used in some experiments [46].

11.6 Coupling with Ablation

The values of the mass flow rate of material loss per particle impact can be very high, substantially above the ablation by phase change or heterogeneous reactions. Typically, if we take the example of carbon treated in Sec. 5.8, a mass flux rate of erosion $\dot{m}_e$ equal to the one due to physicochemical ablation in the absence of erosion $\dot{m}$ leads to:

- A reduction of 1 to 3% of temperature
- A decrease from 15 to 23% of the total ablation as compared to the sum of ablations taken independently

These values vary linearly with the ratio $(\dot{m}_e/\dot{m})$.

All this, of course, does not include indirectly induced effects:

- The laminar–turbulent transition eventually induced by the creation of a very rough surface.
- The increase of heat flux in turbulent flow, due to the same roughness.
- The increase of heat flux that is the subject of Sec. 11.4. Note, however, the vast uncertainty about this. If the reason given for this phenomenon, namely the rebound of particles, is correct, models from wind tunnel tests are irrelevant. Indeed, we saw that the high-speed impact created a large number of ejecta of a size close to the incident particle.

For the record, the particles are injected into the flow in too small a quantity to have any effect by physico-chemical reactions [47].

The rare cases of comparison of the models described here with flight tests on a nose of graphite, however, give respectable results—approximately 20% on recession for stagnation point [48]. But these comparisons are made a posteriori. The precision that can be expected in a calculation is probably much lower than this value.

11.7 Discussion

It seems necessary to go through all the problems outlined in this chapter because great uncertainty exists about the origin of phenomena and their possible effects.

1. In the area of impact ejecta particles on the flow:
   - What is the origin of the phenomena of distortion of the shock?
   - Do these distortions have an effect on the wall heat flux?
   - Does the turbulence have an effect, and if so, what is the mechanism involved?
2. In the area of impacts:
   - Does the cracking of the material have an effect of minor importance in all cases, including composite materials?
   - Is the part of gas created negligible, even in the case of carbon during sublimation, and thus which low-temperature variation significantly alters the phenomenon?
   - What is the impact of changes in temperature on the thermo-mechanical properties of carbon?
3. Is the protective effect of ejecta on the incident particle important?

Note that the answer to these questions either validate or invalidate hot or cold wind tunnel tests.

Finally, we take care to note that the phenomena that have been described are related to an impact on a homogeneous material. Things get notably more complicated when dealing with composite materials, for which mechanical phenomena play an important role (Fig. 11.18).

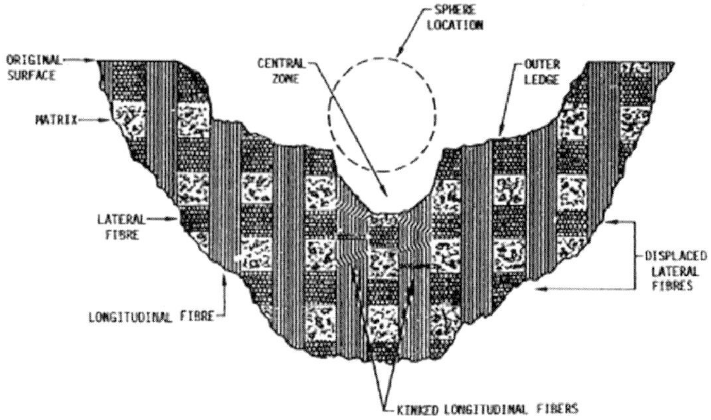

Fig. 11.18 Impact on carbon–carbon material; part of the material is removed by spallation (49).

References

[1] Perko, H. A., Nelson, J. D., and Green, J. R., "Review of Martian Dust Composition, Transport, Deposition, Adhesion, and Removal," *ASCE Conference Proceedings of the 8th International Conference on Space and 5th International Conference on Robotics*, edited by Laubscher, B. E.et al., Vol. 203, No. 40625, ASCE, Albuquerque, New Mexico, 2002.

[2] Goody, R. M., and Yung, Y. L., *Atmospheric Radiation. Theoretical Basis*, 2nd ed., Oxford University Press, Oxford, UK, 1989.

[3] Lushnikov, A. A., and Kulmala, M., "Nucleation Burst in a Coagulating System," *Physical Review E*, Vol. 62, No. 4, 2000, pp. 4932–4939.

[4] Snook, K., MacKay, C., and Cantwell, B., "Analysis of Physical Properties of Dust Suspended in the Mars Atmosphere," NASA Ames-Stanford Consortium NCC-5166, NTRS 19980237262, Aug. 1998.

[5] Forget, F., Hourdin, F., Fournier, R., Hourdin, C., Talagrand, O., Collins, M., Lewis, S. R., Read, P. L., and Huot, J.-P., "Improved General Circulation Models of the Martian Atmosphere from the Surface to Above 80 km," *Journal of Physical Research-Planet*, Vol. 104, 1999.

[6] Hu, Y., Vaughan, M., McClain, C., Behrenfeld, M., Maring, H., Anderson, D., Sun-Mack, S., Flittner, D., Huang, J., Wielicki, B., Minnis, P., Weimer, C., Trepte, C., and Kuehn, R., "Global Statistics of Liquid Water Content and Effective Number Density of Water Clouds over Ocean Derived from Combined CALIPSO and MODIS Measurements," *Atmospheric Chemistry and Physics Discussions*, Vol. 7, 2007, pp. 4065–4083.

[7] Korolev, A. V., Isaac, G. A., Strapp, J. W., Kober, S. G., and Barker, H. W., "In Situ Measurements of Liquid Water Content Profiles in Midlatitude Stratiform Clouds," *Quarterly Journal of the Royal Meteorological Society*, Vol. 133, 2007, pp. 1693–1699.

[8] Pruppacher, H. R., and Klett, J. D., *Microphysics of Clouds and Precipitation*, Atmospheric Science Library, Kluwer Academic Publishers, Dordrecht, The Netherlands, 1978.

[9] Heymsfield, A. J., Bansemer, A., and Field, P. R., "Properties of Tropical and Midlatitude Ice Cloud Particle Ensembles: Observations and Parameterizations," *American Meteorological Society, 11th Conference on Cloud Physics*, Ogden, Utah, June 2002.

[10] Nicolet, W. E., Wool, M. R., Laub, B., and Jaffe, N. A., "Hydrometeor/Shock Layer Interaction Study," *Interim report, Passive Nosetip Technology (PANT) Program, SAMSO-TR-74-86*, Vol. XIX, June 1975.
[11] Saurel, R., Petitpas, F., and Berry, R. A., "Simple and Efficient Relaxation Methods for Interfaces Separating Compressible Fluids, Cavitating Flows and Shocks in Multiphase Mixtures," *Journal of Computational Physics*, Vol. 228, 2009, pp. 1678–1712.
[12] Theofanous, T. G., and Li, G. J., "On the Physics of Aerobreakup," *Physics of Fluids*, Vol. 20, No. 5, 2008.
[13] Simons, R. A., "Aerodynamic Shattering of Ice Crystals in Hypersonic Flight," *Physical Sciences Incorporated Report AD-A019 517, Advanced Reentry Aeromechanics*, Vol. II, April 1975.
[14] Maxey, M. R., and Riley, J. J., "Equation of Motion for a Small Rigid Sphere in a Non-uniform Flow," *Physics of Fluids*, Vol. 26, No. 4, 1983, pp. 883–889.
[15] Henderson, C. B., "Drag Coefficients for Spheres in Continuum and Rarefied Regimes," *AIAA Journal*, Vol. 14, No. 6, 1976, pp. 707–708.
[16] Walsh, M. J., and Henderson, C. B., "Comment on Drag Coefficient of Spheres in Continuum and Rarefied Flows," *AIAA Journal*, Vol. 15, No. 6, 1977, pp. 893–895.
[17] Waldman, G. D., and Reinecke, W. G., "Particle Trajectories, Heating, and Breakup in Hypersonic Shock Layers," *AIAA Journal*, Vol. 9, No. 6, 1971, pp. 1040–1048.
[18] Finson, M. L., Pirri, A. N., Nebolsine, P. E., Simons, G. A., and Wu, P. K. S., "Advanced Reentry Aeromechanics. Final Report," *Physical Sciences Incorporated Report AD-A052 744*, Jan. 1978.
[19] Pfeiffer, E., "Experimental and Theoretical Studies of Mars Dust Effects," *High Performance Space Structure System GmbH Report MDUST-HPS-RP-001*, July 2007.
[20] Fleener, W. A., and Watson, R. H., "Convective Heating in Dust-Laden Hypersonic Flows," *AIAA Paper 1973–761, Eighth Thermophysics Conference*, Palm Springs, CA, July 1973.
[21] Holden, M. S., "Studies of the Effects of Transitional and Turbulent Boundary Layers on the Aerodynamic Performance of Hypersonic Re-entry Vehicles in High Reynolds Number Flows," *Air Force Office of Scientific Research, AFSOR-TR-79-0125*, Dec. 1978.
[22] Minucci, M. A., Toro, P. G. P., Chanes, J. B., Jr., and Pereira, L., "Introduction to Physical Gas Dynamics Investigation of a Laser-Supported Directed-Energy 'Air Spike' in Mach 6.2 Air Flow. Preliminary Results," AIAA Paper 2001–0641, 2001.
[23] Fomin, V. M., Maslov, A. A., Malmuth, N. D., Fomichev, V. P., Shashkin, A. P., Korotaeva, T. A., Shiplyuk, A. N., and Pozdnyakov, G. A., "Influence of a Counterflow Plasma Jet on Supersonic Blunt-Body Pressures," *AIAA Journal*, Vol. 14, No. 1, 2000, pp. 10–17.
[24] Ganiev, Y. C., Gordeev, V. P., Krasilnikov, A. V., Lagutin, V. I., Otmennikov, V. N., and Panasenko, A. V., "Aerodynamic Drag Reduction by Plasma and Hot-Gas Injection," *Journal of Thermophysics and Heat Transfer*, Vol. 14, No. 1, 2000, pp. 10–17.
[25] Chauvat, Y., Moschetta, J.-M., and Gressier, J., "Shock Wave Numerical Structure and the Carbuncle Phenomenon," *International Journal for Numerical Methods in Fluids*, Vol. 47, No. 8–9, 2005, pp. 903–909.
[26] Burghart, G. A., "Evaluation of the AEDC Dust Erosion Tunnel Heat Augmentation Data," *Science Application Incorporated, SAI-72-527-LA*, Sept. 1972.
[27] Courtney, J. F., MacMillen, L. D., Wencer, R. S., and Zuieback, J. M., "Heating Augmentation Phenomenology Studies," Report DNA 3407F, 1974.
[28] Dunbar, L. E., Courtney, J. F., and MacMillen, L. D., "Heating Augmentation in Erosive Hypersonic Environment," *AIAA Journal*, Vol. 13, 1972, pp. 908–912.
[29] Feszty, D., Badcock, K. J., and Richards, B. E., "Driving Mechanisms of High-Speed Unsteady Spiked Body Flows, Part 1: Pulsation Mode," *AIAA Journal*, Vol. 42, No. 1, 2004, pp. 95–106.

[30] Feszty, D., Badcock, K. J., and Richards, B. E., "Driving Mechanisms of High-Speed Unsteady Spiked Body Flows, Part 2: Oscillation Mode," *AIAA Journal*, Vol. 42, No. 1, 2004, pp. 107–113.

[31] Canton-Desmeuzes, C., "Stagnation Point Heat Fluxes Augmentation During Cloud Crossing," *Congrès AAAF*, Arcachon, France, AAAF, March 2003.

[32] Hove, D. T., and Shih, W. C. L., "Reentry Vehicle Stagnation Point Heat Transfer in Particle Environments," *AIAA Paper 1977–93, 15th Aerospace Science Meeting*, Los Angeles, CA, Jan. 1973.

[33] Hermann, W., and Wilbeck, J. S., "Review of Hypervelocity Penetration Theories," *International Journal of Impact Engineering*, Vol. 5, 1987, pp. 307–322.

[34] Wolf, C. J., Nardo, C. T., and Dahm, T. J., "Coupled Erosion/Ablation of Reentry Materials," *Interim Report: Passive Nosetip Technology (PANT) Program*, Vol. XXII, SAMSO-TR-74-86, July 1975.

[35] Zukas, J. A., Nicholas, T., Swift, H. F., Greszczuk, L. B., and Curran, D. R., *Impact Dynamics*, Krieger Publishing Company, Malabar, FL, 1992.

[36] Barré, S., Rochais, D., Tallaron, C., Jurion, M., and Goyhénèche, J.-M., "Comportement thermomécanique de composites texturés en environnements extrêmes," *CHOCS*, No. 34, 2007, pp. 72–83.

[37] Gehring, J. W., Jr., "Engineering Considerations in Hypervelocity Impact," *High-Velocity Impact Phenomena*, edited by Kinslow, R., Academic Press, New York, 1970, pp. 463–514.

[38] Rand, J. L., "Hypervelocity Impact Effects. Semi-annual Progress Report," NASA Report CR 119189, March 1971.

[39] Swain, C. E., "The Effect of Particle/Shock Layer Interaction on Reentry Vehicle Performance," *AIAA Paper 1975–734, 10th Thermophysics Conference*, Denver, CO, May 1975.

[40] Smith, R. H., "Investigation of Crater Growth and Ejecta Cloud Resulting from Hypervelocity Impact of Aluminum Spheres on Thick Aluminum Targets," Technical Report AFML-TR-68-175, June 1968.

[41] Evans, N. J., Shahinpoor, M., and Arhens, T. J., "Hypervelocity Impact: Ejecta Velocity, Angle and Composition," Geological Society of America, Special Paper 293, 1994, pp. 93–101.

[42] Gault, D. E., and Heitowit, E. D., "The Partition of Energy for Hypervelocity Impact Craters Formed in Rock," *Proceedings of the 6th Hypervelocity Impact Symposium*, NASA-TM-X-57428, April–May 1963.

[43] Gault, D. E., Quaide, W. L., and Oberbeck, V. R., "Impact Cratering Mechanics and Structures," *Shock Metamorphism of Natural Materials*, edited by French, B. M., and Short, N. M., MonoBook Corporation, Baltimore, MD, 1968, pp. 87–99.

[44] Ferri, F., Giacomuzzo, C., Pavarin, D., Francesconi, A., Bettella, A., Flamini, E., and Angrilli, F., "Hypervelocity Impact Experiments to Study Craterization and Catastrophic Fragmentation of Minor Bodies of the Solar System," Seventh Workshop on Catastrophic Disruption in the Solar System, Alicante, Spain, June 2007.

[45] Swift, W. R., Moser, D. E., Suggs, R. M., and Cooke, W. J., "An Exponential Luminous Efficiency Model for Hypervelocity Impact into Regolith," *Meteoroids: The Smallest Solar System Bodies*, NASACP-2011-216469, Breckenridge, CO, May 2010.

[46] Canton-Desmeuzes, C., "Analyses of Particles Erosion Tests on Graphite Samples," *Congrès AAAF*, AAAF, Arcachon, France, March 2003.

[47] Wassel, A. T., and Mills, A. F., "Combustion of Carbon Particles in a Hypersonic Turbulent Boundary Layer," *Journal of Spacecraft and Rockets*, Vol. 11, No. 12, 1974, pp. 803–807.

[48] Neuner, G. J., Wool, M. R., and Berry, R. A., "Nosetip Analysis Using the EROS Computer Code," *Interim Report: Passive Nosetip Technology (PANT) Program*, SAMSO-TR-74-86, Vol. XVIII, June 1975.

[49] Evans, A. G., Adler, W. F., and Chesnutt, J. C., "Impact Damage in Carbon-Carbon Composite," Rockwell International Science Center Technical Report SC5076.2TR, Jan. 1977.

Chapter 12 Testing and Specific Test Facilities

- Describe plasma jets
- Measure conductivity

12.1 Models Used in Reentry

Multiple models are used in the calculations of reentry problems. A given database is relative to the type of modeling used. Some examples:

- The trajectory can be calculated in one of two ways:
  - Directly by integration of forces and moments calculated from wall results in an overall calculation [1].
  - From an aerodynamic database obtained by a series of calculations and approximations of the forces and moments according to various parameters (Mach number, Reynolds, altitude, incidence, etc.). This second method is generally used [2].

  In both cases it is necessary to give:
  - Initial reentry conditions: altitude, velocity, and angle relative to local horizontal plane, angular position, and rotation
  - Atmosphere profile $T = f(z)$, $p = f(z)$, where z is the geopotential altitude, and eventually the winds
  - Massic and inertial characteristics of the vehicle
- Ablation can be obtained:
  - From the data of species created by heterogeneous (with the wall) or homogeneous (in the boundary layer) reactions and corresponding reaction kinetics, accompanied by a numerical description of blowing.
  - With the simplifying assumptions described in Secs. 4.3, 4.4, and 8.7.1, by an expression $\mathcal{B}' = f(T, p, \mathcal{B}'_g, \rho_e u_e C_H)$, accompanied by the enthalpy of gas at the wall. The effect of boundary layer exchanges are generally ignored, reducing this expression to $\mathcal{B}' = f(T, p, \mathcal{B}'g)$. At a lower level of physical description, one uses a relationship $\mathbf{v}_a = f(T)$ with an apparent ablation enthalpy Δh, eventually constant. It this case, the blowing is obtained from a correlation.

When conditions of reentry lead to laminar–turbulent transition, it is most often a consequence of material surface state obtained in laminar flow, which can be described by:
  - Its physical geometry, in the case of the most advanced methods (see Chapter 6)
  - An "equivalent" sand grain roughness for the standard engineering models (see Chapter 7)

  In turbulent flow, increased exchanges in the boundary layer lead to the same data type as the real geometry or equivalent sand grain roughness.
- Transfers of mass and energy in the material are described in two possible ways:
  - For a nonpyrolyzable material, the data of density ρ_s, enthalpy h_s (or specific heat C_{ps}), and conductivity tensor $\mathbf{\Lambda}_s$.
  - For pyrolyzable material, the general problem requires a lot of data. With the most common assumptions (see Chapter 8), these data are reduced to virgin and pyrolyzed densities ρ_v and ρ_p, pyrolysis law $\partial\rho/\partial T = f(T, \rho)$, enthalpy of pyrolysis gas $h_g(T)$, and conductivity tensors of virgin $\mathbf{\Lambda}_v$ and pyrolyzed $\mathbf{\Lambda}_p$ materials.

  Some data for various materials are present in the databases as TPSX.*
- Add to this list the thermodynamic and transport properties (enthalpies, diffusion coefficients, viscosities, conductivities):
  - Globally in the case of equilibrium thermodynamics
  - For each species when they are treated explicitly
  - Per degree of freedom in the case of nonequilibrium thermodynamic environments that require more data of vibrational relaxation times (or relaxation rate Z^{rot})
- Finally, if the calculation of radiative transfer is needed, it should be added to the previous list (see Chapter 10):
  - Either an amount of data describing the absorption and, in the case of thermodynamic nonequilibrium, emission.
  - Or those for the approximate description of the spectra.
  - For materials it should be added as a phase function in the general case.

This chapter is devoted to specific test methods for accessing these data. We therefore exclude all information regarding the general test facilities of physics used in the field.

12.2 Plasma Jets

At first, many tests were made on ghost materials in "cold" wind tunnels. Some examples are given in Sec. 1.3.2. The subsequent development of plasma jets has made this type of test a curiosity, perhaps wrongly. But the

* National Aeronautics Space Administration, *Material Properties Database*, http://tpsx.arc.nasa.gov [retrieved 30 April 2009].

rise of plasma jet facilities, their better control, and the desire of engineers to get closer to the real conditions made these facilities the standard reference. We talk of plasma jets, although relatively modest temperatures (2000 to 6000 K) obtained at nozzle outlet make it somewhat difficult to justify the word *plasma*. Specifically, the volume fraction of electrons can be neglected in the assessments of thermodynamic and transport properties. This is obviously not true in the generator itself, where temperatures of the electric arc can reach or exceed 20,000 K.

12.2.1 Description and Performance of a Plasma Jet

A plasma jet is a small wind tunnel compared to cold wind tunnels, in which air (or another carrier gas) is heated by an electric discharge of high intensity (several thousand amperes). The electrical power required is a few megawatts to tens of megawatts. (See Table 12.1. The first panel describes facilities for the tests at low pressure and therefore equipped with an extractor. The notation $xx \times xx$ in the table gives the dimension's minor axis and the major axis of a semi-elliptical nozzle. Some values shown are the maximum performance possible, and never used in practice.) The operating time ranges from a few seconds to continuous. The gas is heated to temperatures ranging from 15,000 to over 20,000 K. There are two types of technologies:

- Huels generators using a natural arc (Fig. 12.1). Air is injected into the vortex to stabilize the arc. The magnetic coil is designed to rotate the arc foot on the anode to limit its wear.
- Generators using a constricted arc with segmented electrodes. This allows the arc to be distributed over a greater length, which decreases the instabilities inherent in this type of medium as well as the surface density of current on the electrode, so the wear and pollution are decreased. Performance, which is superior to that of Huels generators, comes at the price of greater complexity due to independent management of fluids (gas injection, water cooling) of each segment (Fig. 12.2; the cooling pipes of each segment give this facility its remarkable aspect).

Performances are given by three parameters:

- The gas mass flow d
- The plenum pressure p_s
- The bulk enthalpy or its value reduced by the quantity $h_{ref} = \mathcal{R}T_0/\overline{\mathcal{M}}(T_0)$, about 87 kJ kg^{-1} for air

These three parameters are measured as follows:

- The mass flow is measured by a flowmeter.
- The stagnation pressure is assimilated to the static pressure measured in the chamber, the velocity in this region being very low.

Table 12.1 Plasma Jet Facilities: Performances and Uses (3–5)[†]

Facility	Operator	Installed Power (MW)	Maximum Enthalpy (MJ/kg)	Plenum Pressure (10^5 Pa)	Gas Flow ($kg \cdot s^{-1}$)	Mach Nozzle	Φ Nozzle (mm)
Scirocco	CIRA	70	25	16	0.1–3.5	12	900–1950
IHF	NASA	60	47	10	—	5.5–7.5	203 × 813
PTF	NASA	20	35	10	0.5–1	5.5	102 × 457
AHF	NASA	20	33	10	2.5	2.5–12	76–1067
HEAT H2	AEDC	42	7	3.5	1–5	3.4–8.3	230–1070
Simoun	Astrium	6	15	14	—	4.5–5	50
L3K	DLR	5	25	100	0.2	5–10	100–400
HEAT H1	AEDC	30	15	90	0.2–4	1.8–3.5	19–76
HEAT H3	AEDC	70	20	115	2–10	1.8–3.5	28–102
JP200	Astrium	20	15	60	—	1.7–2.4	—
JPHP	Astrium	8	5	130	—	1.6	15

Acronyms can be found in this book's nomenclature.
[†] Arnolds Engineering Development Complex, *High-Enthalpy Arc-Heated Facilities at AEDC*, http://www.arnold.af.mil/library/factsheets/factsheet.asp?id=13894 [retrieved 18 May 2009].

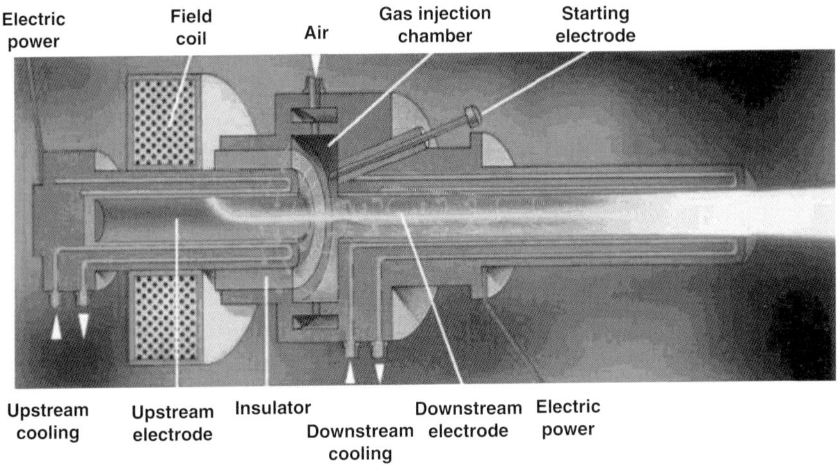

Fig. 12.1 Huels plasma jet (6).

- The bulk enthalpy can be measured in two ways:
  1. The calorimetric method by performing an energy balance: electric energy consumed minus energy dissipated in the cooling circuit. (This presupposes the measurements of water temperature and rate.)
  2. The "aerodynamic" method, in which an inviscid flow in thermodynamic equilibrium is assumed. (This hypothesis is justified by the pressure chamber, which is always greater than atmospheric

Fig. 12.2 Segmented electrodes plasma jet (7).

pressure.) The flow at the nozzle throat of known area S_c is sonic and the plenum pressure allows one to calculate all quantities at the throat by the isentropic expansion/compression relations. One can deduce the flow $\mathrm{d} = \rho_c a_c S_c$. The calculation of density ρ_c and speed of sound a_c is made with an equation of state of real gas type. This method leads to the expression given by Winovich [8] for air in thermodynamic equilibrium (in units of the international system)

$$h = 4.55 \times 10^{-2} \left(\frac{\mathrm{d}}{S_c p_s} \right)^{-2.519} \tag{12.1}$$

It should be noted that there is a stress-related operating maximum allowable heat flux on the inner wall of the generator. This constraint is expressed by the following empirical law [9]:

$$h_{\max} p^{n}_{s_{\max}} = C^{ste} \tag{12.2}$$

with $n \simeq 0.4$. The constant for the second member of the equation is characteristic of the facility, especially the cooling system.

Various facilities are described in Table 12.1. An important distinction must be made. Some facilities work at low static pressure and are equipped with an extractor. Others, restricted to high pressures, run at ambient pressure (see Sec. 12.2.4).

12.2.2 Adequacy of the Material Chacterization

Note that the performance of the generator described in the preceding section depends only on the characteristics upstream of the nozzle throat, and is therefore intrinsic to the facility. The user needs to have representative flow defined by:

- A minimum size of the sample or model
- A total enthalpy
- A stagnation pressure or a static pressure for the sample and a local Mach number

Take, for example, the objects shown in Fig. 1.5 in Chapter 1. At 40 km altitude, three of them (STS-1, Apollo-4, and Stardust) have a speed close to $2000\ \mathrm{m \cdot s^{-1}}$. With the expansion–compression relations for an ideal gas, we can calculate a stagnation pressure of about 5.8×10^5 Pa. To reach this value in a plasma jet is not a problem (see Fig. 12.1). For the ballistic vehicle, the same calculation made with a speed of $7000\ \mathrm{m \cdot s^{-1}}$ provides an enormous stagnation pressure: 2.6×10^9 Pa! It is obviously impossible to achieve a fully representative test. One thus reduces the need and admits a low Mach number and keeps the pressure high enough to maintain high fluxes. This is obtained by reducing the divergence of the nozzle, so the size of the model cannot exceed 0.7–0.8 times the nozzle output in order for the facility to work well.

From simple considerations, one can show that the sample size is directly related to the plant power W. If the efficiency thereof is r (decreasing from 90 to 50% when increasing the bulk pressure), it follows

$$rW = h\mathrm{d} \tag{12.3}$$

Combined with Eqs. (12.1) and (12.2), this equation provides

$$S_{c_{\max}} h_{\max}^{0.603} = C^{ste} \frac{rW}{p_s^{0.5}} \tag{12.4}$$

In this equation, the right-hand side characterizes the installation. If one wants to double the size of the sample at constant stagnation pressure, one must quadruple S_c, which will result in reducing the enthalpy by a factor 10. This shows that there is little leeway in the choice of test domain for a given facility. One can also interpret this relationship by saying that the maximum size of the sample (for a given test environment) will be proportional to $W^{1/2}$.

12.2.3 Characteristics of the Jet

The flow leaving the nozzle is measured by uncooled sensors doing a quick scan of the flow:

- A Pitot probe for stagnation pressure: This measurement does not pose any particular problem. A sample survey is given in Fig. 12.3.
- A stagnation flux probe, from which the enthalpy is inferred through approached relationships, for example, that of Fay and Riddell (see Sec. 4.5.2): This type of survey can have problems. Indeed, the flow is rarely homogeneous; the temptation of designers is to focus energy in the vicinity of the axis to increase the enthalpy performance. This is not always the case: we see in Fig. 12.3 the measured flux, relatively homogeneous, leading to enthalpies close to the average generator enthalpy. This is far from true in a great number of the facilities [10, 11].

The observed enthalpy gradients led to the development of nonstandard measurements. Among these are probes for the direct measurement of the enthalpy [13], which highlighted problems with interpreting fluxmeter results in high pressure facilities [11].

More recently, optical emission spectroscopy has been used [10] for calibration, but their utilization is too cumbersome for routine use. Moreover, their interpretation needs a calculation of the flow. The numerical approach was introduced by Nicolet [9] in fluid mechanics calculations with 1-D radiative transfer. The approach has been generalized more recently [14], but the difficulty of this type of calculation (see Chapter 10) remains an obstacle to its development.

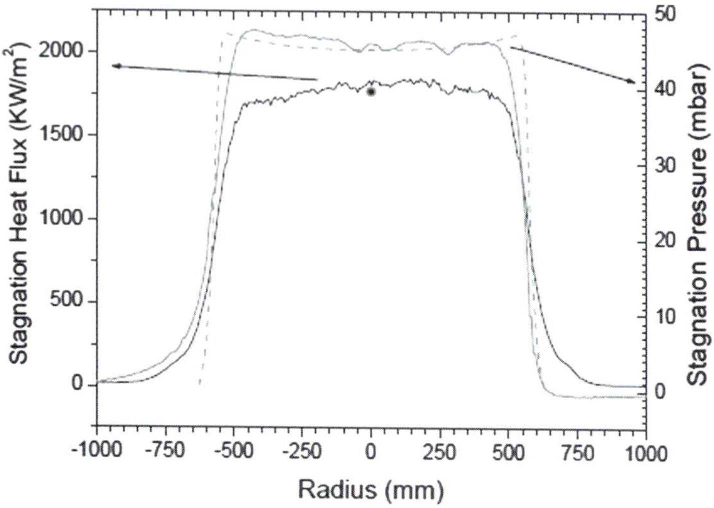

Fig. 12.3 Flow characterization: stagnation pressure (Pitot tube) and stagnation heat flux (fluxmeter) (12).

12.2.4 Experimental Configurations

We can distinguish two types of test configuration. In the first, directly issued from wind tunnel tests' traditional layout, the sample is in front of the nozzle. It is axisymmetric (Fig. 12.4) or flat (Fig. 12.5). The first configuration is used in particular for the characterization of carbon materials for high pressures. In this case, perturbation in the nozzle exit due to a static pressure different from ambient pressure limits the volume of flow usable for testing. This is not a major problem because the large ablations need to compensate for the distance between sample and nozzle output to keep the stagnation point a fixed distance from the nozzle. This is achieved by a servo system using an optical barrier.

We cite a special case of the first type of arrangement. This is a nozzle delivering a pressure gradient in the output axis (flared nozzle) [16]. In this type of test, the sample goes back to the nozzle, and the laminar–turbulent transition is measured by observing the indentation of the specimen or by measuring the temperature.

In the second type of test, the material is part of the assembly. In this configuration, the nozzle exit is flat (2-D nozzle or semi-elliptical). Many variants exist:

- The material is placed at the exit of the nozzle, aligned with it or forming a dihedral. When the test is done at ambient pressure, one can choose the generator bulk pressure and Mach number to obtain a static pressure p_0 in the flow leaving the nozzle, and thereby generate a slip line, minimizing the disturbances. However, this limits the scope of use of the facility.

Fig. 12.4 Test in stagnation point with seeding of particles, CO_2 atmosphere (15).

- The material is one of the faces of a duct whose other sides are made of copper and cooled (Fig. 12.6). This configuration creates problems for the diagnosis of the sample. On the other hand, measurements (pressure, flow) are feasible on the faces of copper walls.

To conclude this summary description of the experiments, we mention the possibility in some facilities (HEAT H1, Simoun, L3K) for injecting solid particles into the flow to simulate the crossing of clouds of clay particles

Fig. 12.5 Test in stagnation point for a dihedral sample (7).

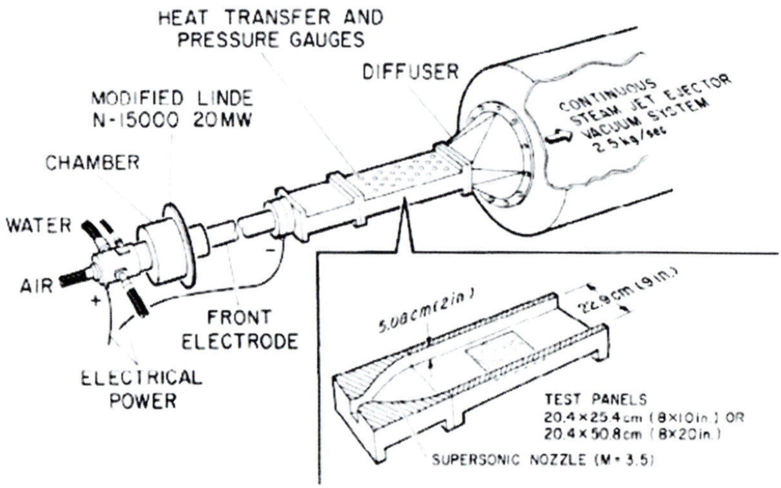

Fig. 12.6 Arcjet duct experiment; flow characterized by static pressure and wall heat flux at different points (17).

present at certain times of the year in the Martian atmosphere. Note that the particles reach a moderate speed (1000 to 2000 $m \cdot s^{-1}$), which poses a problem of representativeness of the impact (see Chapter 11). In addition, the development speed of these particles is incomplete: their lack of speed compared to gas is over 100 $m \cdot s^{-1}$ [5]. In the laboratory frame, the wake of the particle precedes it. This phenomenon may disturb the flow in terms of turbulence.

12.2.5 Measurements

In many cases, one replaces the model with an artifact, generally in cooled copper, equipped with different sensors (pressure port, flowmeter, etc.). This, of course, assumes a good reproducibility of the test or at least an opportunity to readjust the quantities measured by changes in the operating point of the generator.

The most common measures are:

- Profiles using cameras (Fig. 12.7).
- Temperature by pyrometer or infrared (IR) camera. It should be noted that operating such a quantitative image requires knowledge of the thermo-optical properties of the material temperature (sometimes at temperatures for which there is no calibration possibility) and at high incidence from the surface of the material (constraint of observation). Figure 12.8 gives an example of an IR image.

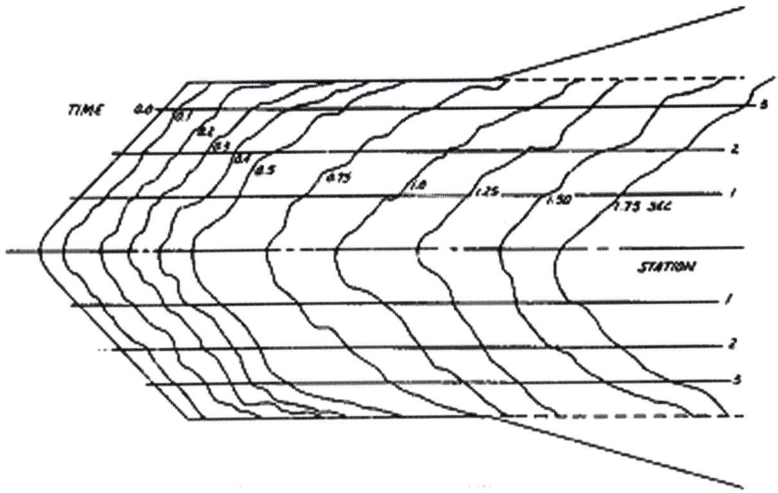

Fig. 12.7 The recession of a specimen in a plasma jet (18).

12.2.5.1 Ablation Asymmetry

Observed ablation asymmetries in flight test (shown previously in Fig. 7.20) are likely issued from asymmetry of laminar–turbulent transition: the various meridians become turbulent at different times. No numerical simulation of this phenomenon has been described. Plasma jet experiments allow it into evidence. Figure 12.9 shows ablation on various meridians. These measures clearly demonstrate the phenomenon. The representativeness of this type of test, however, is obliterated by the pollution of the flow by the turbulence of the jet. However, it is hoped that this effect is limited when the laminar–turbulent transition is dominated by the roughness (see Sec. 7.4).

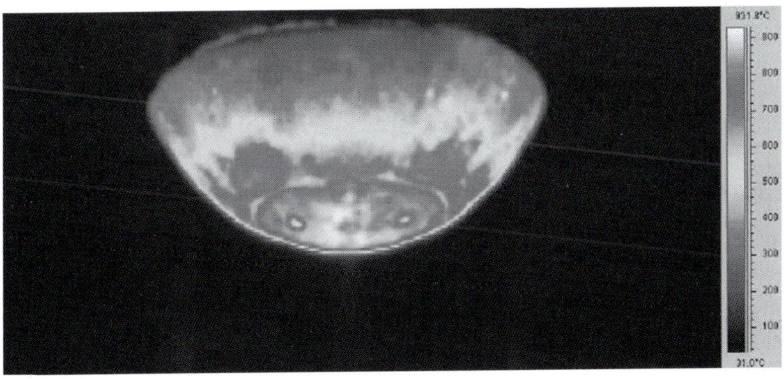

Fig. 12.8 Infrared image of a specimen in a plasma jet (19).

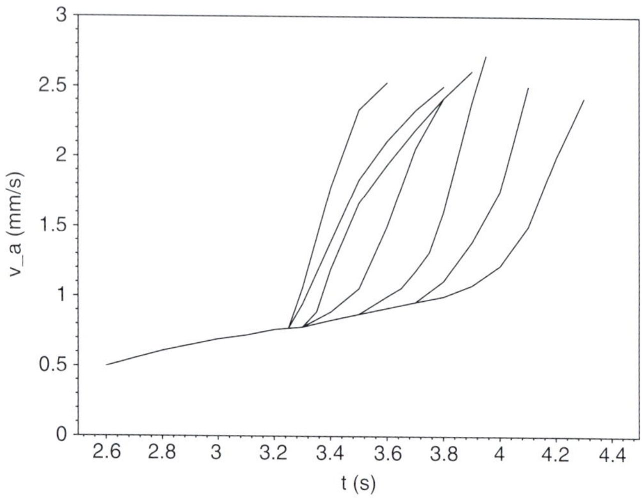

Fig. 12.9 Asymmetry of ablation in a plasma jet experiment, measured at mid-radius on various meridians (6).

12.3 Radiative Facilities

There are methods for delivering significant energy to surfaces in order to perform representative experiments: typically we can get some megawatts per square meter to a few tens of megawatts per square meter on surfaces of a few centimeters squared. These facilities are the arc image furnaces and CO_2 gas lasers. The former have a continuous spectrum. The lasers emit in the vicinity of 9.4 or 10.6 μm in a spectral range not absorbed by water vapor in air. These resources have also been used in ways similar to plasma jets for backup heating of the sample in a cold wind tunnel or low enthalpy generator.

Of course, the use of such a facility requires knowledge of the thermo-optical properties of the sample surfaces; assessment is not always obvious at high temperature. We will look subsequently at an important use of this type of facility: the measurement of material thermal properties.

12.3.1 Methods for Identification of Conductivity

There are many methods for measuring conductivity. For applications of interest here, we distinguish quasi-static methods and dynamic methods. In the first case, the sample is heated to a temperature T, and then one makes a small perturbation of the temperature to obtain the conductivity. Various methods are possible, but we cannot go into detail because they are classics in thermal metrology. We may oppose dynamic measurements that seem particularly suited to varying environments (e.g., during pyrolysis). In the latter case, the experimental setup consists of a sample equipped with thermocouples whose positions are measured by radiography. In general, one tries to place them in a 1-D configuration to simplify, as much as possible, the process of interpretation. Because of ignorance of the thermo-optical properties of the surface, it is difficult to take it as a boundary condition. We prefer to use the thermocouple closest to the hot wall as a "pilot," that is, to impose its measure as a condition of a Dirichlet problem. Note that this has the effect of restricting the temperature range in which the result is identified.

The inverse thermal problem is to calculate the conductivity from the knowledge of temperatures at various points; this is an ill-posed problem in the sense of Hadamard. This means that a small variation in the measurement (e.g., experimental error) causes a sharp variation in the result [20]. This problem is overcome by adding a regularity constraint on the solution; for example, imposing a local maximum on the derivatives of $\lambda(T)$.

Two types of methods are used to solve the problem:

- Minimize in the least squares sense, with given constraints, the difference between the measures $T_{mes_{ij}}$ taken at time i on thermocouple j. Solving the problem will go through the calculation of partial derivatives $\partial T_{ij}/\partial \lambda_i$ (temperature sensitivity to the conductivity). This calculation can be done by finite differences or more sophisticated methods using direct calculation of gradients of the function.
- One can also notice that the discretized 1-D heat equation can be treated by a filtering of Kalman type. This method is indeed highly effective in this type of problem, the regularization solution being naturally performed [21].

We can go a step further by posing as new unknowns the thermocouple positions and identify (small) differences between the a priori measured value and the most likely value that minimizes the gap between calculation and measurements.

12.3.2 Static Characterization Example

We take as an example a well-documented test on carbon–phenolic resin, TWCP [22]. The virgin material has a density of 1368 $kg \cdot m^{-3}$ and a porosity close to zero. The pyrolyzed material at very high temperature (either in oven at 2500 K, or at the plasma jet with a surface temperature of about 2000 K) has a density of 1184 $kg \cdot m^{-3}$ and a total porosity of about 20%. Note that, given the structural transformations of the material beyond 1600 K (Sec. 8.2.1.2) not taken into account in models, the pyrolysis temperatures used are too high.

Figure 12.10 shows quasi-static measurements made on the virgin material in the main axes. Note the decrease of conductivity after 500 K, which, statically, is the beginning of pyrolysis for low temperature increase. This poses a problem in the standard models, which use a mixing law for the material during pyrolysis. The conductivity of the pyrolyzed material being stronger

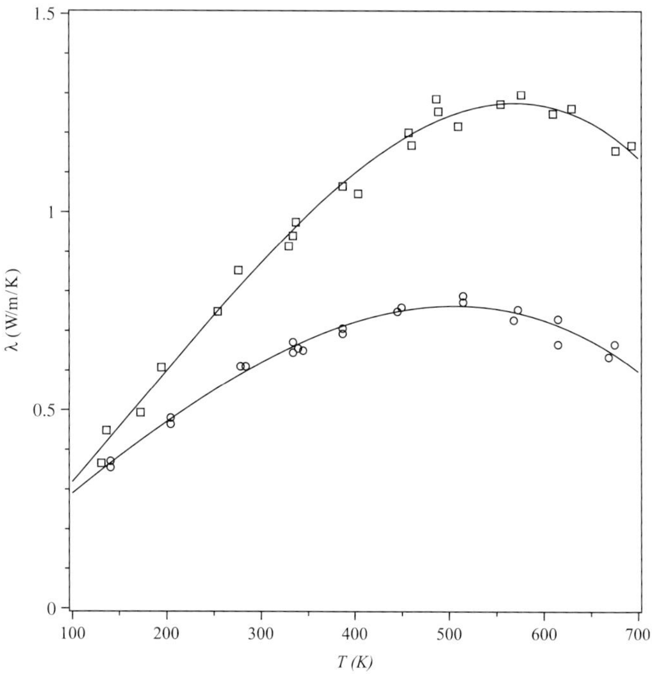

Fig. 12.10 Conductivity of a carbon–resin virgin material [22].

than that of virgin (Fig. 12.11), such a law leads to an increase in calculated conductivity, which is in contradiction with observation. (In Fig. 12.11, note the arbitrary curve corresponding to the cross symbol is an estimate with radiation, the true conductivity of the solid varying slowly with temperature.)

Measurements on the pyrolyzed material show differences with the temperature used for pyrolysis. As can be predicted, the material prepared in the oven is more conductive, because of the higher graphitization. This is confirmed by measurements on the same samples by gradually lowering the temperature: the values (for both types of packaging) are close to the measurements on material issued from the oven. This confirms the assertion from the beginning, that it is better to precondition the material at a low enough temperature.

One can add a comment on the curves of Fig. 12.11. We might expect a decrease of conductivity at high temperatures, such as is found with polycrystalline graphite. However, the high porosity leaves room for heat transfer by radiation, the apparent conductivity varying in T^3 (Sec. 10.4). This effect is also very strong on materials with very low density, such as PICA [24]. Measurements on variants of more or less dense material show:

- An increase in conductivity with density at low temperature, related to the conduction in the solid

- A decrease in conductivity at high temperature (from $\simeq$ 1000 K), due to the decrease in porosity and thus the effect of radiation

12.3.3 Dynamic Characterization Example

Various references [25, 26] give detailed calculations on a TWCP-type material, but only with samples that have undergone pyrolysis at moderate temperature, about 1500 K. For standard models ignoring the structural changes at high temperature ($T > 1600$ K), it is desirable to subject the sample to a representative temperature rise. In the example, the tests are performed on a furnace using an arc lamp image and delivering 8 $MW \cdot m^{-2}$. It is not desirable to use surface temperature measurements due to ignorance of thermo-optic properties; therefore, the hottest thermocouple serves as a reference. Its maximum temperature reached 2100 K and the second-highest was 1500 K.

Detailed analysis highlights some interesting points:

- The sensitivity analysis $\partial T_{ij}/\partial \lambda_i$ shows that under these conditions, we cannot identify the conductivity beyond approximately 1600 K

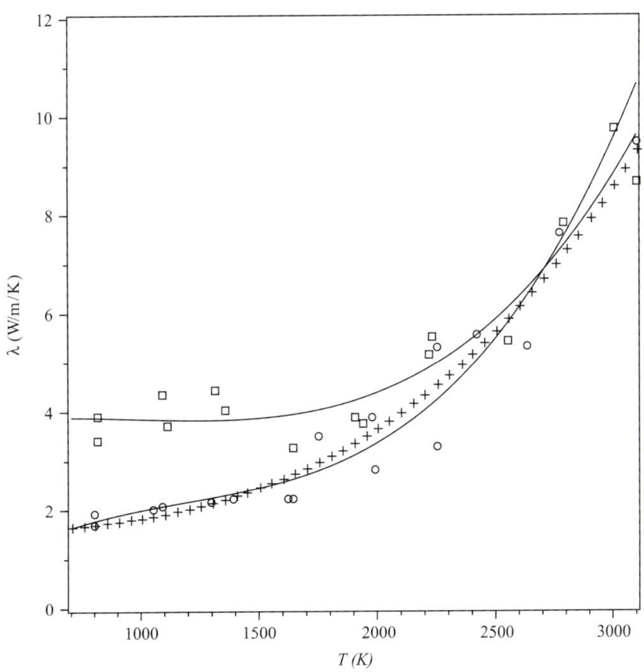

Fig. 12.11 Conductivity of a pyrolyzed carbon–resin (quasi-static measurements along the principal axes) (22, 23).

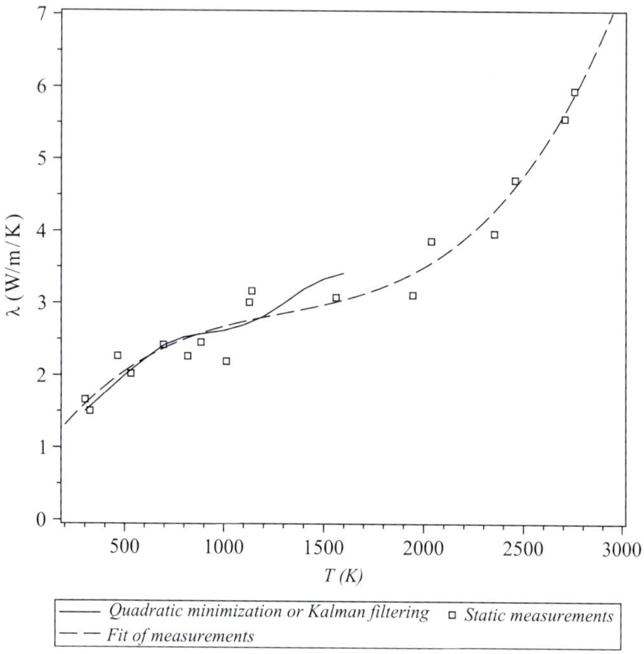

Fig. 12.12 Conductivity of a pyrolyzed carbon-resin (material tilted 20 deg) (21, 26).

(Fig. 12.12; the results obtained by both numerical methods are almost the same).

- Increasing the number of thermocouples did not improve the result.
- It is possible to identify a positioning error at each sensor; this induces a marked improvement on residues: the differences between measurement and calculation shows a decrease from 27 K to 4 K. But modeling and experimental errors permit us to think that this may be an ad hoc result. The method artificially compensates for various errors.

The results obtained by this method are consistent with those from a static characterization (Fig. 12.12).

12.3.4 Averaged Properties

The measurements on the virgin material being made in quasi-static conditions, they cannot go beyond the beginning of pyrolysis in the same conditions, about 700 K. Pyrolysis with a high rate of rise in temperature requires the extrapolation of these values. In the absence of knowledge about the composition and state of division of the material during its thermal degradation, different assumptions are used to calculate the conductivity (Fig. 12.13):

- Both parts, virgin and pyrolyzed, are in parallel. This gives an arithmetic average that is a higher limit

$$\lambda = \zeta\lambda_v + (1 - \zeta)\lambda_p \tag{12.5}$$

ζ is the advancement rate defined by Eq. (8.31).

- Both parts are in series, leading to a harmonic mean that is a lower limit

$$\lambda = \left[\frac{\zeta}{\lambda_v} + \frac{1-\zeta}{\lambda_p}\right]^{-1} \tag{12.6}$$

- Both parties are randomly mixed (2-D case), leading to a geometric mean

$$\lambda = \lambda_v^{\zeta}\,\lambda_p^{1-\zeta} \tag{12.7}$$

Many authors prefer a different approach to the problem, which consists of choosing one of these laws a priori and interpreting a dynamic experience

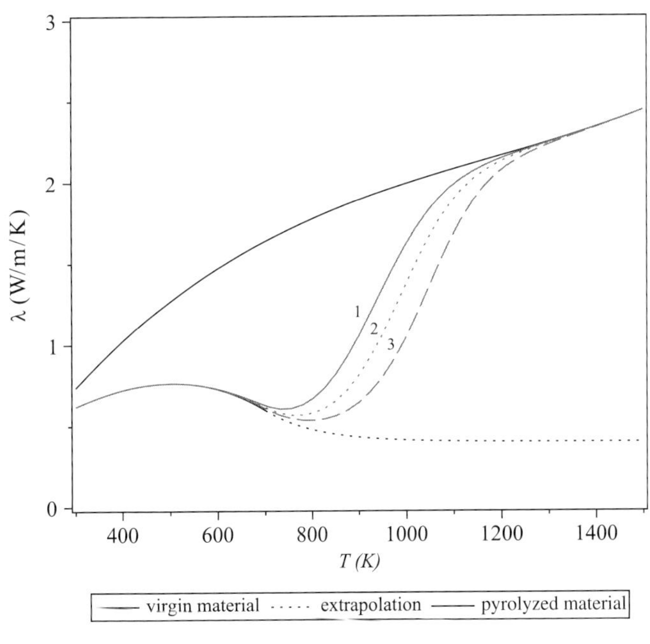

Fig. 12.13 Average conductivity calculated from the data of Figs. 12.10 and 12.11: (1) arithmetic mean, (2) geometric mean, and (3) harmonic mean. Pyrolysis test at 20 K · s^{-1}.

considered representative. This approach generally leads to gratifying but somewhat illusory results in which the apparent conductivity obtained contains all the modeling errors of pyrolysis.

12.4 Ablation Measurements

The following sections describe a number of measurements of ablation or "front" of pyrolysis. The list of sensors described is not complete, and we found measurements based on both similar principles [27] and different ones, such as capacitive measurement for a dielectric material like PTFE [28].

12.4.1 Estimating from Temperature Measurements

The reverse thermal calculation, as was used in Sec. 12.3.3, can estimate the temperature anywhere, at any time. We can also apply this method on ablated material by giving the temperature at any point of a hypothetic nonablative material. The problem is thus reduced to a free boundary problem coupled by means of the embedded surface temperature $T_W = f(\mathbf{x}, t)$, which is a known function. The surface $\mathbf{x}$ is an unknown in the problem. The solution to this problem is more or less complex depending on the methods of resolution used in the fluid medium. An example is given in [20]. The accuracy of the restitution of the surface is strongly related to the temperature gradient in the vicinity thereof, thus the conductivity of the material and the conduction heatflux.

12.4.2 Ultrasonic Measurements

The ultrasonic measurement of ablation is possible on a nonpyrolyzable material. The accuracy is affected by the significant opening of the beam emitted by a sensor of small size and by dispersion of velocity with frequency. However, it works well on cold material, with a signal processing that could be considered simple [29]. The problem is complicated at high temperatures, where measurements of transmission properties in the material are difficult.

In pyrolyzable materials, ultrasonic signal reflection occurs preferentially on the "front" of pyrolysis. It does not have a clear definition, the gradient being dependent on the flow of conduction (see Sec. 8.2.1.3).

12.4.3 Gamma Radiation Measurements

An early method of measuring ablation was by sensing radiation from a radio source; this was at a time when standards in the handling of sources were fairly liberal. This type of measurement can be separated into two methods:

- In the first method, intrusive, the source is included in the material as an insert. Each source (or set of sources with an insert) is subject to a detector through a collimator [30]. In the case of metals, a possible (even probable) liquid oxide is formed during ablation, which will migrate downstream.
- The second method, nonintrusive, uses the Thomson backscatter from a source located near the detector [31]. This method has the disadvantage of requiring more active sources, and thus a protection system (detector and assemblers) is very important.

The measurement error due to statistical fluctuation of the signal, which in any case is low level, is on the order of several percent.

12.4.4 Optical Measurements

Figure 12.14 shows the principle of a device capable of measuring temperature and detecting the ablation: the COTA (temperature and ablation optical sensor) [32]. The principle is simple: a coated cavity (in this case of a glassy carbon) allows the measurement of temperature downhole with good accuracy. Indeed, the hot parts that are not facing the detector send it very little energy (emission at grazing incidence). The cladding has the added advantage of protecting the measurement from gases produced by ablation or pyrolysis. The clearing of the cavity is accompanied by a discontinuity on the signal.

This device is intrinsically intrusive, but simulations of the local perturbation show that it is low, even on a silica-type material resin. For the latter, which has a large difference in conductivity with the sheath, it was shown that a set of sensors allowed the ablation speed to return to better than 20% [32].

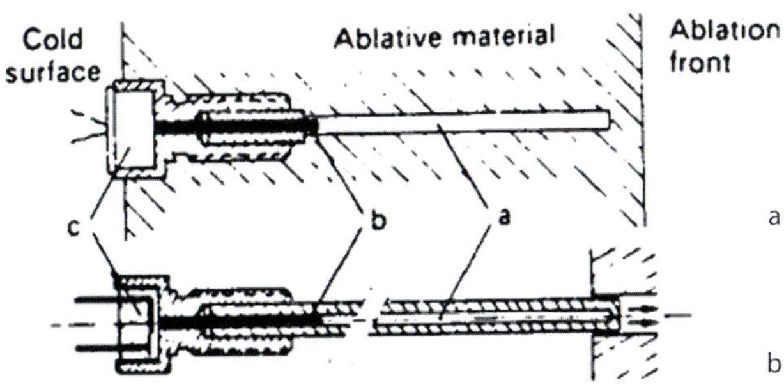

Fig. 12.14 Temperature and ablation optical sensor (32).

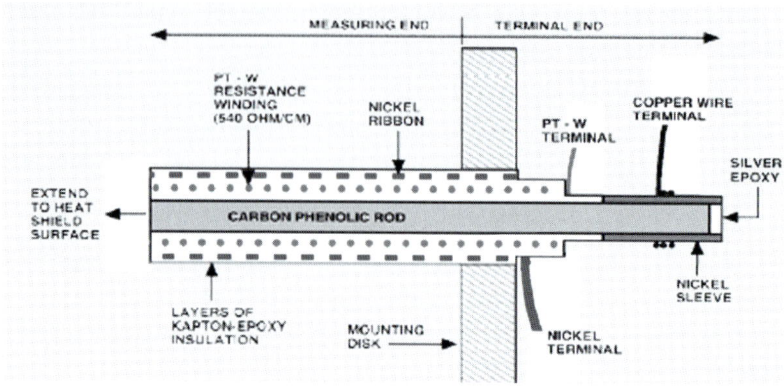

Fig. 12.15 ARAD sensor (33).

12.4.5 Electrical Conductivity Measurements

The Analog Resistance Ablation Detector (ARAD) sensor [33] measures the electrical resistance of a bar of material inserted into a wall (Fig. 12.15). This device makes it possible to access an estimate of the "front" of pyrolysis, with all that this notion implies, discussed in Sec. 12.4.2. This type of sensor therefore requires careful calibration in a plasma jet.

12.4.6 Benchmark Experiments

The development of approximate methods often requires having a reference experiment for calibration codes. These are difficult to obtain and are therefore not very numerous. The following sections contain some of them, although we do not claim that this list is exhaustive.

12.4.6.1 Plasma Jet Experiments

There are some tests on graphite or TWCP that can serve as a reference:

- Tests on a spherocylindrical specimen 12.7 mm in diameter made of various graphites, performed at various facilities, used pressures ranging from 0.035 atm to 15 atm and enthalpies from 2 $\mathrm{MJ \cdot kg^{-1}}$ to 35 $\mathrm{MJ \cdot kg^{-1}}$ [34]. (The high enthalpies are associated with low pressures.) Most of these tests correspond to laminar flow. Ablation is very important and induces significant changes in the geometry during the test. We will retain only the tests at low pressure (0.035 to 0.6 atm) and high enthalpy that give consistent results on various materials. This will avoid idle discussions about the possible erosion of specimen of thermomechanical origin.
- Tests on a spheroconical specimen ($R = 19.1$ mm, $\theta = 10$ deg) were performed at the IHF facility at a pressure of 0.75 atm and enthalpy of

27 $MJ \cdot kg^{-1}$ [35]. Under these conditions, ablation is slow and physical analysis possible for a fairly well-defined geometry.
- Tests were performed on TWCP material specimens with a spherocylindrical diameter of 25.4 mm [36]. Various ply orientations were studied. The tests were carried out in nitrogen or air atmosphere at pressures of 0.07 to 11 atm and enthalpies from 2.6 to 25.5 $MJ \cdot kg^{-1}$. As in the graphite trials, it seems preferable to select for comparison with the calculation the tests made at low pressure and high enthalpy for abstract problems of spallation, real in some of these experiences.

Note that it is not always necessary to have full test results to learn from them (see Sec. 5.10).

12.4.6.2 Flight Tests

A number of flight tests were used to provide specific experiences of reference for understanding the flow properties and wall quantities (Table 12.2).

Few baseline tests are available regarding the measurements on ablative thermal protection materials. Regarding space probes and excluding noninstructive surveillance measurements (Pioneer, Viking), only two experiments are noteworthy:

- Pathfinder, equipped with nine thermocouples in the SLA thermal protection
- Galileo, equipped with ARA sensors in the TWCP material

However, there are a number of experiments from which we can extract valuable data, especially Apollo 4 and Stardust, which were the subject of

Table 12.2 Reference Flight Tests (37–43)

Mission	Objectives	Instrumentation	Wall
RAM C-III	Density and electronic temperature Blackout	Langmuir probes (rack) Wall electrostatic probes Reflectometer S band	Be/PTFE
Re-entry F	Wall heatfluxes Laminar–turbulent transition Pressures (base)	Thermocouples, static pressure probes	Graphite/Be
PAET	Radiation	Radiometers	Be
Fire II	Wall heatfluxes Radiation	Calorimeters Radiometers 0.3–0.6 μm band	Be
Apollo 4	Radiation	Radiometers	Avcoat
Pacemaker	Ablation	Thermocouples Ablatometers	TWCP

postmortem analysis (Fig. 8.12). Also found in the literature are the results of experiments with recovery on materials such as PTFE [44], phenolic nylon [45], Avcoat [46], and TWCP [47].

References

[1] Dubroca, B., Duffa, G., and Leroy, B., "High Temperature Mass and Heat Transfer Fluid-Solid Coupling," AIAA Paper 2002–5180, *11th AIAA/AAAF Meeting, "Space Planes and Hypersonic Systems and Technologies,"* Orléans, France, Sept. 2002.

[2] Gallais, P., *Atmospheric Re-Entry Vehicle Mechanics*, Springer-Verlag, Berlin, 2007.

[3] Crawford, R., "Aerodynamic and Aerothermal Facilities. II: Continuous Flow High Enthalpy Facilities," in *Methodology of Hypersonic Testing*, Von Karman Lecture Series 1993–03, VKI, Rhode Saint Genese, Belgium, Feb. 1993.

[4] De Filippis, F., Del Vecchio, A., Martucci, A., Trifoni, E., Marraffa, L., Savino, R., and Paterna, D., "70 MW Plasma Wind Tunnel Upgrades for ESA Aurora TPS Testing," *Fourth International Planetary Probe Workshop*, Pasadena, CA, June 2006.

[5] Matthews, R. K., "Materials/Structures Testing," in *Methodology of Hypersonic Testing*, Von Karman Lecture Series 1993–03, VKI, Rhode Saint Genese, Belgium, Feb. 1993.

[6] Duffa, G., "Ablation," CESTA monograph, Commissariat a l'Energie Atomique, Le Barp, France, Nov. 1996.

[7] Smith, D. M., and Younker, T., "Comparative Ablation Testing of Carbon Phenolic TPS Materials in AEDC-H1 Arcjet," AIAA Paper 2005-3263, *AIAA/CIRA 13th International Space Planes and Hypersonics Systems and Technologies, Capua*, Italy, May 2005.

[8] Winovich, N., "On the Equilibrium Sonic-Flow Method for Evaluating Electric-Arc Air-Heater Performance," NASA TN-D-2132, March 1964.

[9] Nicolet, W. E., Shepard, C. E., Clark, K. J., Balakrishnan, A., Kesselring, J. P., Suchsland, K. E., and Reese, J. J., Jr., "Analytical and Design Study for a High-Pressure, High-Enthalpy Constricted Arc Heater," AEDC TR-75-47, July 1975.

[10] Park, C., Raiche, G. A. II, Driver, D. M., Olejniczak, J., Terrazas-Salinas, I., Hightower, T. M., and Sakai, T., "Comparison of Enthalpy Determination Methods for Arc-Jet Facility," *Journal of Thermophysics and Heat Transfer*, Vol. 20, No. 4, 2006, pp. 672–679.

[11] Wassel, A. T., and Courtney, J. F., "Environmental Characteristics and Experimental Limitations of Arc-Heated Reentry Test Facilities," Spectron Development Laboratories Report 76–6048, June 1976.

[12] Purpura, C., De Filippis, F., Barrera, P., and Mandanici, D., "Experimental Characterisation of the CIRA Plasma Wind Tunnel SCIROCCO Test Section," *Acta Astronautica*, Vol. 62, No. 6–7, 2008, pp. 410–421.

[13] Smith, R. T., Mac Dermott, W. N., and Giltinan, T. L., "A Transient Enthalpy Probe for the Calibration of High Heat Flux Ablation Facilities," AEDC TR-74-116, Jan. 1975.

[14] Sakai, T., "Computational Simulation of High-Enthalpy Arc Heater Flows," *Journal of Thermophysics and Heat Transfer*, Vol. 21, No. 1, 2007, pp. 77–85.

[15] Desai, P. M., and Qualls, G. D., "Stardust Entry Reconstruction," *Journal of Spacecraft and Rockets*, Vol. 47, No. 5, 2010, pp. 736–740.

[16] Scaggs, N. E., and Stetson, K. F., "Experimental Studies of the AFFDL Flared-Nozzle Ablation Simulation Technique," AIAA Paper 78-776, 1978.

[17] Sepka, S. A., Kornienko, R. S., and Radbourne, C. A., "Testing of SLA-561V in NASA Ames Turbulent Flow Duct with Augmented Radiative Heating," 10th AIAA/ASME Joint Thermophysics and Heat Transfer Conference, June 2010.

[18] DiChristina, V., Howey, D., Schmidt, H., and Ziering, M., "Thermomechanical Erosion of Ablative Plastic Composites," Technical Report AFML-TR-71-153, June 1971.

[19] Cardone, G., Tortora, G., and del Vecchio, A., "IR Thermographic Measurements of Temperatures and Heat Fluxes in Hypersonic Plasma Flow," *Proceedings of the 5th European Symposium on Aerothermodynamics for Space Vehicles* (ESA SP-563), edited by Danesy, D., ESA, Noordwijk, The Netherlands, Nov. 2004.

[20] Guilpart, B., Duffa, G., and Legendre, J., "Problèmes inverses en thermique," *CHOCS*, No. 19, Sept. 1998, pp. 11–23.

[21] Lambelin, J.-P., Lasserre, J.-P., and Ducamp, V., "Identification of High Temperature Pyrolysed Carbon Phenolic Composite Conductivity," *Proceedings of Eurotherm Seminar 68 "Inverse Problems and Experimental Design in Thermal and Mechanical Engineering,"* edited by ENSMA/LET, Poitiers, France, March 2001, pp. 359–366.

[22] Engelke, W. T., Pyron, C. M., Jr., and Pears, C. D., "Thermal and Mechanical Properties of a Nondegraded and Thermally Degraded Phenolic Carbon Composite," NASA Report CR-896, Oct. 1967.

[23] Pradere, C., Batsale, J.-C., Goyhénèche, J.-M., Pailler, R., and Dilhaire, S., "Thermal Properties of Carbon Fibers at Very High Temperature," *Carbon*, Vol. 47, No. 3, 2009, pp. 737–743.

[24] Tran, H. K., Johnson, C. E., Rasky, D. J., Hui, F. C. L., Hsu, M.-T., Chen, T., Chen, Y. K., Paragas, D., and Koyabachi, L., "Phenolic Impregnated Carbon Ablators (PICA) as Thermal Protection Systems for Discovery Missions," NASA Technical Memorandum 110440, April 1997.

[25] Ducamp, V., "Transferts thermiques dans un matériau composite carbone-résine," Ph.D. thesis, University of Bordeaux, France, 1, No. 2505, April 2002.

[26] Lasserre, J.-P., Epherre, J.-F., and Ducamp, V., "Identification et modélisation de la conductivité thermique à haute température d'un composite carbone résine pyrolysè," Congrès SFT, Nantes, France, May 2001.

[27] LeBel, P. J., and Russel III, J. M., "Development of Sensors to Obtain In-Flight Ablation Measurements of Thermal Protection Materials," NASA Report TN D-3686, Nov. 1966.

[28] Noffz, G. K., and Bowman, M. P., "Design and Laboratory Validation of a Capacitive Sensor for Measuring the Recession of a Thin-Layered Ablator," NASA Technical Memorandum 4777, Nov. 1996.

[29] McGunigle, R. D., and Jennings, M., "Ultrasonic Ablation Recession Measurement System," *Advances in Test Measurement*, edited by Instrument Society of America, Vol. 12, Proceedings of the Twenty-First International Instrumentation Symposium, Philadelphia, Pennsylvania, May 1975, pp. 19–24.

[30] Bunker, S. N., and Armini, A. J., *"Many Ray Nosetip Ablation Sensor Design,"* 23rd International Instrumentation Symposium, Las Vegas, Nevada, May 1977.

[31] Droms, C. R., Langdon, W. R., Robison, A. G., and Entine, G., "Heat Shield Ablation Sensor Utilizing CdTe Gamma Detectors," *IEEE Transactions on Nuclear Science*, Vol. 23, No. 1, 1976, pp. 498–500.

[32] Cassaing, J.-J., Balageas, D. L., Deom, A. A., and Lestel, J.-C., "Ablation and Temperature Sensors for Flight Measurements in Reentry Bodies Heat Shields," First European Symposium on Aerothermodynamics for Space Vehicles, ESTEC, Noordwijk, The Netherlands, June 1991.

[33] Martinez, E., Oishi, T., Centner, R., Gorbonov, S., and Venkatapathy, E., "In-Space Propulsion Program. Advanced Sensor Project. Current Developments in Future Planetary Probe Sensors: Update 2004," http://thermo-physics.arc.nasa.gov/fact_sheets/PWS_Sensors.pdf [retrieved 4 June 2008].

[34] Maahs, H. G., "Ablation Performance of Glasslike Carbons, Pyrolytic Graphites, and Artificial Graphite in the Stagnation Pressure Range 0.035 to 15 Atmospheres.," NASA Technical Note TN D-7005, Dec. 1970.

[35] Chen, Y.-K., Milos, F. S., Reda, D. C., and Stewart, D. A., "Graphite Ablation and Thermal Response Simulation Under Arc-Jet Flow Conditions," AIAA Paper 2003-4042, 36th AIAA Thermophysics Conference, Orlando, Florida, June 2003.
[36] Sutton, K., "An Experimental Study of a Carbon-Phenolic Ablation Material," NASA Technical Report D-5930, Sept. 1970.
[37] Cauchon, D. L., "Radiative Heating Results from the Fire II Flight Experiment at a Reentry Velocity of 11.4 Kilometers per Second," NASA Report TM X-1402, July 1967.
[38] Davies, C., and Arcadi, M., "Planetary Mission Entry Vehicles. Quick Reference Guide. Version 3.0," NASA SP-2006-3401, 2006.
[39] Dillon, J. S., and Carter, H. S., "Analysis of Base Pressure and Base Heating on a 5-Half-Angle Cone in Free Flight Near Mach 20 (Reentry F)," NASA Technical Report TM X-2468, Jan. 1972.
[40] Dunn, G. J., and Eagar, T. W., "Calculation of Electrical and Thermal Conductivities of Metallurgical Plasmas," *Welding Research Council (WRC) Bulletin*, No. 357, Sept. 1990.
[41] Ried, R. C., Jr., Rochelle, W. C., and Milhoan, J. D., "Radiative Heating to the Apollo Command Module: Engineering Prediction and Flight Measurement," NASA Technical Report TM X-58091, April 1972.
[42] Walton, T. E., Jr., and Witte, W. G., "Flight Test of Carbon-Phenolic on a Spacecraft Launched by the Pacemaker Vehicle System," NASA Technical Memorandum TM X-2504, March 1972.
[43] Wright, M. J., Bose, D., Palmer, G. E., and Levin, E., "Recommended Collision Integrals for Transport Properties Computations, Part 1: Air Species," *AIAA Journal*, Vol. 5, No. 4, 1962, pp. 380–386.
[44] Winters, C. W., Witte, W. G., Rashis, B., and Hopko, N., "A Free-Flight Investigation of Ablation of a Blunt Body to a Mach Number of 13.1," NASA Report TN D-2354, July 1954.
[45] Witte, W. G., "Flight Test of High-Density Phenolic-Nylon on a Spacecraft Launched by the Pacemaker Vehicle System," NASA Technical Memorandum TM X-1910, Nov. 1969.
[46] Graves, R. A., Jr., and Witte, W. G., "Flight-Test Analysis of Apollo Heat-Shield Material Using the Pacemaker Vehicle System," NASA Technical Note TN D-4713, Aug. 1968.
[47] Walton, T. E., Jr., and Witte, W. G., "Flight Test of Carbon-Phenolic on a Spacecraft Launched by the Pacemaker Vehicle System," NASA Report TM X-2504, March 1972.

Chapter 13 An Example: Apollo

13.1 Thermal Protection Design Requirements

The Apollo program is an excellent example of a system designed for a super-orbital reentry. Its importance in the field covered in this book is great because the reentry of the Command Module encapsulates the variety of difficulties—ablation and pyrolysis in high enthalpic flow with radiation problems—that have been described in this book, with all the accompanying issues in terms of modeling, material characterization, and ground or flight testing. As such, it has been the source of many theoretical, numerical, and experimental tools in our field. Furthermore, the Avcoat material used in the Apollo program, or a similar one, is being considered for the Orion Multi-Purpose Crew Vehicle (MPCV) [1].

The thermal protection material used in Apollo was designed with specifications given from numerical simulations of trajectories. The Command Module's lift-over-drag lies between 0.3 and 0.4 for an equilibrium trim angle of attack close to 30 deg, ensured by an offset of center of gravity. The reference reentry has an initial velocity of about $11\ \mathrm{km \cdot s^{-1}}$, corresponding to a return to lunar orbit [2]. Some examples of reentry conditions are given later in this chapter in Table 13.2. The aerodynamic characteristics derived from actual flight tests can be found in [3]. Codes for solving the Navier–Stokes equations were nonexistent at the time of conception; the concept of the boundary layer on an axisymmetric equivalent body was used to assess the exchanges in the boundary layer. These calculations are made from the current flow lines and Eulerian characteristics along them (see Fig. 13.1). This calculation is performed using approximate methods, usually based on a pressure distribution given by a Newtonian method or by wind tunnel or flight test results. This method neglects the transverse component of velocity in the boundary layer. It therefore implies a low divergence of the streamlines on the body surface and is valid only on nondetached flow. The equations obtained are those of the boundary layer with modification of the metric, which is deduced from the pressure by approximate solution of the Euler equations on the surface [4].

These assessments help to give a maximum value of stagnation point heat flux of about $5.3\ \mathrm{MW \cdot m^{-2}}$, with $1.7\ \mathrm{MW \cdot m^{-2}}$ for the radiative part. Part of both modes of heating varies with location and time [5].

The sizing of the thermal protection is based on simple methods like correlation validated by detailed studies using a boundary layer code called

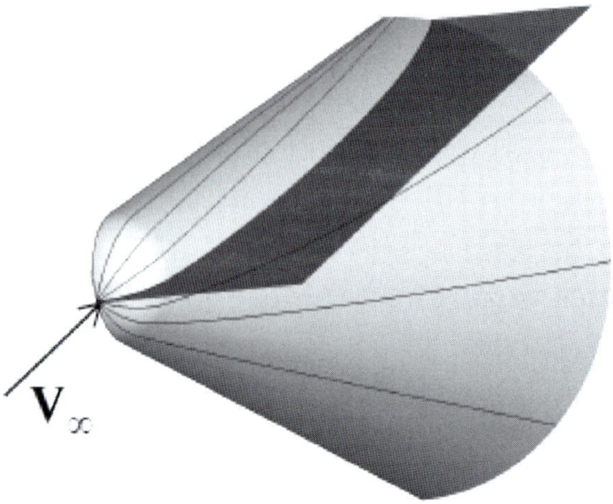

Fig. 13.1 Approximate streamlines on a sphere-cone (6).

BLIMP, whose direct use was not feasible, given the computers available when these studies were performed. This code takes into account a nonequilibrium chemical environment and an accurate description of diffusion by splitting (see Sec. 2.3.3). It is capable of taking into account complex wall conditions to simulate the effects of ablation and pyrolysis [4]. This possibility has helped the development of a correlation for the injection of the type described in Sec. 4.4.3.2.

This method has undergone numerous validations by comparison of heat flux on cold wall wind tunnel experiments.

Note that flight tests were conducted to better understand the aerothermal environment for such reentries, particularly radiation: FIRE I and II [7] and Apollo 4 [8]. This does not include testing of thermal protection, which is discussed later in this chapter.

In these studies, the laminar–turbulent transition is handled by a simple method using a momentum thickness Reynolds number (see Sec. 7.4), which is chosen between 200 and 250, depending on the author.

Moreover, a possible effect of roughness (see Sec. 7.1) of the ablated surface was considered, but this hypothesis has not really been developed [9].

13.2 Avcoat Material

The ablative material is deposited by spraying a composite material made with fiberglass and phenolic resin into a honeycomb carrier of centimetric size (Fig. 13.2). It is actually a family of materials that has evolved since the 5026-22 version of density $1060\ \mathrm{kg \cdot m^{-3}}$ up to the final version 5026-39 of density $530\ \mathrm{kg \cdot m^{-3}}$ by adding microballoons of phenolic resin to the

Fig. 13.2 Avcoat 5026-39 G material (1, 10).

initial composition made of epoxy resin and silica. The final material's mass composition thus consists of the following [4, 11]:

- 25% of a mixture of equal parts of silica fiber (about 6 mm in length) and milled fibers
- 30% of phenolic resin microballoons
- 45% of a polyepoxide novolac resin

The analysis of elemental composition [12] shows the presence of Al, Ca, and B, from which it is possible to calculate the approximate composition of the fiber (Table 13.1). These species are commonly used in silica fibers. Open porosity of virgin material is about 25%.

13.3 Pyrolysis and Gas Flow

13.3.1 Pyrolysis

The structure of the resin is as follows:

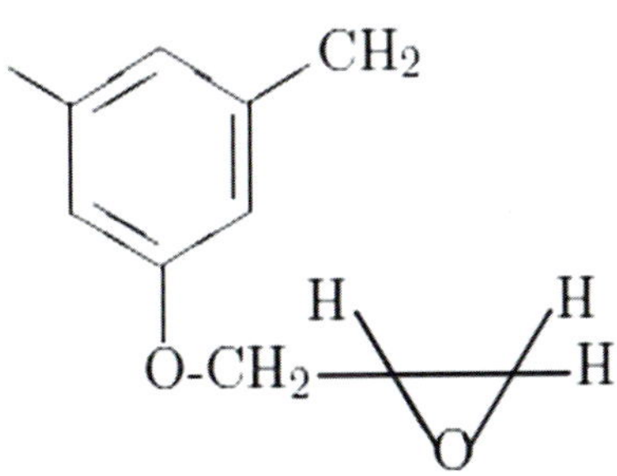

The polymer is a network similar to that of the phenolic resin (see Sec. 8.2), the hydroxyl group being replaced by a group having an epoxy functional group (oxirane). The density of this resin is around $1200\ \mathrm{kg \cdot m^{-3}}$. Its mass loss by pyrolysis is close to 70% [13].

Table 13.1 Composition of Silica Fiber of Avcoat Material

Species	Molar Percentage
Al_2O_3	8.7
CaO	10.1
B_2O_3	8.1

Pyrolysis of a material close to the final material ($\rho = 545\ \text{kg} \cdot \text{m}^{-3}$) was described using a simple law (Eq. 8.15). The reaction order is $n = 2.5$, the activation temperature $T_A = 13{,}600$ K, and the frequency factor $A = 54.9\ \text{s}^{-1} \cdot (\text{kg} \cdot \text{m}^{-3})^{-1.5}$. The pyrolyzed material has a density of $256\ \text{kg} \cdot \text{m}^{-3}$ [12]. It seems that no measurements of high heating rates have been done.

13.3.2 Carbon Deposition

The pyrolytic carbon deposition is evident on tomographies of material (see Fig. 8.12 and [4]). This was recognized very early due to the decomposition of pyrolysis gases. Studies leading to the identification of the hydrogen abstraction C_2H_2 addition (HACA) mechanism (see Sec. 8.2.1.6) were made later, and a more global mechanism was proposed. This is based on the overall reaction $CH_4 \rightarrow C(s) + H_2$ detailed by a Langmuir–Hinshelwood mechanism (Sec. 3.2.1) [14]. This model, leading to a deposit law of the type $\dot{m}_C = f(T, p_{CH_4}, p_{H_2})$ was readjusted on tests. It requires an assessment of the surface area of a medium assimilated to a network of cylindrical pores.

Of course, this model requires a detailed description of the pyrolysis gases, and therefore a heavy calculation. For this reason, an alternate approach was proposed, which assumed the species solid carbon (fiber and pyrolytic deposition) and gas in local thermodynamic equilibrium [15]. Knowing the elemental composition of pyrolysis gas, the solid carbon fraction (the carbon deposited) $\tilde{c}_{C_s}(T, p)$ can be calculated by a heterogeneous thermodynamic equilibrium calculation. This method is amended at low temperatures ($1250\ \text{K} < T < 1500\ \text{K}$) where one performs a simple linear interpolation of $\tilde{c}_C$ in the gas between the value at low temperature and that calculated at 1500 K. This method gives a reasonable estimate of the phenomenon such that it can be used for postflight measurements [15].

13.3.3 Gas Flow

Various levels of modeling (Chapter 8) have been implemented during the program, the most elaborate being a detailed description (multispecies medium, Darcy–Forchheimer's law, gaseous conduction included, analytical description of pyrolytic deposition) made in the CHarring Ablation with Diffusion (CHAD) code [16]. Such a code is of great interest for the study of physical phenomena. It contains a description by volume of the silica-to-carbon reaction.

However, the need for a less detailed description, therefore lighter in time calculation but also less needy in data, has led to the development of models where one assumes negligible permeation time. This type of model is used in the Charring Material thermal response and Ablation (CMA) code still used today.

13.4 Ablation

13.4.1 Plasma Jet Experiments and Experimental Ablation Laws

The material has undergone extensive plasma jet testing to analyze the phenomena in a wide pressure range (800 Pa to 10^5 Pa) and enthalpy (7 to 70 $MJ \cdot kg^{-1}$) [11, 17, 18]. The tests in the presence of air showed fused silica at the surface (Fig. 13.2). This liquid is present in very small amounts in the tests under nitrogen. It may be present in the form of drops on the surface. In the analyses made on samples after testing, the SiC is present in only very small amounts.

The first experiments in the program, which are more sketchy, helped produce an engineering design with the use of a simple model of ablation using an apparent enthalpy of ablation Q^* (see Sec. 4.6). This approach helps to link this material to other wrapped silica-phenolics materials (see Figs. 13.3 and 4.8). Laws of ablation $v_a = f(T_w)$ of the type described in Sec. 9.2.4.3 (Fig. 13.4) also were used.

13.4.2 Physical Models

13.4.2.1 Construction of a $B' = f(p, T_W, B'_g)$ Model

More physical studies were conducted with the Aerotherm Chemical Equilibrium (ACE) code [19] to describe the ablation using the law $B' = f(p, T_W, B'_g)$, as described in Sec. 8.7.1. The effect of the silica–resin honeycomb structure is assumed to be negligible, although the appearance of the ablated area (Fig. 13.2) does not render this hypothesis totally certain. In these attempts to develop a model, secondary species are problematic, especially Al_2O_3 because of its high temperature stability in an oxidizing atmosphere. Its presence is a sticking point for these models, where the presence of a liquid layer migrating to the surface under the effect of boundary layer shear (Sec. 9.1.1) is not taken into account. (The same is true for liquid boron oxide.) Note that we are interested here in the phenomena present at relatively low temperatures, for which the viscosity and therefore

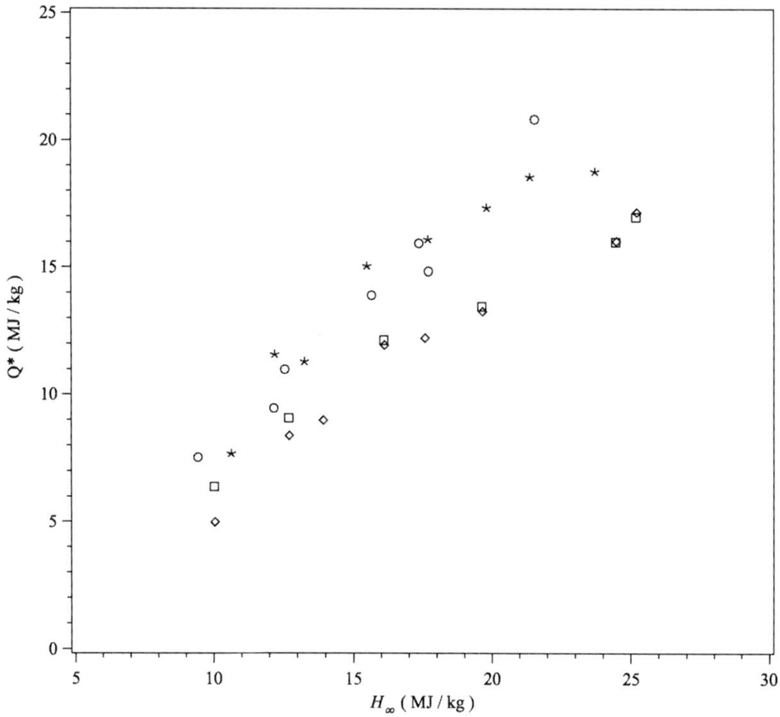

Fig. 13.3 Apparent ablation heat Q^* for Avcoat material (13).

the surface tension of silica (Fig. 9.2) are very high, making unlikely the model of a homogeneous layer developed in Chapter 9.

The model developed is highly dependent on B'_g at low temperature due to the possible presence of hydrocarbons burning at the surface. This aspect of the pyrolysis gas is treated in Sec. 8.2.1.5. At 1500 K the pyrolysis gases lead to surface species such as H_2O, CH_4, and other hydrocarbons, in a medium in chemical nonequilibrium.

The first models were developed with two assumptions: either the presence or absence of liquid silica, the latter being arbitrarily removed in the second hypothesis. This type of model leads to results where the surface consists of SiO_2, SiC, or C, depending on the temperature and the amount of pyrolysis gas. These curves show plateaus corresponding to reactions with oxygen in the boundary layer limited by diffusion. Indeed, part of the oxygen is contained in the silica; the availability of oxygen will differ depending on whether the majority of species formed are SiO, Si, or SiC. An example of the corresponding values are given in Sec. 5.7 and 8.7.1. Figure 13.5 provides an example of such a model showing only two of these plateaus. (The correction in the figure is intended for the cycle of silica on the surface [12].)

In the model just described, the liquid silica, when present, should be eliminated because of the modeling hypothesis. For this, the chemical cycle $(SiO_2)_l \rightarrow SiO \rightarrow SiO_2 \rightarrow (SiO_2)_l$ is treated as a chemical reaction described by the law $B'_{(SiO_2)_l} = p^{-\frac{1}{2}} e^{-\frac{E_a}{RT}}$ based on test results [17, 18].

Comparison of this model with plasma jet and flight tests showed a significant underestimation of ablation and an overestimation of the effect of B'_g. This defect was overcome by assuming that at least part of the pyrolysis gas was discharged through the cracks. This gas is assumed to have no effect on the boundary layer. This assumption is arbitrary, however, even if such cracks are visible on the micrographs of the tested materials.

Note that this model was modified later in an another program to describe more physically the formation of melted silica from SiO [20].

13.4.2.2 Direct Calculation

A simpler model was developed by another team [9, 14]. This model, briefly described in this section, is directly implemented in code without going through the step of establishing an ablation law, as in the previous section. This avoids the error inherent in this approach (Sec. 8.7.1), but the

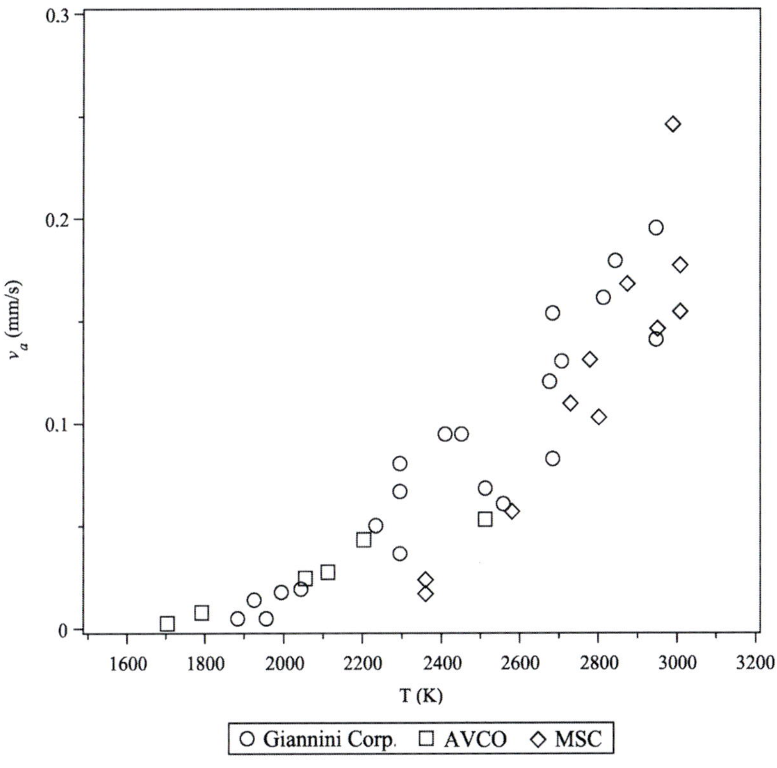

Fig. 13.4 Ablation law $v_a(T)$ for Avcoat material ($\epsilon \simeq 0.5$) (5).

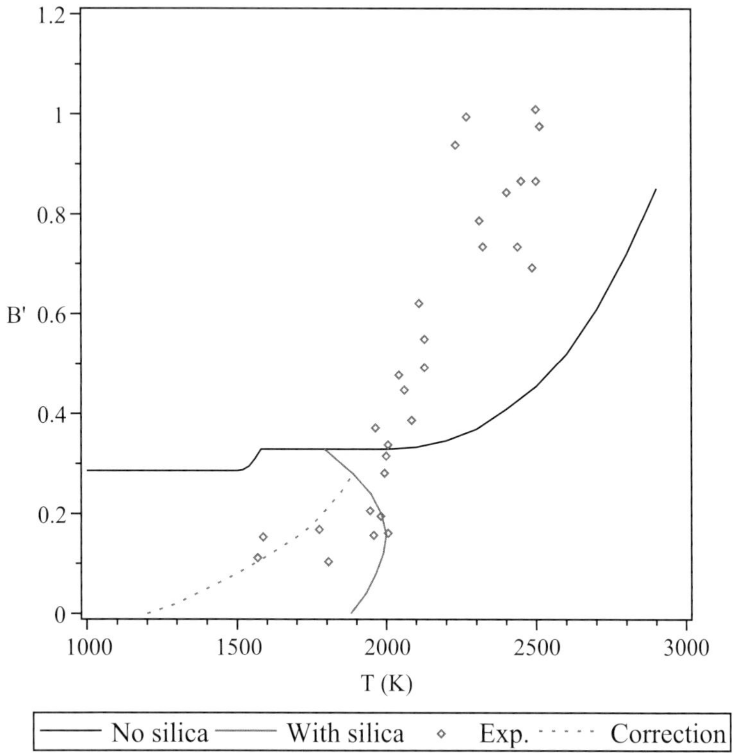

Fig. 13.5 Avcoat material: law $B'(T)$ for $p = 2800$ Pa and $B'_g = 0.6$ (12).

calculation is more time consuming. In this model one takes into account the following:

- Pyrolysis gas combustion.
- A given number of carbon–silica reactions with formation of SiC as an intermediate, aggregated by the reaction

$$(SiO_2)_s + 5C_s + 2O_2 \rightarrow SiO + 5CO$$

  described by an Arrhenius law and leading to a mass flow rate of ablation $\dot{m}_0$.
- The value of $\dot{m}_0$ is reduced to take into account the depletion of oxygen due to the combustion of the pyrolysis gas

$$\dot{m} = \dot{m}_0 - \Lambda\, \dot{m}_g \tag{13.1}$$

  Λ is deduced from experiments.
- Removal of liquid silica through a law giving the flow removed with the assumption of a homogeneous layer (see Sec. 9.1.2) whose coefficients are calibrated on the experience.

In this model there is no reformation of SiO_2 by oxidation of SiO, no sublimation of carbon, and the balance equation used to calculate the wall recession is stationary. The margin of error obtained by comparing with plasma jet test results is about 40% [21], a value quite commonly encountered in this type of problem (see, for example, Fig. 9.7).

13.4.3 Comparison with Flight Tests

13.4.3.1 Ablation

Tests have been made on test vehicles using small ballistic launchers [2, 22]. These tests, performed at high stagnation pressures, have yielded poor results (ablation significantly greater than what was expected) and were forgotten from the moment the test became available on the Apollo system.

Within the flight conditions, measured ablation is on the order of some millimeters and not very reproducible in view of its low value and the likely influence of the honeycomb support (Fig. 13.2). The part of physicochemical phenomena in the energy balance is low in this type of reentry. The re-radiated part of energy is overriding. Given the uncertainty of the emissivity, the diagnosis of an ablation model is difficult. The relative differences can be significant (Table 13.2). A contrario, this phenomenon plays only a minor role in the penetration of the isotherms, which is the dominant phenomenon for thermal protection sizing.

Heat fluxes are those of the station $s/R = 0.9875$ close to the stagnation point. The positions of the stations Y_c and Z_c defined in Fig. 13.6 are given in inches. The station 707 is located on the upwind conical part of vehicle. The calculated ablations are of the model described in Sec. 13.4.2.1 [4]. The representation is in the pitch plane, which is approximately a plane of symmetry.

13.4.3.2 Material Heating

The material was completely characterized in the laboratory [13, 23, 24]:

- Mass loss by pyrolysis at low heating rates
- Combustion enthalpy
- Specific heats and conductivities of samples pyrolyzed at different temperatures, up to 2200 K
- Permeability and Ward constant (measurements made with dry air)
- Spectral reflectance at low temperature

In addition, the properties of the pyrolysis gases have been calculated assuming a medium in local thermodynamic equilibrium.

The evaluation of conduction was done using various methods to generate the thermal conductivity (separating or not the virgin and pyrolyzed phases). These methods are usually verified by comparison with experiments

Table 13.2 Ablation on different Apollo flights; reentry conditions are given at 121.9 km

Flight	V ($m \cdot s^{-1}$)	Angle (deg)	$\dot{q}_{max}$ ($MW \cdot m^{-2}$)	Measurement Station	Measured Ablation (mm)	Calculated Ablation (mm)
AS202	8291	−3.53	0.94	$Z_C = 71,\ Y_C = 0$	2.3	1.5
				$Z_C = 71,\ Y_C = 33$	$\simeq 0$	0.4
AS501 Apollo 4	10,735	−6.93	4.83	$Z_C = 71,\ Y_C = 0$	5.0/6.2	5.1
				$Z_C \simeq -60,\ Y_C = 0$	$\simeq 0/0.5$	4.2
				707	$\simeq 0/1.3/2.5$	3.2
AS502 Apollo 6	9610	−5.85	2.38	$Z_C = 71,\ Y_C = 0$	2.5	4.5
				707	1.2/1.8	1.8

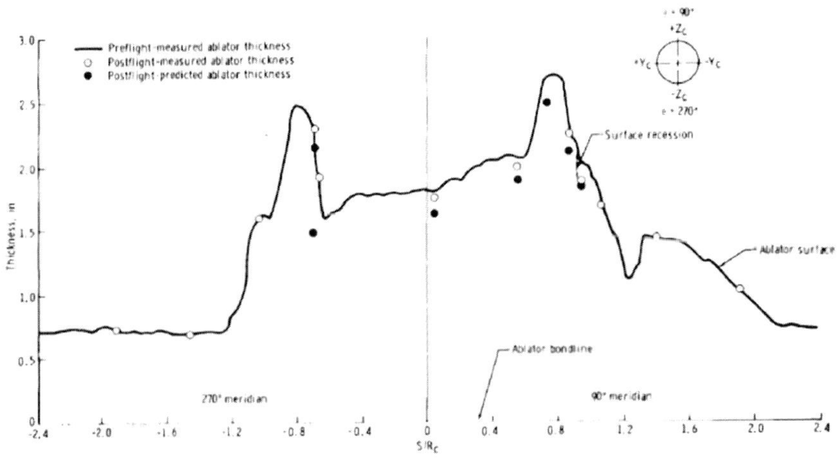

Fig. 13.6 Measurements of ablated thicknesses on Apollo 4 after reentry (5).

using as a boundary condition the thermocouple near the surface, and sometimes readjusted by identification (Sec. 12.3.1). The results (Fig. 13.7) show a rather satisfactory agreement with all measurements, thus validating the models describing the pyrolysis and heat transfer in the material. The representation in the figure is in the pitch plane, which is approximately a plane of symmetry.

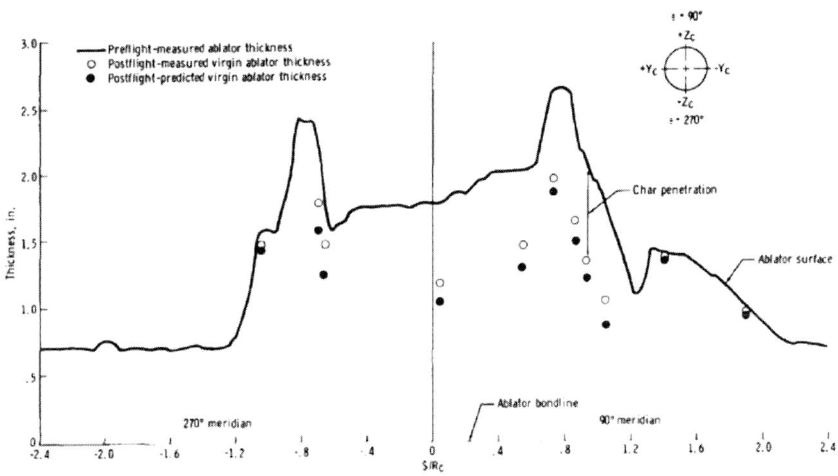

Fig. 13.7 Internal measurements of Apollo 4 thermal protection after reentry; region of pyrolysis initiation (5).

13.5 Radiation

13.5.1 Radiation in the External Flow

Radiation takes a part equivalent to convection, a few $MW \cdot m^{-2}$. The calculations made use approximate methods described in Sec. 10.2.2, with correction terms to take into account fluid–radiation coupling, which is relatively low in this case [25]. These methods are validated by comparison with existing codes, allowing more accurate calculations (RAD/EQUIL code [26]). This is, however, limited

- From an aerodynamic viewpoint: No possibility of considering a medium out of local thermodynamic equilibrium.
- On radiative transfer: The method uses an integration over the frequencies and solves a problem with a space cut into two quadrants.
- From a database viewpoint: The spectrum is treated in a very limited number of bands (two or three) and is incomplete in the ultraviolet region, a problem justified by the absence of experimental results.

All these calculation options would be made possible with the development of numerical methods in the decades after the Apollo program. Despite these limitations, the comparison with experiment is satisfactory in terms of total flux (difference with measurements $< 20\%$) and in shock tube or flight radiometer measurements: Fire I, Fire II [7], and Apollo 4 [4].

13.5.2 Emissivity

The emissivity retained for the material is 0.5[1] or 0.6, the latter value being deduced from simultaneous measurements of the flux emitted in a spectral band and across the whole spectrum [27]. But we saw in Chapter 9 that the liquid layer on the surface could significantly change the energy emitted by the wall, and that the apparent emissivity varies with test conditions (with the temperature gradient). This is a possible source of error because we have seen that this term plays a major role in energy balance (Sec. 8.7.3). In this case, the high surface tension of liquid can cause the presence of drops at the surface. This observation may explain, in part, the difference between the model and the testing regime in plasma jet experiments (Fig. 13.5). Moreover, we note that the emissivity of a cold material previously pyrolyzed is close to 0.95; the measure of this quantity in an ablation test varies considerably with the wavelength [13].

Given the importance of this quantity in the ablation phenomenon, there may be insufficient knowledge of its value. This ignorance is partially offset by ablation models that are not completely analytical, and then partially correcting the error in ad hoc coefficients of approximate laws used in the models.

---

[1] TPSX Material Properties Database. http://tpsx.arc.nasa.gov.

References

[1] Kowal, J. T., "Thermal Protection System (Heatshield) Development. Advanced Development Project," JSC Commercial Human Space Flight Symposium, Houston, Texas, Oct. 2010.

[2] Pavlovsky, J. E. and St. Leger, L. G., "Apollo Experience Report. Thermal Protection Subsystem," NASA Tech. Rep. TN D-7564, Jan. 1974.

[3] Hillje, E. R., "Entry Aerodynamics at Lunar Return Conditions Obtained from the Flight of Apollo 4 (AS-501)," NASA Tech. Rep. TN D-5399, Oct. 1969.

[4] Bartlett, E. P., Abbott, M. J., Nicolet, W. E. and Moyer, C. B., "Improved Heatshield Design Procedures for Manned Entry Systems. Part II: Application to Apollo." NASA Tech. Rep. CR-108689, June 1970.

[5] Curry, D. M. and Stephens, E. W., "Apollo Ablator Thermal Performance at Superorbital Entry Velocities," NASA Tech. Rep. TN D-5969, Sept. 1970.

[6] Theisinger, J. E., Braun, R. D. and Clark, I. G., "Aerothermodynamic Shape Optimization of Hypersonic Entry Aeroshells," *AIAA Paper 2010–9200, 13th AIAA/ISSMO Multidisciplinary Analysis Optimization Conference*, Fort Worth, Texas, Sept. 2010.

[7] Cauchon, D. L., "Radiative Heating Results from the Fire II Flight Experiment at a Reentry Velocity of 11.4 Kilometers per Second," NASA Tech. Rep. TM X-1402, July 1967.

[8] Ried, R. C., Jr., Rochelle, W. C. and Milhoan, J. D., "Radiative Heating to the Apollo Command Module: Engineering Prediction and Flight Measurement," NASA Tech. Rep. TM X-58091, April 1972.

[9] Gaudette, R. S., del Casal E.P. and Crowder, P. A., "Charring Ablation Performance in Turbulent Flow. Volume I: Analytical and Experimental Studies." NASA Tech. Rep. CR-65736, Sept. 1967.

[10] Feldman, J. and Stackpoole, M., "Characterizing Ablator Properties. Review of Some Techniques and Challenges," Fourth Ablation Workshop, Albuquerque, New Mexico, March 2011.

[11] Crouch, R. K. and Walberg, G. D., "An Investigation of Ablation Behaviour of AVCOAT 5026/39M over a Wide Range of Thermal Environments," NASA Tech. Rep. TM X-1778, April 1969.

[12] Bartlett, E. P. and Anderson, L. W., "Further Studies of the Coupled Chemically Reacting Boundary Layer and Charring Ablator. Part II: An Evaluation of Ablation Maechanisms for the Apollo Heatshield Material." NASA Tech. Rep. CR-92472, Oct. 1968.

[13] AVCO, "Apollo Heatshield Phase I Monthly Progress Report," NASA Tech. Rep. CR-128000, Sept. 1962.

[14] Gaudette, R. S., del Casal, E.P., Halstead, D. W. and Deriugin, V., "Analysis of the Apollo Heat Shield Performance. Volume I – Analytical Methods," NASA Tech. Rep. CR-99633, March 1969.

[15] Kendall, R. M., Bartlett, E. P., Rindal, R. A. and Moyer, C. B., "An Analysis of the Coupled Chemically Reacting Boundary Layer and Charring Ablator, Part. I, Summary Report," NASA Tech. Rep. CR-1060, June 1968.

[16] Halstead, D. W., Gaudette, R. S., del Casal E.P. and Deriugin, V., "Analysis of the Apollo Heat Shield Performance. Volume II – CHAD Computer Program," NASA Tech. Rep. CR-99634, April 1969.

[17] Schaefer, J. W., Flood, D. T., Reese, J. J. and Clark, K. J., "Experimental and Analytical Evaluation of the Apollo Thermal Protection System Under Simulated Reentry Conditions. Part I: Program Description and Presentation of Results," NASA Tech. Rep. CR-99634, July 1967.

[18] Schaefer, J. W., Flood, D. T., Reese, J. J. and Clark, K. J., "Experimental and Analytical Evaluation of the Apollo Thermal Protection System Under Simulated Reentry Conditions. Part II: Analysis of Results," NASA Tech. Rep. CR-65818, July 1967.

[19] Kendall, R. M., "An Analysis of the Coupled Chemically Reacting Boundary Layer and Charring Ablator. Part V: A General Approach to the Thermochemical Solution of Mixed Equilibrium-Nonequilibrium, Homogeneous or Heterogeneous Systems." NASA Tech. Rep. CR-1064, June 1968.
[20] Wool, M. R., Moyer, C. B., Powars, C. A. and Rindal, R. A., "Ablative Response of a Silica Phenolic to Simulated Liquid Propellant Rocket Engine Operating Conditions," NASA Tech. Rep. CR-72701, June 1971.
[21] Colony, R., del Casal E.P. and Gaudette, R. S., "Charring Ablation Performance in Turbulent Flow. Volume II: Computer Program." NASA Tech. Rep. CR-65807, Nov. 1967.
[22] Graves, R. A., Jr. and Whitte, W. G., "Flight-Test Analysis of Apollo Heat-Shield Material Using the Pacemaker Vehicle System," NASA Tech. Rep. TN D-4713, Sept. 1964.
[23] Ihnat, M. E., "Evaluation of the Thermal Properties of Materials. Volume I: Technical Report," NASA Tech. Rep. CR-65979, Sept. 1966.
[24] Ihnat, M. E., "Evaluation of the Thermal Properties of Materials. Volume II: Data Handbook." NASA Tech. Rep. CR-65980, June 1966.
[25] Johnston, C. O., Gnoffo, P. A. and Sutton, K., "The Influence of Ablation on Radiative Heating for Earth Entry," *Journal of Spacecraft and Rockets*, Vol. 46, No. 3, 2009, pp. 481–191.
[26] Nicolet, W. E., "Advanced Methods for Calculating Radiation Transport in Ablation-Products Contaminated Boundary Layers," NASA Tech. Rep. CR-1656, Sept. 1970.
[27] Pope, R. B., "Measurement of the Total Surface Emittance of Charring Ablators," *AIAA Journal*, Vol. 5, No. 12, 1967, pp. 2285–2287.

Appendix A Approximate Solutions of Stefan–Maxwell Equation

A.1 System Resolution

This method was introduced by Kendall et al. [1, 2]. Let's start from the Stefan–Maxwell linear system [Eq. (2.38)]

$$\nabla x_i = \sum_{k \neq i} \frac{x_i x_k}{\rho \mathcal{D}_{ik}} \left[\frac{\mathbf{J}_k + \mathcal{D}_k^T \nabla(\ell n\, T)}{c_k} - \frac{\mathbf{J}_i + \mathcal{D}_i^T \nabla(\ell n\, T)}{c_i} \right] \tag{A.1}$$

$\mathbf{J}_k$ is the diffusion mass flux resulting from concentration gradients. We neglect the Dufour effect.

We can rewrite this equation by adding the term $k = i$, identically zero

$$\nabla x_i = \sum_{k} \frac{x_i x_k}{\rho \mathcal{D}_{ik}} \left[\frac{\mathbf{J}_k + \mathcal{D}_k^T \nabla(\ell n\, T)}{c_k} - \frac{\mathbf{J}_i + \mathcal{D}_i^T \nabla(\ell n\, T)}{c_i} \right] \tag{A.2}$$

Now we define the quantity

$$\mathbf{J}_k^{\mathbf{c}} = \mathbf{J}_k + \mathcal{D}_k^T \nabla(\ell n\, T) \tag{A.3}$$

Assume that the binary diffusion coefficients can be written as

$$\mathcal{D}_{ik} = \frac{\mathcal{D}_{\mathrm{ref}}}{F_i(T) F_k(T)} \tag{A.4}$$

Equation (A.2) can be written, given Eq. (A.3), Eq. (A.4), and the definition relationship of the average molar mass [Eq. (2.22)], as

$$\nabla x_i = \frac{\overline{\mathcal{M}}^2}{\rho \mathcal{D}_{\mathrm{ref}}} \left[\frac{c_i F_i}{\mathcal{M}_i} \sum_k \frac{F_k \mathbf{J}_k^{\mathbf{c}}}{\mathcal{M}_k} - \frac{F_i \mathbf{J}_i^{\mathbf{c}}}{\mathcal{M}_i} \sum_k \frac{c_k F_k}{\mathcal{M}_k} \right] \tag{A.5}$$

Summing over the index i, we get

$$\sum_i \frac{\mathcal{M}_i}{F_i} \nabla x_i = \frac{\overline{\mathcal{M}}^2}{\rho \mathcal{D}_{\mathrm{ref}}} \left(\sum_i c_i \sum_k \frac{F_k \mathbf{J}_k^{\mathbf{c}}}{\mathcal{M}_k} - \sum_i \mathbf{J}_i^{\mathbf{c}} \sum_k \frac{F_k c_k}{\mathcal{M}_k} \right) \tag{A.6}$$

but

$$\begin{cases} \sum_i c_i = 1 \\ \sum_i \rho_i \mathbf{V}_{D_i} = 0 \Rightarrow \sum_i \mathbf{J}_i^{\mathbf{c}} = 0\,. \\ \sum_i \mathcal{D}_i^T = 0 \end{cases} \tag{A.7}$$

The relation in Eq. (A.6) simplifies to

$$\sum_i \frac{F_i \mathbf{J}_i^{\mathbf{c}}}{\mathcal{M}_i} = \frac{\rho \mathcal{D}_{\text{ref}}}{\overline{\mathcal{M}}^2} \sum_i \frac{\mathcal{M}_i}{F_i} \nabla x_i \tag{A.8}$$

By introducing this expression in Eq. (A.5), we get

$$\nabla x_i = \frac{c_i F_i}{\mathcal{M}_i} \sum_k \frac{\mathcal{M}_k}{F_k} \nabla x_k - \frac{\overline{\mathcal{M}}}{\rho \mathcal{D}_{\text{ref}}} \frac{F_i \mathbf{J}_i^{\mathbf{c}}}{\mathcal{M}_i} \sum_k \frac{c_k F_k}{\mathcal{M}_k} \tag{A.9}$$

We set

$$\begin{cases} \mu_2 = \sum_i \dfrac{\mathcal{M}_i x_i}{F_i} = \overline{\mathcal{M}} \sum_i \dfrac{c_i}{F_i} \\ Z_i = \dfrac{\mathcal{M}_i x_i}{F_i \mu_2} = \dfrac{\overline{\mathcal{M}} c_i}{F_i \mu_2} \\ \mu_1 = \sum_i x_i F_i \\ \mu_3 = \sum_i \dfrac{c_i}{F_i} \dfrac{\mathrm{d}F_i}{\mathrm{d}T} \end{cases} \tag{A.10}$$

Z_i is analogous to a mass or volume fraction, indeed

$$\sum_i Z_i = \sum_i \frac{\mathcal{M}_i x_i}{F_i \sum_k \dfrac{\mathcal{M}_k x_k}{F_k}} = \frac{\sum_i \dfrac{\mathcal{M}_i x_i}{F_i}}{\sum_k \dfrac{\mathcal{M}_k x_k}{F_k}} = 1 \tag{A.11}$$

By multiplying the expression for Z_i by μ_2 and deriving, we get

$$\frac{\mathcal{M}_i}{F_i} \nabla x_i - \frac{\mathcal{M}_i x_i}{F_i^2} \nabla F_i = \mu_2 \nabla Z_i + Z_i \nabla \mu_2 \tag{A.12}$$

Bringing the expressions from Eqs. (A.10) and (A.12) into Eq. (A.9), the flux $\mathbf{J}_i^{\mathbf{c}}$ is written

$$\mathbf{J}_i^{\mathbf{c}} = -\frac{\rho \mathcal{D}_{\text{ref}}}{\mu_1} \left[\frac{\mu_2}{\overline{\mathcal{M}}} \nabla Z_i + \frac{Z_i - c_i}{\overline{\mathcal{M}}} \nabla \mu_2 + c_i \left(\frac{1}{F_i^2} \frac{\mathrm{d}F_i}{\mathrm{d}T} - \mu_3 \right) \nabla T \right] \tag{A.13}$$

This expression simplifies in the case where F_i is independent of T

$$\mathbf{J}_i^{\mathbf{c}} = -\frac{\rho \mathcal{D}_{\text{ref}}}{\mu_1 \mathcal{M}} [\mu_2 \nabla Z_i + (Z_i - c_i) \nabla \mu_2] \tag{A.14}$$

If we subtract the thermal diffusion term, which is written as follows (see Sec. A.3):

$$\mathcal{D}_i^T = \frac{\rho \mathcal{D}_{\text{ref}} \mu_2}{2\mu_1 \overline{\mathcal{M}}} (Z_i - c_i) \tag{A.15}$$

and introducing the flux due to the concentration gradient [Eq. (A.3)] it follows:

$$\mathbf{J}_i = \rho \mathbf{V}_{D_i} = -\frac{\rho \mathcal{D}_{\text{ref}}}{\mu_1 \overline{\mathcal{M}}} [\mu_2 \nabla Z_i + (Z_i - c_i) \nabla \mu_4] \tag{A.16}$$

with

$$\mu_4 = \ell n(\mu_2 T^{\frac{1}{2}}) \tag{A.17}$$

We can express ∇Z_i as follows. Derive the relations in Eq. (A.10) for Z_i and μ_2

$$\begin{cases} \dfrac{\nabla Z_i}{Z_i} = \dfrac{\nabla \overline{\mathcal{M}}}{\overline{\mathcal{M}}} + \dfrac{\nabla c_i}{c_i} - \dfrac{\nabla \mu_2}{\mu_2} \\[2ex] \dfrac{\nabla \mu_2}{\mu_2} = \dfrac{\nabla \overline{\mathcal{M}}}{\overline{\mathcal{M}}} + \dfrac{\overline{\mathcal{M}}}{\mu_2} \sum_i \dfrac{\nabla c_i}{F_i} \end{cases} \tag{A.18}$$

From the two previous relations, we obtain

$$\frac{\nabla Z_i}{Z_i} = \frac{\nabla c_i}{c_i} - \frac{\overline{\mathcal{M}}}{\mu_2} \sum_i \frac{\nabla c_i}{F_i} \tag{A.19}$$

Hence the new term of Eq. (A.16):

$$\mathbf{J}_i = -\frac{\rho \mathcal{D}_{\text{ref}}}{\mu_1 F_i} \left[\nabla c_i - \frac{\overline{\mathcal{M}} c_i}{\mu_2} \sum_j \frac{\nabla c_j}{F_j} + \frac{F_i}{\overline{\mathcal{M}}} (Z_i - c_i) \nabla \mu_4 \right] \tag{A.20}$$

This equation shows that in this approximation, the flows are not linear combinations of gradients. Indeed, μ_1 and μ_2 at denominator are themselves linear combinations of these gradients.

A.2 Expression of Multicomponent Diffusion Coefficients

A.2.1 Blottner Approximation

This decomposition is attributed to Blottner [3–5]. Let's write some basic relations that are used in the following demonstration:

$$\frac{1}{\overline{\mathcal{M}}} = \sum_i \frac{c_i}{\mathcal{M}_i} \Rightarrow \nabla\overline{\mathcal{M}} = -\overline{\mathcal{M}}^2 \sum_k \frac{1}{\mathcal{M}_k}\nabla c_k \tag{A.21}$$

$$x_i = \frac{\overline{\mathcal{M}} c_i}{\mathcal{M}_i} \Rightarrow \nabla x_i = \frac{\overline{\mathcal{M}}}{\mathcal{M}_i}\nabla c_i - \frac{\overline{\mathcal{M}}^2 c_i}{\mathcal{M}_i}\sum_k \frac{1}{\mathcal{M}_k}\nabla x_k \tag{A.22}$$

$$\sum_i c_i = 1 \Rightarrow \nabla c_i = -\sum_{k\neq i}\nabla c_k \tag{A.23}$$

We will express the diffusion term due to concentration gradients

$$\mathbf{J}_i = \frac{\rho}{\overline{\mathcal{M}}^2}\sum_{j\neq i}\mathcal{M}_i\mathcal{M}_j D_{ij}\nabla x_j \tag{A.24}$$

So, given the relation in Eq. (A.22)

$$\mathbf{J}_i = \frac{\rho\mathcal{M}_i}{\overline{\mathcal{M}}}\sum_{j\neq i} D_{ij}\nabla c_j - \rho\mathcal{M}_i\sum_{j\neq i} D_{ij}c_j\sum_k \frac{\nabla c_k}{\mathcal{M}_k} \tag{A.25}$$

and given $D_{ii} = 0$, it follows

$$\mathbf{J}_i = \frac{\rho\mathcal{M}_i}{\overline{\mathcal{M}}}\sum_{j\neq i} D_{ij}\nabla c_j - \rho\mathcal{M}_i\sum_j D_{ij}c_j\sum_k \frac{\nabla c_k}{\mathcal{M}_k} - \rho\sum_j D_{ij}c_j\nabla c_i \tag{A.26}$$

or, by inversion of the indices i and j and taking into account Eq. (A.23)

$$\mathbf{J}_i = \frac{\rho\mathcal{M}_i}{\overline{\mathcal{M}}}\sum_{j\neq i} D_{ij}\nabla c_j - \rho\mathcal{M}_i\sum_{j\neq i}\frac{1}{\mathcal{M}_j}\sum_k D_{ik}c_k + \rho\sum_{j\neq i}\nabla c_j\sum_k D_{ik}c_k \tag{A.27}$$

We subtract the term identically zero

$$\rho\mathcal{D}_i\sum_j \nabla c_j = 0 \tag{A.28}$$

with the following, because $\frac{c_k}{\mathcal{M}_k} = \frac{x_k}{\overline{\mathcal{M}}}$:

$$\mathcal{D}_i = \frac{\sum_{k\neq i}\frac{c_k}{\mathcal{M}_k}}{\sum_{k\neq i}\frac{c_k}{\mathcal{M}_k\mathcal{D}_{ik}}} = \frac{1-x_i}{\sum_{k\neq i}\frac{x_k}{\mathcal{D}_{ik}}} \tag{A.29}$$

It follows

$$\mathbf{J}_i = -\rho \mathcal{D}_i \nabla c_i - \rho \sum_{j \neq i} b_{ij} \nabla c_j \tag{A.30}$$

with

$$b_{ij} = \mathcal{D}_i - \frac{\mathcal{M}_i}{\overline{\mathcal{M}}} D_{ij} - \left(1 - \frac{\mathcal{M}_i}{\mathcal{M}_j}\right) \sum_{k \neq i} D_{ik} c_k \tag{A.31}$$

A.2.2 Effective Binary Diffusion Approximation

We will investigate the multicomponent coefficients corresponding to this approximation. Let γ_{ij} be an element of the matrix $(\mathbf{Id} - \mathbf{A})^{-1}$. The solution developed in Sec. 2.3.5 is written

$$\begin{cases} \mathbf{J}_i = \mathbf{J}_i^p - c_i \sum_j \mathbf{J}_j^p \\ \mathbf{J}_j^p = -\rho \dfrac{\mathcal{M}_j}{\overline{\mathcal{M}}} \mathcal{D}_j \sum_k \gamma_{jk} \mathbf{d}_k \end{cases} \tag{A.32}$$

We can rewrite this equation as

$$\mathbf{J}_i = -\rho \sum_j (\delta_{ij} - c_i) \frac{\mathcal{M}_j}{\overline{\mathcal{M}}} \mathcal{D}_j \sum_k \gamma_{jk} \mathbf{d}_k \tag{A.33}$$

or, by reversing the index,

$$\mathbf{J}_i = -\rho \sum_k \left[\sum_j (\delta_{ij} - c_i) \frac{\mathcal{M}_j}{\overline{\mathcal{M}}} \mathcal{D}_j \gamma_{jk} \right] \mathbf{d}_k \tag{A.34}$$

To identify the expression in Eq. (2.43), one must rearrange the bracket

$$\mathbf{J}_i = -\rho \sum_{k \neq i} \left[\sum_j (\delta_{ij} - c_i) \frac{\mathcal{M}_j}{\overline{\mathcal{M}}} \mathcal{D}_j (\gamma_{jk} + \gamma_{ji}) \right] \mathbf{d}_k \tag{A.35}$$

Then

$$D_{ij} = \frac{\overline{\mathcal{M}}^2}{\mathcal{M}_i \mathcal{M}_j} \left[\sum_j (\delta_{ij} - c_i) \frac{\mathcal{M}_j}{\overline{\mathcal{M}}} \mathcal{D}_j (\gamma_{jk} + \gamma_{ji}) \right] \tag{A.36}$$

This expression gives an explicit solution to the first order where, from Eq. (2.67),

$$\gamma_{ij} = (c_i + 1)\delta_{ij} + \frac{x_i \mathcal{D}_j}{\mathcal{D}_{ij}} (1 - \delta_{ij}) \tag{A.37}$$

A.3 Thermal Diffusion

The success of the approximation of the binary coefficient by splitting led researchers to seek a similar expression for $\mathcal{D}_i^T$. The flow of thermal diffusion is sought in the form

$$\mathcal{D}_i^T = \rho c_i \mathcal{D}_{\text{ref}}^T (G_i - \overline{G}_i) \tag{A.38}$$

$\mathcal{D}_{\text{ref}}^T$ is a reference diffusion coefficient (true diffusion coefficient, as opposed to D_i^T, which is actually a mass flow per unit length). Given $\sum_i \mathcal{D}_i^T = 0$, we get

$$\overline{G} = \sum_j c_j G_j \tag{A.39}$$

There is no justification in the general multicomponent case. We consider the following binary case where Eq. (A.38) is written

$$\mathcal{D}_1^T = -\mathcal{D}_2^T = \rho c_1 c_2 \overline{\mathcal{D}}^T (G_1 - G_2) \tag{A.40}$$

We then introduce the binary thermal diffusion coefficient [Eq. (2.42)]

$$\frac{k^T}{x_1 x_2} = \frac{\mathcal{D}_1^T}{\rho c_1 c_2 \mathcal{D}_{12}} \tag{A.41}$$

The latter, given the defining relation of F_i [Eq. (2.81)], is written

$$\mathcal{D}_1^T = \frac{k^T}{x_1 x_2} \rho \mathcal{D}_{\text{ref}} \frac{c_1 c_2}{F_1 F_2} \tag{A.42}$$

By comparing Eq. (A.38) and Eq. (A.42), we get

$$\frac{k^T}{x_1 x_2} = \frac{\mathcal{D}_{\text{ref}}^T}{\mathcal{D}_{\text{ref}}} (G_1 - G_2) F_1 F_2 \tag{A.43}$$

The coefficient of thermal diffusion is almost independent of temperature beyond 1000 K, and depends on $(\mathcal{M}_1/\mathcal{M}_2)$. Now there is an approximate relation between this ratio and (F_1/F_2) [Eq. (2.82)]. The relationship is sought in the form

$$\frac{k^T}{x_1 x_2} = c_t \frac{1 - \dfrac{F_1}{F_2}}{1 - x_1 \left(1 - \dfrac{F_1}{F_2}\right)} \tag{A.44}$$

$c_t \simeq 0.5$ gives a reasonable approximation, although not very precise (about 30% maximum error) [1].

By identifying Eq. (A.43) and Eq. (A.44), it follows

$$\begin{cases} G_1 = F_1^{-1} \\ G_2 = F_2^{-1} \\ \mathcal{D}_{\text{ref}}^T = \dfrac{c_t \mathcal{D}_{\text{ref}}}{x_1 F_1 + x_2 F_2} \end{cases} \tag{A.45}$$

These relations are generalized to a multicomponent system. (The justification for this operation is very difficult to establish [1].)

$$\begin{cases} G_i = F_i^{-1} \\ \mathcal{D}_{\text{ref}}^T = \dfrac{c_t \mathcal{D}_{\text{ref}}}{\sum_j x_j F_j} \end{cases} \tag{A.46}$$

Hence the expression of $\mathcal{D}_i^T$

$$\mathcal{D}_i^T = \frac{c_t \rho c_i \mathcal{D}_{\text{ref}} \left(\dfrac{1}{F_i} - \sum_j \dfrac{c_j}{F_j} \right)}{\sum_j x_i F_i} \tag{A.47}$$

Or, depending on the quantities Z_i, μ_1, and μ_2 defined in Eq. (A.10)

$$\mathcal{D}_i^T = \frac{c_t \mu_2 \rho \mathcal{D}_{\text{ref}}}{\mu_1 \overline{\mathcal{M}}} (Z_i - c_i) \tag{A.48}$$

References

[1] Bartlett, E. P., Kendall, R. M., and Rindal, R. A., "An Analysis of the Coupled Chemically Reacting Boundary Layer and Charring Ablator, Part IV, A Unified Approximation for Mixture Transport Properties for Multicomponent Boundary-Layer Applications," NASA Report CR-1063, June 1968.

[2] Kendall, R. M., Rindal, R. A., and Bartlett, E. P., "Thermochemical Ablation," AIAA Paper 65-642, 1965.

[3] Blottner, F. G., "Chemically Reacting Viscous Flow Program for Multicomponent Gas Mixtures," Sandia Report TR SC-RR-70-754, Dec. 1971.

[4] Blottner, F. G., "Finite Difference Methods of Solution of the Boundary Layer Equations," *AIAA Journal*, Vol. 8, No. 2, 1970, pp. 193–205.

[5] Blottner, F. G., "Viscous Shock Layer at the Stagnation Point with Nonequilibrium Air Chemistry," *AIAA Journal*, Vol. 7, No. 12, 1969, pp. 2281–2288.

Appendix B Approximation of Thermodynamic Properties

B.1 Thermodynamic Properties

Thermodynamic properties* vary significantly with temperature in the range of interest. For example, Fig. B.1 shows the value of the specific heat at constant pressure of CO. The translational part, constant, is $5R/2 \simeq 20.8\ \mathrm{J \cdot mole^{-1} \cdot K^{-1}}$ and is therefore a fraction of the total, especially at high temperature. Here, this quantity has been calculated using spectroscopic constants [1]. This calculation is quite expensive to use in aerothermal codes, so a polynomial approximation was pursued.

B.1.1 Specific Heat, Enthalpy, and Entropy

Thermodynamic data are present in different databases. These are the values at the reference pressure p_0. We use the thermodynamic relations

$$\mathcal{C}_p^0 = \left.\frac{\partial \mathcal{H}}{\partial T}\right|_{p=p_0} = T\left.\frac{\partial \mathcal{S}}{\partial T}\right|_{p=p_0}$$

- JANAF [1] provides these data in tables of numerical values, or, in the case of gases, also as spectroscopic constants.
- Chemkin [2] uses NASA polynomials. These polynomials are

$$\begin{cases} \dfrac{\mathcal{C}_p^0}{\mathcal{R}} = a_1 + a_2 T + a_3 T^2 + a_4 T^3 + a_5 T^4 \\ \dfrac{\mathcal{H}^0}{\mathcal{R}T} = a_1 + \dfrac{a_2 T}{2} + \dfrac{a_2 T^2}{3} + \dfrac{a_4 T^3}{4} + \dfrac{a_5 T^4}{5} + \dfrac{a_6}{T} \\ \dfrac{\mathcal{S}^0}{\mathcal{R}} = a_1 \ell n\, T + a_2 T + \dfrac{a_3 T^2}{2} + \dfrac{a_4 T^3}{3} + \dfrac{a_5 T^4}{4} + a_7 \end{cases} \tag{B.1}$$

They are given on several temperature intervals with simple connection to the order 0. Indeed, the increase in accuracy by the order of approximation is limited by the Gibbs phenomenon (oscillation of the polynomial interpolation curve; see Sec. B.3). Extrapolation of values can lead to large errors.

* Thermodynamic quantities $\mathcal{C}_p$, $\mathcal{H}$, and $\mathcal{S}$ are molar values.

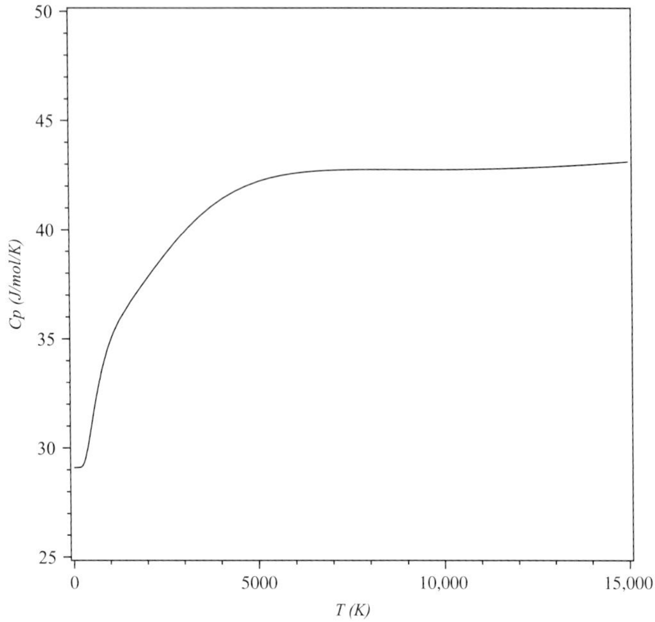

Fig. B.1 Specific heat of CO, $p = p_0$.

- There are various sources using this description [3–5],[†,‡] sometimes using a slightly different definition called the Shomate equation

$$\begin{cases} C_p^0 & = A + BT^* + CT^{*2} + DT^{*3} + E/T^{*2} \\ \mathcal{H}^0 - \mathcal{H}^0_{\text{ref}} & = AT^* + \dfrac{BT^{*2}}{2} + \dfrac{CT^{*3}}{3} + \dfrac{DT^{*4}}{4} - \dfrac{E}{T^*} + F - H \\ \mathcal{S}^0 & = A\ell n\, T^* + BT^* + \dfrac{CT^{*2}}{2} + \dfrac{DT^{*3}}{3} - \dfrac{E}{2T^{*2}} + G \end{cases} \tag{B.2}$$

with $T^* = T/1000$.
- Wilhoit [6] proposed another type of approximation based on the use of partition functions, and which respects the physical evolution of the quantities at high temperatures. Some data of this type are available [3, 7]

† Burcat, A., "Ideal Gas Thermodynamic Data in Polynomial Form for Combustion and Air Pollution Use," http://garfield.chem.elte.hu/Burcat/burcat.html [retrieved 26 Dec. 2009].
‡ National Institute of Standards and Technology, "Gas Phase Thermochemistry Data," http://webbook.nist.gov/chemistry/form-ser.html [retrieved 1 June 2011].

$$
\begin{cases}
\mathcal{C}_p^0 = \mathcal{C}_p^0(0) + [\mathcal{C}_p^0(\infty) - \mathcal{C}_p^0(0)]y^2\left[1 + (y-1)\sum_{i=0}^{n} a_i y^i\right] \\
\dfrac{\mathcal{H}^0 - \mathcal{H}^0(0)}{T} = \dfrac{C^{ste}}{T} + \mathcal{C}_p^0(0) - [\mathcal{C}_p^0(\infty) - \mathcal{C}_p^0(0)]\left[2 + \sum_{i=0}^{n} a_i\right] \\
\qquad \times \left[\dfrac{y}{2} - 1 + \left(\dfrac{1}{y} - 1\right)\ln\left(\dfrac{T}{y}\right)\right] \\
\qquad + y^2 \sum_{i=0}^{n} \dfrac{y^i}{(i+2)(i+3)}\left[\sum_{i=0}^{n} f_{ij} a_i\right] \\
\mathcal{S}^0 = C^{ste} + \mathcal{C}_p^0(\infty)\ \ln T - [\mathcal{C}_p^0(\infty) - \mathcal{C}_p^0(0)] \\
\qquad \times \left[\ln y + \left(1 + y\sum_{i=0}^{n} \dfrac{a_i y^i}{i+2}\right) y\right]
\end{cases}
\tag{B.3}
$$

with

$$
y = \frac{T}{T + B}
$$

and

$$
f_{ij} = \begin{cases} 3 + j & \text{if } i = j \\ 1 & \text{if } j > i \end{cases}
$$

Databases generally use $n = 3$ [3]. $\mathcal{C}_p^0(0)$ and $\mathcal{C}_p^0(\infty)$ are given by statistical physics—$\mathcal{C}_p^0(0) = 7\mathcal{R}/2$ for linear molecules (except hydrogen), and $4\mathcal{R}$ for nonlinear ones. $\mathcal{C}_p^0(\infty) = (3N - 3/2)\mathcal{R}$ for linear molecules, and $(3N - 2)$ $\mathcal{R}$ for nonlinear ones. N is the number of atoms in the species.

One can find many other sources of data [8–10].

B.1.2 Equilibrium Constants

This method is intended primarily for gases for which one can calculate all thermodynamic properties from the spectroscopic constants. Moreover, we note that one can calculate the equilibrium constants compatible with the previous approximations using the Gibbs relation

$$
\Delta(\mathcal{H}^0 - T\mathcal{S}^0) = -\mathcal{R}T\ \ln K_p \tag{B.4}
$$

This quantity represents the change of Gibbs energy in the reaction.

However, we propose to show the relationship between these quantities and the reaction constants. These are given in the form of Arrhenius laws [Eq. (3.8)]

$$
k_{f/b} = AT^B \exp\left(-\frac{T_A}{T}\right) \tag{B.5}
$$

This expression is valid for direct (index f) or retrograde (index b) reactions. At equilibrium, Eq. (3.4) is written

$$k_f \prod_i \left(\frac{\rho x_i}{\overline{\mathcal{M}}}\right)^{\nu'_i} = k_b \prod_i \left(\frac{\rho x_i}{\overline{\mathcal{M}}}\right)^{\nu''_i} \tag{B.6}$$

Or, taking into account the definition of $x_i = \dfrac{p_i}{p}$

$$\frac{k_f}{k_b} = \left(\frac{p}{RT}\right)^{\Delta\Omega} \Pi_k x_k^{\Omega_k} \tag{B.7}$$

where $\Delta\Omega = \sum_i \Omega_i$ is the stoichiometric variation in the reaction. Moreover, the equilibrium constant K_p is defined by

$$K_p = \frac{\Pi_k \left(\dfrac{p_k}{p_0}\right)^{\nu''_k}}{\Pi_k \left(\dfrac{p_k}{p_0}\right)^{\nu'_k}} = \left(\frac{p}{p_0}\right)^{\Delta\Omega} \Pi_k x_k^{\Omega_k} \tag{B.8}$$

Comparing Eqs. (B.7) and (B.8), we get

$$\frac{k_f}{k_b} = K_p \left(\frac{p_0}{RT}\right)^{\Delta\Omega} \tag{B.9}$$

This implies that the equilibrium constants K_p have the same dependency in T as k_f and k_b given by Eq. (B.5), thus

$$\ell n K_p = \alpha_1 + \frac{\alpha_2}{T} + \alpha_3 \, \ell n \, T \tag{B.10}$$

From the Gibbs relationship one can define an equilibrium constant [Eq. (B.12)] for each species so that $\ell n \, K_p = \sum_i \Omega_i \ell n \, K_{p_i}$. This quantity, defined in the next section, has the form given by Eq. (B.10), which is used for the approximation (Fig. B.2).

The example chosen is that of the species N. The approximation restores the original values with great precision for temperatures up to 6000 K, the limit value for JANAF tables [1] (Fig. B.2).

This type of approximation is close to a method proposed by Park [11]:

$$K_p = \exp\left[\sum_{i=1}^{3} A_i \ell n \left(\frac{10^4}{T}\right)^i\right] \tag{B.11}$$

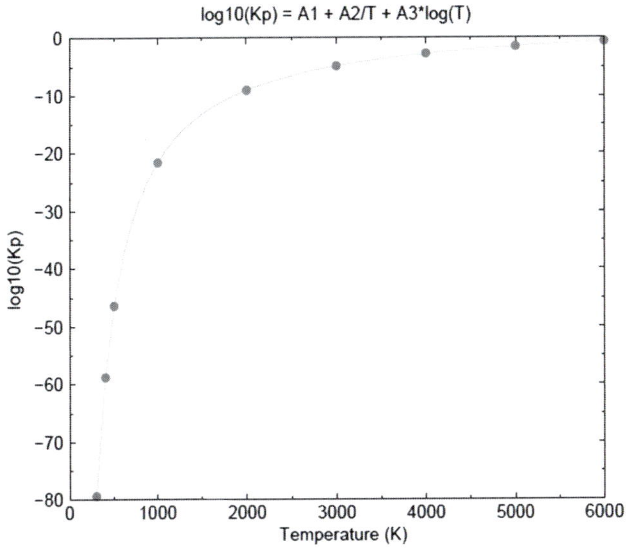

Fig. B.2 Approximation of thermodynamic equilibrium constant for species *N*.

B.2 Homogeneous Description of a Database

B.2.1 Method

The equilibrium constant for each species (the value obtained from a pseudo-reaction of formation involving a reference species of database) is expressed in the functions of enthalpy and entropy by means of the Gibbs relation

$$\ell n\, K_p = \frac{1}{RT}\left[-\Delta\mathcal{H}^0_{f_T} + T\left(\mathcal{S}^0_T - \sum_i \nu_i \mathcal{S}^0_{T_i}\right)\right] \tag{B.12}$$

$\Delta\mathcal{H}^0_{f_T}$ is the enthalpy of formation at temperature T, and index i corresponds to the chemical species involved in the formation of the given species. The enthalpy of formation is

$$\Delta\mathcal{H}^0_{f_T} = \Delta\mathcal{H}^0_{f_{T_{\text{ref}}}} + (\mathcal{H}^0_T - \mathcal{H}^0_{T_{\text{ref}}}) - \sum_i \nu_i(\mathcal{H}^0_T - \mathcal{H}^0_{T_{\text{ref}}})_i \tag{B.13}$$

A sufficient condition to verify Eqs. (B.10) and (B.12) is

$$\mathcal{H}^0 = \beta_1 + \beta_2 T + \beta_3 T^2 + \beta_4 T\, \ell n\, T \tag{B.14}$$

where $\mathcal{H}^0$ is obtained from the partition function Q via the expression

$$\mathcal{H}^0 = RT^2 \frac{\partial\, \ell n\, Q}{\partial T} + RT \tag{B.15}$$

The partition function is a minima written as follows:

$$\ell n Q = \delta_1 T + \frac{\delta_2}{T} + \delta_3 \ell n\, T + \delta_4 (\ell n\, T)^2 \tag{B.16}$$

The same approach for the entropy gives

$$\ell n Q = \gamma_1 + \gamma_2 T + \frac{\gamma_3}{T} + \gamma_4 \ell n\, T + \gamma_5 (\ell n\, T)^2 \tag{B.17}$$

This expression includes one more term than the previous one, so this last form must be used.

B.2.2 Generalization

The thermodynamic constants are generally not compatible with kinetics reactions of experimental origin. The expression in Eq. (B.10) must be generalized

$$\ell n Q = \gamma_1 + \sum_{j=1}^{m} \gamma_{2_j} T^j + \sum_{j=1}^{m} \frac{\gamma_{3_j}}{T^j} + \sum_{j=1}^{m} \gamma_{4_j} (\ell n T)^j \tag{B.18}$$

The derivative of the partition function, necessary for calculations of various thermodynamic quantities, is written

$$\frac{\partial \ell n Q}{\partial T} = \sum_{j=1}^{m} j \gamma_{2_j} T^{j-1} + \gamma_{31} \ell n T - \sum_{j=1}^{m-1} \frac{\gamma_{3_j}}{j T^{j-1}} + \sum_{j=1}^{m} \frac{j \gamma_{4_j}}{T} (\ell n T)^{j-1} \tag{B.19}$$

The order m needed to achieve a given accuracy depends on the species and the interval desired for temperature. For example, N_2, for which the fundamental state is sufficient to describe the molecule up to high temperatures, is easier to approximate than C_2.

Nitrogen in the range 150–15,000 K is used as an example. The accuracy is good once the fourth order (Table B.1), the maximum error, is located at

Table B.1 Approximation Errors on Thermodynamic Constants (in %)

Species, Approximation	Error on Q	Error on $\mathcal{H}^0$	Error on $\mathcal{S}^0$
N_2, Eq. (B.17)	2	13	3.5
N_2, 2nd order	0.4	4	1
N_2, 3rd order	0.1	1	0.2
N_2, 4th order	0.01	0.1	0.06
C_2, 2nd order	2.5	30	7
C_2, 3rd order	0.2	0.5	0.5
C_2, 4th order	0.03	0.4	0.1

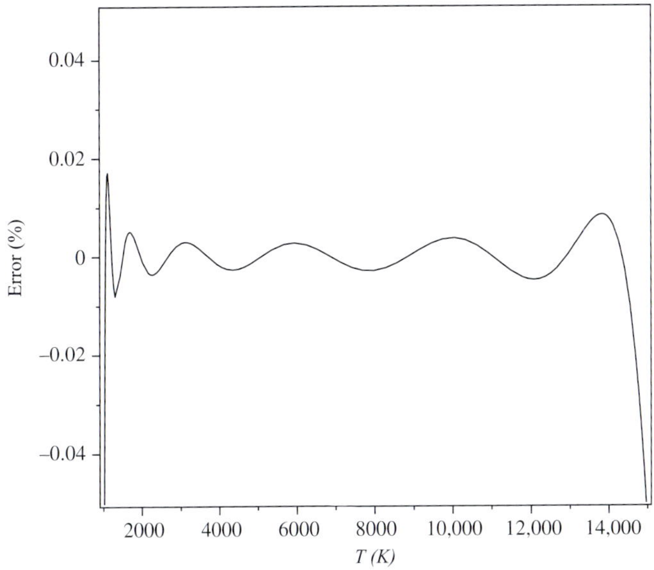

Fig. B.3 Error on approximation of internal energy for species N_2; interval 150–15,000 K.

the boundaries. Again, the extrapolation is very undesirable. Note that we could increase the lower limit approximation, which weighs heavily on the precision (Fig. B.3).

B.2.3 For Solids and Liquids

In the case of solids or liquids, you can always set the same type of approximation, even if the partition function loses its meaning. In these cases (this is also sometimes true in the case of gas) the specific heat at constant pressure C_p is given as a table or a NASA polynomial. Therefore, the enthalpy must be recalculated

$$\frac{\mathcal{H}^0}{\mathcal{R}} = \sum_{j=1}^{m} j\gamma_{2,j} T^{j+1} - \sum_{j=1}^{m} \frac{j\gamma_{3,j}}{T^{j-1}} + \frac{5}{2}T + T\sum_{j=1}^{m} j\gamma_{4,j}(\ell n\, T)^{j-1} \tag{B.20}$$

Then the expression of $\mathcal{C}_p$

$$\frac{\mathcal{C}_p^0}{\mathcal{R}} = \sum_{j=1}^{m} j(j+1)\gamma_{2,j} T^j + \sum_{j=1}^{m} \frac{j(j-1)\gamma_{3,j}}{T^j} + \gamma_{41} + \frac{5}{2} + 2\gamma_{42}$$
$$+ \sum_{j=2}^{m-1} [j\gamma_{4,j} + j(j+1)\gamma_{4,j+1}](\ell n\, T)^{j-1} + m\gamma_{4,m}(\ell n\, T)^{m-1} \tag{B.21}$$

The approximation of C_p will give the coefficients contained in this expression. The coefficients $\gamma_{i,j}$ are deduced immediately.

References

[1] Chase, M. W., Jr., Davies, C. A., Downey, J. R., Jr., Frurip, D. J., MacDonald, R. A., and Syverud, A. N., "JANAF Thermochemical Tables, 3rd ed.," *Journal of Physical Chemistry*, Supplements 1 & 2, 1985.

[2] Kee, R. J., Dixon-Lewis, G., Warnatz, J., Coltrin, M. E., Miller, J. A., and Moffat, H. K. "CHEMKIN III: A Fortran Computer Code Package for the Evaluation of Gas-Phase, Multicomponent Transport Properties," Sandia National Laboratories Report SAND86-8246B, March 1998.

[3] Burcat, A., Dixon-Lewis, G., Frenklach, M., Gardiner, W. C., Jr., Hanson, R. K., Salimian, S., Troe, J., Warnatz, J., and Zellner, R., *Combustion Chemistry*, edited by Gardiner, W. C., Jr., Springer-Verlag, New York, 1984.

[4] Gupta, R. N., Yos, J. M., Thomson, R. A., and Lee, K., "Review of Reaction Rates and Thermodynamic and Transport Properties for an 11-Species Air Model for Chemical and Thermal Nonequilibrium Calculations to 30,000 K," NASA Technical Report RP 1232, 1990.

[5] McBride, B. J., Gordon, S., and Reno, M. A., "Coefficients for Calculating Thermodynamics and Transport Properties of Individual Species," NASA Technical Memorandum 4513, 1993.

[6] Wilhoit, R. C., *Thermodynamics Research Center Current Data News*, Vol. 3, No. 2., 1975.

[7] Frenkel, M., Kabo, G. J., Marsh, K. N., Roganov, G. N., and Wilhoit, R. C., "Thermodynamics of Organic Compounds in the Gas State," in 2 volumes, edited by the Thermodynamics Research Center, Texas A&M System, College Station, TX, 1994.

[8] Barin, I., *Thermochemical Data of Pure Substances*, in two parts, John Wiley & Sons, New York, 1989.

[9] Gurvich, L. B., Veytz, I. V., and Alcock, C. B., *Thermochemical Data of Individual Substances*, 4th ed., in 5 volumes, CRC Press, Boca Raton, FL, 1994.

[10] Knacke, O., Kubaschewsky, O., and Hesselmann, K., *Thermochemical Data of Inorganic Substances*, in two parts, Springer-Verlag, New York, 1991.

[11] Park, C., "On Convergence of Computation of Chemically Reacting Flows," AIAA Paper 85-0247, AIAA 23rd Aerospace Science Meeting, Reno, NV, Jan. 1985.

Appendix C System with Variable Elemental Composition

In Sec. 3.1.4 we showed how to construct an equation of state from a pseudo-system of reactions

$$\frac{\mathrm{d}\rho_i}{\mathrm{d}t} = \dot{\omega}_i(c_k, T) \tag{C.1}$$

with

$$\dot{\omega}_i = \sum_j \Omega_{ij} \left[k_{f_j} \prod_{k=1}^{n_e} \left(\frac{\rho c_k}{\mathcal{M}_k} \right)^{\nu'_{kj}} - k_{b_j} \prod_{k=1}^{n_e} \left(\frac{\rho c_k}{\mathcal{M}_k} \right)^{\nu''_{kj}} \right] \tag{C.2}$$

The given internal energy constraint applies to any point in a flow calculation.* There are other problems, on the boundary conditions upstream and at the wall. Computer codes also use speed of sound and the specific heats. These problems will be dealt with later in this appendix.

C.1 Wall Conditions

In the case of wall boundary conditions, temperature is given. We want to impose pressure p_w. For this, we solve the following system:

$$\frac{\partial \rho_i}{\partial t} = \dot{\omega}_i + \frac{\overline{\mathcal{M}}}{\tau \mathcal{R} T_0} \left(\frac{c_i \overline{\mathcal{M}} p_w}{\mathcal{M}_i} - p_i \right) \tag{C.3}$$

τ is an arbitrary time constant, chosen nearly equal to the characteristic times for the chemical reactions. At convergence, $\partial \rho_i / \partial t = 0$ leads to $\sum_i \partial \rho_i / \partial t = 0$, so $p = p_w$.

Upstream Conditions

In this case, the pressure and density are given. In aerodynamics, the atmosphere (whatever that is) behaves like a perfect gas. The temperature calculation is trivial. However, it is desired for the code to be general. To

* Thermodynamic quantities C_p, C_v, e, and h are mass values.

do this, and by analogy with the previous cases, we will solve the system (when necessary)

$$\begin{cases} \dfrac{\partial \rho_i}{\partial t} = \dot{\omega}_i \\ \dfrac{\partial p}{\partial t} = \frac{1}{\tau}(p_w - p) \end{cases} \tag{C.4}$$

As elsewhere

$$\frac{\partial p}{\partial t} = \frac{\partial}{\partial t}\left(\sum_i \rho c_i \frac{\mathcal{R}T}{\mathcal{M}_i}\right) = \left(\sum_i c_i \frac{\mathcal{R}}{\mathcal{M}_i}\right)\rho\frac{\partial T}{\partial t} + \rho\mathcal{R}T\sum_i \frac{1}{\mathcal{M}_i}\frac{\partial c_i}{\partial t} \tag{C.5}$$

The equation is written on the temperature

$$\rho\mathcal{R}\frac{\partial T}{\partial t} = \frac{1}{\tau}(p_w - p) - T\sum_i \frac{\mathcal{R}}{\mathcal{M}_i}\dot{\omega}_i \tag{C.6}$$

C.3 Pressure and Sound Velocity

C.3.1 Principle of the Method

Among the other variables, pressure and sound velocity play a special role in the codes. They are given by [1]

$$\begin{cases} \mathrm{d}p = \chi \mathrm{d}\rho + \kappa \mathrm{d}(\rho e) \\ a^2 = \chi + \kappa h \end{cases} \tag{C.7}$$

The quantities χ and κ are obtained by derivation of the first equation with respect to ρ and internal energy e

$$\begin{cases} \chi = \left.\dfrac{\partial p}{\partial \rho}\right|_e - \dfrac{e}{\rho}\left.\dfrac{\partial p}{\partial e}\right|_\rho \\ \kappa = \dfrac{1}{\rho}\left.\dfrac{\partial p}{\partial e}\right|_\rho \end{cases} \tag{C.8}$$

The calculation of partial derivatives $\left.\frac{\partial p}{\partial e}\right|_\rho$ and $\left.\frac{\partial p}{\partial \rho}\right|_e$ is done as follows. We start from the pressure given by the state law

$$p(\rho, e) = \sum_i \rho_i \frac{\mathcal{R}T}{\mathcal{M}_i} \tag{C.9}$$

Hence we deduce

$$\begin{cases} \left.\dfrac{\partial p}{\partial e}\right|_\rho = \sum_i \rho \left.\dfrac{\partial c_i}{\partial e}\right|_\rho \dfrac{\mathcal{R}T}{\mathcal{M}_i} + \sum_i \rho_i \dfrac{\mathcal{R}}{\mathcal{M}_i}\left.\dfrac{\partial T}{\partial e}\right|_\rho \\ \left.\dfrac{\partial p}{\partial \rho}\right|_e = \sum_i \dfrac{\mathcal{R}T}{\mathcal{M}_i}c_i + \sum_i \rho\left.\dfrac{\partial c_i}{\partial \rho}\right|_e \dfrac{\mathcal{R}T}{\mathcal{M}_i} + \sum_i \rho_i \dfrac{\mathcal{R}}{\mathcal{M}_i}\left.\dfrac{\partial T}{\partial \rho}\right|_e \end{cases} \tag{C.10}$$

Moreover, given the relation of definition of C_{V_i} and by derivation of the relation $e = \sum_i c_i e_i$

$$\sum_i \frac{\partial c_i}{\partial e}\bigg|_\rho e_i + C_{V_f} \frac{\partial T}{\partial e}\bigg|_\rho = 1 \tag{C.11}$$

with

$$C_{V_f} = \sum_i c_i C_{V_i} \tag{C.12}$$

We will use the relation $\dot{\omega}_i(\rho_k, T) = 0$, true at convergence. By derivation we get

$$\sum_k \frac{\partial \dot{\omega}_i}{\partial \rho_k}\bigg|_T \frac{\partial \rho_k}{\partial \rho}\bigg|_e + \frac{\partial \dot{\omega}_i}{\partial T}\bigg|_\rho \frac{\partial T}{\partial \rho}\bigg|_e = 0 \tag{C.13}$$

with

$$\frac{\partial \rho_k}{\partial \rho}\bigg|_e = c_k + \rho \frac{\partial c_k}{\partial \rho}\bigg|_e \tag{C.14}$$

Equations (C.11) and (C.13) constitute a linear system and can be written as follows:

$$\begin{vmatrix} \rho \dfrac{\partial \dot{\omega}_i}{\partial \rho_k}\bigg|_T & \dfrac{\partial \dot{\omega}_i}{\partial T}\bigg|_\rho \\ e_k & C_{V_f} \end{vmatrix} \begin{vmatrix} \dfrac{\partial c_k}{\partial \rho}\bigg|_e \\ \dfrac{\partial T}{\partial \rho}\bigg|_e \end{vmatrix} = \begin{vmatrix} -\rho \sum_k \dfrac{\partial \dot{\omega}_i}{\partial \rho_k}\bigg|_T \\ 0 \end{vmatrix} \tag{C.15}$$

Its resolution allows the calculation of the terms needed in the first equation of system [Eq. (C.10)]. However, there is a difficulty to which we will return in the next section.

Similarly

$$\sum_k \frac{\partial \dot{\omega}_i}{\partial \rho_k}\bigg|_T \frac{\partial \rho}{\partial e}\bigg|_\rho + \frac{\partial \dot{\omega}_i}{\partial T}\bigg|_\rho \frac{\partial T}{\partial e}\bigg|_\rho = 0 \tag{C.16}$$

with

$$\frac{\partial \rho_k}{\partial e}\bigg|_\rho = \rho \frac{\partial c_k}{\partial e}\bigg|_\rho \tag{C.17}$$

Hence the system equivalent to Eq. (C.15)

$$\begin{vmatrix} \rho \dfrac{\partial \dot{\omega}_i}{\partial \rho_k}\bigg|_T & \dfrac{\partial \dot{\omega}_i}{\partial T}\bigg|_\rho \\ e_k & C_{V_f} \end{vmatrix} \begin{vmatrix} \dfrac{\partial c_k}{\partial e}\bigg|_\rho \\ \dfrac{\partial T}{\partial e}\bigg|_\rho \end{vmatrix} = \begin{vmatrix} 0 \\ 1 \end{vmatrix} \tag{C.18}$$

Note that the matrices of Eqs. (C.15) and (C.18) are the same. Its elements are calculated analytically from the expression of $\dot{\omega}_i$ given by Eq. (5.2)

$$\begin{cases} \left.\dfrac{\partial \dot{\omega}_i}{\partial T}\right|_\rho = \mathcal{M}_i \sum_j \Omega_{ij} \left[\dfrac{\partial k_{f_k}}{\partial T} \prod_k \left(\dfrac{\rho_k}{\mathcal{M}_k} \right)^{\nu'_{kj}} - \dfrac{\partial k_{b_k}}{\partial T} \prod_k \left(\dfrac{\rho_k}{\mathcal{M}_k} \right)^{\nu''_{kj}} \right] \\ \left.\dfrac{\partial \dot{\omega}_i}{\partial \rho_k}\right|_T = \mathcal{M}_i \sum_j \Omega_{ij} \left\{ k_{f_j} \left[\prod_{k \neq i} \left(\dfrac{\rho_k}{\mathcal{M}_k} \right)^{\nu'_{kj}} \right] \dfrac{\nu'_{ij}}{\mathcal{M}_i} \left(\dfrac{\rho_i}{\mathcal{M}_i} \right)^{\nu'_{ij}-1} \right. \\ \qquad \left. - k_{b_k} \left[\prod_{k \neq i} \left(\dfrac{\rho_k}{\mathcal{M}_k} \right)^{\nu'_{kj}} \right] \dfrac{\nu''_{kj}}{\mathcal{M}_i} \left(\dfrac{\rho_i}{\mathcal{M}_i} \right)^{\nu''_{ij}-1} \right\} \end{cases} \tag{C.19}$$

The reaction constants are derived from

$$\begin{cases} k_* = A_* T^{B*} \exp\left(-\dfrac{T_{A*}}{T} \right) \\ \dfrac{\partial k*}{\partial T} = \dfrac{k*}{T} \left(B* + \dfrac{T_{A*}}{T} \right) \end{cases} \tag{C.20}$$

* is any index. In fact, given the renormalization constants of reaction, the derivatives are

$$\begin{cases} \text{if } \dfrac{K_p}{(\mathcal{R}T)^{\Delta\Omega}} > 1 \text{ then } \dfrac{\partial k_f}{\partial T} = 0 \text{ and } \dfrac{\partial k_b}{\partial T} = (\mathcal{R}T)^{\Delta\Omega - 1} \left[\dfrac{\Delta\Omega}{K_p} - \mathcal{R}T \dfrac{\partial(\ell n K_p)}{\partial T} \right] \\ \text{if } \dfrac{K_p}{(\mathcal{R}T)^{\Delta\Omega}} < 1 \text{ then } \dfrac{\partial k_b}{\partial T} = 0 \text{ and } \dfrac{\partial k_f}{\partial T} = (\mathcal{R}T)^{-\Delta\Omega} K_p \left[\dfrac{\partial(\ell n K_p)}{\partial T} - \dfrac{\Delta\Omega}{\mathcal{R}T} \right] \end{cases} \tag{C.21}$$

C.3.2 Complete System

The method described in the previous section cannot be applied directly for reasons that we explain here. First, we define the following:

- F is defined by the vector $F_i = \mathrm{d}/\mathrm{d}t(\rho c_i/\mathcal{M}_i)$, the first member of Eq. (C.1).
- ϖ is the vector of rate $\dot{\omega}_k$ of each reaction, defined by Eq. (C.2).

The stoichiometric matrix Ω has dimension $(n_e \times n_r)$ and rank n_r (based on the choice of pseudo-reactions).

The equation $F = \Omega\varpi$ [Eq. (C.1)] can be solved only if F is in the image of Ω, then

$$F \cdot \vec{e}_i = 0 \tag{C.22}$$

where $\vec{e}_i$ belongs to the kernel of Ω. Hence the equations closing the system, which express the conservation elements (see Table 3.2)

$$\sum_j [A_j] e_{ij} = \sum_j \frac{\rho c_j}{\mathcal{M}_j} e_{ij} = C^{st}, \quad i = 1, n_e - n_r + 1 \tag{C.23}$$

Differentiating with respect to ρ we obtain the closure relations

$$\sum_j \frac{1}{\mathcal{M}_j} \frac{\partial c_j}{\partial \rho} e_{ij} = 0, \qquad i = 1, n_e - n_r + 1 \tag{C.24}$$

These relations will be substituted for the corresponding terms on lines and columns $n_r + 1$ to n_e of the matrix defined by Eq. (C.15) or Eq. (C.18).

Take the example of reaction 3 of Table 3.3. We have

$$\dot{\omega} = k_f \frac{\rho_{N_2}}{\mathcal{M}_{N_2}} \frac{\rho_{O_2}}{\mathcal{M}_{O_2}} - k_b \frac{\rho_{NO}^2}{\mathcal{M}_{NO}^2} \tag{C.25}$$

Then

$$\begin{cases} \dfrac{\partial \rho_{O_2}}{\partial t} = -\mathcal{M}_{O_2} \dot{\omega} \\ \dfrac{\partial \rho_{N_2}}{\partial t} = -\mathcal{M}_{N_2} \dot{\omega} \\ \dfrac{\partial \rho_{NO}}{\partial t} = 2\mathcal{M}_{NO} \dot{\omega} \end{cases} \tag{C.26}$$

Suppose the reaction is unbalanced, for example, $k_f \simeq 0$, because of low temperature, then

$$\begin{cases} \dfrac{\partial \dot{\omega}}{\partial(\rho_{N_2})} \simeq 0 \\ \dfrac{\partial \dot{\omega}}{\partial(\rho_{O_2})} \simeq 0 \\ \dfrac{\partial \dot{\omega}}{\partial(\rho_{NO})} = -2k_b \dfrac{\rho_{NO}}{\mathcal{M}_{NO}^2} \end{cases} \tag{C.27}$$

The latter term is very large. The matrix in Eq. (C.15) is written

$$\begin{vmatrix} 0 & 0 & 2\rho\mathcal{M}_{O_2} k_b \dfrac{\rho_{NO}}{\mathcal{M}_{NO}^2} & 0 \\ 0 & 0 & 2\rho\mathcal{M}_{N_2} k_b \dfrac{\rho_{NO}}{\mathcal{M}_{NO}^2} & 0 \\ 0 & 0 & -4\rho\mathcal{M}_{O_2} k_b \dfrac{\rho_{NO}}{\mathcal{M}_{NO}^2} & 0 \\ e_{O_2} & e_{N_2} & e_{NO} & C_V \end{vmatrix} \tag{C.28}$$

This matrix is not invertible. We take the method described at the beginning of this section

$$A = \begin{vmatrix} 1 \\ 1 \\ -2 \end{vmatrix}, \quad A^T = |1, 1, -2| \tag{C.29}$$

$$\text{Ker } A^T = \{x + y - 2z = 0\} \tag{C.30}$$

The solution, of dimension 2, is

$$\vec{e}_1 = \begin{vmatrix} 2 \\ 0 \\ 1 \end{vmatrix}, \quad \vec{e}_2 = \begin{vmatrix} 0 \\ 2 \\ 1 \end{vmatrix} \tag{C.31}$$

Hence the new matrix

$$\begin{vmatrix} 0 & 0 & 2\rho\mathcal{M}_{O_2}K_b\dfrac{\rho_{NO}}{\mathcal{M}_{NO}^2} & 0 \\ \dfrac{2\rho}{\mathcal{M}_{O_2}} & 0 & \dfrac{\rho}{\mathcal{M}_{NO}} & 0 \\ 0 & \dfrac{2\rho}{\mathcal{M}_{N_2}} & \dfrac{\rho}{\mathcal{M}_{NO}} & 0 \\ e_{O_2} & e_{N_2} & e_{NO} & C_V \end{vmatrix} \tag{C.32}$$

This matrix is no longer a problem.

C.4 Total Specific Heats

From the relation defining the internal energy $e = \sum_i c_i e_i$, one can obtain the total specific heat at constant volume

$$C_V = \sum_i c_i C_{V_i} + \sum_i e_i \frac{\partial c_i}{\partial T}\bigg|_\rho = C_{V_f} + \sum_i e_i \frac{\partial c_i}{\partial T}\bigg|_\rho \tag{C.33}$$

The last term of the equation is unknown. We will search for a different form of the specific heat. From the relationship

$$\mathrm{d}T = \frac{\partial T}{\partial \rho}\bigg|_e \mathrm{d}\rho + \frac{\partial T}{\partial e}\bigg|_\rho \mathrm{d}e \tag{C.34}$$

it follows

$$\mathrm{d}e = \frac{\partial T}{\partial e}\bigg|_\rho^{-1} \mathrm{d}T - \frac{\partial T}{\partial \rho}\bigg|_e \frac{\partial T}{\partial e}\bigg|_\rho^{-1} \mathrm{d}\rho \tag{C.35}$$

By identification with the thermodynamic relationship

$$de = C_V dT + \alpha d\rho \tag{C.36}$$

where $\alpha = \left.\frac{\partial e}{\partial \rho}\right|_T$. We obtain

$$C_V = \left.\frac{\partial T}{\partial e}\right|_\rho^{-1} \tag{C.37}$$

This latter quantity was estimated earlier, in Eq. (C.18).

The calculation of C_p is a bit longer. From

$$\begin{cases} de = C_V dT + \alpha d\rho \\ dp = \chi d\rho + \kappa d(\rho e) \end{cases} \tag{C.38}$$

we deduce

$$\begin{cases} d(\rho e) = \rho C_V dT + (\alpha\rho + e) d\rho \\ d\rho = dp - \frac{\chi + \kappa(\alpha\rho + e)}{\rho C_V \kappa} dT \end{cases} \tag{C.39}$$

Furthermore, we have

$$d(\rho h) = \rho dh + h d\rho = d(\rho e + p) = dp + \rho C_V dT + (\alpha\rho + e) d\rho \tag{C.40}$$

then

$$d\rho = \frac{1}{\chi + \kappa(\alpha\rho + e)} [dp - \kappa\rho C_V dT] \tag{C.41}$$

And finally, referring to Eq. (C.40)

$$\rho dh = \rho C_V \left[1 - \frac{\kappa\left(\alpha\rho - \frac{p}{\rho}\right)}{\chi + \kappa(\alpha\rho + e)} \right] dT + \left[1 + \frac{\alpha\rho - \frac{p}{\rho}}{\chi + \kappa(\alpha\rho + e)} \right] dp \tag{C.42}$$

we identify C_p

$$C_p = C_V \left[\frac{\chi + \kappa h}{\chi + \kappa(\alpha\rho + e)} \right] \tag{C.43}$$

Reference

[1] Dubroca, B., and Morreeuw, J.-P., "An Extension of Roe's Approximate Riemann Solver for the Approximation of Navier-Stokes Equations in Chemical Nonequilibrium Cases," *Computing Methods in Applied Science and Engineering*, edited by Glowinski, R., Nova Sciences, New York, 1992, pp. 345–370.

Appendix D Homogenization of an Inhomogeneous Rough Surface

D.1 Smooth Inhomogeneous Surface

We will first treat the case of a smooth surface consisting of two materials of different reactivities. We assume the following restrictive assumptions:

- Steady flow of an incompressible laminar boundary layer, whose velocity profile, independent of heterogeneous reactions, is given a priori (Blasius profile). Reactive chemical species are therefore assumed in the minority, and the vertical component of velocity is neglected.
- All diffusion coefficients are equal.
- The heterogeneous chemical system is reduced to a single irreversible reaction of order 1 on the species i.
- The calculation is performed in an elementary cell with periodic boundary conditions (length L) following the direction of the flow s. The flow is parallel to the wall. This hypothesis is discussed in Sec. D.2.1. The conditions at the upper end are given as $c_j(y = h) = C^{ste}\ \forall j$.

The system then reduces to (see Chapter 2)

$$\begin{cases} \nabla \cdot (uc_j - \mathcal{D}\nabla c_j) = 0 \\ c_j(x + L) = c_j(x) \\ \mathbf{n} \cdot (\mathcal{D}\nabla c_i)\big|_p = -\dfrac{\dot{\omega}_i}{\rho} \end{cases} \tag{D.1}$$

with

$$\frac{\dot{\omega}_i}{\rho} = \frac{k\theta p_i}{\rho} = \kappa(x)c_i \tag{D.2}$$

where the reactivity $\kappa = (k\theta \mathcal{R}T/\mathcal{M}_i)$ is a discontinuous function of s valued κ_f and κ_R on media f and R, respectively.

One seeks the solution $\mathcal{C}_j$ for the homogenized problem that satisfies the system in Eq. (D.1) in which the surface condition is written

$$\mathbf{n} \cdot (\mathcal{D}\nabla \mathcal{C}_i)\big|_p = -\kappa_{eq}\mathcal{C}_i = -\langle \kappa c_i \rangle \tag{D.3}$$

$\langle f \rangle$ is the moment of order 0 (average) of f.

The perturbation $c_j' = c_j - C_j$ of the solution C_j, of average $\langle c_j' \rangle = 0$, satisfies again the system in Eq. (D.1), with the condition on the surface

$$\mathbf{n} \cdot (\mathcal{D}\nabla c_i')|_p = \kappa C_i + \kappa c_i' - \langle \kappa (C_i + c_i') \rangle \tag{D.4}$$

So by introducing the difference of reactivity $\kappa' = \kappa - \langle \kappa \rangle$

$$\mathbf{n} \cdot (\mathcal{D}\nabla c_i')|_p = \kappa' C_i - \kappa c_i' - \langle \kappa' c_i' \rangle \tag{D.5}$$

We search for the solution as follows:

$$c' = a(x, y)c_j + \mathbf{b}(x, y) \cdot \nabla c_j \tag{D.6}$$

This ansatz represents a local linear perturbation of concentration profile. The unknowns a and $\mathbf{b}$ satisfy the following systems:

$$\begin{cases} \nabla \cdot (ua - \mathcal{D}\nabla a) & = 0 \\ a(x+L) & = a(x) \\ a(y=h) & = 0 \\ \mathbf{n} \cdot (\mathcal{D}\nabla a)|_p & = -\kappa' - \kappa s + \langle \kappa' s \rangle \\ \int_\Omega a \mathrm{d}V & = 0 \end{cases} \tag{D.7}$$

$a(y = h) = 0$ results from the condition imposed in $y = h$. The last equation imposes equal masses for original and homogenized problems

$$\begin{cases} \nabla \cdot (\mathbf{u} \cdot \mathbf{b} - \mathcal{D}\nabla \cdot \mathbf{b}) & = 0 \\ \mathbf{b}(x+L) & = \mathbf{b}(x) \\ \mathbf{b}(y=h) & = 0 \\ \mathbf{n} \cdot (\mathcal{D}\nabla \mathbf{b})|_p & = -\kappa \mathbf{b} + \langle \kappa' \mathbf{b} \rangle \\ \int_\Omega \mathbf{b} \mathrm{d}V & = 0 \end{cases} \tag{D.8}$$

The latter system admits the trivial solution $\mathbf{b} = 0$.

The system in Eq. (D.7) must be solved numerically. Equivalent reactivity is then calculated from Eq. (D.3)

$$\kappa_{eq} = \langle \kappa \rangle + \langle \kappa' a \rangle \tag{D.9}$$

D.2 Rough Inhomogeneous Surface

D.2.1 Flow over Rough Wall

The first step in the calculation is to replace the rough surface with a smooth surface with ad hoc condition limits, leading to approximately the same flow. The aim of this is to not have to describe the details of the flow at the scale of roughness. For this, we wish to establish a boundary condition called "wall law" or "jump condition." This is a relationship giving a slip rate imposed on a hypothetical smooth surface that will be set at a height δ relative to the physical domain. We can illustrate this with an example in modeling of turbulence where we use the following law [1]:

$$v(\delta) = v_\tau \left[\kappa^{-1} \ell n \left(\frac{\rho v_\tau \delta}{\mu} \right) + B(l_\delta) \right] \tag{D.10}$$

where $v_\tau = [(\mu/\rho)(\partial v/\partial y)]|_{y=\delta}^{1/2}$ is the wall shear velocity and $\kappa \simeq 0.41$ is the Von Karman constant. The value of B depends on the roughness height l_δ, obtained from experimental correlations.

In practice, one can apply this boundary condition on the surface itself because the definition of altitude above a rough surface is poorly defined.

Theoretical results were obtained by scaling in 2-D surfaces for compressible flows [2], or for incompressible flows in the case of low Reynolds numbers [3]. The latter case allows for full results. Its development will be done with the same assumptions as in the previous section (see Fig. D.1). We suppose no injection at wall.

The stationary equations to solve for $\mathbf{V}$ in the domain Ω are the following (see Sec. 2.2):

$$\begin{cases} \rho\mathbf{V} \cdot \nabla\mathbf{V} - \mu\Delta\mathbf{V} + \nabla p = 0 & \text{in} \quad \Omega \\ \nabla \cdot \mathbf{V} = 0 & \text{in} \quad \Omega \\ \mathbf{n} \cdot \mathbf{V}_p = 0 & \text{on the wall} \end{cases} \tag{D.11}$$

Speed $\mathcal{U}$ and pressure $\mathcal{P}$, approximate values in Ω_{eq}, differ from exact quantities of a small amount of some $\epsilon = (l_\delta/h)$, where h is the boundary layer thickness. They therefore satisfy the same system of equations in calculating the volume Ω_{eq}.

We define the quantities $\mathbf{V}' = \mathbf{V} - \mathcal{U}$ and $p' = p - \mathcal{P}$, for which we will seek the solution in the local domain containing the roughness Ω_R. For

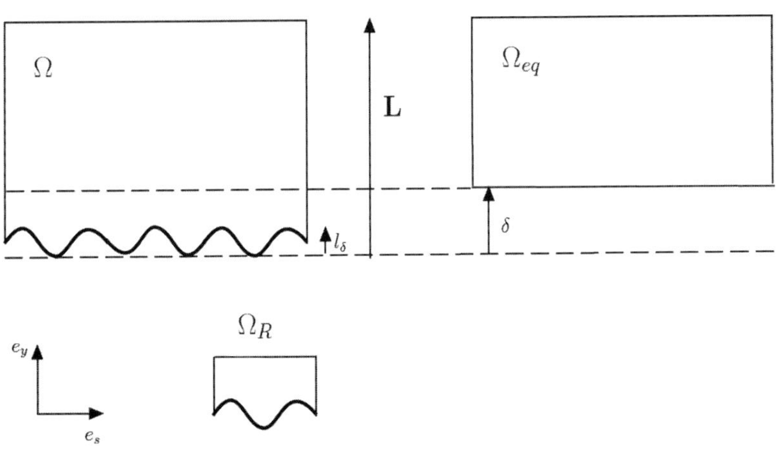

Fig. D.1 Calculation domains on rough walls and notations.

this, we define $\mathcal{U}$ in this domain by linear extrapolation of its value in the domain Ω_{eq}

$$\mathcal{U} = (y - \delta)\frac{\partial \mathcal{U}_s}{\partial y}\mathbf{e_y} \tag{D.12}$$

It is also necessary to impose an upper boundary condition on $y = \delta$. The most natural condition to impose is the stress tensor, then shear and pressure.

This gives, to order 0, the following system:

$$\begin{cases} -\mu\Delta\mathbf{V}' + \nabla p' = 0 & \text{in } \Omega_R \\ \nabla \cdot \mathbf{V}' = 0 & \text{in } \Omega_R \\ \mathbf{V}' + \mathcal{U} = 0 & \text{on the wall} \end{cases} \tag{D.13}$$

The solution will be sought in the form of periodic functions in s. [It was inspired by the form of Eq. (D.12).]

$$\begin{cases} \mathbf{V}' = \dfrac{\partial \mathcal{U}_s}{\partial y}(\mathbf{d} + l_\delta \mathbf{e_y}) \\ p' = \pi \dfrac{\partial \mathcal{U}_s}{\partial y} \end{cases} \tag{D.14}$$

Substituting into Eq. (D.13), we get

$$\begin{cases} -\mu\Delta\mathbf{d} + \nabla \pi = 0 & \text{in } \Omega_R \\ \nabla \cdot \mathbf{d} = 0 & \text{in } \Omega_R \\ \mathbf{d} + y\mathbf{e_y} = 0 & \text{on the wall} \end{cases} \tag{D.15}$$

This system must be solved numerically.

It is shown [3] that $\mathbf{d}$ reaches exponentially to a constant vector. Far from the wall we have $\mathbf{V}' = 0$, which implies that, at the top of the domain Ω_R.

$$\mathbf{d}(\delta) = -\delta\mathbf{e_y} \tag{D.16}$$

This equation allows one to calculate δ.

It is possible to calculate the order 1 using slowly varying functions with s. This eliminates the assumption of strict periodicity. The results are not significantly different numerically from the solution of order 0, which conversely validates the hypothesis of periodicity that can appear rather severe.

D.2.2 Equivalent Reactivity

We must now solve numerically the mesoscopic local problem [Eq. (D.1)] to find the mass fractions of various species and the average mass flux (vertical, taking into account the periodicity condition) $\langle J \rangle$ of reactive species.

It is further noted that the calculation in Sec. D.2.1 is consistent with an approximation of the first order for the diffusion, obtained by extension

$$J_R = \frac{\mathcal{D}c_i}{h - l_\delta} \tag{D.17}$$

The equivalent conductivity is deduced

$$\kappa_{\text{eq}} = \frac{\frac{\langle J \rangle}{C_I}}{1 - \frac{\langle J \rangle}{J_R}} \tag{D.18}$$

The numerical results show that the effective reactivity is mainly dependent on the reaction–diffusion system, and that convection plays no role until Reynolds numbers are high enough.

The comparison between these results and a direct calculation allows us to see an agreement up to a domain where the assumption of decoupling of fields is hardly respected.

Finally, we note that a direct analysis of stationary flow (Sec. 6.5.2) gives the main results presented here. The homogenization method, however, is fundamental to further study of the effect of roughness (see Sec. 7.1.2.4).

References

[1] Chassaing, P., *Turbulence en Mécanique des Fluides*, Editions Cépaduès, Paris, 2000.

[2] Carrau, A., Gallice, G., and Le Tallec, P., "Taking into Account Surface Roughness in Computing Hypersonic Re-entry Flows," *Computing Methods in Applied Sciences and Engineering*, edited by R., Glowinsky, Nova Science, New York, 1991, pp. 331–343.

[3] Achdou, Y., Pironneau, O., and Valentin, F., "Effective Boundary Conditions for Laminar Flows over Periodic Rough Boundaries," *Journal of Computational Physics*, Vol. 147, 1998, pp. 187–218.

Appendix E Mass Loss by Pyrolysis

E.1 Sample Size

The phenomenon of pyrolysis in the phenolic resin composites used for reentry are very violent. The temporal gradient can reach several hundred Kelvins per second. The characterization of this phenomenon is therefore very difficult. Indeed, unless one is able to heat the sample in volume (which is difficult), use of very small samples permits one to minimize the temperature gradient in the material. It is this problem that we will develop here. We will use 1-D situations to obtain orders of magnitude of the problem. Heating will be calculated using the isotherm approximation

$$\bar{\rho} C_p e \beta = \dot{q} \tag{E.1}$$

where e is the thickness of the sample submitted to flux $\dot{q}$, $\beta = \mathrm{d}T/\mathrm{d}t$ is a parameter of experience, $\bar{\rho}$ is the mass density, and C_p the total specific heat (see Sec. 8.4.2).*

The spatial gradient of temperature in the material is given by

$$\frac{\Delta T}{e} \simeq \frac{\dot{q}}{\lambda} \tag{E.2}$$

The constraint expressing the quality of the result will be given by $(\Delta T/T) < \varepsilon$.

By solving, we get

$$\beta = \frac{\alpha \epsilon T}{e^2} \tag{E.3}$$

where $\alpha = \lambda / \bar{\rho} C_p$ is the diffusivity of the material.

So the thickness required to achieve the goal regardless of the temperature

$$e = \left(\frac{\alpha \varepsilon T}{\beta} \right)^{\frac{1}{2}} \tag{E.4}$$

The minimum value is obtained for low temperatures (T and α minimum)

$$e_{\min} = \left(\frac{\alpha \epsilon T_{\min}}{\beta} \right)^{\frac{1}{2}} \tag{E.5}$$

* All densities used in this section are apparent densities $\bar{\rho}$, that is to say, mass divided by the total volume solid plus pores.

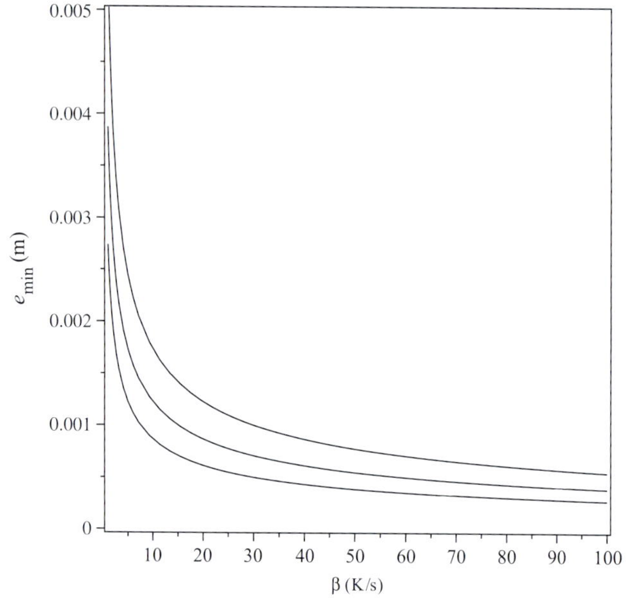

Fig. E.1 Maximum sample size vs. temperature time gradient.

Figure E.1 shows the results for a carbon resin and for various constraints ε, giving the quality of measurement. It was found that the maximum allowable size decreases very rapidly with the desired β. This value is only slightly dependent on the quality criterion (curves correspond to $\varepsilon = 5\%$, 10%, and 20%). For a representative value of $100\ \mathrm{K \cdot s^{-1}}$ already obtained in some experiments [1, 2], it is necessary in all cases to use millimetric samples (or less). For example, the laboratory SoRI [3] uses powders of a diameter less than 75 μm. However, the description of the experiment is not clear on the thickness sample formed with the powder. Inhomogeneities in temperature are recorded and used to estimate error-bound results. ONERA uses millimetric specimens in experiments [1, 2]. Given these values, $100\ \mathrm{K \cdot s^{-1}}$ is an ultimate value used in the laboratory whose results should be viewed with caution.

E.2 Activation Energies

Assume for now that each peak mass loss is separated from the preceding and the following. Each solid species will decompose irreversibly under a law of first order

$$\frac{\mathrm{d}\bar{\rho}_i}{\mathrm{d}t} = -A_i \bar{\rho}_i \exp\left(-\frac{E_{A_i}}{RT}\right) \tag{E.6}$$

The density used in this equation is the mass density (mass per component unit volume of material). We define the temporal gradient of temperature

$$\frac{d\bar{\rho}_i}{dt} = \frac{d\bar{\rho}_i}{dT}\frac{dT}{dt} = \beta\frac{d\bar{\rho}_i}{dT} \tag{E.7}$$

and then the mass loss equation

$$\frac{d\bar{\rho}_i}{\bar{\rho}_i} = -\frac{A_i}{\beta}\exp\left(-\frac{E_{A_i}}{RT}\right)dT \tag{E.8}$$

whose solution is

$$\ell n\left(\frac{\bar{\rho}_i}{\bar{\rho}_{0_i}}\right) = -\frac{A_i T}{\beta}\exp\left(-\frac{E_{A_i}}{RT}\right) + \frac{A_i E_{A_i}}{\beta R} E_1\left(\frac{E_{A_i}}{RT}\right) \tag{E.9}$$

where E_1 is the exponential integral defined by

$$E_1(x) = \int_1^{\infty} \frac{e^{xy}dy}{y} \tag{E.10}$$

By derivation of Eq. (E.8), we get

$$\frac{d^2\bar{\rho}_i}{dT^2} = -\frac{A_i}{\beta}\bar{\rho}_i \exp\left(-\frac{E_{A_i}}{RT}\right)\left[-\frac{A_i}{\beta}\exp\left(-\frac{E_{A_i}}{RT}\right) + \frac{E_{A_i}}{RT^2}\right] \tag{E.11}$$

You can then find a relationship that meets the abscissa T_{m_i} of the minimum of the derivative of mass loss by writing

$$\left.\frac{d^2\bar{\rho}_i}{dT^2}\right|_{T=T_{m_i}} = 0 \tag{E.12}$$

so

$$\frac{d\left[\ell n\left(\frac{\beta}{T_{m_i}^2}\right)\right]}{d\left(\frac{1}{T_{m_i}}\right)} = -\frac{E_{A_i}}{R} = -T_{A_i} \tag{E.13}$$

The Kissinger relationship in Eq. (E.13) provides a way to measure the activation energies E_{A_i} (or activation temperature T_{A_i}) from the abscissas of the peaks derived from mass loss. To do this we draw $\ell n(1/T_{m_i}^2)$ vs. $1/T_{m_i}$. In Fig. E.2, corresponding to the first pyrolysis peak measured at ONERA [2], we see that the points are properly aligned, with the exception of the value corresponding to $\beta = 100\ \mathrm{K \cdot s^{-1}}$, which is not surprising given the remark made in the previous section.

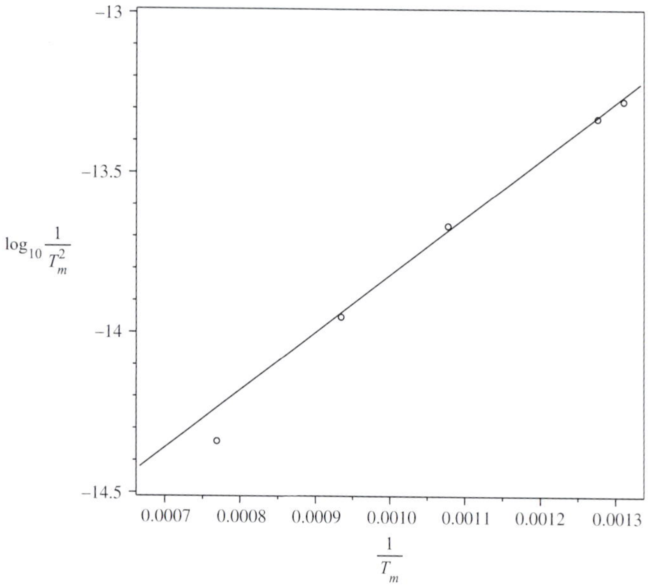

Fig. E.2 Determination of activation temperature (temperatures in K).

The formal solution of Eq. (E.12) is

$$T_{m_i} = \frac{T_{A_i}}{2\mathcal{W}\left[\frac{1}{2}\left(\frac{A_i T_{A_i}}{\beta}\right)^{\frac{1}{2}}\right]} \tag{E.14}$$

where $\mathcal{W}$ is the Lambert function defined by $\mathcal{W}(y)\, e^{\mathcal{W}(y)} = x$.

E.3 Example of Thermogravimetric Analysis

Solve the following example:

$$\begin{array}{ll} \text{Reaction 1} & E_1 \Rightarrow E_2 + \text{gaz} \\ \text{Reaction 2} & E_2 \Rightarrow \text{gaz} \end{array}$$

The system describing these reactions is as follows:

$$\frac{d\bar{\rho}_1}{dT} = -\frac{A_1\bar{\rho}_1}{\beta}\exp\left(-\frac{T_{A_1}}{T}\right) \tag{E.15}$$

$$\frac{d\bar{\rho}_2}{dT} = \alpha\frac{A_1\bar{\rho}_1}{\beta}\exp\left(-\frac{T_{A_1}}{T}\right) - \frac{A_2\bar{\rho}_2}{\beta}\exp\left(-\frac{T_{A_2}}{T}\right) \tag{E.16}$$

where $\alpha = 0.5$ is the mass fraction of E_1 turned into gas and $[\beta = (\mathrm{d}T/\mathrm{d}t) = 10\ \mathrm{K} \cdot \mathrm{s}^{-1}]$ is the rise rate of temperature. In the two cases treated, $A_1 = A_2 = 1\ \mathrm{s}^{-1}$. In the first case, $T_{A_2} = 2000$ K and $T_{A_2} = 3000$ K. In this case (Fig. E.3), the exact calculation and that obtained by simple summation have a moderate difference. Indeed, there is a case where the separation of reactions is a reasonable approximation. In the second case, the exact result is considerably shifted towards higher temperatures. In the figure, the squares represent species 1 and 2, the continuous curve the actual density, and the dotted curve that obtained by summing the Arrhenius laws.

The application to the case of phenolic resin shows a great difference between a three-step model and mass loss obtained by summing the corresponding Arrhenius laws (see Fig. E.4, which corresponds to Fig. 8.7 of Sec. 8.2.1.3). The simple addition of the Arrhenius laws (with the same parameters) underestimates the mass loss obtained by the differential system.

E.4 Homogenization

E.4.1 Case of a Single Gas Species

When a material undergoes a phase change (pyrolysis), the laws of evolution must be considered in the change in geometry and in particular the variation of porosity. We will establish here a simplified mass balance assuming that the medium consists of a single fluid.

We will consider a material that we will homogenize on an elementary representative volume (ERV) Y. This material consists of a fluid phase $Y_s(t)$ of volume V_f separated from a solid phase $Y_s(t)$ of volume V_s by an interface $\sum_1(t)$. The normal to the interface will be denoted $\mathbf{n}(x)$ and will be oriented towards the solid fluid (see Fig. E.5). The recession velocity of the

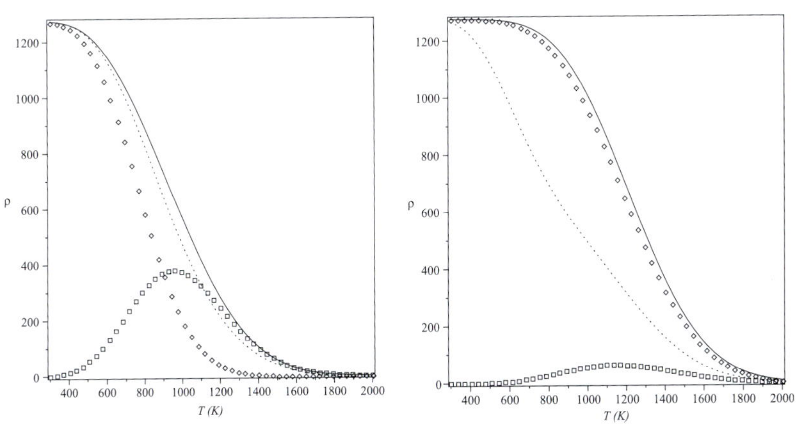

Fig. E.3 Densities of pyrolyzed materials; influence of activation temperatures.

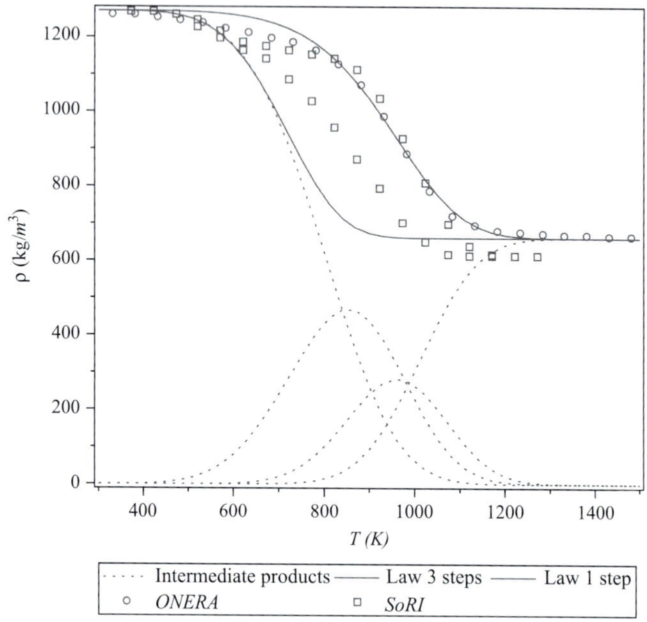

Fig. E.4 Comparison of mass loss models (continuous curves).

interface will be denoted $\mathbf{v}_a$. This ERV is separated from its neighbors by the interface $\sum_2(t)$.

Pyrolysis is interpreted here as an ablation of the inner surface of the material. This is necessary to take into account the concept of porosity. We saw an equivalence between pyrolysis and ablation in Sec. 8.1.2.

E.4.1.1 Gas Phase

Starting from the equation of mass conservation

$$\frac{\partial \rho_g}{\partial t} + \nabla \cdot (\rho_g \mathbf{V}) = 0 \tag{E.17}$$

and integrating on Y_f we get

$$\int_{Y_f} \frac{\partial \rho_g}{\partial t} \mathrm{d}\mathbf{v} + \int_{Y_f} \nabla \cdot (\rho_g \mathbf{V}) \mathrm{d}\mathbf{v} = 0 \tag{E.18}$$

The transport theorem of Leibnitz–Reynolds can be written

$$\int_{Y_f} \frac{\partial \rho_g}{\partial t} \mathrm{d}\mathbf{v} = \frac{\partial}{\partial t} \int Y_f \, \rho_g \, \mathrm{d}\mathbf{v} - \int_{\Sigma_1} \rho_g \mathbf{v}_a \cdot \mathbf{n} \, \mathrm{d}\sigma \tag{E.19}$$

Moreover, the theorem of Green–Ostrogradsky is written

$$\int_{Y_f} \nabla \cdot (\rho_g \mathbf{V}) \mathrm{d}\mathbf{v} = \int_{\Sigma_1} \rho_g \mathbf{V} \cdot \mathbf{n}\, \mathrm{d}\sigma + \int_{\Sigma_2} \rho_g \mathbf{V} \cdot \mathbf{n}\, \mathrm{d}\sigma \tag{E.20}$$

Over $\sum 1,$ the mass conservation is locally written

$$\rho_s \mathbf{v}_a \cdot \mathbf{n} = -\rho_g \mathbf{V} \cdot \mathbf{n} \tag{E.21}$$

Equation (E.18) will thus be written, adding the quantity $\int_{\Sigma_2} \rho_g\ \mathbf{V}\ \mathrm{d}\sigma$, which is identically zero for a nonporous system pyrolysis ($\mathbf{V} = 0$ on the surface)

$$\frac{\partial}{\partial t} \int_{Y_f} \rho_g \mathrm{d}\mathbf{v} + \int_{\Sigma_1} (\rho_s + \rho_g) \mathbf{v}_a \cdot \mathbf{n}\, \mathrm{d}\sigma + \int_{\Sigma_1} \rho_g \mathbf{V} \mathrm{d}\sigma$$
$$+ \int_{\Sigma_1 \cup \Sigma_2} \rho_g \mathbf{V}\, \mathrm{d}\sigma = 0 \tag{E.22}$$

In the second term, we neglect ρ_g compared to ρ_s.

This allows us to define a convection term similar to that obtained in the absence of pyrolysis

$$\int_{\Sigma_1 \cup \Sigma_2} \rho_g \mathbf{V} \mathrm{d}\sigma = \int_{Y_f} \nabla \cdot (\rho_g \mathbf{V}) \mathrm{d}\mathbf{v} \tag{E.23}$$

The homogenized equation is obtained from Eq. (E.22) by dividing by V to make appear

- Porosity $\epsilon = V_f / V$
- Specific surface $S_v = \Sigma_1 / V$

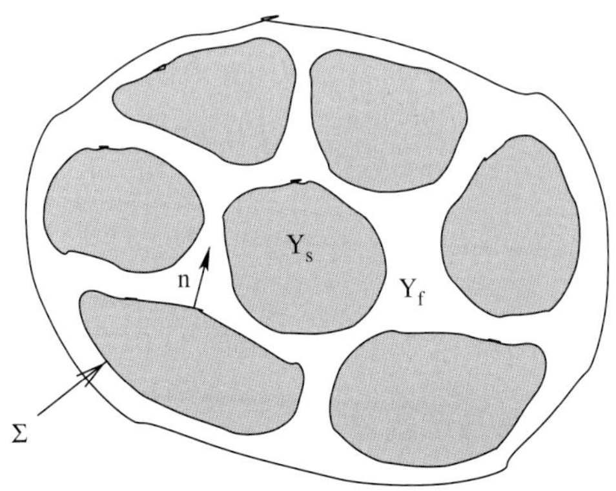

Fig. E.5 Porous material.

This equation is obtained by assuming the different quantities homogeneous in the reference volume

$$\frac{\partial(\epsilon\rho_g)}{\partial t} + \nabla\cdot(\epsilon\rho_g \mathbf{V}) = S_v \dot{m}_s \tag{E.24}$$

In this last equation we introduced the homogenized quantities

$$\begin{cases} \nabla\cdot(\rho_g \mathbf{V}) = \dfrac{1}{\mathrm{V}}\displaystyle\int_{Y_f} \nabla\cdot(\rho_{\mathbf{g}}\mathbf{V})\mathrm{d}\mathbf{v} \\ S_v \dot{m}_s = -\dfrac{1}{\Sigma_1}\displaystyle\int_{\Sigma_1} \mathbf{v}_a \cdot \mathbf{n}\mathrm{d}\sigma \end{cases} \tag{E.25}$$

$\dot{m}_s > 0$ s the average mass flow per unit area on $\sum_1$.

E.4.1.2 Solid Phase

As before, the conservation of the solid phase is written

$$\int_{Y_s} \frac{\partial \rho_s}{\partial t}\mathrm{d}\mathbf{v} = 0 \tag{E.26}$$

so

$$\frac{\partial}{\partial t}\int_{Y_s} \rho_s \mathrm{d}\mathbf{v} - \int_{\Sigma_1} \rho_s \mathbf{v}_a \cdot \mathbf{n}\,\mathrm{d}\sigma = 0 \tag{E.27}$$

Homogenizing, we get

$$\rho_s \frac{\partial \epsilon}{\partial t} = S_v \dot{m}_s \tag{E.28}$$

This law is called the Geometric Conservation Law (GCL). It shows the evolution of porosity. Note that this method supposes implicitly that the products of pyrolysis have the same density as the resin.

E.4.2 Case of a Reactive Gas Mixture

The equation for the conservation of the species i is written

$$\frac{\partial \rho_{g_i}}{\partial t} + \nabla\cdot(\rho_{g_i}\mathbf{V}_i) = \dot{\omega}_i \tag{E.29}$$

By reproducing the calculations in Sec. E.4.1, we obtain the desired equation

$$\frac{\partial(\epsilon\rho_{g_i})}{\partial t} + \nabla\cdot(\epsilon\rho_{g_i}\mathbf{V}_i) = \epsilon\dot{\omega}_i + S_v \dot{m}_{s_i} \tag{E.30}$$

By summing over all species ($\sum_i \dot{\omega}_i = 0$, $\sum_i \dot{m}_{s_i} = \dot{m}_s$ and $\sum_i \rho_{g_i} V_i = \rho V$) we find the Eq. (E.24).

References

[1] Bouvet, A., Demange, D., and Hervé, P., "Caractérisation thermique de matériaux thermo- dégradables," *Congrès Francais de thermique SFT2005*, Reims, France, May–June 2005.
[2] Duffa, G., "Ablation," Monograph CESTA, Commissariat a l'Energie Atomique, Le Barp, France, Nov. 1996.
[3] Stokes, E. H., "Permeability of Carbonized Rayon Based Polymer Composites," *Computational Mechanics of Porous Materials and Their Thermal Decomposition, Applied Mechanics Division*, Vol. 136, ASME 1992, pp. 145–156.

Appendix F Water in Phenolic Composite Materials

Resin

Measuring the quantity of trapped water in phenolic resin (in a stationary state) shows an almost linear variation with the partial pressure of water vapor, made dimensionless by saturation vapor pressure (Fig. F.1): the relative humidity *HR*. The maximum weight set is low, at about 4%. This result is probably dependent on the polymerization cycle.

Diffusion in the resin is an extremely slow phenomenon: the apparent diffusion coefficient is $\mathcal{D} = 1.0 \times 10^{-13}\ \mathrm{m^2 \cdot s^{-1}}$ when rising, and about twice as great on the descent. This figure is comparable to that encountered in zeolites. This coefficient shows a temperature dependence described by an Arrhenius law, activation energy being about $58\ \mathrm{KJ \cdot mol^{-1}}$. This low value and the phenomenon of "activation" in temperature show that diffusion is achieved by moving molecules physically linked to the material (as opposed to a Knudsen diffusion).

The expansion of a silica-based composite with moisture content shows an almost linear rise after a period of "incubation." The silica fiber used in this composite is only slightly susceptible to water; it is conceivable that this is an effect linked to the resin alone. This type of response can be interpreted in terms of "free" volume. Up to about 0.2–0.3% mass, water finds micropores ($d < 2$ nm) compatible with the steric size of the molecule. In addition, the setting causes structural modifications resulting in a change of the volume or also a change in the infrared (IR) response {quoted in [1]} or a reversible change in glass transition temperature. This may also explain the increasing diffusion coefficient during the descent, in an absorption–desorption cycle [2]. This is sometimes interpreted in terms of microcracks [1]. One can also consider deformations of the macromolecular network of the type encountered in foams subjected to an external force [3]. The latter explanation is compatible with the reversibility phenomenon observed in the variation of the glass transition temperature. In the latter case, the increase in permeability has no counterpart in terms of porosity, which is a measurable criterion, which seemingly was never verified.

Mechanisms of absorption in free volume should correspond to different characteristics (diffusion rate, activation temperature) of the subsequent

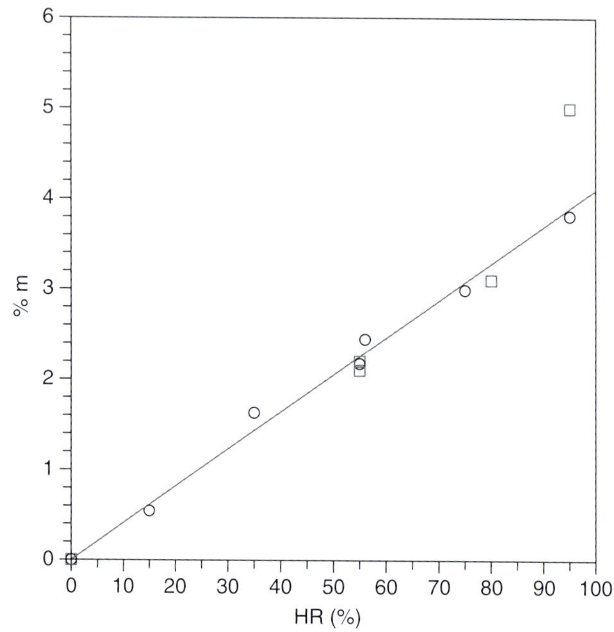

Fig. F.1 Relative mass of water absorbed in the resin.

response. Given the narrowness of the field of functioning for diffusion in free volume and the measurement accuracy, this effect is not observable.

Nuclear magnetic resonance (NMR) measurements do confirm the presence of water molecules in various forms, but are difficult to interpret, perhaps because of the complex structure of a layer of water only a few tenths of a nanometer thick [4].

F.2 Fiber

We focus on the ex-rayon fiber used for carbon-resin materials because of its low thermal conductivity. These fibers have a high oxygen content, typically 4% in mass. Unlike other fibers (ex-PAN, pitch, or silica fibers), they absorb a large amount of water: 15% in mass at 100% *HR* (Fig. F.2).

Figure F.2 presents the results in terms of absorption and desorption. There is a slight difference between the two curves. This can be explained by examining the differential scanning calorimetry (DSC) measurements, which show a binding energy with much lower values for large *HR*: $30\ \text{kJ} \cdot \text{mol}^{-1}$ at 65% *HR* vs. $90\ \text{kJ} \cdot \text{mol}^{-1}$ at 25% *HR*. One can therefore conclude the presence of water in layers up to the condensation for large values of *HR*. This creates a hysteresis phenomenon associated with modified equilibrium pressure by surface tension. Applying the Kelvin equation [5], and assuming cylindrical pores, the differences between the absorption and desorption curves of Fig. F.2 correspond to diameters less than 2 nm.

This phenomenon is called absorption isotherm type V, or Ruthven's isotherm [6]. It can be interpreted as follows:

- Absorption on sites, the phenomenon being limited by the size of the "cages" that can contain molecules. This results in a linear rise followed by saturation.
- Multilayer physical absorption on molecules already stuck in the nanoscale pores (analogous to the BET adsorption), until saturation of macropores.

The apparent diffusion coefficient measured at 20 °C is $6 \times 10^{-16}\, m^2 \cdot s^{-1}$. The variation temperature is unknown.

There are no data on the dilation induced.

F.3 Composite Material

F.3.1 Low-Temperature Outgassing

Measurements made on various types of tape-wrapped carbon phenolic (TWCP) composite [7, 8] show a small difference in absorption/desorption. Assuming additivity of mass absorption by resin and fiber, one can calculate

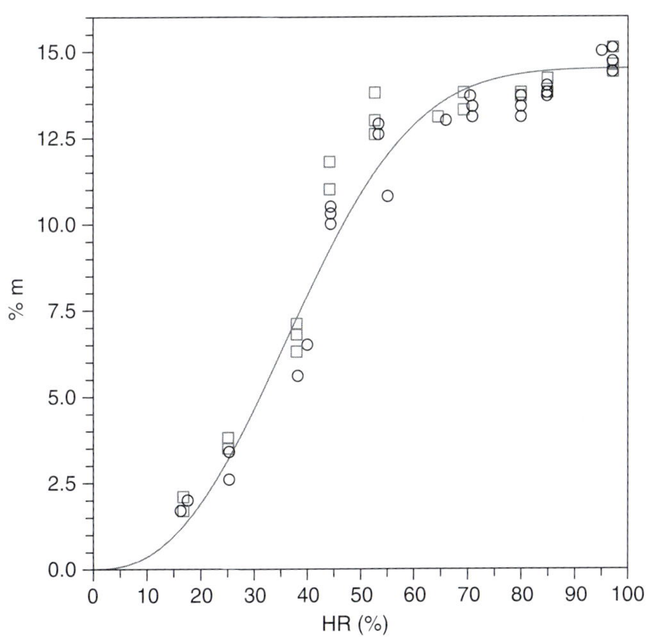

Fig. F.2 Amount of water absorbed relative to mass of the fiber (circles: absorption; squares: desorption).

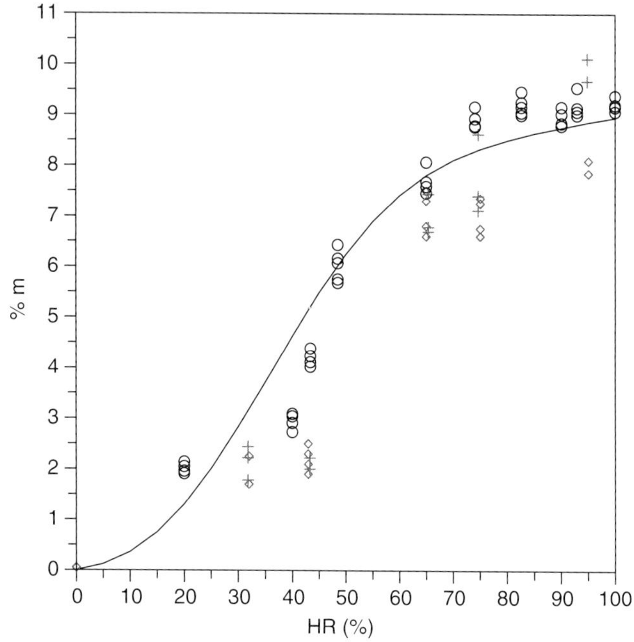

Fig. F.3 Amount of water absorbed on the total mass (8).

the mass absorption from the values measured on both constituents. This gives a curve very similar to the one measured on the composite (Fig. F.3). It simply indicates that all these materials are very similar and they do not have large specific surface porosity at the fiber–matrix interface.

Regarding the diffusion coefficients, the values measured in the composite are, at room temperature, $\mathcal{D} = 4.6 \times 10^{-14}\ \mathrm{m^2 \cdot s^{-1}}$ in the direction of the fabric, and $\mathcal{D} = 1.8 \times 10^{-14}\ \mathrm{m^2 \cdot s^{-1}}$ in the perpendicular direction. These values are consistent with those measured on the fibers, matrix, and composites. Indeed, given the characteristic sizes for the fiber and matrix, and their respective diffusion coefficients, a simple calculation shows that the characteristic time of saturation of the fiber is much shorter than that of the matrix.

$$\frac{t_R}{t_f} \simeq \left(\frac{e_R}{e_f}\right)^2 \frac{\mathcal{D}_f}{\mathcal{D}_R} \simeq 400 \tag{F.1}$$

The resin being the only continuous material in the medium, it pilots the entire process of water uptake.

The diffusion coefficient in the plane of the fabric will then be given simply by the volume fraction of fiber, the ratio $\mathcal{D}_f/\mathcal{D}_R$ being very small: $\mathcal{D}// = V_R \mathcal{D}_R = 4.6 \times 10^{-14}\ \mathrm{m^2 \cdot s^{-1}}$, a value remarkably close to the measured value.

Reference [7] provides values for the expansion of the material. In the absence of data for the fiber, it is difficult to discuss these results. However, we note that this expansion is much larger in the direction perpendicular to the tissue, the expansion of the fiber being much lower than that of the resin.

Of course, the presence of water in the material currently in use is the result of a preliminary study of the life of this material in the phases of manufacturing, warehousing, and so on.

F.3.2 Outgassing at High Heating Rates

Experiments on constrained expansion on a TWCP material were performed and interpreted [9, 10] as follows. It is assumed that water is present in the resin alone; the model does not take into account the fiber. The medium of resin plus water is described as a liquid mixture (model of free volume) at thermodynamic equilibrium. The thermodynamic properties of water are assumed those of the liquid.

Initially, we focus on the expansion of material when porosity is not yet open. From measurements related to the expansion of the dry material, it is possible to know the characteristics of this material. We can then, knowing the volume fraction of water and its expansion coefficient, calculate the measured stress on wet material. There is a good agreement despite a questionable assumption (fiber absent or playing the same role as the resin).

This same approach is used to interpret experiments of free expansion.

This model, coupled with the permeation equation, is then used to predict mass loss in the range $T = [0, 400\ ^\circ\mathrm{C}]$. The result, using a measured permeability ($K = 2.7 \times 10^{-20}\ \mathrm{m}^2$), corresponds fairly well to the measurement. This allows one to predict the internal pressure of the order of 50 bars for experience with a heating of $0.05\ ^\circ\mathrm{C} \cdot \mathrm{s}^{-1}$. However, this model without porosity change with time is difficult to use. Indeed, the permeability materials are very variable, and can even be virtually zero (the detectability limit of the apparatus used for this type of measurement is of the order of $10^{-27}\ \mathrm{m}^2$) [11]. But the calculations [12] show that the internal pressure is very dependent on this value.

It seems likely that this rise in initial pressure leads to a thermomechanical fracturing that can also be observed in the tissue.

After the fracturing, such a medium can be described by a diffusion–permeation model as described in [13]. A calculation assuming no spatial pressure variation in the fractures accredits such an approach.

References

[1] Stokes, E. H., "Equilibrated Moisture Component of Several Carbon Phenolic Composites," *AIAA Journal*, Vol. 30, No. 6, 1992, pp. 1597–1601.

[2] Wong, T. C., and Broutman, L. J., "Water in Epoxy Resins. Part II: Diffusion Mechanism," *Polymer Engineering and Science Journal*, Vol. 25, No. 9, 1985, pp. 529–534.

[3] Aref, H., and Vainchtein, D. L., "The Equation of State of a Foam," *The Physics of Fluids*, Vol. 12, No. 1, 2000.
[4] Polubarinova-Kochina, P. Y., *Theory of Ground Water Movement*, Princeton University Press, Princeton, NJ, 1962.
[5] Atkins, P. W., *Physical Chemistry*, 6th ed., W. H. Freeman & Co., New York, 2000.
[6] Gregg, S. J., and Sing, K. S. W., *Adsorption, Surface Area and Porosity*, Academic Press, New York, 1982.
[7] Stokes, E. H., "Equilibrated Moisture Component of Several Carbon Phenolic Composites," *AIAA Journal*, Vol. 30, No. 6, 1992, pp. 1597–1601.
[8] Stokes, E. H., "Anomalous Swelling Behaviour of FM5055 Carbon Phenolic Composite," *AIAA Journal*, Vol. 31, No. 3, 1993, pp. 584–589.
[9] Sullivan, R. M., and Stokes, E. H., "Porous Media and Mixture Models for Hydrothermal Behaviour of Phenolic Composites," *ASME International Mechanical Engineering Conference and Exposition*, Atlanta, GA, Nov. 1996.
[10] Sullivan, R. M., and Stokes, E. H., "A Model for the Effusion of Water in Carbon Phenolic Composites," Applied Mechanics Division, Vol. 233, Application of Porous Media Methods for Engined Materials, ASME, 1999.
[11] Ducamp, V., "Transferts thermiques dans un matériau composite carbone-résine," Ph.D. thesis, l'Université Bordeaux, No. 2505, Bordeaux, France, April 2002.
[12] McManus, H. L., and Chamis, C. C., "Stress and Damage in Polymer Matrix Composite Materials Due to Material Degradation at High Temperatures," NASA Technical Memorandum 4682, Dec. 1995.
[13] MacManus, H. L., and Tai, D. S., "Advanced Modeling of the Generation and Movement of Gases Within a Decomposing Polymer Composite," *High Temperature Properties and Applications of Polymeric Materials*, edited by Tant M., Cornell, J., and MacManus, H., American Chemical Society, Washington, DC, 1995.

Appendix G Radiative Transfer in a Plane Interface of Silica

The liquid layer surface is treated as a slab. The material at high temperature is supposed to have the absorption coefficient measured at low temperature [1]. This assumption is based on the fact that the structure of this material is, in all cases, a glass.

Boundary conditions are

- At solid–liquid interface

$$I_\nu^+(0) = \epsilon_\nu \mathcal{B}_\nu + (1 - \epsilon_\nu) I_\nu^-(0) \tag{G.1}$$

where $\epsilon_\nu = 0.8$ is the spectral emissivity of a pyrolyzed solid, which will be assumed Lambertian (hypothesis compatible with the resolution method).

- At silica–gas interface

$$I_\nu^-(e) = R_\nu I_\nu^+(e) \tag{G.2}$$

where $R_\nu = \left(1 - 1/n^2\right)^{1/2} \simeq 0.75$ is the reflection coefficient, calculated from index $n = 1.5$ by the Snell–Descartes law.

From the solution for $I_\nu^\pm$, it is possible to calculate the density of leaving flux

$$\dot{q}_{\text{rad}} = \int_0^\infty \pi I_\nu^+ \, \mathrm{d}\nu \tag{G.3}$$

where the apparent emissivity, calculated from the temperature of the external silica–gas interface is

$$\epsilon_{\text{app}} = \frac{\dot{q}_{\text{rad}}}{\sigma T_e^4} \tag{G.4}$$

This quantity is a function of

- The external temperature T_e
- The temperature profile in the boundary layer
- The thickness e of the slab

We have seen (Sec. 9.1.2) that one could express the temperature profile using Eq. (9.8) in which the quantity v_o, in steady state, is connected to v_a by

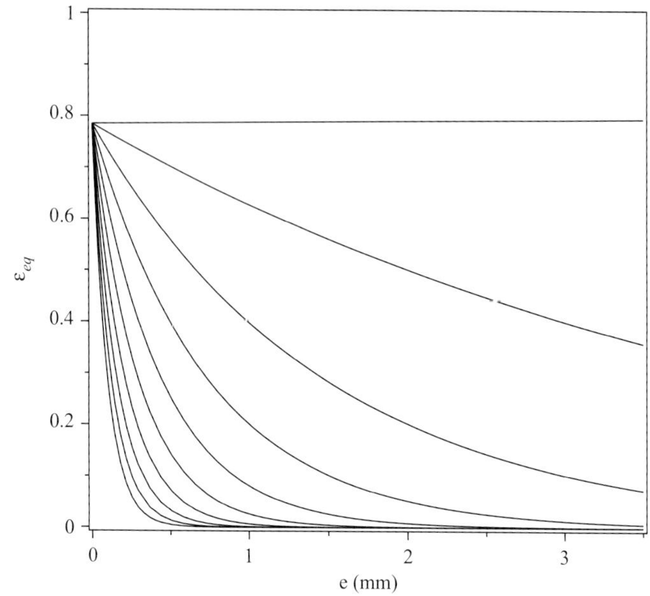

Fig. G.1 Apparent emissivity of silica-resin material, versus thickness of silica, parameterized by ω, proportional to the ablation speed (Eq. (G.5)) representing the temperature profile.

Eq. (9.21). So we are able to express the ϵ_{app} function of T_e, e, and a parameter proportional to the rate of ablation

$$\omega = \frac{v_a + v_0}{\alpha_l} = \left[1 + (1 - 6g)\frac{\rho_p}{\rho_l}\right]\frac{v_a}{\alpha_l} \tag{G.5}$$

In the field we are interested in, there is again a very low dependence on T_e. For a large domain of the parameters, the apparent emissivity is given by the following correlation:

$$\epsilon_{\mathrm{app}} = 0.785\exp[-(3.798\omega - 3.8)e] \tag{G.6}$$

The value of ϵ_{app} can take very low values for large thickness and strong temperature gradients (Fig. G.1). In the calculations, we find values not larger than 0.3. Recall that this quantity has no physical meaning. It is obtained by the desire to return to a classic problem of opaque medium.

Reference

[1] Pascal, P., "Silicium," *New Presentation of Mineral Chemistry*, Vol. VIII, No. 2, Masson, Paris, France, 1965.

INDEX

Note: Page numbers with f represent figures. Page numbers with t represent tables.

(*Continued*)

(*Continued*)

SUPPLEMENTAL MATERIALS

To download supplemental material files, please go to AIAA's electronic library, Aerospace Research Central (ARC), and navigate to the desired book's landing page for a link to access the materials: arc.aiaa.org. Enter the password

reentry

and follow the directions provided.

A complete listing of titles in the AIAA Education Series is available from ARC. Visit ARC frequently to stay abreast of product changes, corrections, special offers, and new publications.

AIAA is committed to devoting resources to the education of both practicing and future aerospace professionals. In 1996, the AIAA Foundation was founded. Its programs enhance scientific literacy and advance the arts and sciences of aerospace. For more information, please visit www.aiaafoundation.org.